WHISKER TECHNOLOGY

Whisker Technology

ALBERT P. LEVITT Editor

Army Materials and
Mechanics Research Center
Watertown, Massachusetts

WILEY-INTERSCIENCE, a Division of John Wiley & Sons, Inc.

New York · London · Sydney · Toronto

Library of Congress Catalogue Card Number: 75-114917

SBN 471 53150 2

Printed in the United States of America

10 9 8 7 6 5 4 3 2 1

Contributors

W. B. Campbell, Associate Professor, Department of Ceramic Engineering, The Ohio State University, Columbus, Ohio

Edward Friedman, Research Engineer, Solid Mechanics Laboratory, Re-entry and Environmental Systems Division, General Electric Company, Philadelphia, Pennsylvania

F. D. George, Supervisor, Solidification Facility, United Aircraft Research Laboratories, East Hartford, Connecticut

ing. J. A. Herzog, Senior Scientist, Air Force Materials Laboratory, Advanced Metallurgical Studies Branch, Dayton, Ohio; presently Filament and Fiber Research Consultant, Darmstadt, West Germany.

Robert H. Kelsey, Staff Scientist, P. R. Mallory & Co., Laboratory for Physical Science, Burlington, Massachusetts

H. E. LaBelle, Jr., Manager, Saphicon Division, Tyco Laboratories, Inc., Waltham, Massachusetts

F. D. Lemkey, Senior Research Scientist, United Aircraft Research Laboratories, East Hartford, Connecticut; now on a leave of absence at University of Oxford, Oxford, England

Albert P. Levitt, Staff Advisor for Composites, Army Materials and Mechanics Research Center, Watertown, Massachusetts; presently Supervisor, High Temperature Materials Group, at the Center

R. L. Mehan, Metallurgist, Space Sciences Laboratory, General Electric Company, King of Prussia, Pennsylvania

A. I. Mlavsky, Director, Corporate Technology Center, Tyco Laboratories, Inc., Waltham, Massachusetts

B. Walter Rosen, Consultant, Space Sciences Laboratory, General Electric Company, King of Prussia, Pennsylvania; now President, Materials Sciences Corporation, Fort Washington, Pennsylvania

M. J. SALKIND, Chief, Advanced Metallurgy, United Aircraft Research Laboratories, East Hartford, Connecticut; now Chief, Structures and Materials, Sikorsky Aircraft, Division of United Aircraft Corporation, Stratford, Connecticut

E. SCALA, Professor of Materials Science and Metallurgical Engineering, Cornell University, Ithaca, New York; formerly Director, Army Materials and Mechanics Research Center, Watertown, Massachusetts

D. M. SCHUSTER, Supervisor of Composite Materials Development, Division 5314, Sandia Laboratories, Albuquerque, New Mexico

WILLARD H. SUTTON, Group Leader, Metallurgy and Ceramics Research, Space Sciences Laboratory, General Electric Company, King of Prussia, Pennsylvania; now Technical Director, Special Metals Corporation, New Hartford, New York

THOMAS VASILOS, Manager, Materials Sciences Department, Avco Systems Division, Avco Corporation, Lowell, Massachusetts

R. S. WAGNER, Member of the Technical Staff, Bell Telephone Laboratories, Murray Hill, New Jersey

STANLEY M. WOLF, Research Metallurgist, Army Materials and Mechanics Research Center, Watertown, Massachusetts

ERNEST G. WOLFF, Senior Staff Scientist, Avco Systems Division, Lowell, Massachusetts; now UNESCO expert, Faculty of Engineering, University of Lagos, Nigeria

Preface

Advances in scientific, industrial, and military products are increasingly dependent on the availability of new and improved structural materials that have high specific strength and modulus and high temperature resistance. In general, any system, device, and component, no matter how sophisticated and advanced, can be further improved by making it out of lighter, stronger, stiffer, and more heat-resistant materials. The great success achieved with glass-fiber-reinforced structures, the availability of new fibers having properties far superior to those of glass fibers, the remarkable theoretical potential of these fibers as strengthening agents, and the ever-increasing need for improved structural materials have stimulated intense interest in advanced fiber-reinforced materials. Since continuous filaments such as boron, carbon, and silicon carbide can be more easily handled, characterized, and fabricated than discontinuous whiskers, the bulk of the fiber composite research and development in the United States has been directed at these continuous filaments.

However, a much smaller, highly significant effort has also gone forward, primarily in the United States and England, with the objective of using the ultrahigh strengths of whiskers (of the order of 1 million psi) in structural composites. This activity utilized the talents of scientists and engineers in a variety of technical fields such as ceramics, metallurgy, plastics, physics, and mechanical and chemical engineering. The important contributions of these specialists to the technology serve to emphasize its interdisciplinary nature and indicate that future achievements can be optimized by such coordinated effort.

Whisker strengths, generally, are about double those attainable in continuous filaments. Sapphire, silicon carbide, graphite, and boron carbide in whisker form also have very high elastic moduli, low densities, and high melting points. Therefore, the theoretical potential of these materials as reinforcing agents exceeds that of any currently available continuous filament at room and elevated temperatures.

The progress made so far has usually been reported in diverse journals and in the proceedings of symposia whose primary emphasis has been on the development of continuous filament-reinforced materials. As a result, there is no single source of information devoted to the state of the art and science of whisker technology.

This book aims to remedy this deficiency and to present, in one volume, a timely summary of progress in all aspects of this fascinating field. The book proceeds from the growth, testing, and properties of whiskers to the mechanics of whisker strengthening, thence to the fabrication of whisker composites and their properties, and finally to a discussion of possible applications of whisker-reinforced composites.

Where appropriate, the author or authors of each chapter discuss the limitations in our current knowledge and examine the problems that are still to be resolved. I hope that the book will serve as a useful reference to the practicing materials scientist and engineer as well as the student of materials science, and that it will stimulate further research and development in this challenging and promising field.

I express my gratitude and deep appreciation to the contributing authors, experts in their respective fields, for their cooperation and many helpful suggestions. I also gratefully acknowledge the interest and encouragement of the Army Materials and Mechanics Research Center in this undertaking.

ALBERT P. LEVITT

Newton Centre, Massachusetts
January 1970

Contents

WHISKER TECHNOLOGY

Albert P. Levitt

1 Introductory Review

INTRODUCTION

Whiskers are metallic and nonmetallic filamentary single crystals that have ultrahigh strength (of the order of 1 million psi) and micron-sized widths or diameters. Whiskers—so called by reason of their appearance—have become the subject of intense study in recent years because of their remarkable characteristics. Analytical studies indicate that low-density, high-modulus, high-melting-point whiskers such as B, B_4C, C, Al_2O_3, SiC, and Si_3N_4 have great potential as reinforcing agents for metals and plastics. Experimental verification of these studies using model systems, new insights into their growth and properties, and new techniques for handling and incorporating them into metallic and nonmetallic matrices are paving the way for the development of a new class of improved high-strength, light weight, whisker-reinforced composite materials.

It is the purpose of this book to present the current status of the use of whiskers in advanced structural composites, including the achievements, the challenging problems, and the exciting possibilities of these fascinating materials.

In the interest of standardization of terms in the field of whisker technology, this book adopts the definitions of fiber and whisker presented by Rauch, Sutton, and McCreight [1]:

Fiber—Any material in an elongated form such that it has a minimum length to a maximum average transverse dimension of 10:1, a maximum cross-sectional area of 7.9×10^{-5} in.^{2} (corresponds to a circular cross-section of 0.010 in. in diameter), and a transverse dimension of 0.010 in. (The term fiber is used generically to describe all types of materials that fit the definition: i.e., filaments, wires, whiskers, etc.)

Whisker—Any material that fits the definition of a fiber and is a single crystal.

HISTORICAL DISCUSSION AND REVIEW

Observations on whiskers were reported as early as 1574 by L. Erker [2]. He noted hairlike or whiskerlike growths emerging from copper and silver sulfide ores. Eighty-seven years later, Robert Boyle [3] in his *Sceptical Chymist* compared the growth of silver whiskers from rocks to the growth of grass, implying that they might be related phenomena. Silver whiskers were subsequently produced experimentally by heating silver sulfide in air. Although some studies of these laboratory curiosities continued to be made, they became the subject of closer scrutiny in the 1940s when the growth of metal whiskers from metal surfaces was found to be the cause of short circuits in miniaturized communication equipment. The objective of this study was to suppress these undesirable growths. In 1952 whiskers came under intense study after Herring and Galt [4] of the Bell Telephone Laboratories experimentally determined the strength of tin whiskers to be an order of magnitude greater than the strength of ordinary tin, so high, in fact, that they were close to their theoretical limits. This discovery suggested that the theoretical estimates of the strengths of solids were correct and that solids in whisker form have strengths that approach theoretical values. This

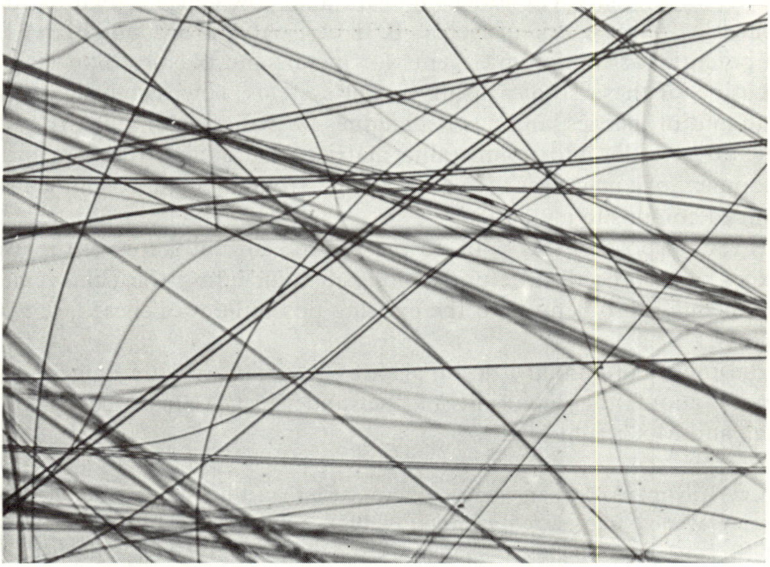

Figure 1.1 Al_2O_3 whiskers. $l/d \approx 10{,}000$, diameter range 1–10 μ (400×) (Courtesy General Electric Space Sciences Laboratory).

TABLE 1.1 PROPERTIES OF CERAMIC WHISKERS [1]

Material	M.P. (deg F)	Density (lb/cu in)	Tensile Strength (psix10^3)	Modulus of Elasticity (psix10^6)
Al_2O_3	3700	0.143	4000	62
BeO	4660	.103	1900	50
B_4C	4400	.091	2000	70
SiC	4870	.115	3000	70
Si_3N_4	3450	.115	2000	45
Graphite	6600	.060	2850	102

work stimulated renewed interest in the growth and properties of whiskers; since then a wide variety of whiskers, both metals and nonmetals, have been grown and their strengths measured, with results comparable to those of Herring and Galt.

The ultrahigh strength of whiskers is attributed to their crystalline perfection (internal and surface) and small dimensions [5, 6], which minimize the occurrence of the defects that are responsible for the low strength of materials in bulk form. Figure 1.1 shows some alumina whiskers enlarged. Table 1.1 presents the properties of high-melting-point ceramic whiskers, and Figure 1.2 graphically compares the room-temperature strengths of

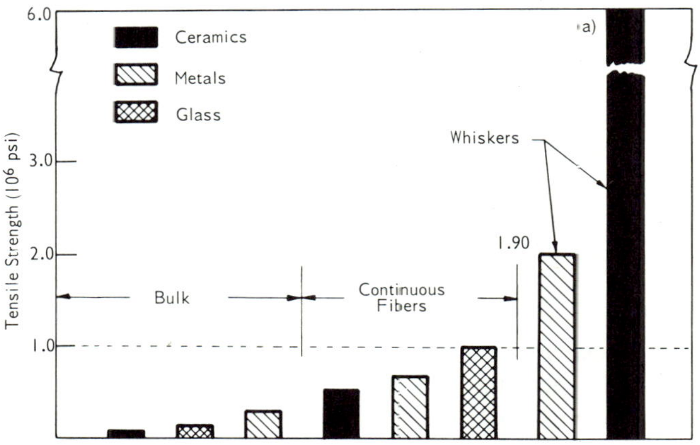

Figure 1.2 Strength of bulk materials, continuous fibers, and whiskers. (a) Strengths exceeding 6 million psi have been observed on silicon carbide whiskers grown by Philips Laboratories, Eindhoven.

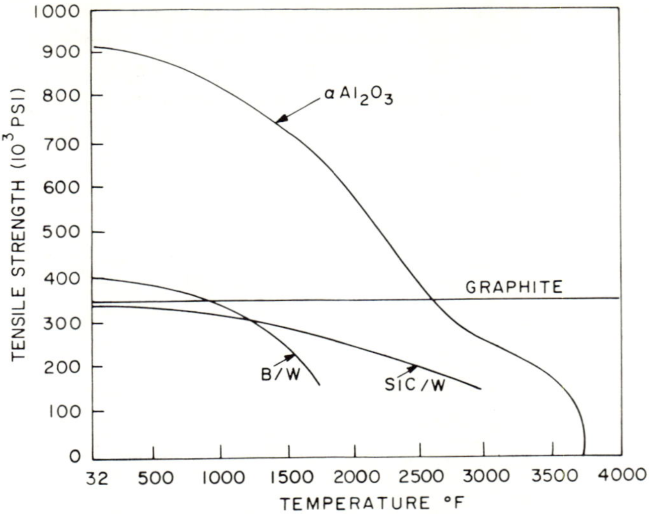

Figure 1.3 Strength retention at elevated temperature of various commercially available fibers (Ref. [1], Courtesy Air Force Materials Laboratory).

materials in bulk, filament, and whisker form. It can be seen that whiskers are many times stronger than their bulk, polycrystalline counterparts and approach theoretical strengths that range from 0.03 E to 0.17 E [7], where E is the modulus of elasticity.

Whiskers also have relatively high strengths at elevated temperatures. Figure 1.3 compares the elevated-temperature strength retention of α-Al_2O_3 whiskers with currently available, continuous, polycrystalline filaments. It can be seen that the α-Al_2O_3 is stronger than boron on tungsten and silicon carbide on tungsten at all temperatures and that it is superior to graphite filaments up to 2600°F.

It is these extremely high strengths together with high elastic moduli, low densities, and high melting points that make whiskers so attractive as strengthening agents for metals, plastics, and ceramics.

It should be noted that the experimental values for whisker strength reported in Table 1.1 are representative values that are used for illustrative purposes only. Since whiskers of a given material may have different growth morphologies, corresponding to growth in various crystallographic directions, a characteristic strength and modulus of elasticity for each type of growth can be expected. The properties of both metallic and nonmetallic whiskers are the subject of current study in which experimental techniques and equipment are continuously being refined and improved. Therefore, these reported values will probably change.

The fascinating properties of whiskers and their great potential as strengthening agents have stimulated extensive research into them with an attendant burgeoning literature. In 1956, Hardy [8] presented a comprehensive review of theoretical and experimental studies in metal whiskers during the past two hundred years. Additional reviews of whisker growth mechanisms and techniques followed, including those by Nabarro and Jackson [9], Brenner [10], Baskey [11], Salkind [12], Coleman [13], and Levitt [14]. Whiskers have been grown by a variety of techniques which include stress-induced growth, evaporation-condensation, vapor-phase reaction, growth from the melt, electrolytic growth, growth from solution, and directional solidification of binary eutectics.

This book confines its discussion to those techniques of whisker growth that show promise of being adapted to relatively low-cost mass production. Thus, in Chapter 2, Campbell describes whisker growth by vapor-phase reactions. His studies indicate that significant cost reductions are probable by the scale-up of the process he describes.

The discovery by Wagner and Ellis [15] of the vapor-liquid-solid mechanism of crystal growth has tied together and explained earlier observations of whiskers growing with molten droplets on their tips. In Chapter 3, Wagner describes this extremely important, fascinating, and widespread mechanism of whisker growth. This discovery has provided us with new insight into a crystal growth mechanism that is operative for many types of whiskers. Again, Wagner describes a very promising technique for the low-cost production of whiskers.

The growing interest in whiskers in the United States and Europe is shown by the increasing number of whisker suppliers. Some of them are listed in Table 1.2, together with properties and prices [1]. As costs are reduced and techniques improved for utilizing whiskers, the number of suppliers will probably increase to meet the increased demand. The most recent entry in this table is Tyco Laboratories, Inc. (Waltham, Mass.), which has succeeded in growing continuous whiskers of aluminum oxide (see Chapter 4). The Norton Research Corporation is currently developing a process originated by the Phillips Laboratories, Eindhoven, for the growth of high-quality silicon carbide whiskers at a projected cost of $10/lb or less. Some silicon carbide whiskers grown by this process are shown in Figure 1.4.

One of the major barriers to realizing the great potential of whiskers as reinforcing agents is the great difficulty associated with classifying and handling these small fibers. This problem has been greatly alleviated by the development of a method for growing "long wool" whiskers which lend themselves well to elutriation and fabrication techniques. As will be seen in Chapter 9, remarkable progress has been made in economically handling and preparing whiskers for incorporation into metal matrices.

TABLE 1.2 PROPERTIES OF COMMERCIALLY AVAILABLE HIGH-MODULUS WHISKERS [1]

	Supplier	Diam. (μ)*	Density (lb/in^3)	Tensile Strength (10^3 psi)	Elastic Modulus (10^6 psi)	Strength/Density (10^6 in.)	Modulus/Density (10^6 in.)	Cost† ($/lb)
Al_2O_3	Thermokinetic Fibers, Inc	1-10	0.143	600-3500	80-150	4.2-24.5	560-1050	7,500
Al_2O_3	Horizons, Inc	0.5-2.0	0.143	1000-3000	60-70	6.9-21	420-490	8,500
SiC	Thermokinetic Fibers, Inc	1-10	0.116	2000-6000	80-150	17.2-51.8	690-1300	11,500
SiC	Carborundum Co	0.5-3.0	0.116	3000‡	70	25.8	600	200
Si_3N_4	Thermokinetic Fibers, Inc	1-10	0.115	700-1500	40	6.1-13	348	-
Al_2O_3	Tyco Laboratories, Inc	38-500	0.143	370-625	67	2.5-4.4	470	**

*1 micron = 0.000039 inch
†Based on supply and demand, therefore subject to major changes
‡Bend strength
**$10 per foot for 0.010-inch-diameter (256 μ) continuous whiskers

Fabrication would, of course, be greatly simplified if whiskers could be grown as continuous filaments. This exciting and very promising breakthrough has recently been achieved at Tyco Laboratories. Working under an Air Force Contract [16], Mlavsky and LaBelle succeeded in growing continuous whiskers from the melt. They describe this fascinating development in Chapter 4. This work may well bridge the gap between whisker technology and continuous filament technology.

As noted earlier, the year 1952 marked the beginning of accelerated effort in whisker studies. Although their small sizes make them extremely difficult to handle and test, techniques were developed for doing so and mechanical property measurements were made and published on a variety of whiskers [7, 17]. However, it is unlikely that the data thus far obtained are representative of the average whisker strength in a given batch of whiskers. The reason is that the testing process requires great skill and patience of the investigator who, in order to ensure a successful test, generally selects one of the best-looking whiskers whose properties may vary widely from the average properties of the batch. In Chapter 5, Kelsey discusses the problems associated with mechanical testing of whiskers and describes the relative merits and limitations of various testing devices.

Mehan and Herzog, in Chapter 6, present a comprehensive and critical review and analysis of all the pertinent data available to date on the mechanical properties of ceramic and metal whiskers. They discuss the room- and elevated-temperature mechanical properties of ceramic and metal whiskers; the relative advantages of the former as reinforcing agents over the latter; the variation of tensile strength with cross-sectional area (size effect); elevated temperature effects; and the limitations in our understanding

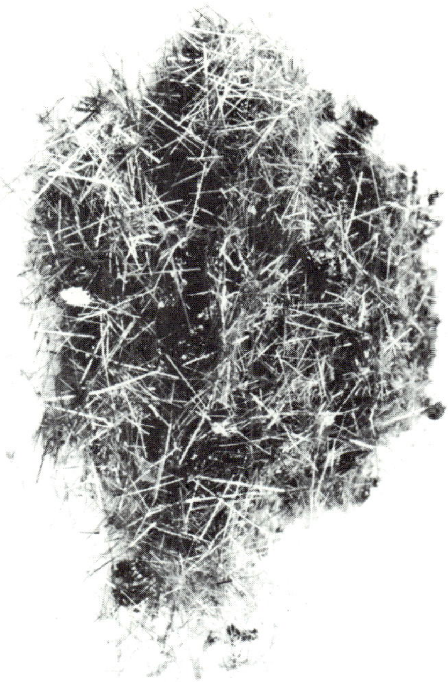

Figure 1.4 Silicon carbide whiskers l/d range 200 to 1000, diameter range $\frac{1}{8}$ to $\frac{1}{2}$ in. strength range 1×10^5 to 1×10^6 psi (5 ×) (Courtesy Norton Research Corp.).

of the mechanical behavior of whiskers. Additional data on properties of whiskers extracted from unidirectionally solidified eutectics are given by Salkind, Lemkey, and George in Chapter 10.

The concept of developing efficient, high-strength, lightweight structures by strengthening a relatively weak low-modulus matrix by way of incorporating into it stronger, higher-modulus fibers has been most successfully applied to continuous glass filaments in a resin matrix. This is because, in addition to their relatively low cost, continuous glass filaments lend themselves well to fabrication by filament winding techniques, which are currently being adapted to the newer continuous filaments such as boron, Borsic,* silicon carbide, and carbon.

* Trade name of United Aircraft Corporation's silicon carbide-coated boron filaments.

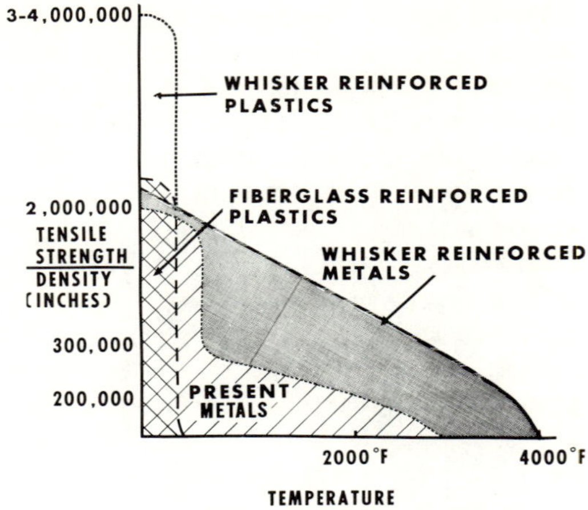

Figure 1.5 Potential of Al_2O_3 whisker-reinforced materials compared to presently available materials (Courtesy General Electric Space Sciences Laboratory).

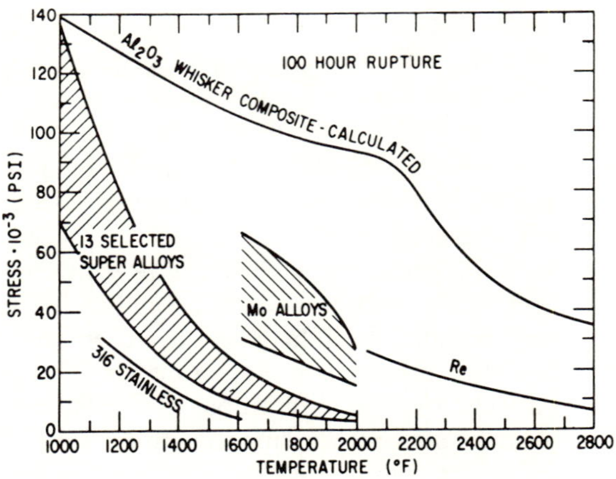

Figure 1.6 Comparison between the calculated unidirectional strength of a composite containing 50 vol% Al_2O_3 whiskers and the strongest metals currently available (Courtesy of American Society for Metals).

On the other hand, the quantitative extension of this concept to discontinuous filaments is relatively recent. One of the earliest analyses of discontinuous fiber strengthening was presented by Cox [18] in 1952, the same year that Herring and Galt reported their discovery of the ultrahigh strength of tin whiskers. Cox's paper pointed out that short, discontinuous fibers can strengthen a weaker matrix and their short lengths can be accounted for as a reduction in their elastic moduli. He experimentally verified his analysis with paper fibers in a resin matrix.

Coleman [19] in 1958 presented an analysis indicating that the strength of a bundle of parallel fibers is less than the mean strength of the individual fibers. Therefore, the fibers or whiskers used in a composite should have uniformly high strengths if a high bundle strength is to be achieved.

In 1957, 1958, and 1959, Hoffman [20–22] discussed the potential of whiskers as reinforcing agents and pointed out their possible advantages over conventional materials for aerospace applications. Assuming that whiskers can be mass-produced into usable form, he calculated that reductions to one-fifth of the weight of structures made of conventional materials appear to be attainable.

Then, in 1959, McDanels, Jech, and Weeton [23], using short tungsten fibers in titanium alloy and copper matrices, demonstrated experimentally that discontinuous fibers can significantly strengthen a weaker, lower-modulus matrix. As these results were communicated to the scientific and engineering communities, effort in this area began to accelerate.

Sutton and his associates at the General Electric Company conducted theoretical and experimental studies aimed at developing new and improved whisker-reinforced metals. They calculated and graphically presented the potential of Al_2O_3 whisker-reinforced materials compared to currently available materials shown in Figure 1.5 [24]. These projections assume (*a*) 50 vol% whisker additions, (*b*) alumina whiskers having average tensile strengths of 1×10^{-6} psi, (*c*) unidirectional orientation and uniform spatial distribution of whiskers in the matrix, and (*d*) good bonding between matrix and whisker so that the load can be effectively transferred from the former to the latter. Brenner's theoretical calculations [7] of the 100-hour rupture stress versus temperature for an Al_2O_3 whisker-reinforced composite also indicated that such a composite is far superior to any currently available alloys (Fig. 1.6).

Dow [25] analyzed the relationship between interfacial shear stress and the stress distribution in discontinuous fibers for elastic and plastic deformation of the matrix. His results in the elastic case were in agreement with those of Cox. He concluded that the matrix must undergo inelastic (elastic and plastic) deformation in order to transmit appreciable load into the fiber. Kelly and Tyson in 1963 [26] presented the concept of critical fiber length

l_c and fiber transfer ratio l/l_c and derived an important relationship between the average fiber stress $\bar{\sigma}$ and the l/l_c. This relationship provides a means for quantitatively accounting for the effect of fiber length on composite strength. In Chapter 7, Rosen and Friedman discuss these concepts and extend their work of 1964 [27] on the mechanics of composite strengthening to discontinuous fibers.

The interface is critically important in fiber composites, especially whisker composites, since it is the region through which the matrix transfers its load to the fibers. Wetting of the ceramic whiskers by the matrix, followed by interfacial bonding, is essential for fabrication of sound composites by liquid infiltration. These requirements are much more difficult to satisfy with metal matrix systems than with resin matrix systems, since metals generally do not wet ceramics. Basic studies of metal-ceramic interfacial wetting and bonding [28–31] have provided a better understanding of the factors affecting these phenomena. The sessile drop studies of Sutton and Feingold [32] and Wolf, Levitt, and Brown [33] have characterized two systems of importance as structural composites: nickel/nickel alloy-sapphire, and aluminum/aluminum alloy-sapphire, respectively. The current status of whisker matrix bonding and interaction is presented and evaluated by Levitt and Wolf in Chapter 8.

Starting in 1960, Sutton and his associates [34] at the General Electric Company [35] conducted studies that yielded a number of significant advances in the development of noneutectic whisker-reinforced metals. The results included the following:

1. Demonstration of the need to coat Al_2O_3 whiskers to achieve fiber matrix bonding.
2. The growth of high-quality Al_2O_3 whiskers (long wool) free from dendrites and debris.
3. The development of an air-elutriation process for classifying whiskers.
4. Development of an automatic tape-making machine.
5. The demonstration of significant whisker strengthening at room and elevated temperatures. The strength of silver was increased fivefold at room temperature, and much of this strength was maintained at a temperature of about 98% of the melting point of silver. The strength of pure aluminum was increased sixfold and that of epoxy resins tenfold. These results are truly impressive and have provided the incentive for further work in this field.

Studies by Raynes and Kelsey [36] (Al_2O_3 in nickel chromium alloys), Hahn et al. [37] (Al_2O_3 in aluminum and nickel alloys), and Barr [38] (SiC in aluminum) also contributed to the interest in this field. Sutton, in Chapter 9, describes in detail techniques for the fabrication of whisker-reinforced composites.

Research in whiskers has by no means been limited to the United States. Significant progress in the development of whisker-reinforced composites has been made in England by Parratt and his associates at the Explosives Research and Development Establishment. In 1966, Parratt [39] reported on the development of a semiautomated technique for the growth, classification, and incorporation of aligned silicon nitride whiskers in metal and resin matrices. The fiber classification technique has since been improved and refined to the point where a short fiber classification unit is being marketed as a laboratory tool for fiber reinforcement work.

Kraft's and Albright's [40, 41] observations and experimental studies at United Aircraft Research Laboratories on the directional solidification of eutectic alloys provided the basis for the development of a new class of aligned whisker-reinforced composites. Such composites are particularly attractive because the solidification technique used in their formation automatically solves the major problems of whisker growth, spatial distribution, stability, and bonding. Although the volume fraction of the reinforcing fibers is limited by the eutectic composition, systems such as Ni-$NiAl_3$, Ta-Ta_2C, and Nb-Nb_2C, have already shown remarkable improved properties over the as-cast alloys. In Chapter 10, Salkind, Lemkey, and George discuss the historical events leading to the development of these remarkable composites and then present, in detail, the characteristics of currently important systems. They also describe the properties of extracted fibers produced by unidirectional solidification to prove that they are actually ultrahigh-strength whiskers.

Before any new materials can be fully and rationally utilized in the design of a system, their properties must be fully determined. Whisker composites, being relatively new and available only in very limited amounts, have yet to be fully characterized and evaluated. Schuster and Scala in Chapter 11 discuss the mechanical properties of currently available whisker and fiber composites and point out, by means of photoelastic techniques, the factors associated with fiber geometry and spacing that affect composite strength. Their early observations [42] of the stress concentrations around the tips of fibers indicated the need for fibers having rounded or tapered ends to minimize such stress concentrations.

Finally, Vasilos and Wolff present their views on the potential applications of whisker composites. While the greatest potential of these composites lies in aircraft and aerospace applications where high strength, heat resistance, light weight, and stiffness are at a premium, breakthroughs that would greatly reduce the cost of the whiskers themselves and of the techniques to fabricate them into composites could truly broaden their applications to the point where they would enjoy wide military, aerospace, and industrial applications.

REFERENCES

[1] H. W. Rauch, Sr., W. H. Sutton, and L. R. McCreight, "The Fabrication, Testing and Application of Fiber-Reinforced Materials: A Survey," Space Sciences Laboratory, General Electric Company, Technical Report AFML-TR-68-162, September 1968.

[2] L. Erker, *Treatise on Ores and Assaying*, 1574; 2nd ed., 1580. Translated by A. G. Sisco and C. S. Smith, University of Chicago, 1951, p. 177.

[3] R. Boyle, *The Sceptical Chymist*, Everyman's Edition, London, 1661, p. 193.

[4] C. Herring and J. K. Galt, *Phys. Rev.*, **85** (1952), 1060.

[5] G. W. Sears, *Acta Met.*, **1,** 458 (1953); **3,** 367 (1955).

[6] R. F. Register, Thesis, University of Pennsylvania, 1963.

[7] S. S. Brenner, "Factors Influencing the Strength of Whiskers," Chapter 2, *Fiber Composite Materials*, ASM, 1965.

[8] H. K. Hardy, *Progr. Metal Phys.*, **6** (1956).

[9] F. R. N. Nabarro and P. J. Jackson, "Growth of Crystal Whiskers, A Review," in *Growth and Perfection of Crystals*, edited by R. H. Doremus, B. W. Roberts, and David Turnbull, John Wiley and Sons, New York, 1958.

[10] S. S. Brenner, *The Art and Science of Growing Crystals*, edited by J. J. Gilman, John Wiley and Sons, New York, 1963.

[11] R. H. Baskey, "Fiber Reinforcement of Metallic and Non-Metallic Composites, Phase I—State of the Art and Bibliography of Fiber Metallurgy," ASD Interim Report 7-924(I), February 1962.

[12] M. J. Salkind, "Whisker and Whisker Strengthened Composite Materials," Watervliet Arsenal Technical Report WVT RR 6315, October 1963.

[13] R. V. Coleman, "Growth and Properties of Whiskers," *Met. Rev.*, **9,** 35 (1964).

[14] A. P. Levitt, "Recent Advances in Whisker Technology," Army Materials and Mechanics Research Center Monograph, MS 64-08, November, 1964.

[15] R. Wagner and W. C. Ellis, "The Vapor-Liquid-Solid Mechanism of Crystal Growth and its Application to Silicon," *Trans. Met. Soc. AIME*, **233** (June 1965), 1057.

[16] H. E. LaBelle, Jr. and A. I. Mlausky, *Nature*, **216,** 5115 (1967), 574.

[17] R. L. Mehan and E. Feingold, "Room and Elevated Temperature Strength of α-Al_2O_3 Whiskers and Their Structural Characteristics," *J. Metals*, **2,** 2 (1967), 235–270.

[18] H. L. Cox, "The Elasticity and Strength of Paper and Other Fibrous Materials," *Brit. J. Appl. Phys.*, **3,** 72 (1952).

[19] B. D. Coleman, "On the Strength of Classical Fibres and Fiber Bundles," *J. Mech. and Phys. of Solids*, **7** (January–March 1958), 60–70.

[20] G. A. Hoffman, "The Structural Exploitation of the Strength of Whiskers," The Rand Corp. Paper, August 1957, p. 1149.

[21] G. A. Hoffman, "A Look at Whiskers," *Astronautics*, **3** (August 1958), 31–33.

[22] G. A. Hoffman, "Future Possibilities in Fibered Materials," Proc. 6th Sagamore Ordnance Materials Research Conference on Composite Materials and Composite Structures, August 1959.

[23] R. W. Jech, D. L. McDanels, and J. W. Weeton, "Fiber Reinforced Metallic Composites," Proc. 6th Sagamore Ordnance Materials Research Conference, August 1959.

[24] W. A. Sutton, "Sapphire Whiskers for Structural Reinforcement," General Electric Company, MSD, Dept. No. R62 SD994, December 1962.

[25] N. F. Dow, "A Study of Stresses Near a Discontinuity in a Filament-Reinforced Composite Metal," General Electric Company, Report R 63SD61, August 1963.

[26] A. Kelly and W. R. Tyson, "Fiber Strengthened Materials in High Strength Materials," V. F. Zachay, Ed., John Wiley and Sons, New York, 1965, pp. 578–602.

[27] B. W. Rosen, "Mechanics of Composite Strengthening," Chapter 3, *Fiber Composite Materials*, ASM, 1965.

[28] W. D. Kingery, "Role of Surface Energies and Wetting in Metal-Ceramic Sealing," *Bull. Am. Ceram. Soc.*, **35** (1956), 108–111.

[29] C. R. Kurkjian and W. D. Kingery, "Surface Tension at Elevated Temperatures, III, Effect of Cr, In, Sn and Ti on Liquid Nickel Surface Tension and Interfacial Energy with Al_2O_3," *J. Phys. Chem.*, **60** (1956), 961–963.

[30] B. C. Allen and W. D. Kingery, "Surface Tension and Contact Angles in Some Liquid Metal-Solid Ceramic Systems at Elevated Temperatures," *Trans. AIME*, **215** (1959), 30–37.

[31] W. M. Armstrong, A. C. D. Chaklader, and J. F. Clarke, "Interface Reactions Between Metals and Ceramics, I. Sapphire-Nickel Alloys," *J. Am. Ceram. Soc.*, **45** (1962), 115–118.

[32] W. H. Sutton and E. Feingold, *Role of Interfacially Active Metals in the Apparent Adherence of Nickel to Sapphire—Materials Science Research*, Vol. 3, Plenum Press, New York, 1966, pp. 577–611.

[33] S. M. Wolf, A. P. Levitt, and J. Brown, "Whisker-Metal Matrix Bonding," *Chem. Eng. Progr.*, **62,** 3 (March 1966).

[34] W. H. Sutton, A. Talento, and J. Chorné, "Development of Composite Structural Materials for High Temperature Applications," Final Report, Contract NOw-60-0465d, February 1965.

[35] R. L. Mehan, C. A. Bruch, and E. Feingold, "Evaluation of Sapphire Wool and its Incorporation into Composites of High Strength," Technical Report, AFML-TR-66-357, December 1966.

[36] B. C. Raynes and R. H. Kelsey, "Studies of the Reinforcement of Metals with Ultra High Strength Fibers (Whiskers)," Horizon, Inc., Final Report, Contract NOw62-0235-c, October 1962.

[37] H. Hahn, A. P. Divecha, P. Lare, and B. Denneim, "Research on Whisker Reinforced Metal Composites," Final Report Contract NOw-65-0387-8, April 1966 and subsequent final reports under Navy contracts May 1967, July 1968.

[38] H. N. Barr, "Investigation of Silicon Carbide Whisker Reinforced Metallic Matrix Composites," Final Technical Report AFML-TR-67-296, September 1967.

[39] N. J. Parratt, "Whisker Reinforced Plastics and Metals," *Chem. Eng. Progr.*, **62,** 3 (March 1966).

[40] R. W. Kraft, U.S. Patent 3,124,452, March 10, 1964.

[41] R. W. Kraft and D. Albright, *Trans. AIME*, **221** (1961), 95.

[42] D. M. Schuster and E. Scala, "The Mechanical Interaction of Sapphire Whiskers in a Birefringent Matrix," *Trans. AIME*, **230** (1964), 1635–1640.

W. B. Campbell

2 Growth of Whiskers by Vapor-Phase Reactions

INTRODUCTION

Whisker growth has become a necessary and important phase of whisker technology since the whisker changed from a laboratory curiosity to a most desirable structural component in recent years. The high tensile strength of these micron-size diameter, single crystals stimulated exploratory studies of their character, behavior, and utilization. Considerations of quality, availability, and cost were minor until the utility of these materials was proven and the need for large whisker-reinforced structures was recognized. As in most other areas of crystal growth, production techniques were modeled after those of the pioneering investigators; worthy improvements were incorporated but production remained dependent upon a scale-up approach. Techniques such as evaporation-condensation, stress induction, controlled-melt chilling, electrolytic deposition, and chemical transport or decomposition, were found to have low scale-up efficiencies with respect to amount and quality of product and production time. Of these older batch operations, controlled-melt chilling and chemical transport have enjoyed limited successes in whisker production. The former technique has been adapted for *in situ* formation of whiskers during the fabrication of structural components (details of this development are described in Chapter 10). Chemical transport processes have been adequate in providing masses or mattes of whiskers in large growth chambers and with long growth times; the process is economically feasible but the products require separation or further processing before utilization in a composite.

The practical application of growth theories and the control of major process parameters have received limited attention in the development of whisker-growth processes. The development of the necessary supersaturation gradient is entirely dependent upon the prevailing temperature gradient in most production systems and leads to inherent control difficulties in multiple or complex reaction systems. In the simplest case, the temperature gradient approach becomes more difficult and less controllable as the growth chamber increases in dimensions. Isothermal growth conditions circumvent this problem but require other means of supersaturation control. Similarly, a steady-flow system reduces the influence of concentration gradients and changes in concentration. Thus an isothermal, steady-flow, whisker-growth system reduces the number of critical control variables.

This chapter is intended to provide a qualitative understanding of techniques that involve vapor-phase reactions and produce metastable growth species. The condensation of these species through either homogeneous or heterogeneous nucleation produces the desired morphological product. The establishment of initial growth conditions for isothermal systems from theoretical calculations is presented and compared with experimental observations. Some indication of the effect of growth environment on the strength of whiskers is presented as experimental evidence.

Vapor-phase reaction represents one of the two more recently developed techniques for whisker production; the other, the vapor-liquid-solid mechanism (VLS), is discussed in Chapter 3. Older techniques have been well described in previous publications and are not discussed in detail here. The effect of vapor growth conditions on morphology is presented in principle by Langham and Mason [1] and the reader is referred to this work for experimental evidence.

Thermodynamic and kinetic considerations establish the limits for the practical application of whisker growth. From thermodynamics we shall obtain a first approximation for the feasibility of a process and also some implication as to the kinetics involved. In other words, we shall define the limiting thermodynamic conditions that must be satisfied before reaction kinetics are of consequence. The vapor-phase reaction system developed for the continuous production of Al_2O_3 whiskers is presented as an example of predicting initial conditions, controlling major variables, and obtaining high-quality products in significant quantities.

BACKGROUND

Whisker Growth Theories

In general, the driving force for whisker growth arises from either recrystallization (primary) or a decrease in supersaturation. The growth rate of tin whiskers was found to be proportional to the pressure applied to an electroplated coating—an observation which indicates that the driving force for whisker growth is the relaxation of strain in the metal, that is, primary recrystallization [2]. Franks postulated that whisker growth will occur on a free surface of a metal in compression, or internally strained, provided that plastic flow is prevented and that the temperature is high enough to permit rapid diffusion [3]. Whisker growth caused by a decrease in supersaturation can be explained by the classical crystal growth theory when an atom or molecule is pictured as condensing at a step on the growing crystal surface. Theory requires that each step sweep the growth surface before a new step is nucleated on the completed layer. The creation of a new step on a perfect crystal surface requires an increase in the surface energy of the crystal; consequently, a critical supersaturation is required for continued growth. Although Mason discusses the effect of supersaturation on the growth rate and morphology of ice crystals, his work presents and supports the effects of supersaturation in a vapor system. Since ice structures are similar to other oxide structures, this work provides a useful qualitative analogy [4]. It has been shown experimentally that crystals grow at supersaturations far below the critical values [5]. Frank concluded from this observation that real crystals are not perfect but contain permanent growth steps in the form of crystal imperfections [6]. (Whisker growth has been attained in the presence and the absence of surface imperfections.)

At higher supersaturations, macrocrystal growth occurred and preferential growth velocities in certain crystal directions were characteristic. Preferential growth forms may be expected at supersaturations below those required for euhedral morphology. Sears postulated that anisotropic growth observed in certain materials is due to the absence of imperfections in specific crystal directions [7]. According to this hypothesis, whisker growth can occur by (*a*) having nuclei that contain dislocations parallel to only one crystal direction and (*b*) supersaturation less than that required for euhedral development. Sears demonstrated the validity of (*b*) for a number of materials; however, the requirement for dislocation orientation has not been illustrated conclusively [8].

Although most whiskers are straight, a variety of shapes has been observed in the same batch. Nonlinear shapes have been attributed to dislocation propagation and impurity effects, although kink angles may frequently correspond to twin-plane orientations. Webb and Forgeng reported a twist in alumina with an estimated Burgers vector equal to, or some multiple of, the lattice unit [9–10]. The occurrence of whisker polytypes has been observed but has shown little correlation to the step- or unit-cell height [11]. The magnitude of impurity effects has been questioned because most whiskers have not been of exceptionally high purity [12].

Sears postulated a whisker-growth model for the impingement of atoms onto a whisker surface, temporary adsorption of the impinging atoms, and diffusion of atoms along the lateral surface to a sink at the whisker tip; those atoms not reaching the tip being removed by reevaporation [13, 14]. Both Sears and Gomer have derived appropriate steady-state growth laws to describe the experimentally observed asymptotic and initial growth behavior of whiskers [15]. Within appropriate limits, the general steady-state growth law derived by Dittmar and Neumann reduces to the laws of Sears and Gomer [16]. Ruth and Hirth presented a steady-state solution for the whisker-growth problem which showed that for most physically meaningful cases, the steady-state condition should be fulfilled both for initial and terminal stages of whisker growth [17]. The violation of boundary conditions at the tip, rather than the inability of atoms to reach the tip, was reported to be responsible for a steady-state breakdown.

WHISKER GROWTH PRACTICES

At the present time more than thirty elements and fifty compounds have been grown in whisker form. A variety of methods have been developed and used to obtain large numbers of whiskers for subsequent studies of their physical and chemical properties. For composite utilization, the selection of whisker-growth techniques depends on the influence of the growth conditions on the abundance of product, the ultimate strength, uniformity of product and product properties, surface and internal perfection, and chemical composition. Although the interplay of the preceding factors is well recognized, they are delineated to facilitate the subsequent discussion of whisker-growth practices.

Evaporation-Condensation

Sublimation or evaporation of a source material, mass transport through the vapor phase, and condensation at the growth site causes whisker growth under low supersaturations. It is common practice to establish the necessary concentration gradient by a controlled temperature gradient; additional diffusional control of supersaturation being attained by the introduction of an inert gas. Partial pressure control of the supersaturation has also been achieved in evacuated systems.

The application of this method to the growth of metallic whiskers has been widespread, whereas the growth of oxide whiskers has been limited by the high melting points and low vapor pressures that cause experimentation to be very difficult, if not impossible. The sizes and lengths of metal whiskers grown by condensation are usually found to be proportional to the vapor

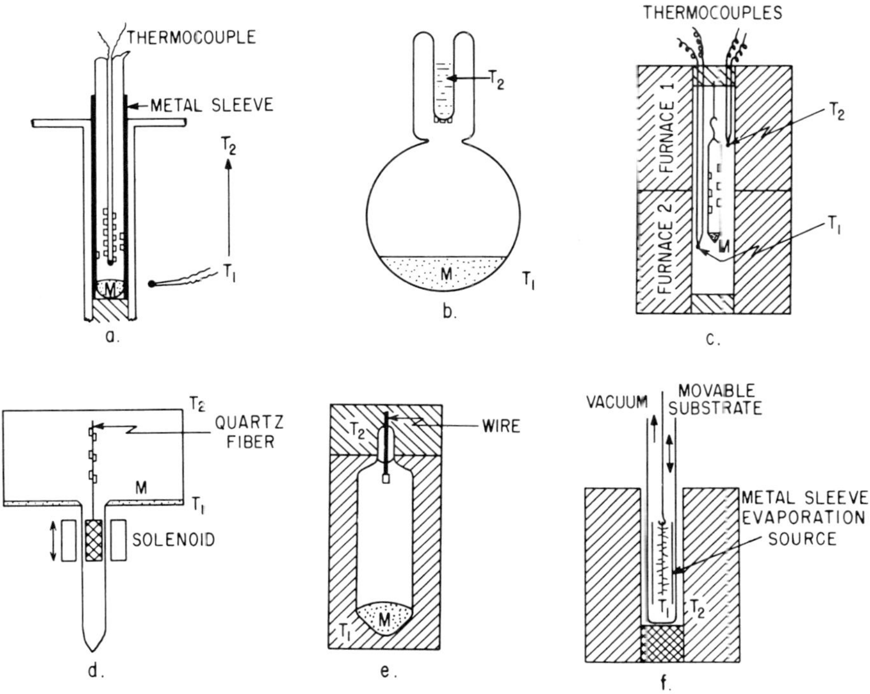

Figure 2.1 Growth cells for the condensation of metal vapors.

pressure of the metal at the melting point. Mercury, zinc, and cadmium are typical metals that have high vapor pressures at the melting point and can be grown easily in a simple thermal gradient system. Systems used for this type of growth are shown in Figure 2.1, and simply consist of a growth chamber with a controlled thermal gradient between the source and the growth site.

In early work, Coleman and Sears [18] found a substantial enhancement of both the population and the length of zinc whiskers when a pure inert gas atmosphere was maintained in the growth chamber. According to Coleman [19], various pressures of gases such as helium, neon, argon, krypton, nitrogen, and hydrogen have been found effective in establishing a diffusional controlled supersaturation at the growth sites of both zinc and cadmium whiskers. Although hydrogen cannot be considered an inert gas, it is often required to maintain reduced conditions and to permit metal growth in oxide growth chambers. At pressures near 10^{-6} torr, Sears [20] observed the whisker growth of metals with a wide range of vapor pressures, and made an extensive study of the growth of mercury whiskers [21]. Melmed and Gomer [22] have grown a number of metal whiskers in both vacuum and inert gas atmospheres (Table 2.1), which suggests that the supersaturation at the growth site is critical for whisker growth of the materials studied.

In practice, to grow cadmium and zinc whiskers, for example, each metal is triple distilled at 10^{-6} torr and spectrographically pure argon is introduced into the growth chamber to establish a pressure between 1 and 1000 mm. When the temperature of the source is held as much as 80°C above the melting point, whiskers are observed to grow in temperature regions ranging from slightly below the melting point to as much as 100°C below. For any specific experiment, the geometry of the growth chamber and the thermal gradient determine the exact pressure and temperature conditions. Whiskers grown by this method vary in length from fractional millimeters to several centimeters and in size from fractional microns to forty microns. Growth times range from one hour to several days. Price [23] determined that cadmium whiskers grew in a supersaturation range of 0.4 to 2.0 and had the same orientation as zinc. Silver, potassium, copper, iron, platinum, gold, aluminum, germanium, nickel, barium, and other metals have been grown by this method.

The experimental difficulties involved in working at high temperatures have limited the application of evaporation-condensation methods to refractory oxides. Although it is not unusual to observe whisker growth in regions of high temperature within furnaces, it is quite difficult to obtain sufficient control for realistic measurements; also, the effect of ambient atmosphere must be considered. Kliman [24] established the evaporation-condensation growth of aluminum oxide in the system shown in Figure 2.2. Aluminum

TABLE 2.1 GROWTH OF METAL WHISKERS

Metal	Source Temperature, °C	Substrate Temperature, °C
Ni	1049–1271	700–900
Cu	900	600–700
Fe	1198	661–753
Pt	1616	750–960
Ti	1247	765
Au	1092–1111	557–630
Al	—	500–630
Ba	—	550–600
Ge	900	750–800
V	—	800–850
Ag	800	600–650

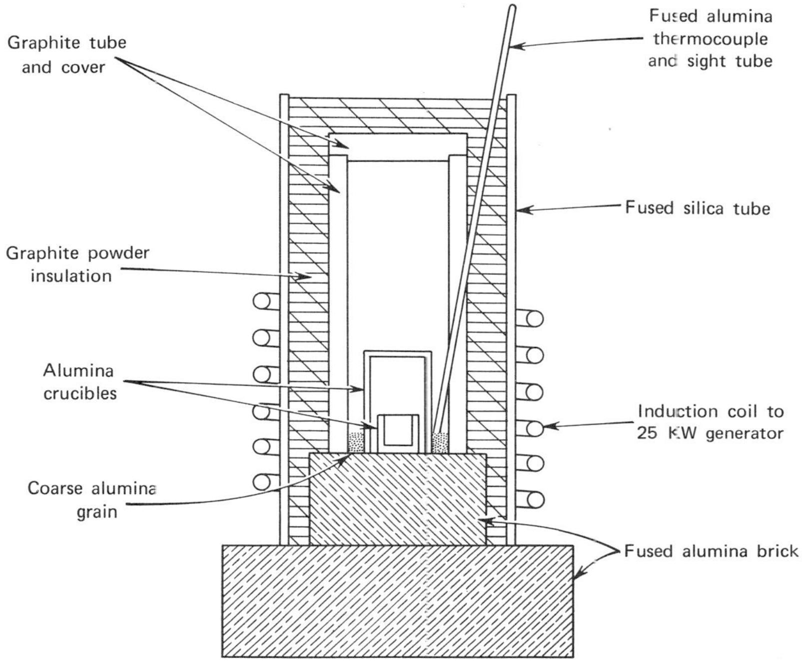

Figure 2.2 High-temperature induction furnace (From Kliman [30]).

oxide whiskers formed in the cooler portion of the growth chamber when the source material was heated to 1950°C.

Evaporation-condensation methods have been less successful for many materials because of the small whisker size, the low population, extended growth times, and experimental difficulties at higher temperatures. In practice this simple method is confined to those materials which grow readily and/or to those studies which require a very limited number of whiskers.

Chemical Reduction

Reaction between the source material and an ambient gas phase to produce a volatile species, vapor transport to the growth site, and condensation results in whisker growth under low supersaturations. In practice the hydrogen reduction of metal salts has been widely used for the growth of metal whiskers; for oxides, wet hydrogen or reductive metals have been utilized. The apparatus for chemical reduction is shown in Figure 2.3 and usually consists of a tube furnace containing either a mullite, alumina, or quartz tube, a ceramic boat

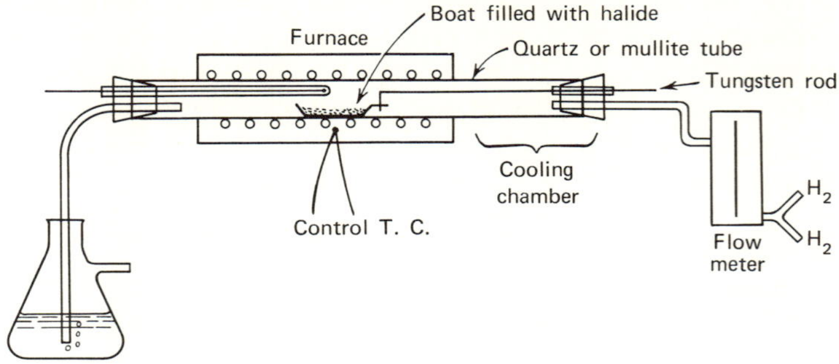

Figure 2.3 Whisker growth apparatus.

filled with the source material, and a controlled gas-flow system. The whiskers grow on the sides and top of the boat in relatively large, downy deposits. Growth times may range from 0.5 to 20 hours depending on the exact conditions and the material. Diameters ranging from submicron to one mm and lengths up to 10 cm have been observed.

Brenner [25] reported the conditions for reduction growth of metal whiskers as shown in Table 2.2 and presented a compilation of conditions for other metals in a later review article [26]. In general, the best reduction temperature is near or slightly above the melting point of the source material. Webb and Forgeng [27] reported the growth of aluminum oxide whiskers by passing wet hydrogen over a boat filled with molten aluminum or $TiAl_3$. At source temperatures between 1300 and 1450°C and operating times of

TABLE 2.2 REDUCTION GROWTH CONDITIONS

Metal	Halide	Temperature Range °C	Optimum Temperature	Maximum Length mm	P_0 Halide, mm Hg	P_0 Metal, mm Hg
Cu	CuCl CuBr CuI	430–850	650	50	10.4	2.23×10^{-10}
Ag	AgCl AgI	700–900	800	10	0.14	3.71×10^{-5}
Fe	$FeCl_2$ $FeBr_2$	730 760	—	20	59	2.07×10^{-10}
Ni	$NiBr_2$	740	—	2	68	8.18×10^{-15}
Co	$CoBr_2$	650	—	3	8	—
Pt	$PtCl_x$	800	—	3	Decom.	7.23×10^{-16}
Au	$AuCl_x$	550	—	2	Decom.	8.58×10^{-15}

2 to 20 hours, relatively large deposits surrounding the charges were produced. The partial pressure of water increased from 1.2×10^{-4} atm to 3.6×10^{-3} atm with passage through the growth chamber and was attributed to hydrogen reduction of the porcelain furnace tube. Larger deposits were found around the intermetallic charges than around the pure aluminum. Whiskers with lengths from 1 to 30 mm and hexagonal diameters of 3 to 50 μ were obtained.

Minor variations of the Webb and Forgeng technique have been used to produce useful quantities of whiskers for composite fabrication. The scale-up of this method presents problems of gas and moisture control, temperature uniformity in large chambers, batch operations, and relatively long growth times. These problems, in addition to the intrinsic requirement for product separation and classification, are indicative of the limits of quantity production by this method; conversely, the present state of the art is a tribute to the ingenuity of workers in this field.

With respect to the present status of quantity production by chemical reduction techniques, it is well to consider some of the history of these developments. Toward the end of the last decade the evaluation of whiskers and whisker-reinforced composites suggested the great advances to be achieved in material properties, and studies were initiated with the intention of determining the factors pertinent to whisker exploitation. An early report of the quantity growth of aluminum oxide whiskers was presented by Talento [28]. He used a system shown schematically in Figure 2.4, wherein hydrogen gas was passed through a humidifier before passing over a 99% alumina boat charged with aluminum metal. A typical deposit formed over the surface of the metal is shown in Figure 2.5. The whisker type and size varied within the charge as well as between charges and other growth forms such as ribbons and powders were observed. Whisker lengths up to 4 cm and diameters down to 0.5 μ were reported; growth conditions are shown in Table 2.3. Growth times ranged from 0.5 to 10 hours at a constant furnace pressure of 3 cm Hg.

A. Gatti extended the observations of Webb and Forgeng on the reaction of hydrogen with porcelain refractory furnace tubes and boats; Brenner [29] reported these observations and suggested consideration of other reductions in the system. A porcelain boat filled with aluminum metal was heated to 1250°C in a flowing stream of hydrogen (dew point, −55°C) and some of the metal was oxidized to form a “cocoon” of whiskers over the top of the boat. Whiskers grew profusely at this temperature only in a porcelain boat and subsequent analysis showed 2 to 3% silicon in the aluminum oxide whiskers ranging in diameter from a few microns up to 100 μ and in lengths up to 1.5 cm. Brenner suggested that the work of Tombs and Welch [30], which showed an appreciable amount of SiO vapor in equilibrium with SiO_2, H_2,

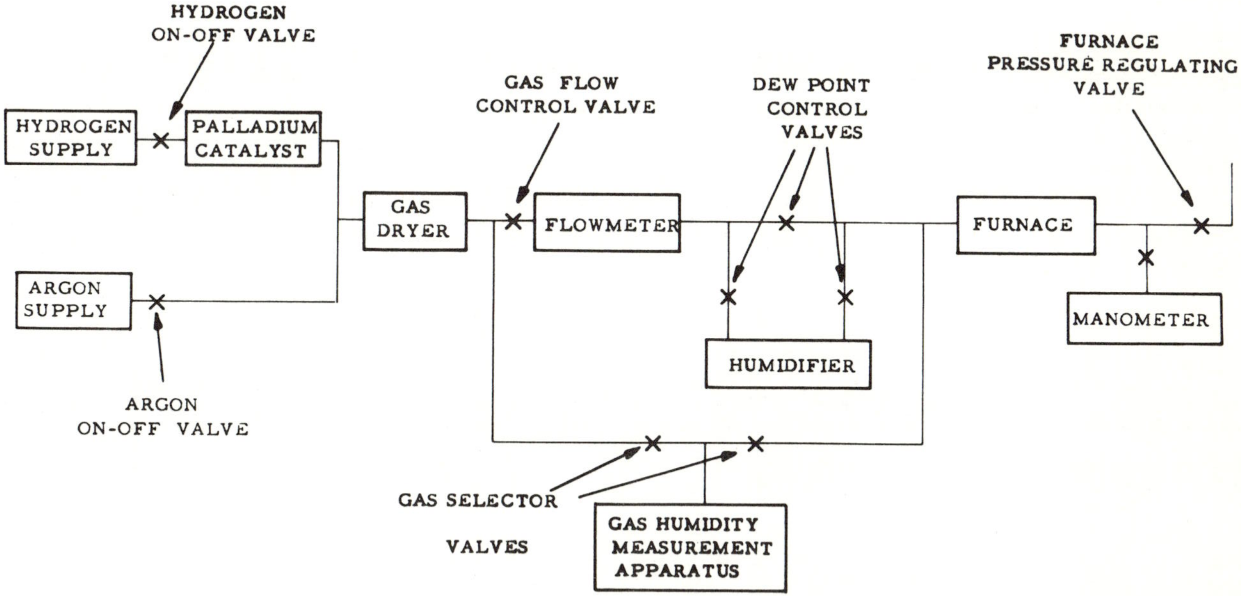

Figure 2.4 Schematic diagram of alpha-alumina whisker-growing furnace system.

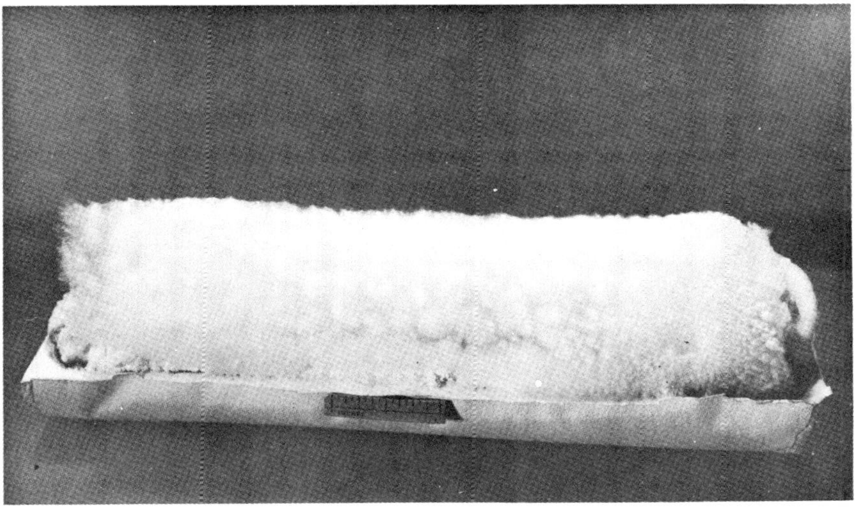

Figure 2.5 Photograph of an 8 in. × 2 in. alumina boat containing wool growth as taken from furnace.

and H_2O at 1250°C, may indicate an alternate reaction for whisker growth to be,

$$2Al(g) + 3SiO(g) \rightarrow Al_2O_3(s) + 3Si(s)$$

Later developments in the reduction method for aluminum oxide have utilized dry hydrogen and silica-rich refractory boats and chambers. The two major overall reactions were found to be [31]:

$$H_2(g) + SiO_2(s) \rightarrow H_2O(g) + SiO(g)$$
$$4Al(l) + 3H_2O(g) + 3SiO(g) \rightarrow 2Al_2O_3(s) + 3H_2(g) + 3Si(l)$$

This modification has been reported to be a substantial improvement in controlling the production growth [32]. It appears that better production control is also achieved in the dry system. Barber [33] reported, however, the growth of aluminum oxide whiskers in "a growth chamber of pure recrystallized alumina" by reaction of molten aluminum with oxygen of uncertain origin. Although he attempted to exclude water vapor and oxygen from the chamber, he suspected the presence of the former by the odor of H_2S in the growth chamber and reasoned this to be formed by residual hydrogen and sulfur in the aluminum. In this instance, it is also likely that the metal reduced the oxide. The apparatus is shown in Figure 2.6.

Alley, et al. [34] reported the formation of fibrous silicon carbide by the reduction of complex silicates in specially constructed graphite crucibles

TABLE 2.3 CONDITIONS FOR Al_2O_3 WHISKER GROWTH BY HYDROGEN REDUCTION

Factor Studied	Temperature, °F	Hydrogen Flow Rate, cfgh[a]	Hydrogen Dew Point, °F	Growth Appearance	Remarks
Temperature	2370	0.8	−50	Fuzz	Growth produced directly above charge inside boat.
	2550	0.8	−50	Fuzz	Growth produced directly above charge inside boat.
	2730	0.8	−50	Wool and coarse	Coarse growth produced directly above charge inside boat. Wool nucleated on furnace tube wall.
	2910	0.8	−50	Wool and coarse	Coarse growth directly above charge inside boat.
Dew point	2730	0.8	+75	Fine powder	Growth produced on charge inside boat.
	2730	0.8	+25	Fine powder	Growth produced on charge inside boat.
	2730	0.8	−2	Very coarse	Growth produced over charge inside boat.
	2730	0.8	−29	Coarse and wool	Coarse growth produced over charge inside boat. Wool nucleated on furnace tube wall.
	2730	0.8	−50	Coarse and wool	Coarse growth produced over charge inside boat. Wool nucleated on furnace tube in much greater quantities than at −29°F dew point.
	2730	0.8	−75	Fuzz	Growth produced directly over charge inside boat.
Flow rate	2730	0.33	−50	Coarse and wool	Coarse growth produced over charge inside boat. Wool nucleated on furnace tube wall in large quantities.
	2730	0.75	−50	Wool	Wool nucleated on furnace tube wall in large quantities.
	2730	1.50	−50	Coarse and wool	Coarse growth produced over charge inside boat in large quantities. The wool nucleated on furnace tube surface in medium quantities.
	2730	3.00	−50	Very coarse	Very coarse growth produced over charge inside boat.

[a] cfgh = cubic feet of gas per hour.

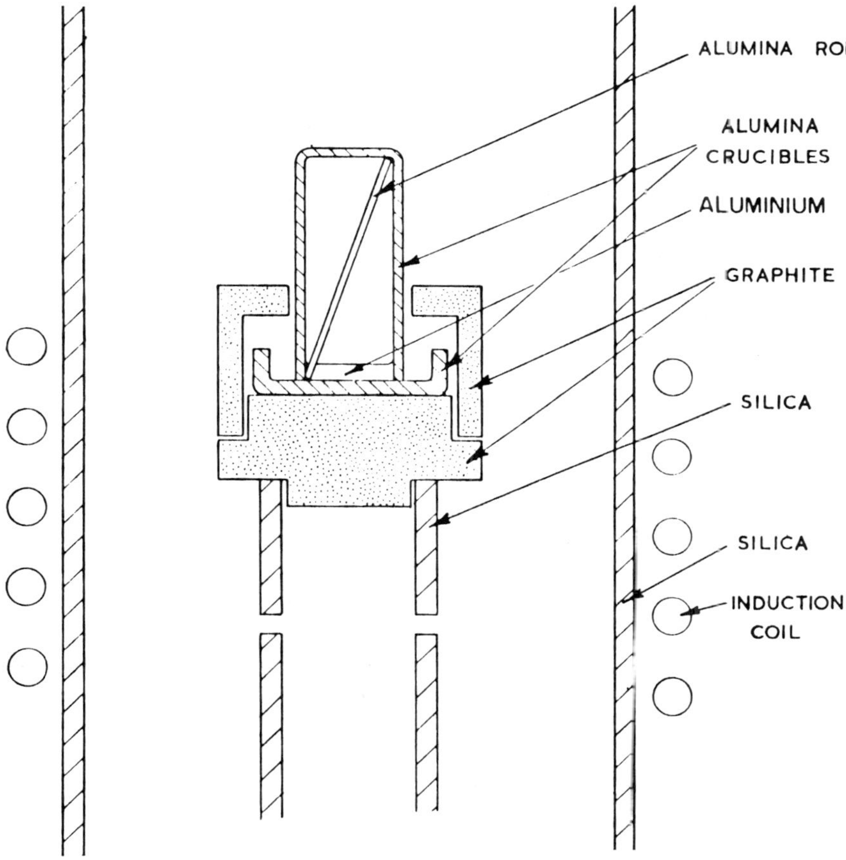

Figure 2.6 Schematic diagram of whisker growth apparatus.

which restricted the escape of gaseous products. Major growth occurred in the thermal gradient between 1300 and 1700°C. Wolff and Coskren [35] used a similar system to form MgO whiskers by tungsten and carbon reduction; they also grew MgO whiskers by wet hydrogen reduction.

Vapor Phase Reaction

Reactions between gaseous species and subsequent condensation at growth sites causes deposition or crystal growth under suitable conditions and supersaturations. Mason [1], Price [23], and other investigators have shown the influence of supersaturation on the growth morphology of many materials. The relative supersaturations associated with the principal growth forms (whiskers, bulk crystals, and powders) can be described qualitatively. A low supersaturation is required for whisker growth whereas a medium supersaturation supports bulk crystal growth; an intermediate supersaturation is evidenced by the growth of dendrites and/or platelets. At higher supersaturations, powders are formed by homogeneous nucleation in the gas phase. Schaffer [36] has reported the vapor-phase reaction growth of ruby and sapphire crystals weighing up to 150 g. Campbell [37, 38] utilized vapor-phase reaction techniques to produce a spectrum of aluminum oxide growth forms.

In practice the gaseous reactants may be mixed in the heated growth chamber and allowed to condense at rates determined by the prevailing conditions, or the reactant and product partial pressures may be controlled to obtain immediate and sustained growth. Substrates may be either stationary with the gases flowing across their surface or small particles (i.e., nuclei) carried in the reacting gas stream. Whisker recovery from stationary substrates is usually a hand operation, whereas recovery of gas-borne whiskers requires a collector in the exhaust side of the gas stream. A schematic of the system for aluminum oxide whisker growth is shown in Figure 2.7. Appropriate source gases, nuclei or substrates, and growth conditions are utilized for the growth of other materials such as silicon nitride, cristobalite, and boron.

For the growth of aluminum oxide whiskers, monomeric aluminum chloride is produced by the controlled chlorination of aluminum metal. Other source gases are carbon monoxide (99.5 %), prepurified hydrogen (99.8 %), and carbon dioxide (99.8 %). Carbon monoxide and excess hydrogen

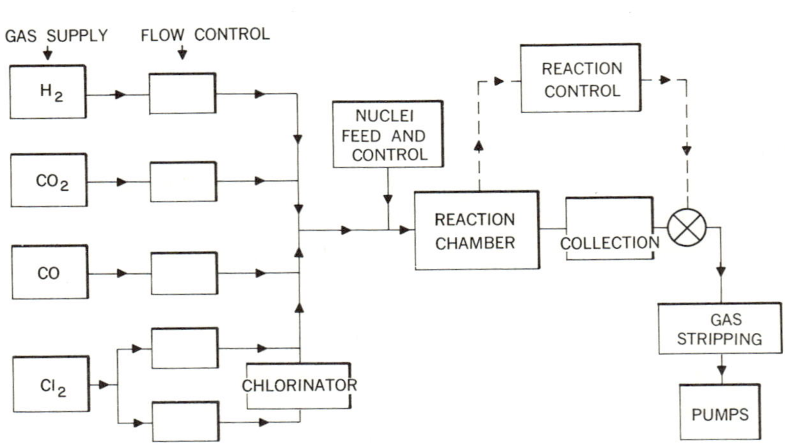

Figure 2.7 Schematic of vapor reaction system for Al_2O_3.

and chlorine are provided to obtain a more controllable reaction environment. The source gases flow through drying columns, calibrated flow meters, and metering valves into the injector system. To control gas flows at standard temperature and pressure conditions, the metering valve and the flow-meter valve in each line are adjusted until the desired flow rate is obtained and the line pressure between the two valves is atmospheric. Appropriate gases are diverted through the nuclei chamber to transport nuclei into the injector. Either a concentric-tube injector or a single-tube injector may be used to inject reactants into the reaction chamber heated by molybdenum elements. Reaction chamber sizes may range from 2.2 to 5.0 cm inside diameter and only materials above 99.7% Al_2O_3 are completely satisfactory. The reaction chamber is continually swept by reactant and product gases which pass into the collector where solid products are removed [37].

Observations and determinations of induction times and axial growth rates were made for aluminum oxide whisker growth on static substrates. The induction time was reported as the time lag between the introduction of the last gas component (H_2) and the observed initiation of growth; the axial growth rate was obtained from the timed translation of the whisker tip across a telescopic field of view. Under isothermal conditions and appropriate gas compositions, induction times were approximately zero. For the gas compositions, similar chamber temperatures, and nonisothermal conditions, induction times up to 15 min were measured. Instant growth conditions produced axial growth rates as high as 4 cm/sec; however, most measured rates were between 0.1 and 1 cm/sec. Under nonisothermal conditions, the observed growth rate was several orders of magnitude lower. The lower range of growth rates for each thermal condition may have been due to an actual difference in rate or the effect of whisker inclination out of the field of view. Qualitatively the growth rate decreased as the induction time increased; in both thermal cases, the growth rate was apparently higher for smaller whisker diameters. The growth rates for gas-borne substrates determined from product dimensions and calculated residence times were the same order of magnitude as those observed for instant growth on static substrates [38].

The gas compositions, temperatures, pressures, and calculated velocities for aluminum oxide whisker growth are shown in Table 2.4 and are applicable to an isothermal zone about 10 cm in length. Increasing the diameter of the chamber from 2.2 cm to 5.0 cm necessitates only a gas velocity adjustment. Vertical-axis reaction chambers improved process control for whisker formation on gas-borne nuclei. Nuclei of varying purity, particle size, and species have been evaluated experimentally; apparently the purity is not critical, the particle size influences the whisker diameter, and the species must be compatible. For species other than aluminum and aluminum oxide, no growth was obtained.

TABLE 2.4 CONDITIONS FOR Al_2O_3 WHISKER GROWTH BY HALIDE OXIDATION

Specimen	Input Gas Flows (lpm) H_2	CO_2	CO	$Cl_{2_{xs}}$	Cl_2 to $AlCl_3$	v(cm/sec)	P_T	Temperature, °C
A. Static Substrate with Induction Time								
1	1.0	0.1	—	—	0.03	950	17 mm Hg	1275
2	1.0	0.2	1.0	—	0.03	4700	10 mm Hg	1750
3	0.9	0.8	0.5	0.06	0.03	4000	14 mm Hg	1500
4	0.9	0.8	0.2	0.02	0.03	3100	12 mm Hg	1500
5	0.9	0.8	0.2	0.02	0.03	3200	12 mm Hg	1600
6	0.9	0.8	0.1	0.02	0.03	7500	5 mm Hg	1565
B. Static Substrate with Instantaneous Growth								
1	0.95	0.80	0.14	0.02	0.03	3100	12 mm Hg	1500
2	0.95	0.90	0.14	0.02	0.04	2900	14 mm Hg	1530
3	0.95	0.80	0.22	0.02	0.03	8000	5 mm Hg	1560
4	0.95	0.80	0.14	0.02	0.03	3100	13 mm Hg	1605
5	0.95	0.95	0.14	0.02	0.04	3300	14 mm Hg	1700
C. Dark Whiskers on Static Substrate								
1	1.0	0.20	—	—	0.07	2500	11 mm Hg	1750
2	1.0	0.10	—	—	0.03	2700	9 mm Hg	1750
D. Gas-Borne Substrates								
Linde A	0.95	0.80	0.49	0.03	0.03	6500	7 mm Hg	1560
Alundum 38B	0.95	0.80	0.60	0.04	0.03	6800	7 mm Hg	1560
Alcoa 408	0.95	0.80	0.22	0.02	0.03	2500	15 mm Hg	1560

When aluminum powder was used as nuclei, conversion to crystalline products was virtually complete; examples of products are shown in Figure 2.8. The morphology of these products suggests a VLS growth mechanism and the absence of crystallographic influence from the nuclei. Examples of whiskers grown on aluminum oxide nuclei are shown in Figure 2.9 [37].

Whiskers of silicon nitride, cristobalite, and boron have been grown to show the feasibility of the vapor-phase reaction method. The conditions of growth are shown in Table 2.5, and silicon nitride products are shown in Figure 2.10.

Whisker Strength and Growth Technique

The whisker has been found to be stronger than its bulk counterpart by several orders of magnitude in a wide variety of elements and compounds.

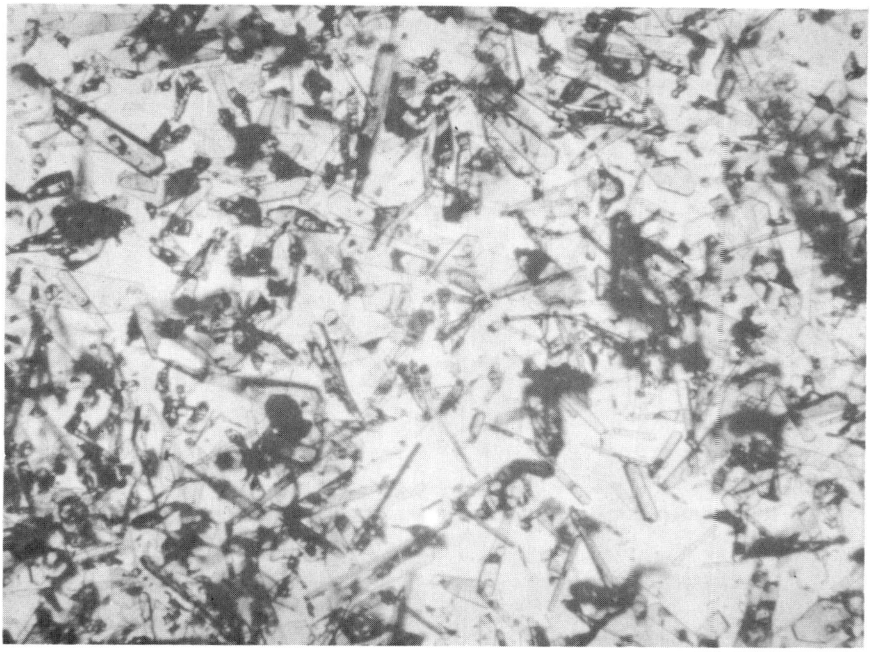

Figure 2.8 Photomicrograph of Al_2O_3 whiskers grown on aluminum nuclei (66 ×).

The high strengths of whiskers have been attributed generally to a high degree of internal and surface perfection. Although values for the strength of the same whisker material may show wide variations and ranges may not be consistent between different investigators, these anomalies are usually regarded as being due to different testing conditions. Only a few investigations have been directed toward the delineation of the variables affecting the strength of whiskers. Studies on the effect of temperature, surface films, and impurities have shown the regions of influence of these variables; however, it is not apparent from the literature that a definitive study has been conducted to determine the strength variations that arise from the conditions of growth.

Certainly the recent paper by Regester, et al. [39] supports the premise that the strength of an aluminum oxide whisker depends on the perfection of the surface. It is also interesting to note that the logarithmic presentation of their data and the line of best fit coincides with previously reported data [38], which was intended to indicate that the dark color of aluminum oxide whiskers grown by vapor-phase reaction under reducing conditions may be

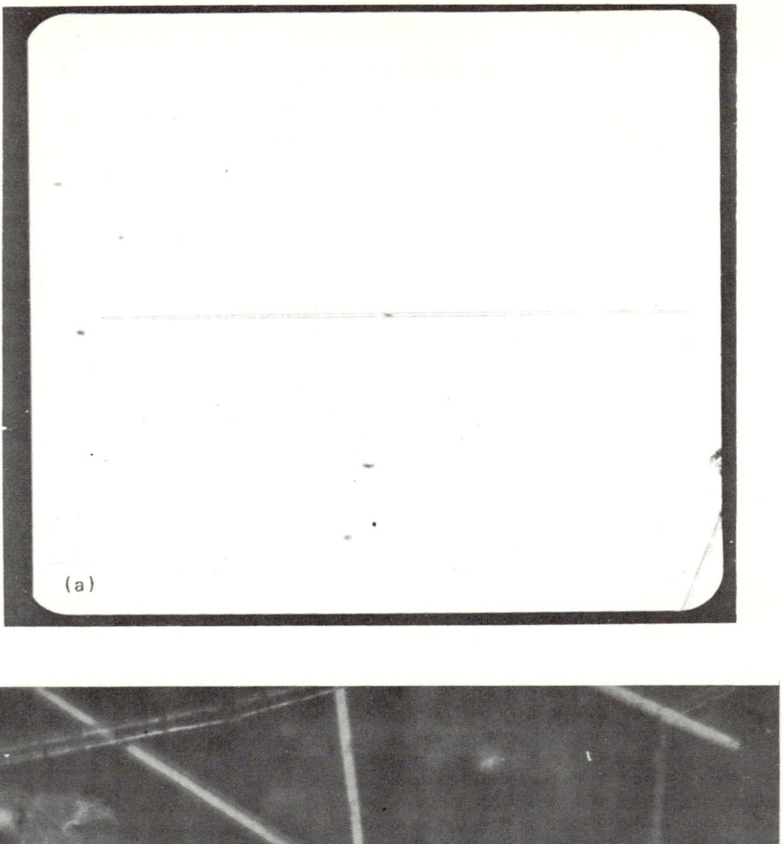

Figure 2. 9 (a) Photomicrograph of Al_2O_3 whiskers grown on Al_2O_3 nucleus (66 ×) (Courtesy Army Materials and Mechanics Research Center). (b) Photomicrograph of Al_2O_3 whiskers grown on Al_2O_3 nuclei (66 ×).

TABLE 2.5 GROWTH CONDITIONS FOR SILICON AND BORON WHISKERS

Specimen	Input Gas Flows (lpm)				v(cm/sec)	P_T	Temperature, °C
	H_2	CO_2	N_2	$SiCl_4$			
SiO_2	0.80	0.20	—	0.02	1700	10 mm Hg	1450
Si_3N_4	0.60	—	0.2	0.02	2100	8 mm Hg	1600
	H_2	BCl_3	$Cl_{2_{xs}}$				
B	1.40	0.02	0.02		7000	4 mm Hg	1700
B	0.75	0.03	0.03		9000	1 mm Hg	1500
B	0.75	0.03	0.04		5000	2 mm Hg	1600

due to nonstoichiometry of the material and may thus have an effect on the strength of the whiskers. Chemical treatments produced no color change in the dark whiskers; however, oxidation at elevated temperatures effected a color change and increased strength. The number of samples for the above studies has been insufficient for confirmation of this postulated effect, although the data indicate a provocative trend. Figures 2.11 and 2.12 present

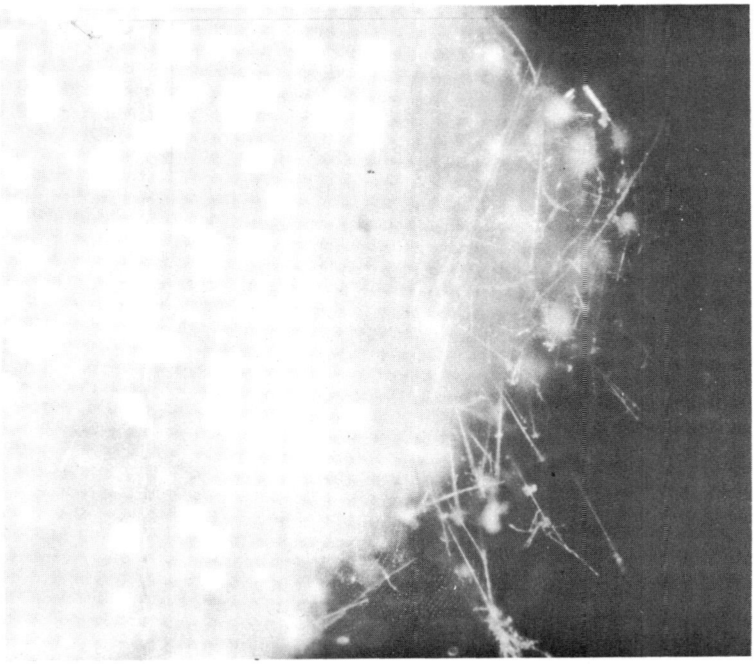

Figure 2.10 Photomicrograph of silicon nitride whiskers (66 ×).

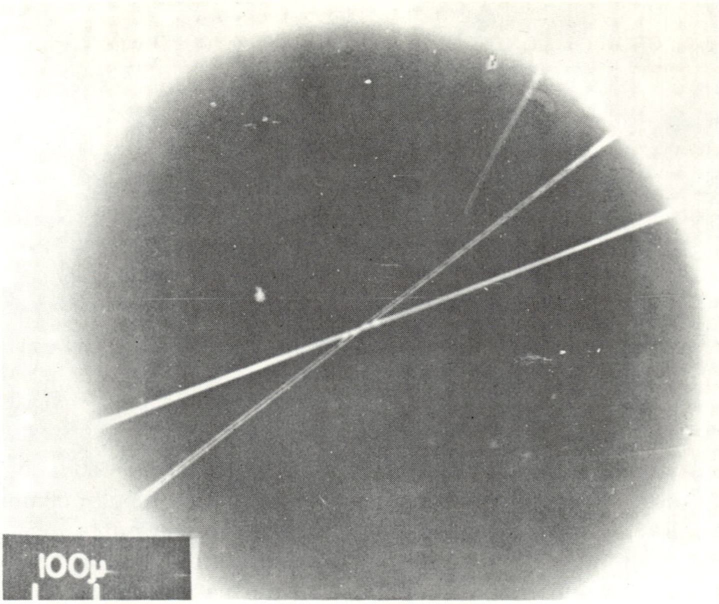

Figure 2.11 Observed color differences in Al_2O_3 whiskers (66 ×).

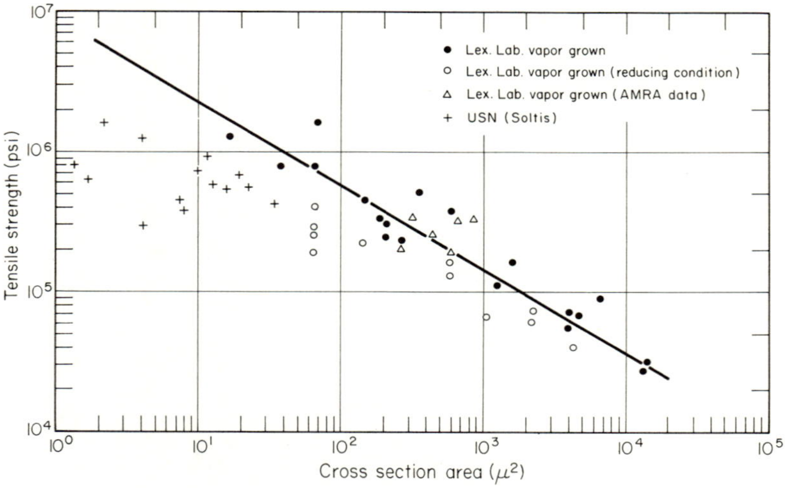

Figure 2.12 Tensile strength of whiskers grown under different conditions.

the data; for a more detailed presentation and analysis of whisker properties see Chapter 6.

PRINCIPLES OF VAPOR-PHASE WHISKER GROWTH

Several basic requirements must be satisfied to utilize a vapor-phase reaction system for whisker growth. The first consideration is the thermodynamics of the reaction at selected temperatures. Secondly, the kinetics of reaction, deposition, and evaporation must not be rate limiting within reasonable limits. When the rate of evaporation is low and the reaction rate substantially exceeds the deposition rate, process control may be achieved by controlling the composition of the reactant gases. When the rates of reaction and deposition are similar, process control may be based on gas velocities or on the compositional control of product and reactant gases. For any system, the most important requirement is that the whiskers have properties equivalent to those of products formed in other systems.

Thermodynamic Considerations

The application of thermodynamic principles to vapor-phase reactions has been utilized and presented by numerous authors mentioned in a preceding section. Thermodynamic consideration should be the first step in developing a vapor-phase whisker-growth system. We should recognize that the thermodynamic representation of an equilibrium system is only as accurate as the identification of the species and temperature, the determination of the thermodynamic activity, and the assigned free energy values. As the initial consideration, thermodynamics establishes the possibility of the selected reaction; in the event that the thermodynamic analysis of a selected system indicates a potential barrier for reaction, the successful development of the system is quite unlikely. It should not be implied that the reaction "will not go"; rather that its slowness will make the system unattractive for experimentation or production.

It has been customary to represent a reaction system by a summary equilibrium reaction. For a first approximation this is satisfactory; however, more insight into the criticality of intermediate species may be gained by detailed analysis. For example, many reports on controlled oxidation by wet hydrogen in the gas stream present summary reactions that identify water vapor as a reactant even at temperatures over 1200°C. Physically, this assumption is not valid because infrared analyses have shown an appreciable dissociation of water vapor at elevated temperatures. Thermodynamically, the free energy values are correct because account was made for the dissociation. The realization that the species identification of free energy values may be misleading in some cases is conducive to the development of a control scheme based on reacting gas mixtures or reducible solid constituents. Feingold and Gatti [31] showed the latter to be consistent with silica constituents in mullite growth chambers serving as oxygen sources.

In practice, a vapor-phase reaction system should be observationally reversible to permit adequate control of reacting gas mixtures. This implies that the introduction of product gases should drive the reaction backward. Reversibility permits the determination of the physical continuity of the selected reaction by indicating the presence of intermediate species. Since a vapor-reaction system involves the growth or solution of material under identical temperature and pressure conditions, the presence of a discontinuous species will be evident.

The formation of aluminum oxide from a halide may be represented by the following summary reaction,

$$2AlCl_3 + 3CO_2 + 3H_2 = Al_2O_3 + 3CO + 6HCl \tag{1}$$

where $\Delta F^\circ_{R\,1800^\circ K} = -81.8$ kcal

In concept, the reaction of hydrogen and carbon dioxide shown in Eq. 2 should provide a controlled partial pressure of oxygen for the halide reaction.

$$3H_2 + 3CO_2 = 3H_2O + 3CO \tag{2}$$

where $\Delta F^\circ_{R\,1800^\circ K} = -14.5$ kcal

When the products of the preceding reaction are mixed with monomeric aluminum chloride, the following intermediate reaction may occur:

$$2AlCl_3 + 3H_2O = Al_2O + 6HCl + O_2 \tag{3}$$

where $\Delta F^\circ_{R\,1800^\circ K} = +124.5$ kcal

Since the reaction free energy is positive for the preceding reaction, the products will be unstable. If we consider the possible deposition reactions and their free energies, we find:

$$Al_2O + O_2 = Al_2O_3; \qquad \Delta F^\circ_{R\,1800^\circ K} = -191.7 \text{ kcal} \tag{4}$$

$$2AlO + \tfrac{1}{2}O_2 = Al_2O_3; \qquad \Delta F^\circ_{R\,1800^\circ K} = -220.2 \text{ kcal} \tag{5}$$

If the reactions shown in Eqs. 2, 3, and 4 are combined, the resultant free energy of reaction is equal to that for Eq. 1.

$$\begin{array}{rll}
3H_2 + 3CO_2 = 3H_2O + 3CO; & \Delta F^\circ_R = -14.5 \text{ kcal} & (2) \\
2AlCl_3 + 3H_2O = Al_2O + 6HCl + O_2; & \Delta F^\circ_R = +124.5 \text{ kcal} & (3) \\
Al_2O + O_2 = Al_2O_3; & \Delta F^\circ_R = -191.7 \text{ kcal} & (4) \\
\hline
2AlCl_3 + 3H_2 + 3CO_2 = Al_2O_3 + 3CO + 6HCl; & \Delta F^\circ_R = -81.8 \text{ kcal} & (1)
\end{array}$$

For equilibrium conditions, the equilibrium constant, or mass action constant, may be calculated from the standard free energy of the reaction by the following relation:

$$K = e^{-\Delta F_R^\circ / RT} \tag{6}$$

Also, the component partial pressures can be related to the equilibrium constant by Eq. 7.

$$K = \frac{P_{CO}^3 P_{HCl}^6 a_{Al_2O_3}}{P_{AlCl_3}^2 P_{CO_2}^3 P_{H_2}^3} \tag{7}$$

The activity of Al_2O_3 may be assumed to be unity if the solid phase is present in the reaction system. The activities of the gaseous components will be equal to their partial pressures.

Supersaturation

For whisker growth, the control of supersaturation is a prime consideration because there is good evidence that the degree of supersaturation determines the prevailing growth morphology. The relative supersaturations associated with the principal growth forms (whiskers, bulk crystals, and powders) have been described qualitatively. A low supersaturation is required for whisker growth whereas a medium supersaturation supports bulk crystal growth. At high supersaturations, powders are formed by homogeneous nucleation in the gas phase.

At equilibrium, a reversible reaction is saturated with respect to each component. If equilibrium is violated, the system will become supersaturated or undersaturated in various components, depending upon the direction of shift. The change in supersaturation should be proportional to the rate constant for the reaction shift. Thus, the supersaturation of selected products in a vapor-phase reaction may be established by controlling component partial pressures under isothermal conditions.

In gaseous systems of constant volume, the supersaturation is proportional to the partial pressure of the growth species.

$$\Sigma_T = \frac{P_a}{P_e} \tag{8}$$

The supersaturation ratio, Σ, is proportional to the partial pressure of the growth species produced by the reaction, P_a, and the equilibrium partial pressure of the growth species, P_e, at temperature T. Supersaturation exists when the ratio exceeds one—the equilibrium ratio. When the ratio is less than one, the imbalance will result in the evaporation of the solid material.

According to the Law of Mass Action, the equilibrium rate of reaction is represented by the equilibrium constant,

$$K_{p_T} = \frac{P_p}{P_r} \tag{9}$$

where K_{p_T} is the equilibrium constant derived from partial pressure relations at temperature T, P_p is the product of the exponential partial pressures of the products, and P_r is the product of the exponential partial pressures of the reactants. In systems where a solid is derived by gaseous reaction, the partial pressures of products and reactants are as follows (activity of solid, $a_s = 1$; partial pressure of gas, P_g):

$$P_p = (P_{g_1})^x(a_s)^n = P_D^d P_E^e \tag{10}$$

and

$$P_r = (P_{g_2})^y = P_A^a P_B^b P_C^c \tag{11}$$

In systems where the desired solid phase is present in excess, the partial pressure of its evaporation species is a constituent of the total reactant and product partial pressures.

The supersaturation ratio defined in Eq. 8 may be derived in terms of thermodynamic quantities by the following calculations. Setting a condition of equilibrium for the gas phase reaction, Eqs. 3 and 4 are combined as follows:

$$2AlCl_3(g) + 3H_2(g) + 3CO_2(g) \rightleftharpoons O_2(g) + 3CO(g) + 6HCl(g) + Al_2O(g) \tag{12}$$

It should be noted that the assumption of equilibrium for the gas-phase reaction does not imply an equivalent condition for the condensation reaction. The equilibrium constant for the gas-phase reaction at a particular temperature is

$$K_{12} = \frac{P_{O_2}P_{CO}^3 P_{HCl}^6 P_{Al_2O}}{P_{AlCl_3}^2 P_{H_2}^3 P_{CO_2}^3} \tag{13}$$

Rearranging in terms of O_2 and Al_2O partial pressures,

$$[P_{O_2}P_{Al_2O}] = \frac{K_{12}P_{AlCl_3}^2 P_{H_2}^3 P_{CO_2}^3}{P_{CO}^3 P_{HCl}^6} \tag{14}$$

For the condensation reaction shown in Eq. 4, the equilibrium constant is

$$K_4 = \frac{a_{Al_2O_3}}{P_{Al_2O}P_{O_2}} = P_{Al_2O}^{-1}P_{O_2}^{-1} \tag{15}$$

Rewriting Eq. 8 in terms of partial pressures of the growth species, we can relate the supersaturation ratio to the equilibrium constants and partial pressures.

$$\Sigma_T = \frac{(P_{O_2}P_{Al_2O})_a}{(P_{O_2}P_{Al_2O})_e} = \frac{K_{12}P^2_{AlCl_3}P^3_{H_2}P^3_{CO_2}}{K_4^{-1}P^3_{CO}P^6_{HCl}} \tag{16}$$

since $K_1 = K_{12}K_4$,

$$\Sigma_T = \frac{(P_{O_2}P_{Al_2O})_a}{(P_{O_2}P_{Al_2O})_e} = K_1\left[\frac{P^2_{AlCl_3}P^3_{H_2}P^3_{CO_2}}{P^3_{CO}P^6_{HCl}}\right] \tag{17}$$

This derivation suggests that the supersaturation ratio, Σ_T, can be controlled by adjusting the partial pressures on the input gases. It should be noted that the total pressure of the system is not included in this relation.

The calculation of the partial pressure of Al_2O from the free energy of reaction for Eq. 4 provides a value of $10^{-11.5}$ atm. Thus, the reaction system becomes supersaturated with respect to alumina when the partial pressure of the growth species exceeds the equilibrium value.

If the temperature, gas composition, and other critical parameters of a steady-flow, vapor-phase reaction system are maintained and only the substrate surface area is increased, one of three effects may be expected; namely, (1) an increase in supersaturation, (2) a decrease in supersaturation, or (3) no change in the balance of the system. The last effect is least probable because the system parameters and the equilibrium constant remain fixed while the mass of the system is increased. A decrease in supersaturation would imply that the condensation rate changes as a function of surface area and might exceed the reaction rate under experimental conditions. The most probable occurrence will be an increase in supersaturation if the vapor pressure of the substrate is maintained by a reaction in which the number of moles of gaseous products is greater than the number of gaseous reactants. Such is the case for aluminum oxide when the pressure of the reacting species is increased by increasing the substrate area in a constant volume, steady-flow system. Under these conditions, the equilibrium is displaced in the direction of volume contraction and aluminum oxide is deposited. This effect would not be expected in a system where $\Delta n = 0$ (i.e., metals).

From the evidence for substrate area effects, the derived expression, Eq. 17, can be modified to include the substrate partial pressure, P_s.

$$\Sigma_T = \frac{(P_{O_2}P_{Al_2O})_a + (P_{O_2}P_{Al_2O})_s}{(P_{O_2}P_{Al_2O})_e} = K_1\left[\frac{P^2_{AlCl_3}P^3_{H_2}P^3_{CO_2}}{P^3_{CO}P^6_{HCl}}\right] \tag{18}$$

The equilibrium partial pressure of the system, P_e, should remain unchanged and dependent upon the initial parameters of the system. The degree to which P_e is exceeded will be the supersaturation reference. Gas compositions must be selected to compensate for the effect of substrate area in steady-flow, vapor-phase reaction systems where product morphology is critical.

Kinetic Considerations

Kinetic theory suggests that heterogeneous nucleation may be limited by one or more of the following factors: (1) the rate of reaction in the vapor phase, (2) the rate of diffusion in the gaseous phase, (3) the nature of the solid-vapor interface, (4) the rate of deposition, (5) the configuration and thickness of the boundary layer, (6) the rate of surface diffusion. The relation between the supersaturation and the equilibrium constant indicates that a vapor process based on a reaction with a high negative free energy and controlled by appropriate gas inputs is experimentally attractive. For example, the volume change usually associated with the gaseous reaction forming a solid product may appear as a pressure fluctuation in a system of constant volume; the magnitude of this volume-pressure change is reduced as the total number of gaseous moles is increased. For different reactions, each producing the desired product, the kinetics of each are related to the respective standard free energies of reaction. The relation shown in Eq. 19 indicates the influence of the reaction free energy on the mass action constant.

$$K_{P_T} = e^{-\Delta F^\circ/RT} = k_f/k_b \tag{19}$$

A negative increase in the reaction free energy will increase the difference between the rate of the forward reaction, k_f, and the rate of the reverse reaction, k_b, under equilibrium conditions; that is, there is an increased tendency toward the formation of the desired product. The equilibrium constant will be increased exponentially by an increased reaction free energy and will produce a large effect on the supersaturation.

The mass action constants for some component reactions may be used to anticipate limiting kinetics in the vapor-phase reaction system. For example, the Arrhenius relation,

$$k = Ae^{-Q/RT} \tag{20}$$

where Q is the activation energy and A is the frequency factor, can be used to predict the effect of temperature on the reaction rate. In most cases, an increase in temperature will increase the rate of the forward reaction. A similar prediction cannot be made for the effect of changing component partial pressures.

VAPOR-PHASE WHISKER-GROWTH OPERATIONS

Reaction Reversibility

If an experimental system is to be controlled by component partial pressures, it is necessary that the summary reaction be reversible. The reversibility of the aluminum oxide reaction is implied in Eq. 1 and in the thermodynamic prediction of the temperature effect.

For the summary reaction shown in Eq. 1, the reversibility and temperature effects were observed experimentally. Under constant pressure and for different temperatures (1100 to 1800°C), a series of runs was made to form vapor reaction products and to subsequently dissolve the initial product by appropriate changes in gas compositions. The rate of product formation was observed to be more rapid than the rate of solution when the changes in gas composition provided partial pressure ratios proportional to the equilibrium ratio. Similarly, high concentrations of product gases increased the rate of solution of aluminum oxide. For comparable gas ratios, the rates of formation and solution increased as the temperature of the reaction chamber increased.

A boat containing fine, aluminum oxide powder apparently decreased the rates of deposition and solution in the reaction chamber. Morphological evidence indicated the decrease in deposition rate to be associated with the growth rate intrinsic to the product. The decreased solution rate was attributed to the uncompensated increase in the volume of aluminum oxide in the system. The rate changes observed for constant vapor-reaction parameters with an increased volume of aluminum oxide suggested the reversibility of the condensation reaction.

The reaction reversibility and the relation between morphology and gas composition supported the applicability of the reaction shown in Eq. 17. It should be noted that discrete amounts of product gases must be introduced with the reactants in order to achieve supersaturation control in a steady-flow system.

Pressure and Velocity Effects

The supersaturation relation, Eq. 17, derived from thermodynamic quantities does not include the total pressure of the system. Several experiments were conducted to determine the validity of the expression for isothermal conditions and different system pressures. At pressures near 1 atm, dendrites and platelets developed from individual whiskers because precise gas control was limited by the flow-metering system. Incrementally decreasing the system pressure while maintaining constant gas compositions enhanced the formation of whiskers. The pressure changes in the steady-flow system produced gas velocities up to 8000 cm/sec. At the higher velocities, whisker growth diminished and reactants were condensed beyond the reaction zone. From these observations it was concluded that supersaturation was insensitive to system pressure and greatly dependent upon reactant gas velocities.

The ability to reverse the reaction and control the degree of supersaturation by selection of gas compositions suggested that the rates of reaction and diffusion in the gas phase were sufficiently high to permit a broad range of velocities in a steady-flow system. No effects detrimental to whisker growth were ascribed to coherent or semicoherent substrate surfaces. A dirty or contaminated surface was observed to retard heterogeneous nucleation. Products were formed at similar rates under the same supersaturation conditions on a wide variety of substrates. Observed variations in growth rates on static substrates which were positioned relative to the injector tip suggested the influence of boundary layer configuration and thickness. It was concluded from these experiments that turbulence in the boundary layer improves the kinetics of heterogeneous nucleation on a static substrate and that the configuration and thickness of the boundary layer is a critical kinetic factor for heterogeneous substrate nucleation.

Control of Whisker Diameters

The effect of nuclei size on growth diameter was determined for whiskers grown on static and gas-borne substrates. The microstructural analysis of three static substrates (high purity, polycrystalline aluminum oxide) provided average grain sizes of 39 μ, 33.5 μ, and 30.5 μ. The respective range in grain size was 7.5 to 100 μ, 11 to 82 μ, and 7 to 71 μ. The average base diameters of whiskers grown on these substrates approximated the average grain diameter in each substrate. Although average values of substrate-grain and whisker-base diameters corresponded, the range in whisker-base diameters was substantially less than the grain size range. This variance probably stems from the variation in the crystallographic orientation of the grains on the substrate surface.

An apparent relation between the whisker-base diameter and the size of gas-borne nuclei was noted. Although no rigorous statistical analysis was made, it was observed that the average base diameter was proportional to the average nucleus size and that the distribution of nuclei and whisker-base dimensions were similar. The proportionality factor was related to the residence time of the particles in the growth chamber.

Substrate Effects on Supersaturation

Following the determination of conditions for vapor-phase whisker growth on static substrates, several attempts were made to grow whiskers on gas-borne nuclei. Examination of the products of each run showed that the aluminum oxide nuclei had been polygonized by three-dimensional growth. Since the same gas composition, temperature, and velocity had produced profuse whisker growth on static substrates, it was evident that the injection of nuclei or the characteristics of the nuclei had changed the

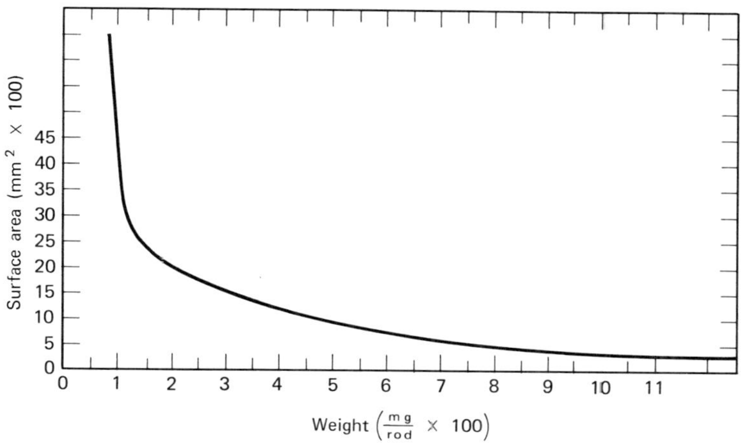

Figure 2.13 The effect of surface area on condensation rate.

supersaturation of the system. Several experiments showed the apparent change in supersaturation to be associated with the volume or surface area of the injected powder. Gas conditions were established for instantaneous whisker growth on an aluminum oxide boat; subsequent runs were made with the boat containing varying amounts of powder. As the amount of powder increased, three-dimensional growth on the boat and powder surface increased and whisker growth decreased. When the gas composition was changed to increase the input flow of product gases, whisker growth was dominant on powder-filled boats. Similar results were observed for powders of different purity and grain size. It was evident from these observations that the nuclei surface area must be considered to establish optimum conditions for whisker growth.

The substrate area was varied in a series of experiments to determine the effect of surface area on product morphology. Ten-minute runs were made with the following gas flows: 0.95 lpm H_2, 0.22 lpm CO, 0.80 lpm CO_2, 0.03 lpm Cl_2, and 0.02 lpm excess Cl_2. A polycrystalline rod of known composition, dimensions, and weight extended from the injector to the exhaust end of the reaction chamber. A reaction-zone temperature of 1560°C was used with a 10 cm flat. The substrate surface area was increased by increasing the number of rods in subsequent runs. Figure 2.13 shows the change in substrate weight per unit area with the increase in substrate surface area. The product morphologies are tabulated in Table 2.6 and agree with the results reported by Sutton [40]. Since residence time, input gas composition, and temperature were held within 1%, the changes in product yield and morphology reflect the effect of substrate surface area on the supersaturation.

TABLE 2.6 PRODUCTS OF SURFACE AREA EXPERIMENTS

Number of Substrate Rods	Whiskers	Platelets	Crystals	Powder
0.1	Whiskers with aspect ratios of 800 to 1	None	None	None
1.0	Abundant fine whiskers with aspect ratios of 200–300 to 1	Few bladelike crystals	None	Grainy and highly flocked powders
2.0	Few stubby and thickened whiskers	Abundant bladelike crystals	Few three-dimensional crystals	Much lightly flocked powder
3.0	Sparse, short, and quite thick whiskers	Abundant flat plates and blades with hexagonal faces on their sides	Abundant hexagonal crystals	Abundant lightly flocked powder

SUMMARY

The theories for whisker growth and their limits of applicability in various systems have been presented. The extension of theory to practice in evaporation-condensation, chemical reduction, and vapor-phase reactions has been considered with particular emphasis on the advantages and limitations that govern the ability of each method to produce abundant, high-strength, uniform whiskers. Because of the promise of a vapor-phase reaction process to establish continuous whisker production, the treatment of a representative system, Al_2O_3, has been detailed in thermodynamic, kinetic, and supersaturation considerations. The applicability of vapor reactions to other material systems has been indicated, as has the ease of process scale-up.

It is obvious that no mention has been made of the present production costs or those anticipated in the future. Suffice it to say that without chemical recirculation of product gases, the process will have an efficiency of 40% and the most optimistic estimate of product cost would be \$15 per pound. The reader should consider this cost figure as being near minimum for large-scale production.

ACKNOWLEDGMENTS

This work was conducted under Contract No. DA 19-020-AMC-0068 (X) and sponsored by the Army Materials and Mechanics Research Center. The aid and assistance of R. C. Folweiler, P. S. Schaffer, B. Albanese, L. Tagliani, and D. W. Jones of Lexington Laboratories, Inc. during the experimentation if gratefully acknowledged. The author is deeply thankful to W. D. Kingery, Massachusetts Institute of Technology, for his stimulating discussions and suggestions.

REFERENCES

[1] E. J. Langham and B. J. Mason, "The Heterogeneous and Homogeneous Nucleation of Supercooled Water," *Proc. Roy. Soc., A*, **247,** 493–504 (1958).
[2] R. M. Fisher, L. S. Darken, and K. C. Carrol, *Acta Met.*, **2,** 368 (1954).
[3] J. Franks, *Acta Met.* **6,** 103 (1958).
[4] J. Hallett and B. J. Mason, "The Influence of Temperature and Supersaturation on the Habit of Ice Crystals Grown from the Vapor," *Proc. Roy. Soc., A*, **247** (1958).
[5] W. Burton, N. Cabrera, and F. C. Frank, *Phil. Trans. Roy. Soc.*, **243,** 299 (1951).
[6] F. C. Frank, *Discussions Faraday Soc.*, **5,** 67 (1949).
[7] G. W. Sears, *Acta Met.*, **3,** 361 (1955).
[8] G. W. Sears, *Acta Met.*, **3,** 367 (1955).
[9] W. W. Webb, R. D. Dragsdorf, and W. D. Forgeng, *J. Appl. Phys. Rev.* **28,** 498 (1957).
[10] W. W. Webb, R. D. Dragsdorf, and W. D. Forgeng, *J. Appl. Phys.*, **29,** 814 (1958).
[11] G. K. Chadha and G. C. Trigunayat, "Optical and X-ray Study of Polytypism In Vapour-Grown Crystals," *Crystal Growth, Proceedings of An International Conference*, Pergamon, New York, 1967.
[12] S. S. Brenner, unpublished data.
[13] G. W. Sears, *Acta Met.*, **3,** 361 (1955).
[14] G. W. Sears, *Acta Met.*, **4,** 268 (1956).
[15] R. Gomer, *J. Chem. Phys.*, **28,** 457 (1958).
[16] W. Dittmar and K. Neuman, *Z. Elektrochem.*, **64,** 297 (1960).
[17] V. Ruth and J. P. Hirth, "The Kinetics of Diffusion Controlled Whisker Growth," AD439497, Nonr 495 (26) (1964).
[18] R. V. Coleman and G. W. Sears, *Acta Met.*, **5,** 131 (1957).
[19] R. V. Coleman, *Met. Rev.*, **9,** 35 (1964), 265.
[20] G. W. Sears, *Acta Met.*, **3,** 7 (1955), 367–369.
[21] G. W. Sears, *Acta Met.*, **1,** 8 (1953), 457–458.
[22] A. J. Melmed and R. Gomer, *J. Phys. Chem.*, **34,** 1802 (1961).
[23] P. B. Price, *J. Appl. Phys.*, **32,** 1746 (1961).
[24] M. I. Kliman, "Formation of Alumina Whiskers," AMRA, WAL TR 371/52 (1962).
[25] S. S. Brenner, *Acta Met.*, **4,** 62 (1956).
[26] S. S. Brenner, "Metals," *The Art and Science of Growing Crystals* Gilman, Ed., John Wiley and Sons, New York, 1963.
[27] W. W. Webb and W. B. Forgeng, *J. Appl. Phys.*, **28,** 12 (1957).
[28] J. A. Talento, "The Growth of Alpha-Alumina Whiskers," presented American Ceramic Society Meeting, 1962 (unpublished).
[29] S. S. Brenner, *J. Appl. Phys.*, **33,** 1 (1962), 33–39.
[30] N. C. Tombs and A. J. E. Welch, *J. Iron Steel Inst.*, **172,** 69 (1952).

[31] E. Feingold and A. Gatti, Tech. Rept. R65SD11, General Electric Space Sciences Lab., 1965.
[32] W. H. Sutton, private communication.
[33] D. J. Barber, *Phil. Mag.*, **10,** 103 (1964), 75.
[34] J. K. Alley, R. C. Johnson, C. Huggins, and H. R. Shell, Bu. Mines Rept. 6220 (1963).
[35] E. G. Wolff and T. D. Coskren, *J. Am. Ceram. Soc.*, **48,** 6 (1965), 279–285.
[36] P. S. Schaffer, *J. Am. Ceram. Soc.*, **48** (1965).
[37] W. B. Campbell, *Chem. Eng. Progr.*, **62,** 3 (1966), 68–73.
[38] W. B. Campbell, "Determination of the Feasibility of Forming Refractory Fibers By A Continuous Process," AMRA CR 63-03/8F (1966).
[39] R. F. Regester, P. D. Gorsuch, and L. Girifalco, *Mater. Res. Std.*, **7,** 5 (1967).
[40] W. H. Sutton, Tech. Rept. R62SD994, General Electric Space Sciences Lab., 1962.

R. S. Wagner

3 VLS Mechanism of Crystal Growth

EXPERIMENTAL OBSERVATIONS LEADING TO THE VLS CONCEPT

Introduction

Crystal growth from the vapor is usually explained by a theory proposed by Frank [1] and developed in detail by Burton, Cabrera, and Frank [2]. In this theory a screw dislocation that terminates at the growth surface provides a self-perpetuating step. Accommodation of atoms at the step is energetically favorable and is possible at much lower supersaturation than required for two-dimensional nucleation. Crystals of a unique form result from anisotropic growth from the vapor—"whiskers," or filamentary crystals. Such crystals have a lengthwise dimension that is orders of magnitude larger than that of the cross section. For most filamentary crystals both the fast-growth direction and directions of slow lateral growth have small Miller indices. The special growth form for a whisker crystal implies that the tip surface of the crystal must be a preferred growth site. Sears [3] proposed that, according to the Frank theory, a whisker contains a screw dislocation emergent at the growing tip. Such an axial dislocation provides a preferred growth site and accounts for unidirectional growth. The hypothesis was extended by Price, Vermilyea, and Webb [4], still implying the presence of a dislocation at the whisker tip. They postulated that impurities arriving at the fast-growing tip face become buried while those arriving on the surface of the slow-growing lateral faces accumulate and thereby hinder growth. These considerations led to a whisker morphology. Axial screw dislocations have been reported for a few filamentary materials lending support to the

Frank mechanism. However, reservations have arisen regarding its universal application. Webb and co-workers [5] examined whisker crystals of nine different metals grown from the vapor and found unequivocal evidence for a single axial dislocation in only one metal—palladium. These considerations have stimulated a careful study [6] of morphology and growth features, structural defects, and the role of impurities in silicon filamentary crystals.

Silicon has several advantages for such an investigation. Whisker crystals can be grown readily by disproportionation of SiI_2 [7–10]. Several independent methods, such as etching, infrared and electron microscopy, and X-ray diffraction are available to study imperfections. It is particularly important that silicon does not deform plastically at room temperature and, therefore, it is unlikely that dislocations are introduced during experimentation.

Preparation of Whiskers and the Thermochemistry of the Process

Whiskers were prepared by reaction of silicon and iodine in a closed quartz tube in a temperature gradient following the procedure described by Greiner [7]. In a typical experiment, a carefully cleaned reaction tube (25 cm long and 140 cm^3 in volume) is charged with approximately 1 g of high-purity silicon (resistivity 2000 ohm cm) and 0.84 g of high-purity iodine. A known amount of a suitable impurity is added, for example, 1 mg of nickel. The reaction tube is subsequently evacuated to a pressure of about 10^{-6} torr, sealed, and placed in a furnace with a temperature profile ranging from 1100 to 800°C. The silicon source material and the impurity are located at the higher temperature region. Crystal growth is observed after about 10 minutes. The quantity, character, and size of whiskers and the region of deposition depend on the impurity added and on the duration of heating. Whisker crystals have been prepared with cross-sectional dimensions as small as 500 Å and as large as 100 μ or more, with lengths to a few centimeters.

The thermochemistry of the process has been developed by Schäfer and Morcher [11] and by Wadja and Glang [12]. The process uses the disproportionation reaction:

$$2SiI_2(g) \rightleftharpoons Si(s) + SiI_4(g)$$

At a high temperature, typically 1050°C, and with pressure of the iodine species above about 80 torr, the reaction proceeds to the left with the production of SiI_2. The SiI_2 moves to a region of lower temperature, typically 900 to 1000°C, where the reaction goes to the right. Silicon is deposited and SiI_4 returns to the higher temperature region to react with additional source silicon. It can be assumed that the chemistry of the transfer process is not affected significantly by the relatively small concentrations of impurities added.

Because the deposition of silicon is by a chemical reaction, there are complexities not encountered with condensation. The reaction could take place either homogeneously in the vapor phase followed by condensation of silicon or heterogeneously at local areas of the substrate or growing crystal. The question of a homogeneous versus a heterogeneous or surface-catalyzed reaction has been considered in detail for the reduction of halides by hydrogen to produce metal whiskers, for example, iron and copper [13–16]. Conclusive experiments show that in these cases a surface-catalyzed reaction is the main source of metal atoms. It seems likely that the disproportionation of SiI_2 is also surface catalyzed.

Role of Impurities

In earlier studies [7] of silicon filamentary crystals, it was found that with arsenic-doped silicon as the source material the addition of small traces of nickel iodide was essential to promote whisker growth. Hydrogen also was present in the reaction tube. Later, Gutowski and Beaver [17] extended the studies of impurities and found that in addition to nickel iodide, iodides of copper, manganese, silver, and cadmium promoted whisker growth when the source material was arsenic-doped silicon. They did not obtain filaments when the impurities were iron, zinc, or magnesium.

The pronounced impurity effect was emphasized in the studies described here when filamentary crystals were not obtained from high-purity source silicon (2000 ohm cm resistivity) and high-purity iodine. The product either consisted of nodules or was a film. If, however, a trace of gold (about 1 mg) was added to the high-purity materials, a copious crop of filamentary crystals resulted. The ratio of filamentary to nodular deposit increased with increasing addition of gold. Transport experiments indicated that under effective temperature and pressure conditions, the amount of transported silicon is independent of the presence and concentration of impurities.

A number of other impurities promoted the growth of filamentary crystals. The impurities that produced extensive filamentary growth under the conditions of the present study were Au, Ni, Pd, Cu, Gd, Mg, and Os. The following impurities did not promote growth: Zn, C, Mn, Sn, and Ge; only a film or a nodular deposit was obtained. When a small amount of oxygen was added intentionally, no transfer occurred. The silicon source material became discolored, probably by an oxide film.

Morphology of Whiskers

Silicon deposits from SiI_2 disproportionation may be of three forms. The first is either a film or nodulelike deposit attached to the walls of the reaction

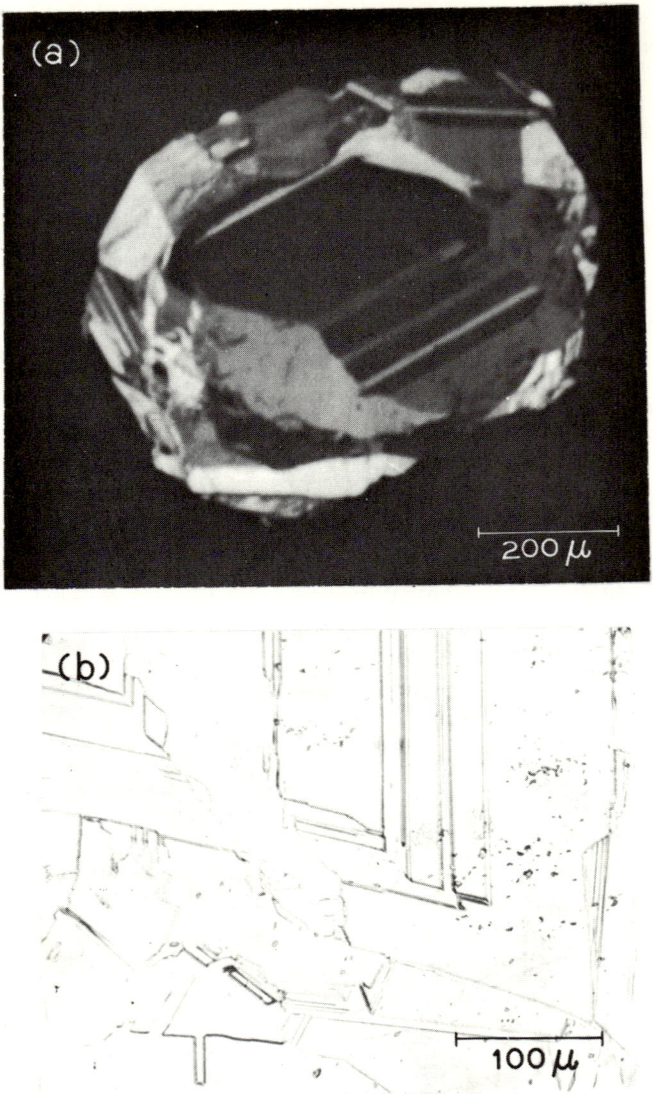

Figure 3.1 Silicon nodular crystal formed in the absence of impurities. (a) A typical nodule showing surface twinning features; (b) a section through a nodule etched to show twin plane markings.

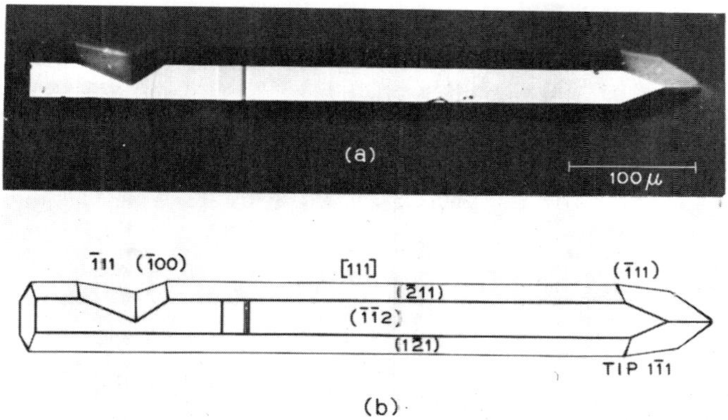

Figure 3.2 Silicon needle morphology: (a) a typical silicon needle showing surface and tip morphology; (b) schematic drawing of the morphology of a ⟨111⟩ needle. The needle is 6-sided with a ⟨111⟩ axis. The main lateral surfaces are {211}.

tube. A nodule and a section through a nodule are shown in Figure 3.1. Nodules are polycrystalline, containing a large number of twin orientations. The second form, in the micron size range, is a hexagonal needle growing in a ⟨111⟩ direction [9, 18] with well-developed {211} lateral faces. Needle crystals prepared from high-purity silicon or from silicon doped with arsenic or boron, and with nickel as the growth-promoting impurity had the same morphology. The third form is a twinned ribbon growing in a ⟨211⟩ direction [19] with a {111} twinning plane parallel to the main faces.

In the present study, 15 needle crystals were examined for growth direction, presence of twinning, and external faces using light microscopy and a Weissenberg X-ray diffraction technique [20]. All needles had a hexagonal cross section, as shown in Figure 3.2a, and the larger ones frequently were tapered. The growth direction was ⟨111⟩ and the lateral faces were {211}, as illustrated in Figure 3.2b. Twinning over the length of the needle was not detected. The terminating faces of a needle were established unambiguously to be {111} type. Sometimes notches were present in the hexagonal column as seen in Figure 3.2a. The surfaces of the notch are {111} and {100} type planes. The gradual change of cross section of larger needles probably occurs through small step planes of these types. A branched needle in Figure 3.3, which occurs rarely, was examined and the entire crystal was found to be a single orientation. The branch resulted from simultaneous growth in another [111] direction from the point of branching.

More than 20 ribbons and blades have been examined by X-ray diffraction. Typical ribbons are shown in Figure 3.4. In all cases, ribbons and blades were twinned crystals, as first reported by Wagner and Treuting [19], with the {111} twinning plane parallel to the {111} main faces. In thin ribbons the

Figure 3.3 A branched needle. The branched and main stems have the same orientation.

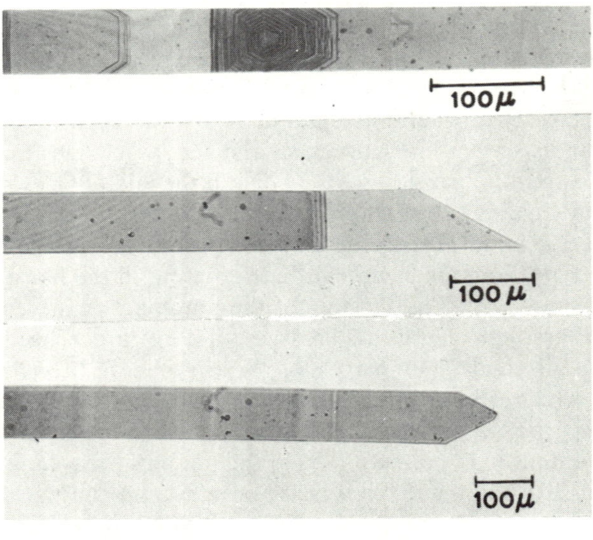

Figure 3.4 Photographs of typical ribbons. The one at the top shows faceting, which frequently occurs. The lower two show growth terminations often observed.

growth direction is ⟨211⟩ but thicker blades obtained after prolonged growth may have other edge directions.

Macroscopic Growth of Crystals

The growth of crystals was viewed in a furnace fitted with an observation window. The temperature gradient could be varied through control of the current in sections of the furnace while the reaction tube could be translated to observe the source material or the site of crystal growth. By raising the temperature in a region of growth, whisker crystals could be made to evaporate.

When the gradient in the furnace was adjusted so that the source material was at about 1100°C and observations were made in a region of lower temperature, where crystal growth could be expected, the first visible occurrence was the development of a number of dark specks on the tube wall. Each small speck rapidly developed into a cluster of fine radiating filamentary crystals (needles and ribbons) that often grew in a few seconds to lengths of a centimeter. A cluster of this type is shown in Figure 3.5. The initial growth was rapid and similar to the leader growth reported by Gordon [21] for hydroquinone from water solution, and by Brenner [22] for copper whiskers produced by hydrogen reduction of CuI. With a steep temperature gradient the initial rate of growth was large; lengthening was so rapid that a person could not consciously observe the progress. After a short period of rapid growth, increase in length ceased or continued only at a slow rate. Thickening

Figure 3.5 Silicon filaments extending radially from a nucleation site.

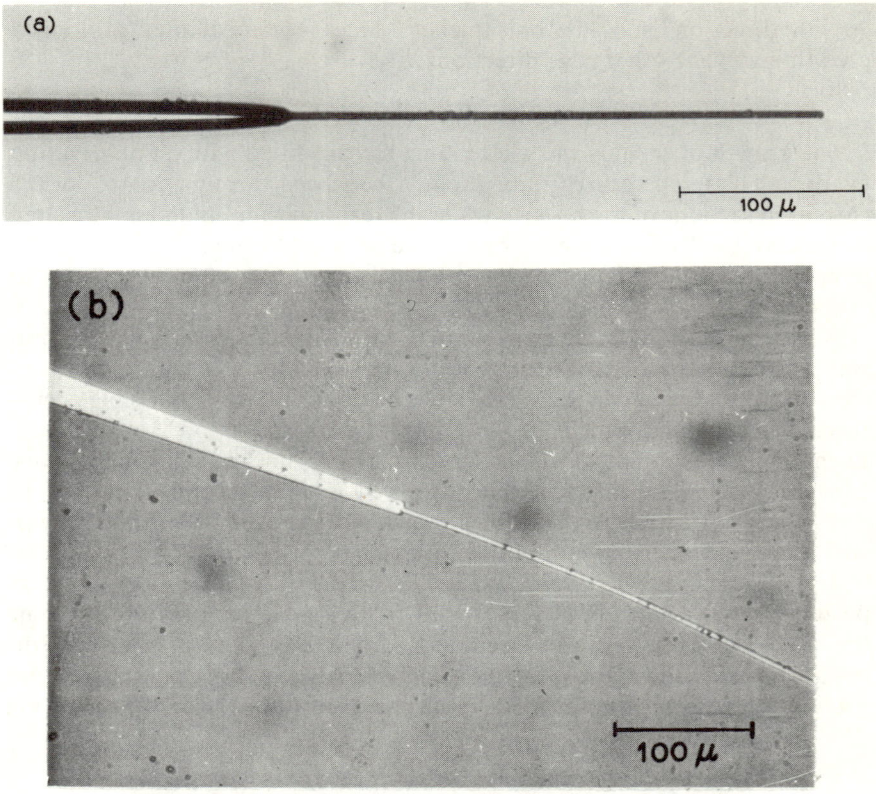

Figure 3.6 Regrowths on needles and blades: (a) needle; (b) blade.

then began. The final product was one of several forms: (1) a very fine clump of woollike needles or ribbons, (2) a thick cluster of ribbons, (3) a thinly grouped cluster of small needles or ribbons, (4) large needles and bladelike growths.

Experiments were performed in which the temperature of the growth region was raised relative to other parts of the reaction tube. Under these conditions the needles and ribbons evaporated back from the tip. On lowering the temperature, evaporation was halted and regrowth began. The new growth was an extension of the parent crystal but with a smaller cross section, as illustrated in Figures 3.6a and b. Quite frequently the author observed a hemispherical termination at the tip of the regrowth region, which appeared liquid.

Orientations of the parent and regrowth regions were determined by X-ray diffraction. Needles regrew in the $\langle 111 \rangle$ direction as a continuation of the original orientation; ribbon regrowth was twinned and also was a continuation of the orientation of the parent twinned crystal. No new orientation was observed. We conclude that the condition necessary for regrowth of the same morphology was present in the solid after a substantial end portion had evaporated. If a dislocation was required for growth, it must have survived the thermal treatment and stresses that accompanied the original growth and partial evaporation.

Abrupt changes in growth direction (kinking) were sometimes observed during growth of $\langle 111 \rangle$ needles. The extension in length beyond the kink definitely established that these crystals grew by addition of material at the tip. The regrowth experiments discussed above also show that silicon whiskers lengthen by adding on at the tip.

Search for Imperfections in Silicon Whiskers

A search was made for imperfections in both needles and ribbons by chemical etching, by examination of needles for the twist proposed by Eshelby [23], and by observation of needles and ribbons of about 4000 Å and smaller cross sections with transmission electron microscopy.

Hexagonal needles of about 50 μ section were cleaved to expose a transverse $\{111\}$ surface and then etched with CP-4,* Superoxol,† W-Ag,‡ or Sirtl§ etch. Etching developed a triangular pit in the center of the needle cross section as seen in Figure 3.7 suggesting the presence of an axial screw dislocation. The etch pattern surrounding the etch pit consists of closed hexagons and is not a spiral. When such a needle was viewed on its side with transmitted infrared light, a narrow dark band was observed extending longitudinally in the crystal. A similar narrow band was observed in a ribbon when viewed with transmitted light normal to a $\{111\}$ major surface. However, after annealing in vacuum at 1350°C for 2 hours, the central, darker band was no longer observed in either needles or ribbons. Likewise, the etching of a cleaved $\{111\}$ surface of a heat-treated needle did not reveal a pit or the hexagonal pattern of Figure 3.7. Rather, a shallow round depression formed in the central region.

It is concluded that the narrow dark bands in needles or ribbons are regions of relatively higher impurity content resulting from the initial rapid growth of the leader, which subsequently thickens. The annealing treatment caused a diffusion of the impurities throughout the specimen and the disappearance of the dark band. These observations emphasize the role of

* 120 cc HF, 200 cc HNO_3, 120 cc CH_3COOH, 2 cc Br.

† Superoxol = $HF:H_2O::1:1:2$.

‡ $HNO_3:HF:5\%\ AgNO_3$ sol. $::1:1:2$.

§ HF:50 g CrO_3 in 100 cc $H_2O::1:2$.

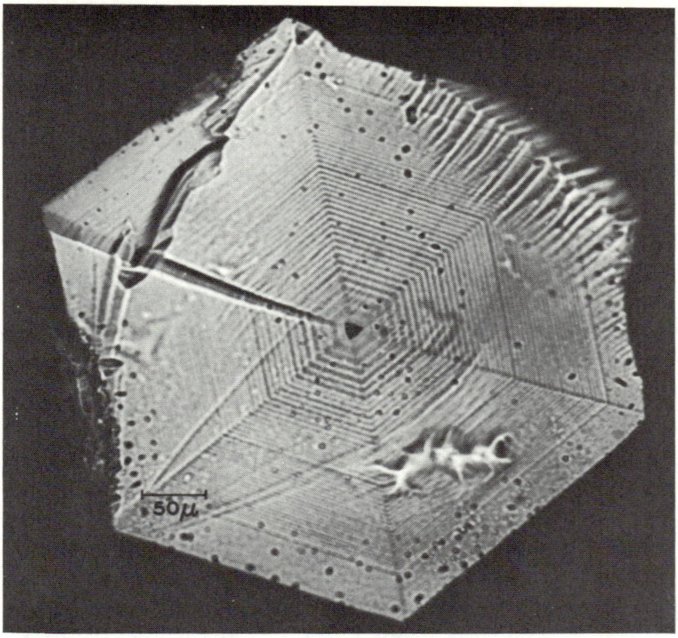

Figure 3.7 The etched cleaved cross section of a hexagonal needle showing a central triangular etch pit and concentric hexagonal markings.

impurities in the growth of these crystals and suggest that the initial rapid and predominately unidirectional growth is related to impurities.

A test for Eshelby twist was made on three hexagonal ⟨111⟩ crystals of the sizes set forth in Table 3.1. The twist α in radians/mm for a single axial screw dislocation is given by the equation $\alpha = b/\pi R^2$ where b is the Burgers vector for the dislocation and R is the "radius" of the crystal. All dimensions are in millimeters. The crystals were examined by the Weissenberg moving-film technique on areas separated by a few millimeters. The calculated

TABLE 3.1 TEST FOR ESHELBY TWIST IN SILICON NEEDLES

Specimen Number	Cross Section (μ)	Calculated[a] Twist (α-deg/mm)	Separation of Irradiated Areas (mm)	Calculated Total Twist (°C)	Observed Twist (°C)
Si92C	10	0.32*	4.5	1.0	0
Si41A	3.5	1.86	2.0	3.7	0
Si41B	2.0	5.7	2.0	11.4	0

[a] Burgers vector is taken as $\frac{1}{3}d_{111} = 3.13$ Å.

total twist for a central screw dislocation for the largest crystal was large enough for easy detection by this method, precise to at least 0.2°. No twist was observed in any of the crystals. Since the whiskers examined were of micron size, it is improbable that a dislocation initially present could migrate out at room temperature due to imposed handling stresses. It is concluded that the crystals did not contain an axial screw dislocation during growth.

Whisker crystals with cross sections from 500 to 4000 Å, prepared with Ni, Au, and Pd as introduced impurities, were examined in the electron microscope. The most striking feature of the small whisker crystals was the constancy of cross section. A given whisker was seldom observed to change in width along its length, which was hundreds of times greater than the width. Cross sections of crystals were revealed by extinction contours in photographs of bent crystals [24]. The cross sections of untwinned whisker crystals were either hexagonal or circular. With gold as an added impurity, the cross section was hexagonal as in Figure 3.8a, and the side faces of the whisker were {110} planes. With nickel as an added impurity, both hexagonal and circular cross sections were observed and, in some cases, the cross section was rounded with flat segments as in Figure 3.8b. The growth axis of untwinned crystals was ⟨111⟩. The twinned crystals in a lot often had larger cross sections than needle crystals; 3000 Å for a twinned specimen compared to 1000 Å for needles. The twin plane usually extended through the narrow dimension of the crystal. The growth direction of twinned crystals was ⟨211⟩ and the broad flat faces were {110}. The narrow faces of the twin were irregular and could not be uniquely identified but were approximately {113}. Twinning could be readily detected in a crystal by tilting the specimen stage in the electron microscope to obtain a difference in extinction contrast between the two components. The twin plane was identified as {111}.

A few crystals were observed to contain many parallel twin boundaries. In general, there were no axial defects in any of the whisker crystals. There were two exceptions: one whisker, which had dislocations in a local area, and a broad, flat, twinned crystal which had a series of parallel dislocations. A few branched whiskers were found. In these, an axial dislocation, if present, would be pinned at the intersection and would be unable to glide out of the crystal. No dislocation was observed in such specimens.

Differences in morphology were found between submicron and micron or larger whiskers. The long axes of both were ⟨111⟩. In the initial lengthening of the leader, {110} side faces most frequently occur but whiskers with circular cross sections or {211} side faces are also found. Larger macroscopic crystals, produced by thickening of these leaders, have {211} side faces only.

A difference in morphology was observed also for twinned crystals. In

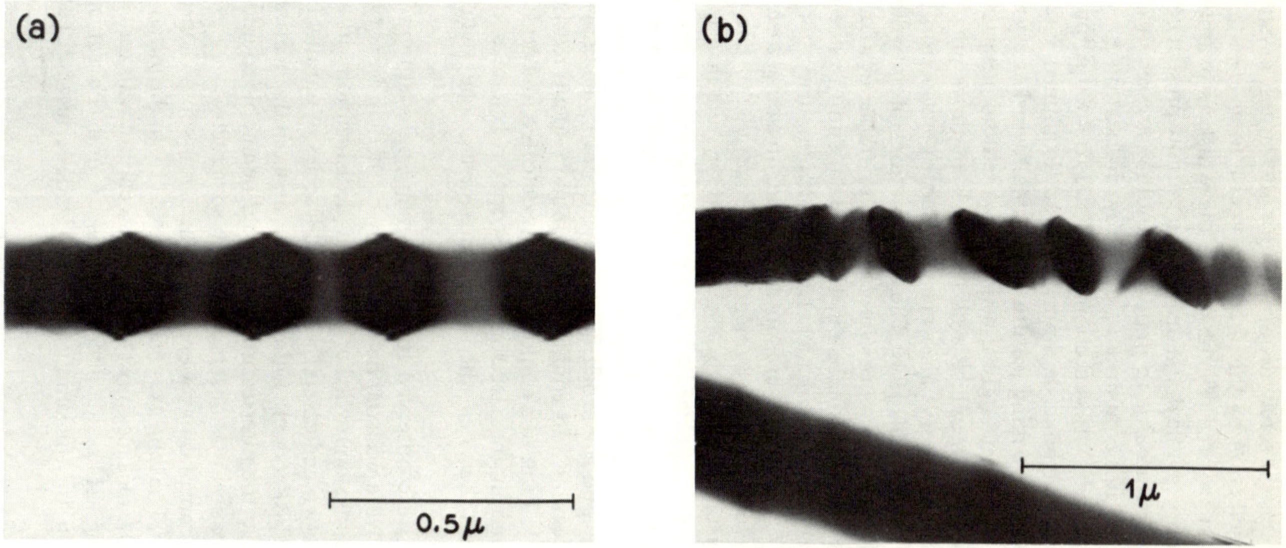

Figure 3.8 Extinction contours revealing the shape of silicon needles. (a) A hexagonal needle; (b) a nearly circular needle.

this case, both small and large crystals had $\langle 211 \rangle$ axes. The submicron twinned crystals have large flat $\{110\}$ faces, normal to the twin plane. The $\{111\}$ external faces, parallel to the twin plane, were not developed but were irregular surfaces which, on the average, were approximately $\{113\}$. The large twinned crystals, formed by the thickening of the smaller crystals, had large flat $\{111\}$ external surfaces parallel to the twin plane.

These differences in morphology support a two-stage growth process. The first stage, in which the whisker lengthens without change in cross section, is very rapid, as discussed earlier. It is believed that the lateral faces of the whisker are determined by surface-energy relationships and the growth mechanism at the tip. These lateral faces probably have a minimum surface free energy. During subsequent thickening of the whiskers in the second stage (a much slower process), the exposed side faces are determined by the relative rates of nucleation and growth of steps on these faces. The exposed faces on the larger crystals need not be equilibrium ones.

Discussions and Conclusions

It has been shown in this study that filamentary crystals of silicon do not contain axial screw dislocations or other line defects during growth. It was concluded from observation of growing whiskers that these crystals lengthen by the addition of new material at the tip. The requirement of certain impurities for filamentary growth and the absence of filamentary growth without these impurities is of the highest significance. Without impurities, silicon is deposited in the form of nodules or films. With certain impurities, silicon whiskers are formed. Whiskers grow rapidly initially as leaders which subsequently thicken. It was found that the impurity concentration in the leader is higher than in the subsequent overgrowth.

Whiskers with submicron diameters have extremely uniform cross sections. Thicker whiskers, however, exhibit relatively large steps on their surfaces. This strongly supports the concept that silicon whiskers grow by a two-stage mechanism; a rapid leaderlike growth in length followed by a subsequent slower growth in thickness. The effect of impurities on whisker growth appears to be most important during the growth of the leader and of lesser importance during epitaxial deposition on the leader.

It was found that needle crystals always grow in a $\langle 111 \rangle$ direction, and that twinned ribbon crystals initially grow in a $\langle 211 \rangle$ direction. The morphology and growth direction of ribbon crystals from the vapor and of "twin dendrites" [25] from a supercooled melt are identical except for the presence of but one twin plane in a ribbon crystal. In twinned dendrites the $\langle 211 \rangle$ growth direction is the mean of the $\langle 111 \rangle$ growth directions of twin components. It is reasonable to conclude that a ribbon crystal also grows in ⟨111⟩ directions, the two ⟨111⟩ directions of the twin components resulting in the observed ⟨211⟩ direction. The ⟨111⟩ direction, however, is the slowest growing one for silicon crystals from the vapor or from the liquid. Wajda and Glang [12], using the silicon disproportionation reaction, studied epitaxial deposition rates on silicon substrates of different surface orientations. The measured growth rates were found to be small for {111} and {100} and increasingly larger for {110} and {211} surfaces. The growth rate apparently increases with decreasing atomic density (increase in roughness) of the deposition plane in accordance with the bonding model. Therefore, the fast ⟨111⟩ growth rate of silicon whiskers cannot be attributed to normal epitaxial growth from the vapor phase.

It must be concluded that the screw dislocation theories of whisker growth cannot explain the experimental results of this investigation. However, the observations support a new concept of whisker growth, the vapor-liquid-solid mechanism described in the next section [26, 27].

THE VAPOR-LIQUID-SOLID MECHANISM

Introduction

The observations made in the preceding section can be summarized as follows:

1. Silicon whiskers are dislocation free.
2. Certain impurities are essential for whisker growth.
3. Whisker growth is a two-stage process: a fast initial extension in length (leader growth) followed by a slow increase in thickness (layer growth). The impurity concentration in the leader is higher than in the layer growth region.
4. The rapid initial extension in length occurs by the addition of material to the tip of the whisker.
5. The main growth direction of silicon whisker is ⟨111⟩, known to be the direction of slow growth from the vapor and from the liquid.
6. Silicon whiskers can be regrown from partly evaporated whiskers without change in morphology and defect structure.
7. A "liquidlike" globule has been observed occasionally at the tip of the whisker crystals during regrowth.

Observations on the growth of silicon whiskers are similar in many respects to those reported on the growth of whisker crystals of other materials [22]. The evidence raised a substantial doubt that silicon whiskers and filamentary crystals of other materials grow by a mechanism based on a dislocation model. Moreover, since impurities are essential to whisker growth, they must be involved fundamentally in the growth mechanism.

In addition, another important phenomenon that has received little attention is the presence of a liquid layer or droplets on the surface of some crystals growing from the vapor. Crystals in which this has been observed include *p*-toluidine [28], MoO_3 [29], ferrites [30], and silicon carbide [31]. The liquid layers or globules were considered to be metastable phases, molecular complexes, or intermediate polymers originating from condensation of the vapor phase. The possibility has been suggested that the halide being reduced is condensed at the tip [32] or adsorbed on the surface [22] of a growing metal whisker. Furthermore, the literature on whiskers discloses many illustrations of rounded terminations at the tip of the whiskers. These appear, for example, on crystals of Al_2O_3 [33, 34], SiC [35], and BeO [36].

The Vapor-Liquid-Solid Mechanism

The growth mechanism discussed in this chapter departs from previous theories in that the layer of liquid in which the crystal material is soluble is situated between the vapor and the growing crystal. The presence of a liquid layer in contact with the growing crystal has important consequences. The surface of the liquid has a large accommodation coefficient and is therefore a preferred site for deposition. The liquid becomes supersaturated with material supplied from the vapor, and crystal growth occurs by precipitation at the solid-liquid interface. The process is illustrated in Figure 3.9

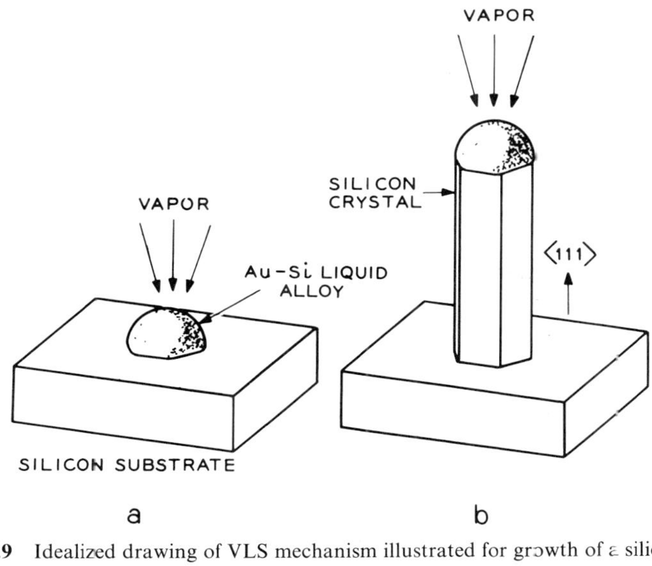

Figure 3.9 Idealized drawing of VLS mechanism illustrated for growth of a silicon crystal.

for the growth of a silicon crystal. Unidirectional growth is the consequence of an anisotropy in solid-liquid interfacial energy. This concept, in which vapor, liquid, and solid phases are involved, is called the vapor-liquid-solid (VLS) [26, 27] mechanism of crystal growth. The mechanism will be discussed first in general and then in detail for the growth of seeded silicon whiskers.

In crystal growth from the vapor, two processes may occur. First, deposition may be directly on a solid substrate in a vapor-solid system; for example, the epitaxial deposition of thin films [37]. The second process, which is the VLS mechanism, occurs in two steps as has been described. First, there is deposition from the vapor directly on a liquid solution in a vapor-liquid system. The second step occurs in a liquid-solid system and is precipitation from the supersaturated liquid solution at the liquid-solid interface.

For a condensation reaction, the growth rate J (atoms/cm^2 sec) of an extensive crystal surface can be estimated from gas kinetics as:

$$J = \alpha\sigma p_o(2\pi mkT)^{-\frac{1}{2}} \tag{1}$$

where α is the accommodation coefficient, $\sigma = (p - p_o)/p_o$ is the supersaturation of the vapor in which p is the vapor pressure and p_o is the equilibrium vapor pressure of the solid at temperature T. α is the fraction of impinging atoms which becomes accommodated on the growing surface. Theory predicts that the value of α depends on the state of perfection of the surface and on σ. Figure 3.10, a modification of one by Hirth and Pound [38],

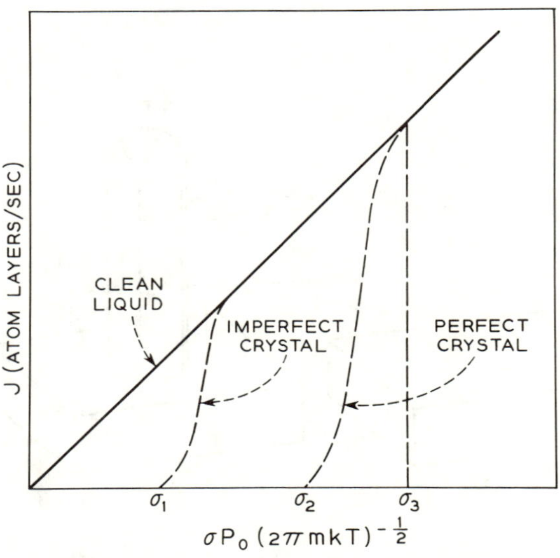

Figure 3.10 Growth rate of a crystal with a surface of unit accommodation (clean liquid), a crystal with an emergent screw dislocation (imperfect crystal), and a perfect crystal.

shows the predicted growth rate for an imperfect crystal surface (one that contains a screw dislocation), a perfect crystal surface, and a surface of unit accommodation coefficient (clean liquid). For the first two cases, the growth rate is essentially zero below a critical supersaturation (σ_1 or σ_2, respectively) and increases to the ideal rate with increasing σ. Nucleation of new layers, within the respective ranges of supersaturation, occurs either by two-dimensional nucleation on the perfect surface or through the agency of an emergent screw dislocation on the imperfect one. The straight line depicts the growth rate for a clean liquid surface (ideal rate). Theory predicts an accommodation coefficient of one for such a surface. A clean, liquid surface is distinctly different from that of a perfect, or imperfect, crystal surface and can be considered as ideally "rough"; that is, composed of ledges and steps only interatomic distances apart [38]. Accommodation sites exist over its entire area. Deposition on a liquid surface, therefore, follows the ideal growth rate law depicted in Figure 3.10, even at extremely small supersaturation.

For the VLS mechanism to operate, the liquid solution must be stable. This condition imposes a limitation on the minimum diameter of a VLS crystal. The stability of a liquid droplet of curvature r in its own vapor depends on the degree of supersaturation σ. The minimum critical radius is given by the equation:

$$r_{\min} = \frac{2\sigma_{LV} V_L}{RT \ln \sigma} \tag{2}$$

where σ_{LV} is the liquid-vapor interfacial energy, V_L is the liquid molar volume and R and T have the usual meanings. For most materials the minimum critical radius, $r_{\min}$, is about 10^{-5} cm for $\sigma = 1.02$. For whiskers with submicron dimensions, stability of the liquid droplet can be obtained by an appropriate increase in supersaturation of all components of the liquid.

Next, we consider deposition from the vapor on a perfect crystal surface with a local area covered by a liquid solution. Such a solution can be readily formed by alloying a suitable solvent agent (impurity) with the crystal substrate. From Figure 3.10 it follows that for a supersaturation smaller than σ_2, the deposition rate on the liquid is ideal, whereas in this range no deposition occurs on a perfect crystal surface. Within the range of $\sigma < \sigma_3$ there is preferred deposition on the liquid as compared to the crystal surface. Similar considerations hold for a liquid on an imperfect crystal surface. It will be shown later that the difference in deposition rates on a liquid and on a crystal surface can be orders of magnitude.

Predictions of Figure 3.10 hold strictly for evaporation-condensation and for homogeneous vapor-phase reactions. For a heterogeneous reaction, the presence of the liquid droplet in the deposition region can have additional effects. The large accommodation coefficient results in a long "lifetime" of the adsorbed gas molecules at the liquid surface. Furthermore, the impurity forming the liquid with the substrate material can act as a catalyst for the heterogeneous reaction. Each factor results in an enhanced deposition rate on the liquid. An additional complexity, particularly in the application of Eq. 1, is the maintenance of chemical equilibrium in a dynamic system.

The liquid solution in contact with the crystalline substrate seeks to maintain the equilibrium composition at a given deposition temperature. However, the liquid solution receives a flux of material from its surface. The resulting concentration gradient in the solution provides the driving force for diffusion from the surface to the liquid-solid interface. The material transport in the liquid is diffusion-controlled and occurs under essentially isothermal conditions.

Requirements for Agent

A suitable solution-forming agent (impurity) for VLS growth must meet a number of requirements: (1) It must form a liquid solution with the crystalline material to be grown at the deposition temperature. (2) From (1) it follows that the distribution coefficient of the agent, $k = C_s/C_l$, must be less than one, where C_s and C_l are the solubilities of the agent in the solid and liquid, respectively, at the deposition temperature. In the absence of a flux of agent from the vapor, the length of a crystal grown by VLS is inversely proportional to the k value. For whisker growth k should be about 10^{-4} or smaller; for large area growth, k values as large as 0.1 can be tolerated. (3) The equilibrium vapor pressure of the agent over the liquid alloy should be small. Evaporation of the agent from the liquid solution does not change the composition, but does change the volume and, therefore, the cross section of the crystal. (4) The agent must be inert to the chemical reaction products, for example, the halogen acid formed in a reaction involving the reduction of a halide. This criterion does not enter into an evaporation-condensation reaction. (5) The interfacial energies—vapor-solid, vapor-liquid, and liquid-solid—play an important role. The wetting characteristics arising from these parameters strongly influence the shape of the growing crystal. A small contact angle is desirable for large area growth; a large contact angle leads to whisker formation. (6) In VLS growth of compound crystals, such as GaAs or SiC, the agent can be one of the components, Ga or Si. (7) In some cases, the agent and the deposition temperature must be selected to avoid the formation of a solid intermediate phase of the agent with one of the constituents of the vapor. (8) For controlled unidirectional growth by VLS, the solid-liquid interface must be well defined crystallographically. The requirement implies that the "growth front" of the crystalline material is faceted. Jackson [39] has shown that only a restricted group, which includes silicon and bismuth, grows faceted from the melts. Faceted growth can be obtained, however, for the group of materials containing most metals, if growth occurs from a dilute solution. At a given deposition temperature, the agent determines the composition of the liquid alloy. By the proper selection of an agent, therefore, the condition for VLS growth, namely growth from a dilute solution, can be met.

Experimental Study of VLS Mechanism

The apparatus for fundamental experiments on VLS and the growth of silicon crystals is shown in Figure 3.11. This apparatus is similar to that of Theuerer [37]. Crystals were grown with a resistance furnace 7 as shown and also with a graphite susceptor and radio-frequency heating. In most instances, the substrate 9 consisted of a single-crystal silicon wafer placed on a fused silica support 8 within the reaction tube 6. In some experiments, when the reaction tube was in a vertical position, the substrate was on a graphite pedestal which served as the susceptor. The substrate temperature was measured with an optical pyrometer or with a thermocouple to within $\pm 10°C$.

A temperature gradient in which the liquid surface is colder than the substrate interior must be avoided for controlled VLS growth. As Pfann [40] has shown, such a temperature gradient gives rise to the phenomenon of temperature-gradient-zone melting (TGZM), in which the liquid alloy moves toward the higher temperature.

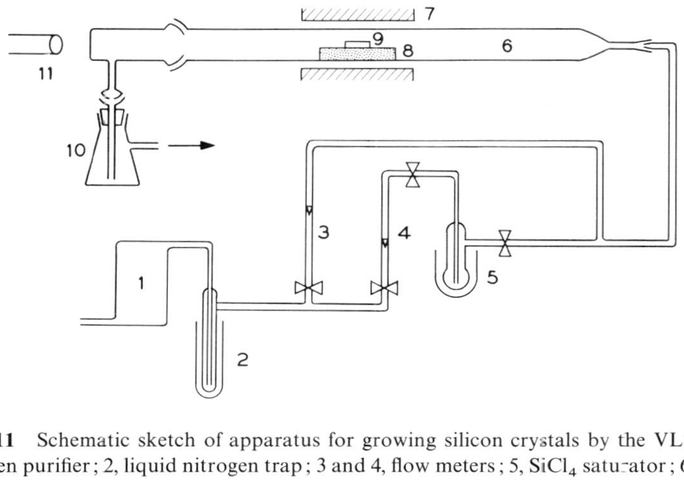

Figure 3.11 Schematic sketch of apparatus for growing silicon crystals by the VLS process. 1, hydrogen purifier; 2, liquid nitrogen trap; 3 and 4, flow meters; 5, $SiCl_4$ saturator; 6, reaction tube; 7, furnace; 8, fused silica support; 9, silicon substrate; 10, exit trap; and 11, telescope.

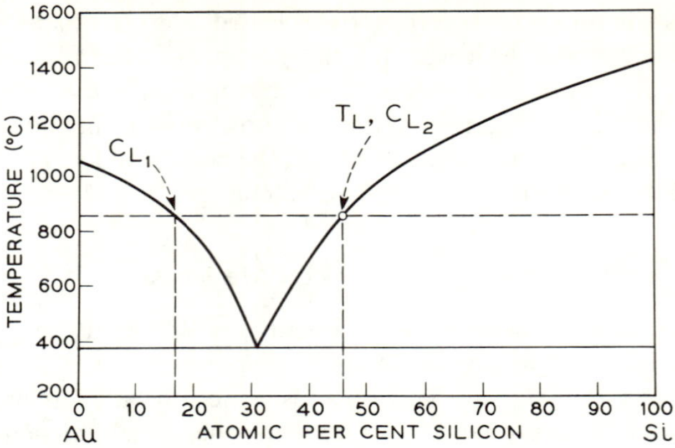

Figure 3.12 Equilibrium relations in the gold-silicon system.

A mixture of H_2 and $SiCl_4$ is introduced into the deposition apparatus following the exact experimental details given later. The chemical processes involved during the hydrogen reduction of $SiCl_4$ and the deposition of silicon on a silicon substrate have been extensively discussed [37, 41]. Experimental evidence suggests that the reduction process occurs at the liquid droplet surface and is heterogeneous and surface catalyzed. For example, no silicon deposit is observed at cold regions in the apparatus.

The VLS process is described for the case of gold as the agent. A small particle of gold is placed on the {111} surface of the silicon wafer. The substrate is subsequently heated to a temperature, T_L in Figure 3.12, to form a gold-silicon liquid alloy. Since the liquid is in contact with an excess of silicon, its composition is the equilibrium one, C_{L_2}. It is well known that silicon dissolves anisotropically during the alloying process. The solid-liquid interface is faceted, being composed of the slow-dissolving or slow-growing {111} crystal planes.

The liquid gold-silicon alloy, of composition C_{L_2}, is a preferential sink for silicon atoms arriving from the vapor. The liquid alloy becomes supersaturated with silicon to a value critical for growth of silicon at the solid-liquid interface. The critical supersaturation or the corresponding liquidus supercooling is probably very small. (It is well known that single crystals of silicon can be grown from the melt free of dislocations at very small supercooling of not more than a few degrees C [42].) The excess silicon from the liquid is deposited on the {111} planes which form the faceted solid-liquid interface. During the early stages the substrate material, which dissolved during the alloying process, will be regrown. Subsequently, the liquid droplet rises from the original substrate surface atop the growing crystal (Fig. 3.9).

The growth direction of the whisker is normal to the (111) surface. Essentially the same process takes place when the agent is gold, platinum, palladium nickel, copper, or silver. The details, such as temperature of deposition, vary because the phase relations of the several elements with silicon are different.

From the discussion it is evident that, for VLS growth, the liquid-forming impurity does not have to be in contact with a substrate. If gold is the agent with silicon available from the vapor, the particle of gold, for example, may be free in space or on an inert support, such as fused silica. As soon as silicon is formed at temperature T_L by the hydrogen reduction of $SiCl_4$, a liquid alloy film of composition C_{L_1} (Fig. 3.12) is formed on the surface of the gold particle. During subsequent deposition of silicon, the liquid layer spreads and thickens until all solid gold is dissolved. From this point on, the composition of the liquid droplet increases from C_{L_1} to C_{L_2}, a higher silicon concentration. Since the deposition is carried out at constant temperature, the liquid droplet saturates with silicon at C_{L_2}. The silicon content of the liquid increases further and supersaturation results. When the supersaturation is large enough, a silicon crystal nucleates either heterogeneously or homogeneously. The liquid is now in contact with solid silicon, and the liquid composition will be again close to the equilibrium value C_{L_2}. VLS crystal growth now proceeds as discussed before. It is proposed that whisker crystals of many substances grow by this procedure. Impurities that satisfy the criteria for a suitable alloying agent are very likely already present in the reaction zone, or are deposited there by vapor transport. For example, the silicon whiskers prepared by the disproportionation of SiI_2 were grown in this way. Impurities intentionally added with the source silicon were transported by evaporation-condensation or other transport processes to the deposition region, giving rise to whiskers which grew by VLS.

Basic experiments were carried out to obtain rates of deposition of silicon on several liquid layers in comparison with the rate on solid silicon. The deposition apparatus was as shown in Figure 3.11 but with the reaction tube in a vertical position. The graphite susceptor served as the pedestal to support the substrate. The deposition was carried out with a hydrogen flow of 1000 cc/min and a $SiCl_4$ to H_2 mole ratio of 0.02. Wafers of three substrate materials, silicon, platinum, and gold, of identical size ($1 \times 1 \times 0.1$ cm) were used. The single-crystal wafers of silicon (about 5 ohm-cm, *n*-type) having (111) faces were etched chemically. Polycrystalline wafers of high purity (99.999%) platinum or gold were chemically cleaned in aqua regia. Silicon was deposited for 10 minutes at constant temperature (± 10°C) and the weight gain of the wafer was measured.

For platinum substrates, at a deposition temperature higher than 850°C, a liquid film formed on the substrate surface within a few seconds after

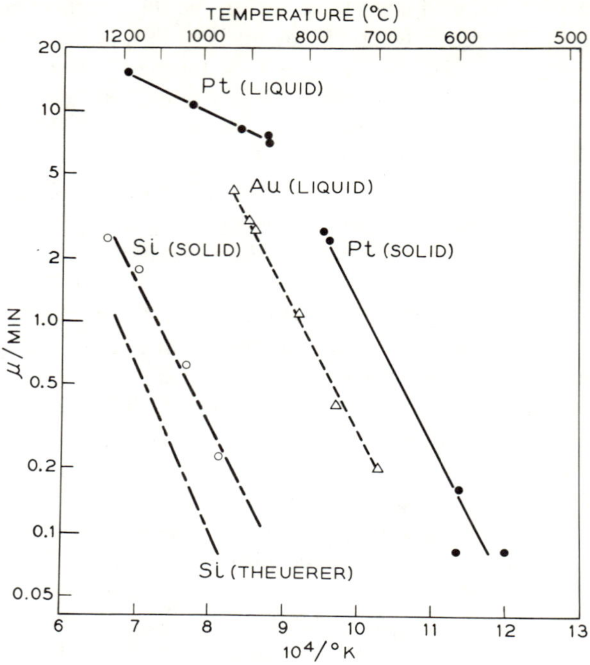

Figure 3.13 Deposition rates of silicon on a silicon {111} surface, on a liquid solution of gold-silicon, and on platinum-silicon.

deposition began. The surface of the liquid alloy was continuous and remained bright and clean during further deposition. For gold substrates, the deposition temperature was kept below 950°C to avoid complete melting within the deposition period. Small droplets of liquid alloy formed on the substrate and subsequently spread over the surface. A clearly recognizable thin solid film was observed to cover most of the liquid. Even at high temperatures it was impossible to avoid formation of the unidentified surface film.

Figure 3.13 shows the measured rates of deposition of silicon in μ/min on the three different substrate materials, as a function of deposition temperature. There is a striking difference in rate for a liquid surface compared to that for a solid silicon substrate. The deposition rate for a silicon substrate is slightly larger than reported by Theuerer [37] although the temperature dependence of the process is about the same. This difference is accounted for by the ratio of area of substrate to that of the reaction tube, which was smaller in the Theuerer experiments.

For deposition on the liquid silicon–platinum alloy, the temperature dependence of the rate was 7.5 kcal/mole, in comparison with about

31 kcal/mole for the silicon substrate. For example, at 900°C, about 60 times more silicon was deposited on the liquid alloy than on the silicon substrate in the 10-min period. Interesting but obscure was the result for deposition on the platinum substrate at temperatures at which a liquid platinum–silicon alloy does not form. The temperature dependence of the rate was about the same as for a silicon substrate but the rate was larger by almost two orders of magnitude. The silicon deposit was polycrystalline with the grain size decreasing with lower deposition temperature. An X-ray diffraction pattern of a silicon deposit on a solid platinum substrate formed at 600°C was similar to a powder pattern for a particle size in the micron range. In the same temperature range, the silicon deposit on a silicon substrate was also highly polycrystalline. Therefore, the enhanced deposition rate on a platinum substrate is not likely due to an imperfect surface structure. An attractive possibility is that platinum diffuses through the silicon deposit at a rate comparable with the deposition rate and acts as a catalyst for the reduction process.

For a gold substrate the deposition rate was about 20 times larger than for a silicon one. There was considerable scatter in the results, perhaps due to the variable surface film mentioned earlier. Only maximum values were used in Figure 3.13.

Summary and Discussion

The new mechanism of crystal growth involving vapor, liquid, and solid phases explains many observations of the effect of impurities in crystal growth from the vapor. The role of impurities is to form a liquid solution with the crystalline material to be grown from the vapor. The solution is a preferred site for deposition from the vapor which causes the liquid to become supersaturated. Crystal growth occurs by precipitation from the supersaturated liquid at the solid-liquid interface. For a "smooth" (faceted) solid-liquid interface, the VLS mechanism produces unidirectional crystal growth perpendicular to the solid-liquid interface.

Crystalline defects, such as screw dislocations, are not essential for VLS growth. However, defects present at the solid-liquid interface may reduce the required supersaturation in the liquid solution during growth. In the case of growth of whiskers, in which the liquid-forming agent is transported in the vapor to the nucleation site, an emergent dislocation could serve as the initial location for the formation of the liquid phase.

The concept of the VLS mechanism has been discussed with reference to the growth of silicon whisker crystals using gold, platinum, nickel, copper, or silver as impurity agent.

The morphologies of silicon crystals grown by VLS are the same, except for details, as those found in earlier studies for crystals grown by the disproportionation of SiI_2 [6]. With the VLS mechanism, dislocation-free crystals and films of silicon have been produced at temperatures several hundred degrees lower than those required in the usual epitaxial methods.

CONTROLLED VLS GROWTH OF SILICON CRYSTALS

Introduction

It is the objective of this section to describe in detail the experimental procedure for controlled VLS growth of silicon crystals [43]. The author has also successfully applied the same principles for controlled growth of germanium crystals [44]. It is reasonable to assume that the general principles outlined in this chapter can be used for VLS growth of any material. Three factors should be emphasized. First, great care must be taken to keep undesired impurities from the deposition region. Second, a temperature gradient may be either advantageous or adverse during VLS growth, depending on geometry. Third, the supply of material from the vapor must be kept within certain limits determined by the transport process through the liquid alloy zone.

Sample Preparation and Deposition Apparatus

Single-crystal silicon wafers preferably with $\{111\}$ main faces are used as substrates because the usual VLS growth direction for this material is $\langle 111 \rangle$. An exact orientation of the substrate surface is not necessary for subsequent VLS growth. Even a $\{211\}$ surface may be used. The VLS crystals grown on such a substrate are then inclined by 19.47° with respect to the $\langle 211 \rangle$ substrate normal. The number of the dislocations in the substrate crystal has no observable effect on VLS growth. Dislocation-free crystals were successfully grown on dislocation-free and on plastically deformed crystals containing up to 10^7 dislocations/cm^2. The growth of silicon whiskers from dislocations that terminate at the substrate surface has never been observed. It was found that for silicon, the growth of whisker crystals only occurs by the VLS mechanism on the site of the liquid alloy.

The silicon substrate crystals (1 to 100 ohm-cm, *n*- or *p*-type) were prepared using the following procedure: (1) The main faces are ground with 600 grit Carborundum. (2) The wafers are washed in acetone and toluene to remove any traces of mounting wax. (3) They are next washed in Alconox solution. (4) The wafers are etched for 5 minutes in 1:4 (HF, HNO_3) solution followed by an etch in 1:2:6 (HF, acetic, HNO_3) for 5 minutes. Subsequently, the wafers are rinsed in deionized filtered water and dried at 110°C.

The gold (99.999% purity) used for alloying may be prepared by filing, slicing a fine wire or sheet, or by evaporation through a mask. Of these the sliced wire or sheet is the least contaminated material. The gold pieces were cleaned in acetone and toluene and boiled in Alconox solution. This was followed by an etch in aqua regia for a few minutes. The particles were rinsed in deionized, filtered water and dried. In most cases the gold pieces were placed either manually or with the aid of a mask on the silicon substrate.

The apparatus used for controlled VLS growth of silicon has been described before. In the present work, two different heating systems were used. In one, the reaction tube of silica was placed inside a resistance furnace. With such a system, a rather long isothermal deposition zone can be obtained, and an adverse temperature gradient avoided. The other system consisted of a graphite tube, serving as an *r-f* susceptor, placed inside the reaction tube. The graphite tube was first baked at 1200°C and subsequently coated with silicon. It was found that frequently numerous fine silicon whiskers grew inside the graphite tube. The whisker crystals were removed from the tube by scraping, and more silicon was deposited. This procedure was repeated until no more whisker crystals were observed. For both heating systems, the silicon substrate crystal was placed either on a silicon or silica support situated in the isothermal deposition zone. Great care was taken to keep the whole deposition apparatus as clean as possible. The hydrogen used for the experiment was purified by a palladium diffuser. The temperature in the deposition zone was controlled to within ± 2°C. The $SiCl_4$ to hydrogen mole ratio used in the experiment ranged from 0.01 to 0.1, with a total hydrogen flow rate of 300 to 1000 cc/min. It should be noted that etching of the silicon substrate may occur, even for an extremely small hydrogen to $SiCl_4$ mole ratio, if an isothermal reaction zone is used. In such a system the ratio of HCl to unreacted $SiCl_4$ increases along the deposition zone. Therefore, the proper location of the silicon substrate for subsequent VLS growth must be found empirically; it depends strongly on the geometry of the reaction zone, the total gas flow, and hydrogen to $SiCl_4$ mole ratio. The silicon substrate was placed in a downstream region where no silicon deposit occurred on a clean silica substrate. This deposition region is also useful for VLS growth of silicon crystals on silicon substrates partly masked with SiO_2.

Alloying and Initial VLS Growth

The most difficult and least understood aspects of VLS growth are the alloying process of the impurity agent with the substrate and the initial stages of VLS growth, involving the rise of the liquid droplet from the substrate. It has been found that alloying gold with the silicon substrate is difficult to control. The temperature at which the liquid alloy forms may vary from 600°C to well over the melting point of gold (1063°C). For freshly cleaned gold and silicon, the alloying temperature is about 600 to 800°C. An

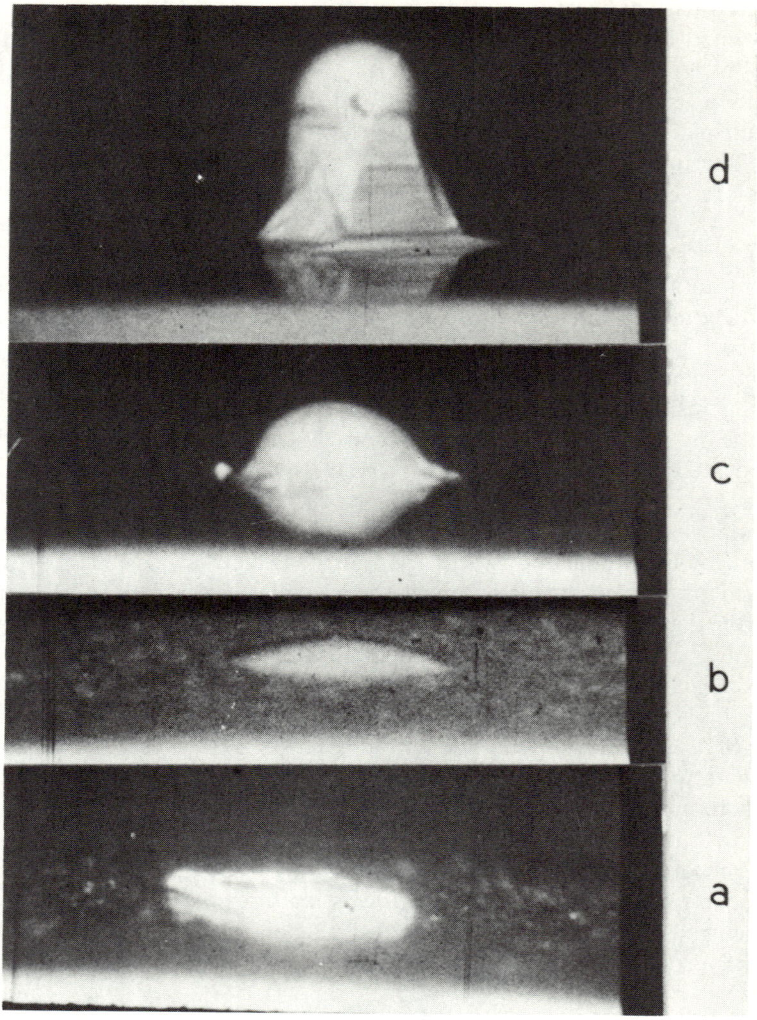

Figure 3.14 Time-lapse sequence of alloying and initial VLS growth: (a) gold particle on silicon substrate at room temperature; (b) liquid droplet of gold-silicon alloy at 1050°C; (c) and (d) after 10 and 30 minutes respectively of growth at 1050°C (mag. 100×).

oxide layer on the silicon substrate may prevent alloying at the gold-silicon eutectic temperature (370°C).

The shape of the liquid droplet, determined by the liquid-solid contact angle, depends on the rate of temperature increase during alloying. We have not found a satisfactory procedure to control the spreading of the liquid over the substrate surface. It is suggested that the spreading may be controlled to some extent by masking the substrate around the intended alloy region with an SiO_2 layer. This would result in preferred dissolution of silicon from the substrate interior and also would control the shape of the droplet. The observed contact angle at 1050°C is about 10° to 20° which would indicate good wetting of the substrate by the liquid. However, this contact angle is most likely not the equilibrium contact angle but rather a consequence of the dissolution kinetics.

The small contact angle remains constant or decreases slightly when $SiCl_4$ is introduced into the carrier gas. However, sometimes a sudden increase in contact angle is observed. The sequence of events during alloying and the initial VLS growth is shown in Figure 3.14. Figure 3.14a shows a piece of gold on the silicon substrate before alloying; in Figure 3.14b the temperature has been increased to about 1050°C. Figure 3.14c, photographed after about 10 minutes of VLS deposition, shows a marked increase in contact angle. As a consequence, the diameter of the growing crystal decreases during the initial stage of VLS growth. Finally, in Figure 3.14d, photographed after about 30 minutes of VLS growth, the contact angle has become stabilized. From this point on, there is no more change and the diameter of the growing crystal remains constant if the deposition temperature is held constant. The gradual increase in contact angle results in the formation of a tapered section at the base of a VLS crystal as shown in Figure 3.14d. At the substrate surface, the crystal is either 6- or 3-sided, depending on deposition temperature and time. At temperatures higher than about 1050°C, the VS deposition is considerable at the base, resulting in a three-sided form.

The liquid alloy is very sensitive, in the first stage of VLS growth, to a temperature gradient or oversupply from the vapor phase. Apparently the droplet is not in a stable configuration. Too much silicon vapor may result in a partial breakup of the liquid into smaller droplets, each new droplet then giving rise to VLS growth as shown in Figure 3.15. Once the crystal grows beyond the tapered section, it is rather stable with respect to changes in temperature and deposition rate. The effect of a temperature gradient or deposition oversupply on the stability of a liquid droplet is discussed in detail in the following sections of this chapter.

Vapor Supply

At steady-state VLS growth, the same number of silicon atoms which enters the liquid from the vapor are deposited at the solid-liquid (S-L) interface. This flux of silicon atoms diffuses through a concentration difference ΔC across the liquid alloy of thickness L. The value of ΔC is an important factor for the stability of the liquid alloy. For the following discussion we assume that convective mixing in the liquid is negligible and that we have a disklike shape of the liquid alloy. Figure 3.16 shows a section of the silicon

Figure 3.15 Breakup of liquid alloy during initial VLS growth giving rise to formation of main crystal surrounded by fine whiskers (mag. 50X).

liquidus, Si_L, and the deposition isotherm T_E. With no vapor deposit, the composition of the liquid will be C_E, the equilibrium concentration. For steady-state VLS growth, the composition of the liquid next to the gas interface is C_L and is C_g at the solid-liquid interface. C_g, which may be very small, is the supersaturation required for nucleation and growth of new crystal layers.

The concentration gradient in the liquid disk is given by

$$\frac{\Delta C}{L} = \frac{V}{D} \tag{3}$$

where V is the speed of advancement of the S-L interface and D is the diffusion coefficient of silicon in the liquid alloy. From the phase diagram (Fig. 3.16) it follows that for a given V, D, and L, the concentration difference in the liquid, $\Delta C = C_L - C_E$ (atom fraction), corresponds to a liquidus supercooling ΔT. Homogeneous nucleation theory [45] predicts a ΔT_{max} of about 300°C for pure Si or a liquid Si alloy. For $\Delta T > 300$°C, the supercooling is high enough to nucleate silicon crystal by homogeneous nucleation in the liquid alloy. This sets an upper limit for ΔC during VLS growth, ΔC can be about 0.25 for $T_E = 950$°C. It has been found experimentally that this consideration is extremely important in order to avoid instability of the liquid alloy by nucleation. For example, for $L = 10^{-2}$ cm. $D = 5 \times 10^{-5}$ cm^2/sec, and $T_E = 950$°C, the critical growth velocity $V_{max} \sim 3.5$ mm/hr.

The VLS growth rate V(cm/sec) for a given total gas flow (cm^3/sec), a $SiCl_4/H_2$ mole ratio m_R, and for a reaction tube of cross section A(cm^2) can be estimated as

$$V = \frac{fxm_RxM}{AxV_0x\rho} \tag{4}$$

where M = molecular weight of $SiCl_4$, V_0 = standard volume of perfect gas, and ρ = density of solid silicon. In Eq. 4 is assumed the use of a dynamic

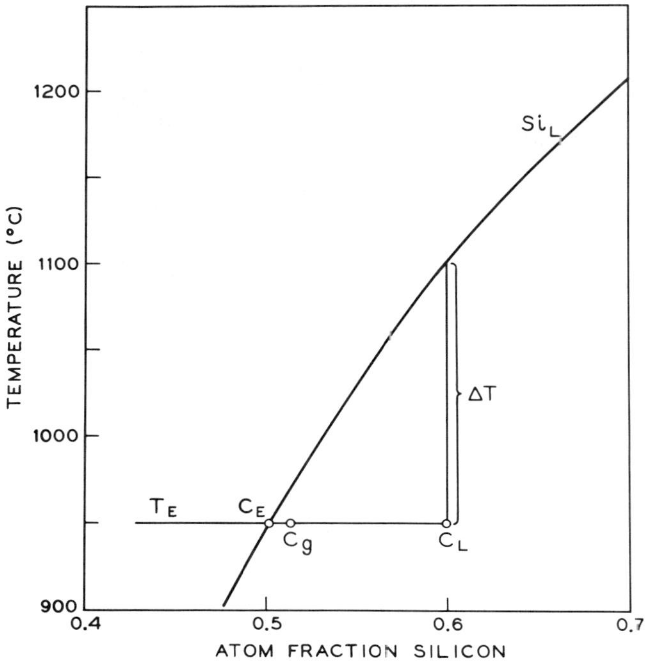

Figure 3.16 Section of the Au-Si phase diagram.

Figure 3.17 Uncontrolled VLS growth resulting from nucleation of silicon crystals in liquid alloy (mag. 125X).

system of high flow rate, a condensation coefficient of unity for the liquid alloy, and the absence of the disproportionation reaction. It is also assumed that no surface migration of Si atoms on the side faces of the growing crystals to the liquid occurs. For $f = 6\ \text{cm}^3/\text{sec}$, $m_R = 0.014$, and $A = 5\ \text{cm}^2$, the calculated VLS growth rate $V = 0.3$ mm/hr. For these conditions, in the temperature range 950 to 1100°C, the observed growth rate is only slightly larger than the calculated value.

Both Eqs. 3 and 4 are only crude approximations of the true state of affairs. Convective mixing of the liquid alloy, which may be considerable when using *r-f* heating, will increase the critical velocity V. A more static reaction system (slow flow rate) will increase V because of higher impingement rate of $SiCl_4$ molecules. The reader is referred to a more rigorous calculation of the vapor-phase deposition rate of silicon given by Shepherd [46]. Finally, ΔT_{max} may be much smaller than 300°C if heterogeneous nucleation sites are present in the liquid alloy or on the surface of the liquid.

It is evident from the previous discussion that a sudden decrease in deposition temperature T_E during VLS, results in a sudden increase in ΔC and therefore ΔT. During the transient, the ΔT can become large enough for homogeneous or heterogeneous nucleation. Figure 3.17 depicts a silicon crystal that was originally grown at 1000°C. During growth the temperature was suddenly reduced to 600°C for about 30 seconds and subsequently raised to 950°C. The breakup of the liquid alloy is dramatic, giving rise to the formation of numerous small crystals which are in random orientation with respect to the main crystal.

Effect of Temperature Gradient on VLS Growth

Pfann [40] has shown that a liquid alloy zone in contact with solid solvent material can migrate through the solvent material under the influence of a temperature gradient. This phenomenon, called temperature-gradient zone melting (TGZM), arises because of continuous melting and freezing of solvent at the hotter or colder section of the liquid zone. The liquid alloy droplet on a silicon substrate comprises such a system if a temperature gradient is imposed. If the gradient is such that the substrate interior is hotter than the substrate surface, the liquid alloy will migrate into the substrate until no more liquid is exposed to the vapor. It is evident that VLS growth is impossible under such conditions. If the temperature gradient is along the substrate surface, the alloy droplet may move along the substrate towards a higher temperature region. The silicon entering the liquid from the vapor will be preferentially deposited at the colder side of the droplet. This gives rise to the buildup of silicon on the substrate surface in the form of a ramp. On a (111) substrate, the ramp surface can be an extremely smooth (111) face. An additional consequence of a lateral temperature gradient can be that the liquid droplet may break up into smaller droplets. During TGZM on the substrate surface, the liquid-solid contact angle becomes nonuniform, being smallest at the hot end of the liquid zone. This can lead to partial breakup of the liquid droplet. This is shown in Figure 3.18, where a large lateral temperature gradient was imposed during growth. The initial position of the alloy droplet was close to A, where the largest crystal grew. The droplet broke up during migration, giving rise to smaller crystals. The smallest droplet migrated to B. The liquid droplet can rise from the substrate surface

Figure 3.18 Adverse effect of a lateral temperature gradient on VLS growth. A single alloy droplet, initially at position A, broke up into smaller droplets which migrated towards a higher temperature region. All droplets but the smallest one at position B gave rise to VLS crystals (mag. 19X).

even with a lateral temperature gradient, if the VLS growth rate is larger than the rate of TGZM.

From the preceding section it follows that a temperature gradient in which the vapor-liquid interface is hotter than the S-L interface would be of great advantage during VLS growth. The liquid supercooling, ΔT as shown in Figure 3.16, could be reduced by an amount equal to the temperature difference across the liquid alloy. By this means the stability of the liquid could be greatly improved.

Control of Diameter

The diameter of a VLS crystal is controlled by three factors, namely the initial amount of gold used for alloying, the deposition temperature, and the vapor-solid (V-S) deposit which also depends on the deposition temperature. For a given deposition temperature, the amount of gold determines the diameter of the growing crystal. The volume of the liquid droplet is given by the phase diagram. For a given amount of gold, the diameter of the crystal can be varied over a wide range by simply changing the deposition temperature. Figure 3.19 is a photograph of an end-section of a VLS crystal. The lower part has been grown at about 1050°C, the upper part at about 950°C. Note the gradual decrease in cross section which occurred during the temperature change. It is advisable to reduce or even stop completely the $SiCl_4$ supply and to reduce the temperature gradually. A temperature change that is too rapid may cause a "pinching off" of small droplets from the main body of the liquid, giving rise to the formation of branches. An increase in diameter can be obtained by raising the deposition temperature. In this case it is helpful to increase the $SiCl_4 : H_2$ mole ratio during the gradual temperature increase. This prevents remelting of silicon at the S-L interface which may cause the (111) interface to become unstable.

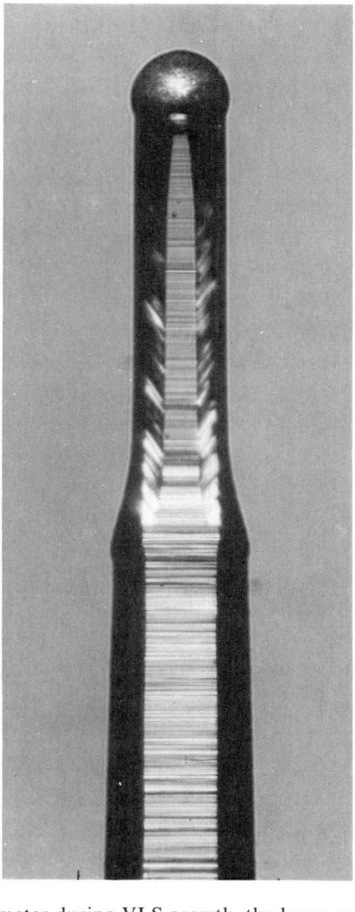

Figure 3.19 Change of diameter during VLS growth, the lower section was grown at 1050°C, the top section at 950°C (mag. 110X).

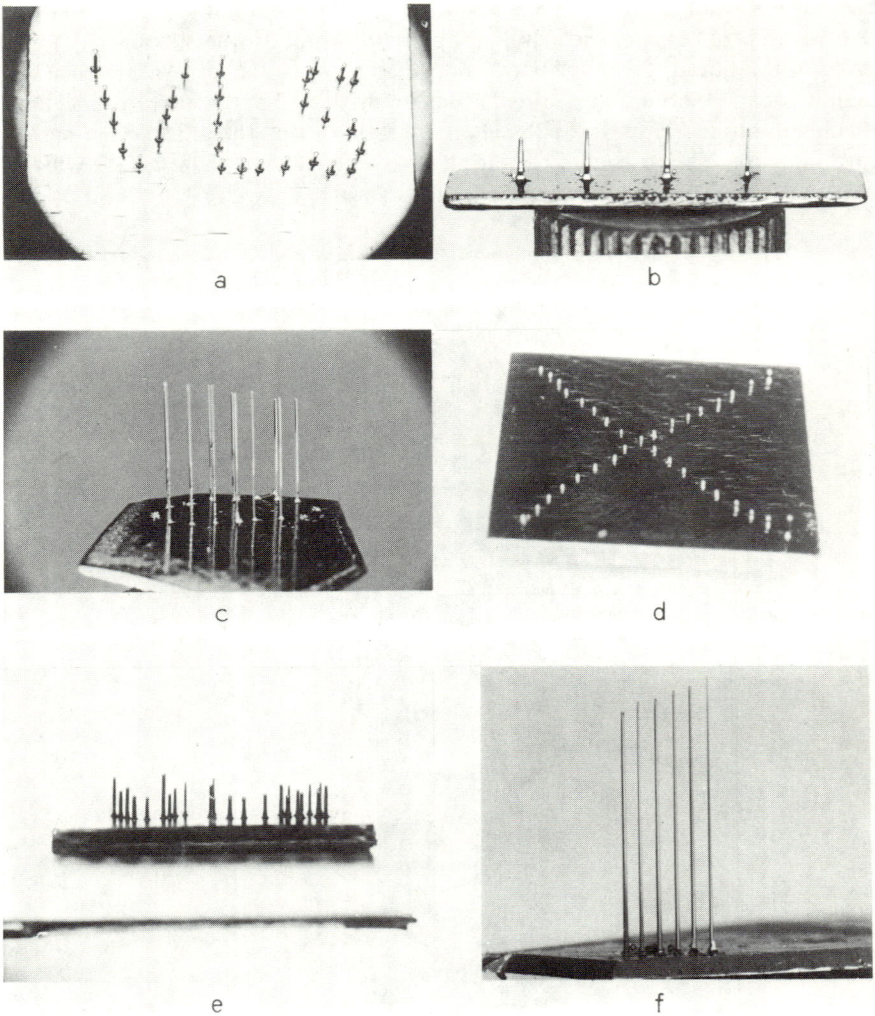

Figure 3.20 Controlled VLS growth of silicon crystals in various arrays of silicon substrates (mag. 3×).

VLS Growth of Crystal Patterns

A goal of controlled VLS growth is the preparation of arrays of silicon crystals on silicon substrates. The growth of such arrays is not very difficult if all the experimental results discussed in the previous chapters are considered in detail. Six growth patterns of different form and dimension are shown in Figure 3.20. Crystals up to 1.5 cm in length have been grown with a maximum diameter of about .3 mm. The smaller the diameter of the crystal, the fewer are the difficulties encountered. For the growth of a crystal pattern, it is very important to use an isothermal reaction zone. A lateral temperature gradient may distort the intended growth pattern by TGZM during the initial stages of growth. The length of the crystals on an individual substrate may vary considerably, depending on the position of the crystals in the reaction tube. In Figure 3.20b, the substrate was situated perpendicular to the gas flow. The outermost crystals were close to the susceptor wall. In Figure 3.20f, the crystal array grew parallel to the gas flow direction. The crystals growing "downstream" are shorter, reflecting the depletion of $SiCl_4$ along the reaction tube. This effect can be eliminated by proper geometrical design of the reaction zone. X-ray measurements, using the Lang technique, have shown that the crystals in an array are "epitaxial" with respect to each other and therefore to the substrate. The observed angular misorientation between the crystals is smaller than 140 sec arc, which is the resolution of the measurement technique employed.

Arrays of crystals have been grown with as many as 30 crystals on a 1 cm × 2 cm substrate. When conditions of cleanliness and proper temperature and gas-phase conditions were observed, one crystal grew on each and every site where a gold particle had been placed manually. Another method, more suitable for production of arrays of many crystals, was to place a mask or template of pyrolytic graphite on the substrate crystal. An impurity particle was placed in each hole in the template and alloying was accomplished with the template in place. Subsequently, the template was removed and the VLS growth begun. Patterns of alloy dots also were made by evaporation of impurity agents through a mask onto the wafer. However, not every alloy dot produced only one crystal. A center crystal normally grew surrounded by fine whiskers of micron dimension. More work will be required to eliminate the breaking up of evaporated alloy dots.

Mass Production of Whisker Crystals

There are two methods that could be used for large-scale production of whisker crystals. The first one utilizes a suitable substrate crystal, whereas the second method circumvents the need for a substrate altogether. Many experiments of the first procedure were performed using either gold or platinum as agent. For example, a silicon substrate crystal was used and a film

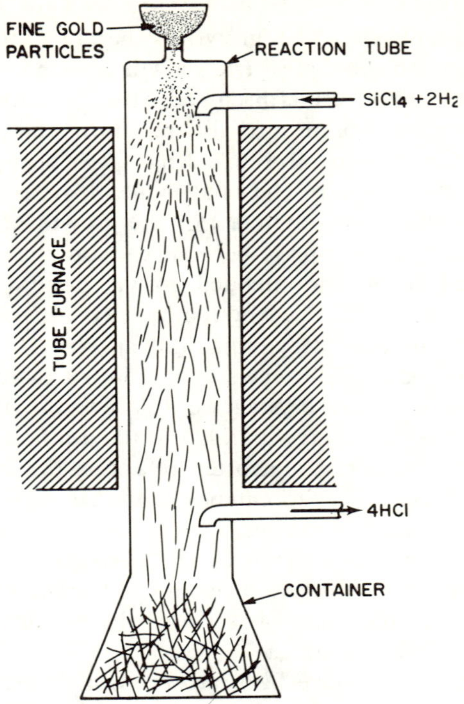

Figure 3.21 Schematic sketch of apparatus for mass production of whisker crystals by the VLS process.

of gold approximately 500 Å thick was evaporated on the $\{111\}$ substrate surface. The substrate was subsequently heated in a deposition apparatus to about 900°C. The gold film alloyed with silicon. However, surface tension caused a breakup of the film and the formation of a multitude of fine alloy droplets of surprisingly uniform size. Upon deposition, each alloy droplet gave rise to VLS growth. The resulting crystals have a rather uniform length and most crystals grow perpendicular to the substrate surface. It was found that the thickness of the impurity film controls both the diameter of the whiskers and their density per unit area. The thinner the film, the smaller the diameter and the higher the density. Silicon whiskers with a length of more than 1 cm and a density of about $10^6/cm^2$ have been prepared by this method. The final product has a velvety appearance. The crystals can be easily removed from the substrate by scraping, for further fabrication. Another possible application would be to fill the space between the whisker crystals with a proper matrix for use as a composite. The substrate crystal, which can be rather thin, can be subsequently removed from the composite.

The second method for whisker production employs the principles outlined in the section on the experimental study of the VLS mechanism, and will be discussed with reference to Figure 3.21. A proper impurity such as gold or platinum in the form of small particles is introduced at the top section of the reaction tube. The particles fall through an extended reaction zone, which is kept at the deposition temperature. VLS growth occurs in this zone as described with reference to Figure 3.12. The growth rate can be adjusted to be many cm/sec for impurity particles smaller than 1000 Å. The final product, a multitude of fine whisker crystals, could be collected at the lower section of the reaction tube.

Summary and Discussion

Silicon crystals can be grown readily by the VLS technique under controlled conditions. The location of growth, the diameter of the crystal, and its ultimate length can be controlled. However, two important points must be considered: the presence of undesired impurities in the deposition region and the stability of the liquid alloy droplet. The first point requires extreme cleanliness of substrate, alloying agent, reaction tube, and carrier gas. The second point demands the absence of undesirable temperature gradients during alloying and VLS growth, and proper adjustment of the total $SiCl_4$ concentration in the carrier gas. A temperature gradient may cause the alloy droplet to migrate into the substrate interior, or along the substrate surface, or may cause the formation of kinking or branching during growth. Proper gas composition is necessary to keep the concentration gradient in the liquid and, therefore, the corresponding liquidus supercooling, small enough to avoid nucleation of silicon crystals in the liquid. It has been shown that the diameter of a VLS crystal depends both on the amount of the alloying agent used and also on the deposition temperature. Crystals with sections of different diameter can be prepared simply by changing the deposition temperature during growth. The growth of arrays of crystals in various patterns has been demonstrated.

DEFECTS IN SILICON CRYSTALS GROWN BY THE VLS TECHNIQUE

Introduction

One objective of this section is to report that silicon crystals of respectable sizes (over 200 μ diameter and length up to 1 cm) can be grown without axial defects such as screw dislocations [47]. The other objective is to emphasize that solidification during VLS growth occurs from a multicomponent liquid solution. Such a system is extremely sensitive to sudden changes in growth rate which may be caused by fluctuations in temperature or vapor supply. Impurity bands [48] can be formed and, in the extreme case, the solid-liquid interface may become nonplanar [49, 50]. Such an interface breakdown usually results in entrapment of the liquid alloy, which subsequently causes the formation of dislocations and stress fields. Many different defects such as dislocations, impurity striations, elastic stress fields, and second-phase regions have been observed. It will be shown that these defects are not necessarily involved in the VLS process, but arise from instability during growth.

General

The hydrogen reduction of $SiCl_4$ was used as the transport reaction and gold was the liquid-forming impurity. Only crystals with a $\langle 111 \rangle$ growth direction, the most common species, were studied. The samples were sectioned either parallel or perpendicular to the growth direction, and etched with the Sirtl etch, for about 1 to 4 minutes. X-ray topographs by the Lang technique [51] were taken with silver radiation.

It was found that certain defects repeatedly occur in specific regions of the crystals which are defined with reference to Figure 3.22. The figure shows an etched section of a dislocation-free silicon VLS crystal that was grown epitaxially on a dislocation-free silicon substrate. There are three regions of interest, A, B, and C. In the figure, the A-region includes the substrate crystal (X), the original substrate surface (S), the vapor-solid deposit (VS), the alloying interface (A), and the tapered section of the crystal grown by VLS. The B-region consists of the part of the crystal with essentially uniform diameter. The C-region starts at the silicon crystal-liquid alloy interface (SA). At this point, the vapor deposition is stopped during crystal growth and the temperature gradually reduced to room temperature. The C-region also contains the regrowth zone (R) of silicon which solidified from the liquid alloy on cooling, and finally the gold-silicon alloy (Au, Si) which forms the hemispherical termination on the crystal. The only defects observed in crystals such as shown in Figure 3.22 are impurity striations (I) which are caused by fluctuations in VLS growth rate. These impurity bands can be eliminated by careful control of both temperature and gas composition during the deposition process.

A-Region

Over 30 crystals and their substrates were sectioned on a plane parallel to the $\langle 111 \rangle$ growth direction, as shown in Figure 3.22. Etching revealed that the A-region of most crystals is imperfect. Crystalline detects can be either generated in this zone or extended into it from an imperfect substrate crystal. Dislocations and stacking faults that terminate at the substrate-liquid alloy interface may continue to grow during the VLS process.

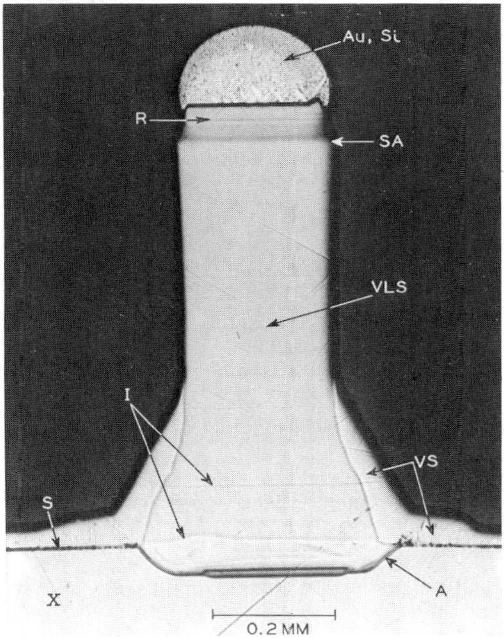

Figure 3.22 Etched section of a dislocation-free VLS crystal grown epitaxially on a dislocation-free silicon substrate. (X, substrate crystal; S, original substrate surface; VS, vapor deposit; A, alloying interface, SA, regrowth interface; R, silicon regrowth; Au, Si, two-phase alloy termination; I, impurity striations.)

Fortunately, the preferred directions of dislocations in silicon are ⟨211⟩ and ⟨110⟩ [52, 53] and stacking faults are parallel to {111} planes [54]. Therefore, all those defects are inclined or parallel to the {111} solid-liquid interface. Defects may continue to grow until they eventually terminate at the side faces of the growing crystals. The extension of dislocations from an imperfect substrate is illustrated in Figure 3.23. The remainder of the crystal, not shown in the photograph, was found to be dislocation free. Note the high density of defects in the VS deposit region on the substrate crystal.

Highly imperfect silicon crystals with a dislocation density of about $10^7/cm^2$ have been used as substrates. Two significant observations were made. Dislocations that terminate at the substrate surface which is not covered by the liquid alloy did not give rise to whisker growth. Furthermore, dislocations continue to grow into the VLS crystal, although there is an abrupt decrease in dislocation density at the alloy interface. The extension

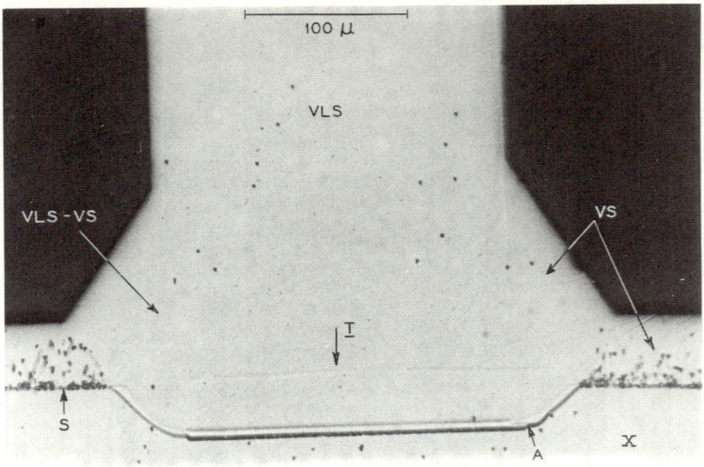

Figure 3.23 Etched section of a VLS crystal grown on an imperfect silicon substrate.

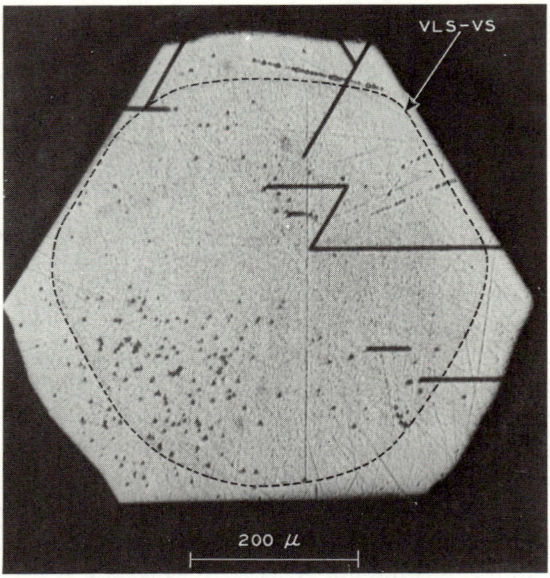

Figure 3.24 Extension of dislocations and stacking faults from an imperfect substrate. (VLS-VS, interface between VLS crystal and vapor deposit.)

of both dislocations and stacking faults from an imperfect substrate is shown in Figure 3.24, which is a section through the A-region perpendicular to the ⟨111⟩ growth direction. The VS-VLS interface is indicated by a dashed line. Note the apparent anisotropy in VS deposition on opposite {211} side faces, which will be discussed later.

The impurity striations I, such as shown in Figure 3.22, reveal successive positions of the solid-liquid interface during VLS growth. It was found that these striations are not always planar, parallel to {111}, in the A-region. It appears that a temperature gradient in the substrate crystal, which may also cause temperature-gradient zone migration and an oversupply of $SiCl_4$ in the vapor, enhances the interface breakdown. An interface instability of this type can result in entrapment (IN) of liquid alloy as shown in Figure 3.25. An island of liquid surrounded by solid silicon may become a powerful source for generation of dislocations (D). Any decrease in temperature will result in regrowth of silicon from the liquid alloy. There is, however, a considerable change in density [55] associated with the freezing process, which results in a buildup of a hydrostatic pressure in the liquid. The resulting stress in the surrounding material may become large enough for generation and motion of dislocations, as shown in Figure 3.25. This process of

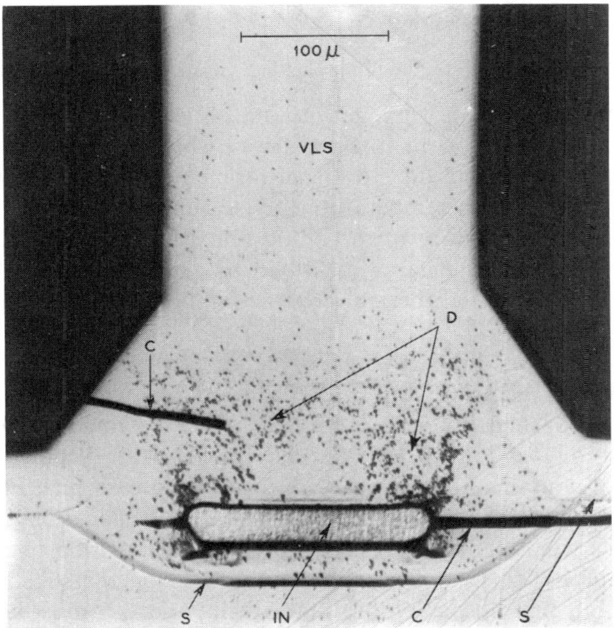

Figure 3.25 Formation of defects due to entrapment of liquid alloy during VLS growth. (IN, Au-Si entrapment; C, cracks; D, dislocation etch pits.)

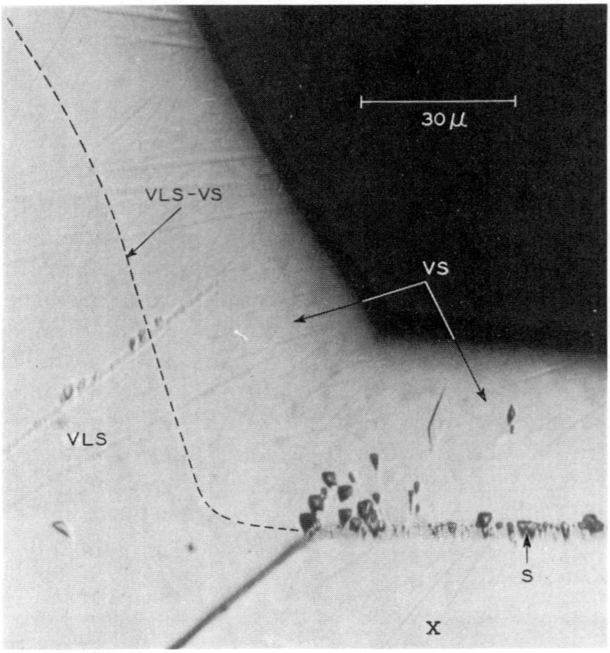

Figure 3.26 Intersection of the alloying, substrate, and VLS-VS interfaces.

defect formation is of a very general nature and can occur in crystals grown from the vapor or melt. For silicon, which is plastic above 500°C, the generation of defects due to an entrapment occurs when a crystal is cooled from the growth temperature to ambient. On cooling from 500°C to the eutectic temperature of the liquid alloy, the pressure may generate cracks (C), Figure 3.25. A stress field will be generated on further cooling because of the differential thermal contraction of silicon and the entrapped two-phase alloy.

Figure 3.26 shows the part of the A-region that contains the intersection of the alloying, substrate, and VLS-VS interfaces. The latter interface is marked by a dashed line. Note that both the alloying and the VLS-VS interface are dislocation free. In contrast, the substrate-VS interface is very imperfect; it contains a dense array of dislocation etch pits. Furthermore, the VS deposit on the crystal grown by VLS is dislocation free. Similar observations were made on crystals grown at a deposition temperature as low as about 700°C. In this case, the VS deposit on the substrate is polycrystalline, but that on the VLS crystal is still perfect. It must be concluded that the surface of a VLS crystal is highly perfect and free of contamination. It is mainly surface contamination that appears to be responsible for imperfect epitaxy of silicon during VS deposition [56–58].

B-Region

It has been mentioned that ⟨111⟩ silicon crystals grown by VLS are either 6- or 12-sided. This interesting change in morphology was further investigated and it was found that these crystals can also be grown with either 24 or only 3 side faces. The morphology of ⟨111⟩ crystals depends on the deposition temperature during VLS growth, as shown in Table 3.2. The number of observed prismatic growth faces increases with decreasing growth temperature. It must be noted that none of these faces is microscopically flat. They consist of arrays of steps which are either perpendicular or oblique to the ⟨111⟩ growth axis. There are three factors that may influence the formation

TABLE 3.2 MORPHOLOGY OF PRISMATIC GROWTH FACES AS A FUNCTION OF VLS DEPOSITION TEMPERATURE

Temperature Range (°C)	Number of Side Faces	Index of Faces
<850	24	{110}, {211}, ~{311}
~850–1000	12	{110}, {211}
~1000–1100	6	{211}
>1100	3	{211}

of the external surfaces: the surface tension of the liquid alloy, the growth kinetics at the solid-liquid interface, and the vapor-solid deposits on the side faces which may occur simultaneously during VLS growth.

If the surface tension of the liquid were the only controlling factor, the crystals would grow with a circular cross section. Since surface tension increases with decreasing temperature, it can be expected that its effect on the shape would be more important at low growth temperatures. Without vapor-solid deposit, the shape of a silicon crystal grown by VLS should be the same or very similar to a crystal grown by the Czochralski method, at low melt undercooling. The physical conditions for both methods are very similar, only the relative motion of solid and melt are reversed. It has been shown for Czochralski growth [59] that the interplay of surface tension and growth kinetics at the solid-liquid interface produces essentially round crystals with faceted side faces. A crystal pulled in a ⟨111⟩ direction has a triangular cross section with rounded corners. The shape of a ⟨111⟩ silicon crystal formed during alloy regrowth, without vapor deposition, is shown later in Figure 3.30a. The cross section is very similar to one prepared by the pulling method [59]. The shapes of crystals grown by VLS at three different deposition temperatures are shown in Figure 3.27. It is concluded

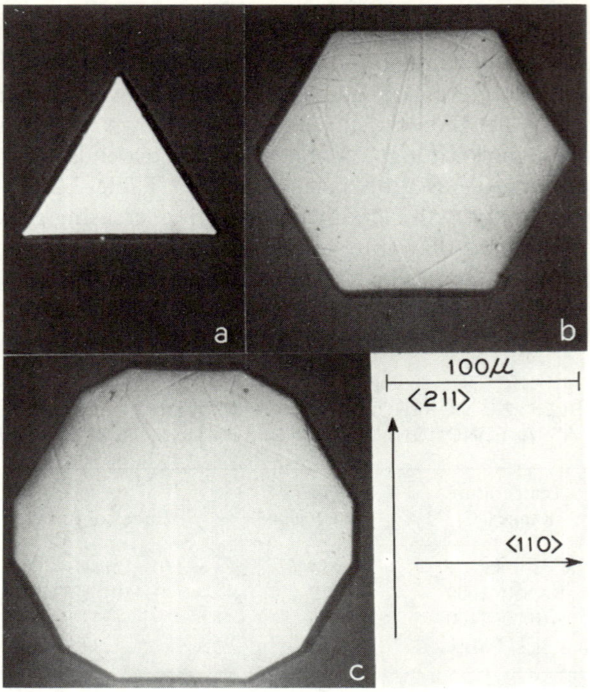

Figure 3.27 Morphology of prismatic side faces of ⟨111⟩ silicon crystals. (a, b, c grown at about 1100°C, 1050°C, and 950°C respectively.)

from a comparison of Figures 3.30a and 3.27 that the formation of the prismatic side faces is due to vapor deposit during VLS growth. Although the VS deposition rate is much smaller than the VLS growth rate for a given temperature, it is still effective in controlling the morphology. Since the difference in the two deposition rates decreases with increasing temperature, the VS deposit will greatly influence the morphology in the high temperature range.

It is well known that the rate of vapor-solid deposit on a {111} silicon substrate crystal is strongly temperature dependent [37]. It is also known that this rate, for constant deposition conditions, depends on the surface orientation of the silicon substrate [60]. This is indirectly confirmed by the temperature dependence of the morphology of the prismatic faces as shown in Table 3.2. For a given VLS growth temperature, the slowest growing faces become stable side faces. The change in morphology of ⟨111⟩ crystals, as a function of temperature, indicates that the rate of vapor-solid deposition increases in the sequence $(\bar{2}11)$, $(2\bar{1}\bar{1})$, (110), (311).

It must be noted, however, that the anisotropy in deposition rate is not that for a microscopically smooth surface of the given crystallographic indices. As stated before, all the prismatic faces consist of steps. The anisotropy in stability of the $(\bar{2}11)$ as compared to a $(2\bar{1}\bar{1})$ faces can only be explained if the steps on these faces have a different nucleation behavior, because both faces have identical atomic bonding.

If a silicon crystal is grown in an intermediate temperature range, the section next to the liquid droplet has the low temperature form, while the remainder of the crystal has the high temperature morphology. By raising the temperature during growth, for example from 900°C to 1050°C, the whole crystal will become 6-sided. The reverse change in temperature, however, does not affect the high temperature morphology; only the part grown at 900°C will become 12-sided.

It was found in numerous crystals that the vapor-solid deposit on the B-region is highly perfect. In particular, the interface between VS and VLS deposit is free of defects, such as dislocation or stacking faults. Even crystals which grew initially by VLS with a diameter of about 0.1 μ and which subsequently thickened by VS deposit to over 100 μ are found to be perfect. The procedure of first growing a thin crystal by VLS and subsequently vapor depositing upon it at low temperature may have practical implications for vapor growth of perfect crystals of macroscopic size.

Over 30 crystals were sectioned perpendicular to the growth direction and etched to reveal crystalline defects. The majority of these crystals was perfect. A few samples contained entrapped Au-Si eutectic such as shown in Figure 3.28a, which is an etched longitudinal section of a $\langle 111 \rangle$ crystal. The islands of eutectic are surrounded by dense networks of dislocation etch pits. An X-ray topograph of the same crystal is shown in Figure 3.28b. The two-phase alloy regions are light because of enhanced absorption. The dark areas around the islands are mainly due to an elastic stress field as shown in the following section. X-ray topographs confirmed the etching results that most silicon crystals grown by VLS are highly perfect in the B-region. The only defects observed are the impurity bands that are caused by growth rate fluctuations.

C-Region

There are two distinctly different alloy regrowth regions which are shown in Figure 3.29. The initial regrowth interface is marked on the photomicrograph by a dashed line. The regrowth in (a) and (b) will be referred to as C1 or C2 region respectively. In C1, which is obtained by slow cooling, the solid-liquid interface remains planar during regrowth. The region is tapered because the volume of the liquid alloy gradually decreases on alloy regrowth. During this process the liquid becomes enriched in gold. The amount of

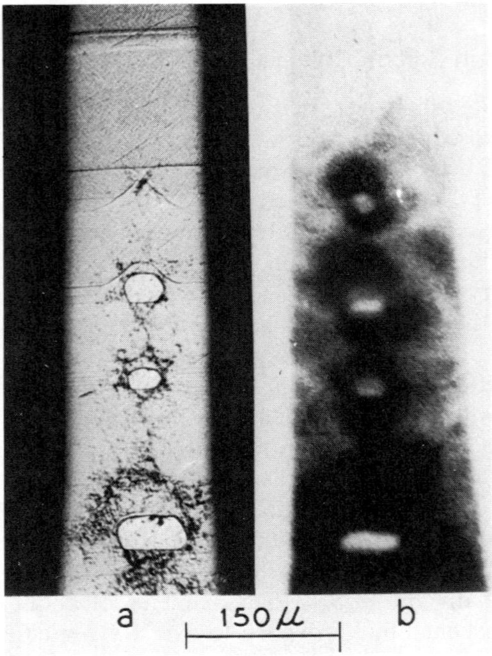

Figure 3.28 Defect formation due to entrapment of liquid alloy. (a, etched section parallel to [111] growth direction; b, X-ray topograph, (111) reflection.)

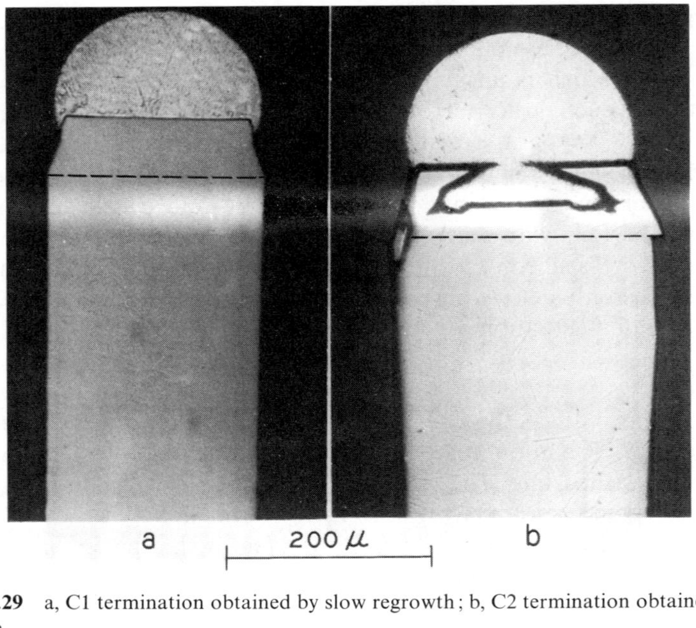

Figure 3.29 a, C1 termination obtained by slow regrowth; b, C2 termination obtained by fast regrowth.

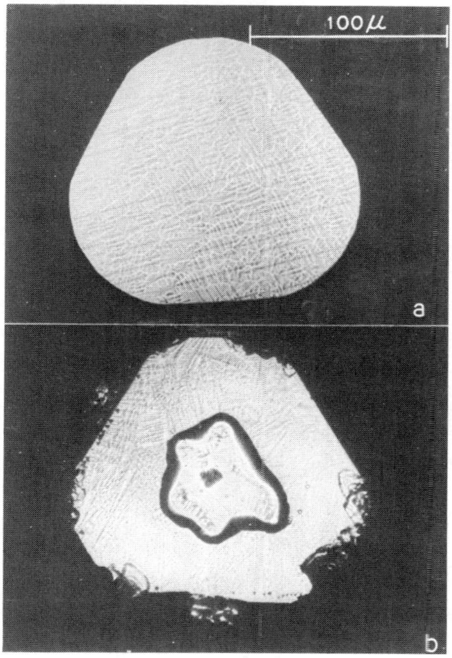

Figure 3.30 Interface between silicon crystal and two-phase alloy. (a and b for the Cl or C2 type, respectively.)

silicon which freezes out on cooling depends both on the original deposition temperature and, to a small extent, on the supercooling that is necessary to nucleate the gold phase [61]. Gold crystals nucleate in the liquid adjacent to the growth interface at or below the eutectic temperature. These crystals will grow predominantly parallel to the interface and will grow in size until the remainder of the liquid has assumed eutectic composition. The rest of the liquid will then solidify as eutectic.

The two-phase alloy comprising the hemispherical cap on top of the silicon crystal can be removed by etching the crystals in aqua regia for about 1 hour followed by ultrasonic cleaning in acetone. This procedure does not damage the main silicon crystal. The exposed-silicon two-phase alloy interface is very flat for the C1 type and parallel to a $\{111\}$ plane (such an etched interface is shown in Fig. 3.30a). The impurity striations form an image of the gold dendrites which grew parallel to the solid-liquid interface. No dislocations or other defects except impurity striations were found in the C1 terminations.

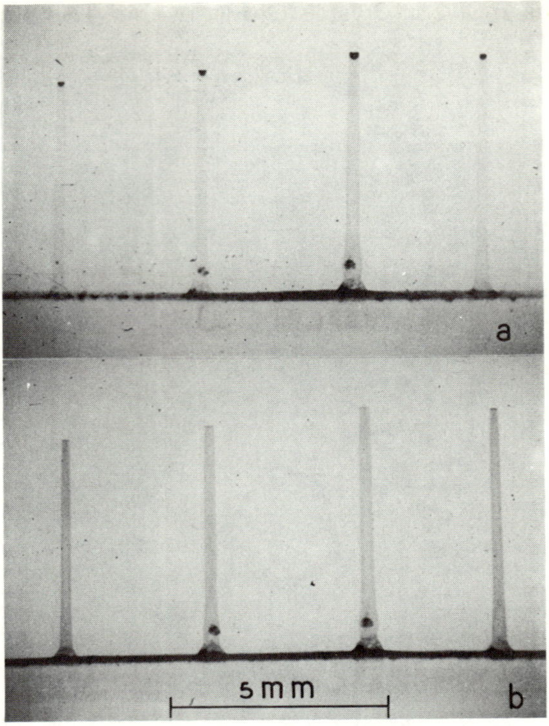

Figure 3.31 X-ray topograph of four silicon crystals and their common substrate. (With (a) and without (b) alloy termination; (111) reflection.)

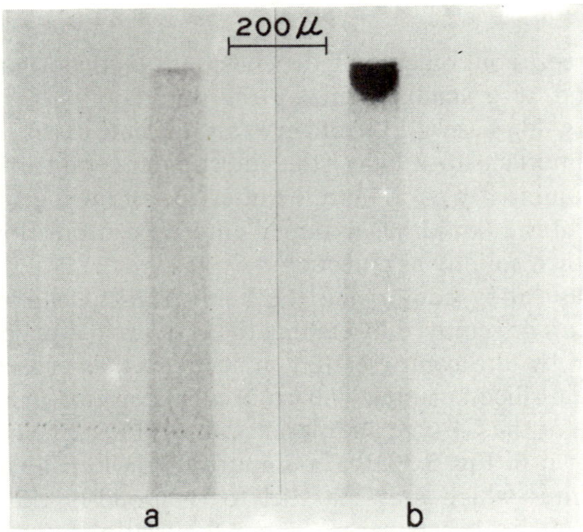

Figure 3.32 Stain pattern for the type Cl termination (b = with and a = without two-phase alloy termination; (111) reflection).

The C2 type regrowth (Fig. 3.29b) is formed on fast cooling from deposition to room temperature. The solid-liquid interface becomes nonplanar during regrowth, which may result in partial or complete entrapment of the liquid alloy. On fast cooling, the outsides of the crystal and of the liquid are colder than the interiors. This causes enhanced regrowth at the outer portion of the growth interface leading to entrapment of the liquid. It should be mentioned that silicon crystals may also nucleate in the liquid alloy during rapid cooling. Therefore, it is possible to entrap liquid above the eutectic temperature. Dislocations can be generated by the process discussed before. Figure 3.30b shows an etched alloy interface for a typical C2 regrowth. Note in comparison with Figure 3.30a that the interface is not planar and that there is a depression in the center region. Dislocation etch pits, in arrays typical for slip pattern, are along the outer edges of the section. Enhanced etching of the rounded corners of the interface indicate that the material is highly strained in this region.

An X-ray topograph of four silicon crystals and their common substrate crystal is shown in Figure 3.31a. Since all crystals are equally well resolved on the topograph, their angular misorientation with respect to each other must be smaller than 140 sec arc, the horizontal divergence of the X-ray beam used for this experiment. The crystals of Figure 3.31 show a partly defective A-region, a perfect B-region, and a defective C-region. A detailed study of the diffraction effect in the C-region gave the following results. Regardless of whether a crystal has a C1 or C2 regrowth region, there is a stress field in the silicon crystal which usually extends beyond the C- into the B-region. This stress field is caused by differential thermal contraction of the two-phase alloy and the silicon crystal during cooling to room temperature. Figure 3.31b shows the topograph of the same crystal array as shown in Figure 3.31a. However, the alloy was removed from the C1 regrowth by etching. The X-ray diffraction contrast due to the strain field has completely disappeared. Figure 3.32, a high magnification photograph, shows the topographs of the crystal C1 region with and without the alloy termination. This result, repeated on many crystals, clearly shows that the strain field due to thermal contraction in a crystal of the C1 type is of elastic nature. The shape of the diffraction effect caused by the strain field depends on the reflection used for the topograph. The {111}, {220}, and {422} topographs are very different in form, which indicates that the strain field is anisotropic.

Topographs of crystals with a C2 termination show that this type contains both an elastic stress field and also a plastically deformed region. Figure 3.33 shows a photograph of such a crystal with alloy tip and the X-ray topograph using {220} reflection. Note that the diffraction contrast again extends beyond the C-region and that its form is different from the one shown in Figure 3.32. The decapitated crystal and its topographs are shown in Figure

Figure 3.33 Strain pattern for the type C2 termination (a, actual crystal; b, X-ray topograph; $(\bar{2}\bar{2}0)$ reflection).

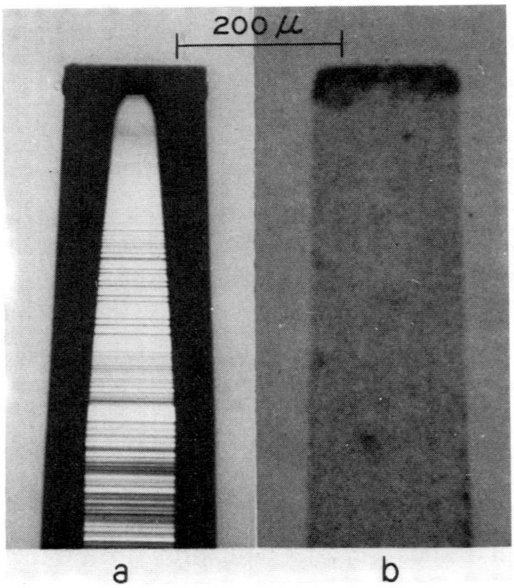

Figure 3.34 a, same crystal as shown in Figure 3.33a, however, the two-phase alloy was removed. b, X-ray topograph, $(\bar{2}\bar{2}0)$ reflection.

3.34. The strain field extending into the B-region has disappeared because it was elastic. In the C2 region, however, there are still defects present that give rise to enhanced X-ray intensity. The regrowth interface of this particular crystal is shown in Figure 3.30b. The stress field in the C-region can be large enough to cause cleavage; the alloy termination and part of the silicon regrowth can be ruptured from the main part of the crystal. This has been observed for crystals with very small diameters.

Summary and Discussion

Highly perfect silicon crystals of macroscopic size have been grown by the VLS technique. This confirms the earlier stipulation that screw dislocations or other crystalline defects are not essential for VLS growth. The main feature of this process is its low activation energy as compared to the usual vapor-solid (VS) process. Furthermore, the liquid alloy protects the growth interface from contamination which may cause defects. The surface of a growing crystal is also perfect and defect-free epitaxy by the VS process can be obtained at low temperature.

There are two requirements for the growth of perfect crystals; a perfect substrate crystal must be used and the various growth parameters must be carefully controlled. Both dislocations and stacking faults can continue to grow from an imperfect substrate crystal. These defects, however, terminate eventually at the side faces of the growing crystal. The second requirement concerns the stability of both the liquid alloy and the solid-liquid interface.

The liquid alloy is always constitutionally supercooled (CS) during VLS growth. This arises because a concentration difference (Δc) is required as driving force for diffusion of material from the liquid-gas to the solid-liquid interface. It has been shown that there exists an upper limit for Δc at which the liquid alloy becomes unstable. Silicon crystals can nucleate in the liquid when the liquidus supercooling $\Delta T = m\,\Delta c$ (m = slope of liquidus) exceeds the critical value for heterogeneous or homogeneous nucleation. For a critical ΔT, the maximum permissible growth rate (v) is inversely proportional to the thickness (L) of the liquid alloy zone. The stability of the liquid with regard to nucleation can be reasonably well controlled for the VLS growth of silicon crystals.

It is surprising that the solid-liquid interface remains stable, parallel to a $\{111\}$ plane, during VLS growth. The gradient of constitutional supercooling $\Delta T/L = mv/D$ (D = diffusion constant of silicon in liquid alloy) is extremely large, even for small growth rates. The assumptions implicit in this equation are a liquid zone of disklike shape and the absence of convective mixing of the liquid. For example, the growth parameter of the crystals investigated here are $v \sim 2.5 \times 10^{-5}$ cm/sec, $L \sim .02$ cm, $m \sim 20°C/.01$

atom fraction, and $D \sim 5 \times 10^{-5}$ cm^2/sec. The gradient of CS at the interface is about 1000°C/cm. However, the impurity striations that delineate interface positions show that the interface is stable under these conditions. Silicon crystals with a diameter of 10^{-5} cm can be grown at a rate of about 1 cm/sec. For this case the gradient is enormous, namely in the order of 10^7 °C/cm. These crystals also grow exactly in a $\langle 111 \rangle$ direction and it must be inferred that the interface is stable. Furthermore, the isoconcentration lines in the liquid are not parallel to the interface because of the geometry of the system. Therefore, the driving force at the interface is nonuniform, being smallest at the center. The stability of the solid-liquid interface during VLS growth can only be explained if the interface kinetics is a stabilizing factor as proposed by Cahn [50]. Interface diffusion [62, 63] must also play an important role to prevent the formation of hopper morphology.

It has been shown that the interface may break down, but not to a dendritic morphology. Instability occurs if a large lateral temperature gradient is imposed. The interface appears to conform closely to the equilibrium temperature isotherm. Parts of the liquid may become entrapped during this process. These entrapments are powerful sources for generation of defects and elastic stress fields. Finally, it was found that the morphology of the prismatic growth faces depends on the deposition temperature. The anisotropy of the vapor-solid deposition rate, as a function of temperature, determines the number and crystallography of the side faces.

MECHANISM OF BRANCHING AND KINKING DURING VLS GROWTH

Introduction

Many different theories [64–68] have been proposed to explain branching and kinking during vapor-phase growth of whisker crystals. All theories assume that the whisker crystals grow by the "screw dislocation" mechanism. Branching and kinking are explained by a variation of the dislocation model involving either a climb process of the dislocation at the whisker's tip or nucleation of a branch at a dislocation intersection. However, there is no reported experimental evidence to justify any of the proposed theories. It appears rather that the observation of growth kinking and branching has lead many investigators to believe that these changes in growth habit can be attributed to the operation of a screw dislocation mechanism in whisker growth. It is the purpose of this chapter to show that branching, kinking, and the growth of curved crystals can also be accounted for by the VLS process [69]. The work described here is an outgrowth of the investigation on the controlled growth of silicon crystals in the form of intricate patterns on silicon substrates. The crystals shown in Figure 3.20 were grown under nearly isothermal conditions using a resistance-heated furnace 10 in. long. Branched or kinked whiskers did not form under these experimental conditions. At this point of the investigation, it was considered worthwhile to take motion pictures during VLS growth. For this objective, the experimental conditions had to be changed drastically. The limited working distance of the microscope required an alteration of the heating system. The resistance furnace was replaced by a 2-in.-long graphite tube which served as susceptor inside the reaction tube. A considerable lateral and axial temperature gradient in the growth zone could not be avoided.

Motion pictures of VLS growth clearly showed that very frequently the crystal branched or kinked during growth. In some experiments all the crystals in an array kinked in the same direction, toward the higher temperature region. Because of the small thermal mass of the graphite susceptor, sudden changes in temperature in the growth zone could be obtained by changes in the *r-f* output of the generator. Such changes, whether intentional or accidental, quite often resulted in branched and kinked whiskers. Similar changes in growth morphology were obtained by sudden changes of the total gas flow through the reaction tube. These observations were studied in detail and the results are discussed in the following sections.

The proposed models are both simple and unique, and the experimental verification is unequivocal. It will be shown that crystalline defects, such as dislocations, are not essential for the branching and kinking process. We discuss first the factors that control the direction of growth of silicon crystals grown by VLS. Four different processes by which branches and growth kinks can be formed are described in subsections that follow. It is shown that controlled "growth shaping" during whisker growth from the vapor is possible.

Experimental

The hydrogen reduction of $SiCl_4$ was used as the transport reaction and gold was the liquid-forming impurity. The following experimental parameters were used in this investigation. The mean temperature of crystal growth was 1000°C. Temperature changes that were introduced intentionally will be discussed in the appropriate sections. A temperature gradient smaller than 3°C/cm was used for the isothermal growth experiments. The gradient was increased to over 40°C/cm for nonisothermal growth. In some experiments, the gradient was further increased by focusing a high-power heat source on the sides of the growing crystals. A total hydrogen flow of 300 cc/min was used with a $SiCl_4$-to-hydrogen mole ratio of .02. Preliminary experiments indicated that controlled doping during VLS growth can yield important information about the growth mechanism. It was found that the dopant, introduced in gaseous form, also preferentially enters the liquid alloy droplet at the tip of the growing crystal. Doping therefore results in clearly identifiable impurity striations which are parallel to the solid-liquid (SL) interface. These interface markers, introduced at given time intervals, permit a careful evaluation of interface transitions during branching and kinking. For the doping experiment, 10 cc/min hydrogen with a mole ratio of .0004 PCl_3 was used. The duration of doping varied from 10 to 30 sec. Doped silicon crystals were sectioned parallel to the growth direction and etched with Sirtl etch for about 1 to 4 min. The maximum diameter of all crystals shown in this chapter is about 250 to 300 μ. The measured rate of VLS growth was about 1 mm/hr. A high-temperature microscope* in conjunction with a movie camera were used to record the growth of individual crystals by time-lapse photography [70]. X-ray topographs by the Lang technique were taken on some samples using silver radiation.

Isothermal VLS Growth

The majority of silicon crystals prepared by the VLS technique, under isothermal conditions, grew in a [111] direction. It has been proposed that this particular growth direction arises because the solid-liquid interface is a single (111) plane. It has been shown that the interface can be very stable during VLS growth; the stabilizing factor is very likely the interface kinetics. To confirm this point, a number of silicon crystals were doped with phosphorus during isothermal growth. The crystals were subsequently sectioned on a plane parallel to the growth direction and etched as shown in Figure 3.35. The pronounced bands indicate the introduction of dopant during growth, whereas the fine bands are produced by fluctuations in growth rate. The figure clearly shows that the solid-liquid interface remains stable, parallel to a (111) plane during isothermal VLS growth. It must be concluded that layer growth occurs on this interface; the alloy droplet therefore advances perpendicular to this face in a [111] direction.

It has been shown that the shape of the solid-liquid interface of a [111] crystal is not circular but triangular with rounded corners. Microscopic observation shows that the solid-liquid contact angle along the meniscus of the droplet is nonuniform during growth, being slightly larger at the rounded corners. However, such a contact angle configuration, having threefold symmetry, stabilizes the liquid droplet on the growing crystal. The effect of gravity on the contact angle was found to be negligible. Crystals with diameters ranging from 50 to about 300 μ were grown in the usual way, vertically upwards. Subsequently, the whole growth apparatus was rotated 90°, and the VLS growth continued in a horizontal direction to about 1 cm length. The crystals were afterwards examined in an optical goniometer and found to be straight. Similarly, X-ray reflections obtained

* American Optical long working-distance microscope.

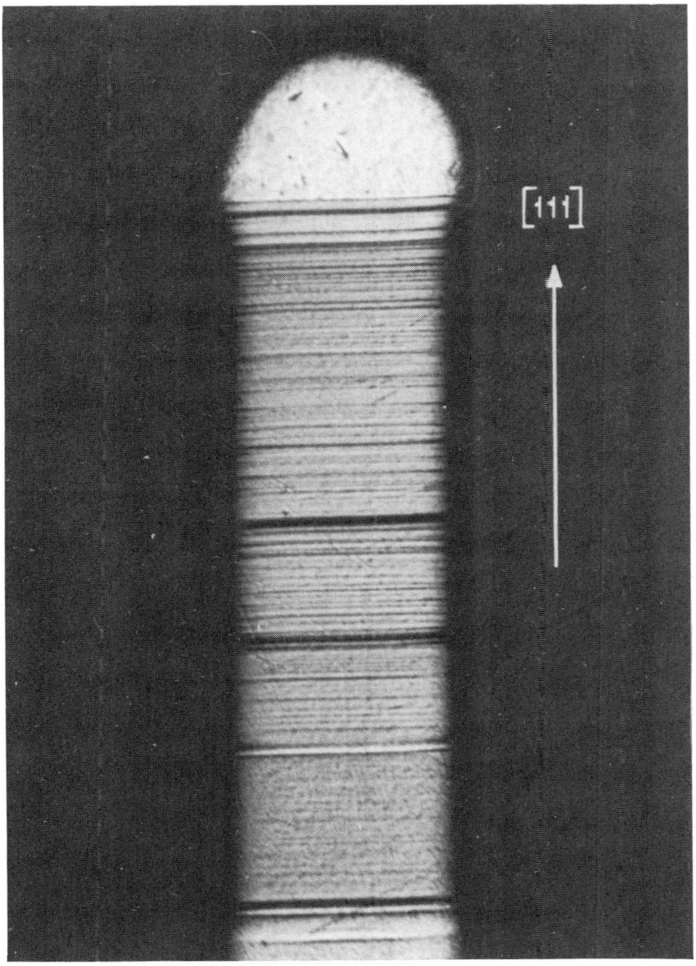

Figure 3.35 Etched section of a [111] silicon crystal doped with phosphorus during isothermal growth (mag. approx. 100×).

with a Weissenberg goniometer showed that the crystals were straight and grew in a [111] direction. It is evident, however, that any force which is strong enough to change the symmetry of the solid-liquid contact angle configuration can influence the direction of growth. Although layer growth still occurs on a (111) plane under isothermal conditions, the resultant growth direction will be controlled by the liquid contact angles.

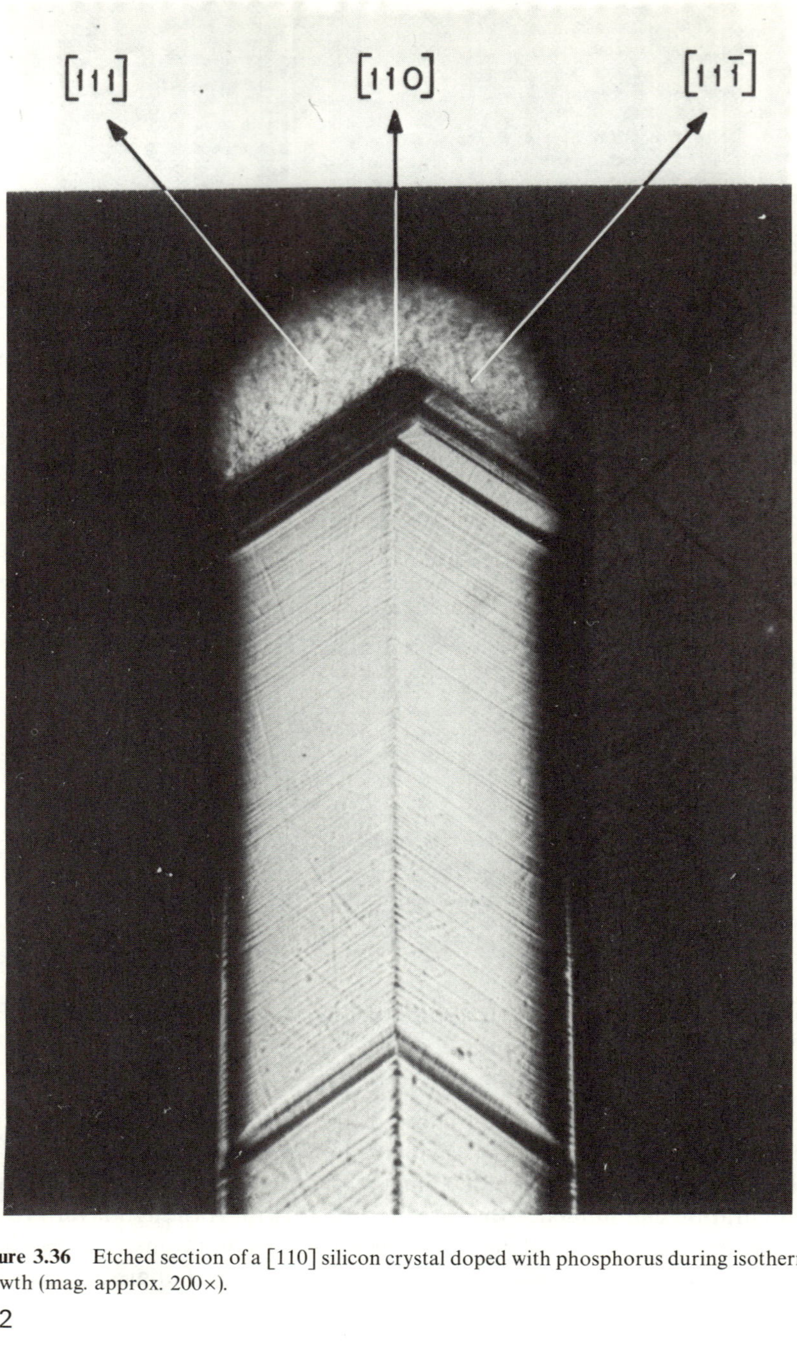

Figure 3.36 Etched section of a [110] silicon crystal doped with phosphorus during isothermal growth (mag. approx. 200×).

A suitable combination of interface morphology and liquid contact angle can give rise to growth directions other than [111]. For example, silicon crystals can be grown in a [110] direction as shown in Figure 3.36. The doping striations show that the interface consists of two {111} planes forming a ridge configuration. If the crystals are grown at about 1050°C, the prismatic side faces are four {111} planes that are parallel to the [110] growth direction. This configuration leads to a twofold symmetry of the contact angles, stabilizing the droplet on the growing crystals. VLS-grown germanium whiskers are frequently observed with the same [110] morphology. In addition, germanium crystals also grow in the [100] direction [44]. In this case, the solid-liquid interface is formed by four {111} planes in the shape of a pyramid. Another morphology has been observed for both germanium and silicon, a twinned crystal growing in a [211] direction [71]. It can be said in conclusion that the solid-liquid interface is not necessarily perpendicular to the direction of growth. The interface morphology determines the direction of growth and unidirectional growth is only possible for a stable contact angle configuration. It will be shown in the following section that a change in interface morphology during VLS growth invariably results in a change in growth direction.

Nonisothermal Growth

The model for branching and kinking during nonisothermal VLS growth [71] is based on the assumption that a lateral temperature gradient can change the morphology of the SL interface and thereby change the direction of crystal growth. The experimental verification of the proposed mechanism will be discussed in this section. The temperature conditions during VLS growth were changed from isothermal to nonisothermal. Time-lapse photography was used to obtain a permanent record of the growth sequence. The crystals were doped with phosphorus before, during, and after completion of the kinking process. Subsequently, the samples were sectioned, polished, and etched on a plane which is co-planar with the axial directions of the kinked crystal. A typical example of such an experiment is shown in Figure 3.37 for a [111] to [$\bar{1}$11] kink. In this case the kinking was caused by exposing the right side of the growing crystal to a higher temperature than the left side.

The value of using controlled doping to study successive interface positions is evident. During isothermal conditions, the crystal grew perpendicular to the (111) plane, as discussed previously. However, the introduction of a lateral temperature gradient caused a change in interface morphology. The ($\bar{1}$11) plane was partly exposed in addition to the original interface. The growth rate perpendicular to the newly formed interface is slower than on the (111) plane because it is in a region of higher temperature. The new

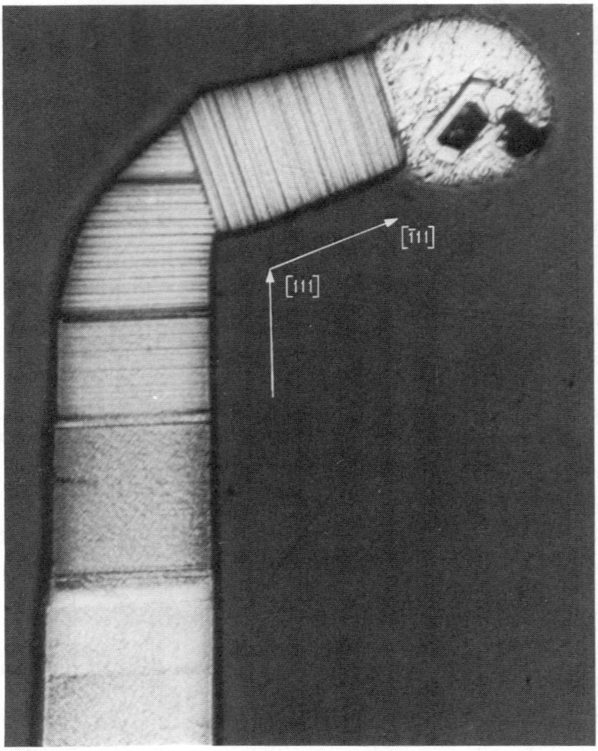

Figure 3.37 Etched section of a [111] to [$\bar{1}$11] kinked silicon crystal doped with phosphorus (mag. approx. 100X).

interface grows in area at the expense of the old one until finally the (111) interface has disappeared. The liquid droplet shifts during this process from the (111) to the ($\bar{1}$11) interface. Further VLS growth proceeds in the [$\bar{1}$11] direction. Figure 3.38 shows an enlargement of a (111) to ($\bar{1}$11) kink. Note that neither the (111) nor the ($\bar{1}$11) interface planes deviate from planarity during the kinking process. The macroscopic growth direction gradually changes from [111] to [$\bar{1}$11] resulting in the growth of a curved crystal.

Kinking under the influence of a temperature gradient was observed for over 100 crystals. About 20 crystals were examined for the presence of crystalline defects in the kink region. Lang topographs were taken and sectioned crystals were etched as shown in Figure 3.37. The majority of the samples did not contain any observable defects in the kink region. A few specimens contained islands of Au-Si entrapments which are known to give rise to the

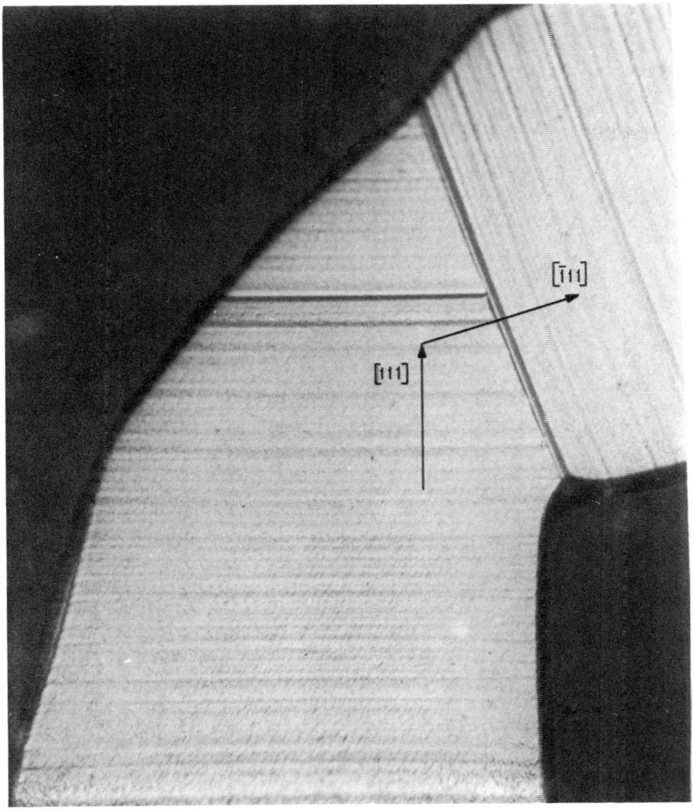

Figure 3.38 Etched section of a [111] to [$\bar{1}$11] kink (mag. approx 250X).

generation of defects. It must be concluded that crystalline defects, such as dislocations, are not essential for the reported kinking process. Entrapment of liquid alloy may occur during kinking with subsequent formation of defects during cooling.

Figure 3.39 shows a Lang topograph of a [110] to [11$\bar{1}$] kink. The crystal was grown on an imperfect (111) substrate. Some dislocations continued to grow from the substrate into the crystal. All dislocations apparently grew out of the crystal at the lower section. The remainder of the sample, including the kink, is dislocation free. Note the occurrence of "Pendellösung" fringes in both the [110] and [11$\bar{1}$] sections of the crystal. The shapes of the fringes in the two sections are different because the cross section of the crystal is not the same in the two regions. The cross section is diamond shaped in the

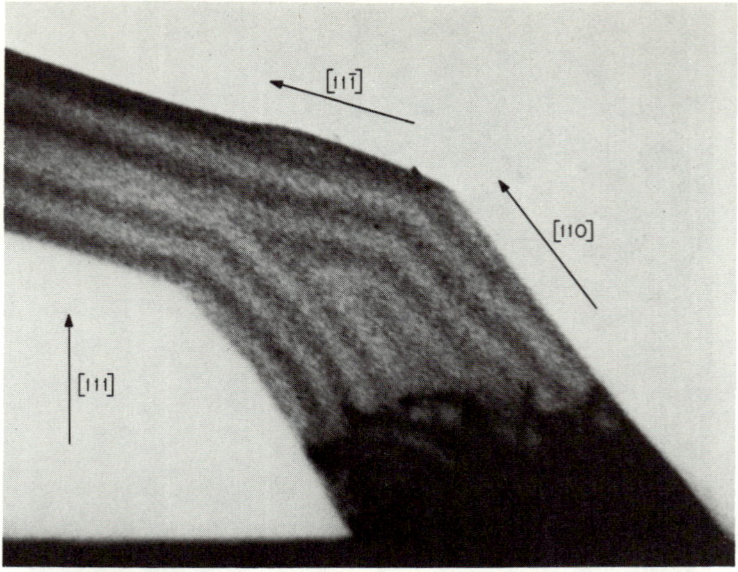

Figure 3.39 (111) X-ray topograph of a [110] to [11$\bar{1}$] kinked silicon crystal (mag. approx. 100X).

[110] region and hexagonal in the [11$\bar{1}$] part. The occurrence of "Pendellösung" fringes in the kink provides additional evidence that defects are not involved in the kinking mechanism.

It is evident from Figure 3.38 that the area of the SL interface changes during the kinking process. The area first increases and finally returns to its original value after completion of kinking. The change in interfacial area imposes a constraint on the contact angle of the liquid alloy droplet. The constraint frequently causes rupture of the droplet, thereby giving rise to the formation of branch crystals as shown in Figure 3.40. The time-lapse sequence, taken at constant magnification, illustrates four different stages during and after the kinking sequence. The crystal was grown on a (111) substrate at about 1050°C. The plane of the photograph is approximately (110). Figure 3.40a shows the trace of the SL interface and the droplet configuration during kinking at a time when the interfacial area is at a maximum. The kinking sequence is almost completed in Figure 3.40b. The new {111} interface has increased considerably at the expense of the old interface. A small droplet, indicated by the arrow, has broken away from the main droplet during the transition. Figure 3.40c shows the interface and droplet configuration after completion of kinking. The small droplet gave rise to the growth of a branch crystal with a [111] growth direction. A later sequence of growth is shown in Figure 3.40d. The effect of VS epitaxy in changing the external morphology can be seen at the base region of the small branch crystal by comparison of Figure 3.40c with Figure 3.40d. Some samples, in which branching occurred during the kinking process, were examined for perfection, both by X-ray topograph and etching. In most cases, the branch and kink were found to be dislocation free.

Generally, it is rather difficult to control the formation of a desired kink configuration under the influence of a lateral temperature gradient. Kinking can be initiated readily; however, there is little control over the shape of the

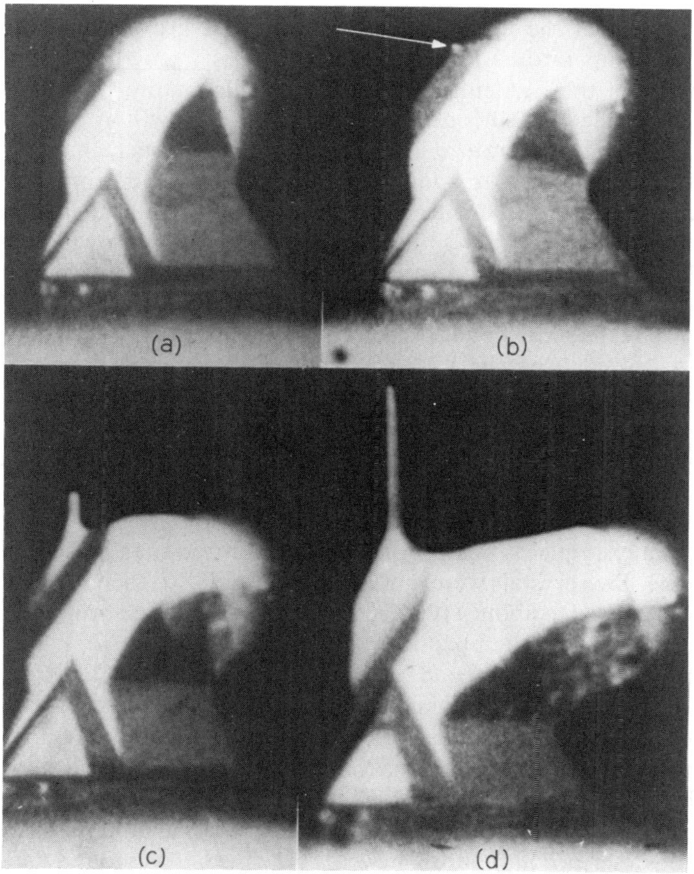

Figure 3.40 Time-lapse sequence of branching and kinking caused by a lateral temperature gradient (mag. approx. 125X).

liquid droplet during the transition process. The orientation of the growing crystal with respect to the isotherm is an important factor, as is evident from Figure 3.37. The preferred orientation, as indicated in the figure, is that in which only one plane, for example the $(\bar{1}11)$ plane, is parallel to the isotherm. If the interface transition occurs by exposing two or even three new $\{111\}$ planes, the droplet invariably ruptures and multiple kinking ("antler" growth) will result. A slow VLS growth rate and a small temperature gradient help to prevent a rapid interface transition which also can cause a rupture of the liquid droplet.

Sudden Increase in Temperature

The volume of the liquid droplet during VLS growth depends on the deposition temperature. This dependency arises from the liquidus phase relationship of the particular alloy system. This feature has been used to change the diameter of silicon crystals during VLS growth by proper changes of the deposition temperature. It has been noted that such changes must be made gradually in order to avoid instability of the liquid droplet. A sudden increase in temperature, for example, from 1000 to 1100°C in 10 seconds, will usually cause instability. The volume of the droplet will increase by about 13% for such a temperature change. The amount of silicon that is required to satisfy the new phase equilibrium cannot be supplied rapidly enough from the vapor-phase reaction. As a result, the necessary amount will be dissolved from the VLS crystal. The SL interfacial area, however, remains that for the 1000°C growth condition. As a consequence, the liquid will bulge over the side of the crystal. It will eventually wet the side faces in order to increase the interfacial area and to satisfy a stable contact angle configuration. The individual droplets will cause formation of branch crystals or kinks as VLS growth continues.

A typical example of this process is shown in the time-lapse sequence in Figure 3.41. Two crystals were initially grown at 100°C, then the temperature was rapidly raised to about 1100°C. Figure 3.41a shows the droplet configuration during the rise in temperature. The solid-liquid contact angle, which is slightly larger than 90° under equilibrium conditions, has noticeably increased. In Figure 3.41b, taken at maximum temperature, the droplet has increased its contact area by wetting the side faces of the crystal. Subsequently, the temperature was reduced to about 1000°C. The next two photographs show the resulting growth morphology at different times. The liquid broke up into smaller droplets giving rise to branch growth. Numerous experiments of the kind described in Figure 3.41 were performed. It was found that the only important factor is a sufficiently large, rapid increase in temperature. The temperature does not have to be returned to its starting value. A large decrease in temperature after the instability has occurred must be avoided

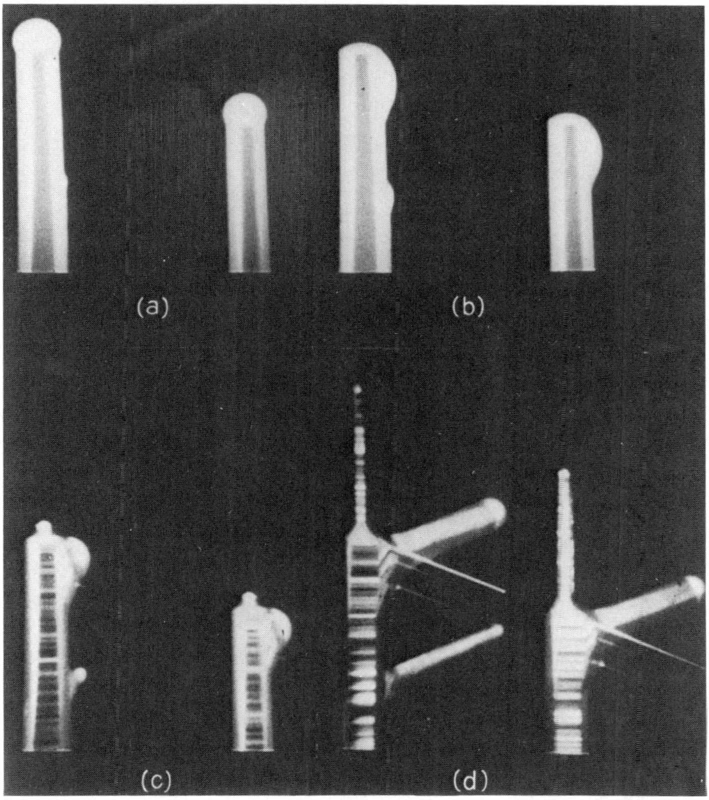

Figure 3.41 Time-lapse sequence of kinking and branching caused by a sudden increase in temperature (mag. approx. 15X).

because this may lead to nucleation of silicon crystals in the droplets, resulting in uncontrolled growth.

The droplet configuration during the rise in temperature was observed visually in some experiments. This provides some control over the resulting droplet configuration. The photograph, Figure 3.42, is an etched $(0\bar{1}1)$ section of a crystal which was kinked by a controlled increase in temperature. Doping was used to mark the position of the SL interface during the growth experiment. Before raising the temperature, the droplet was on top of the [111] crystal which had grown under isothermal conditions. The temperature was carefully increased until the droplet was clinging to the side of the crystal near its tip. The doping striations show that a new SL interface configuration was established. The crystal first grew in approximately a [011] direction

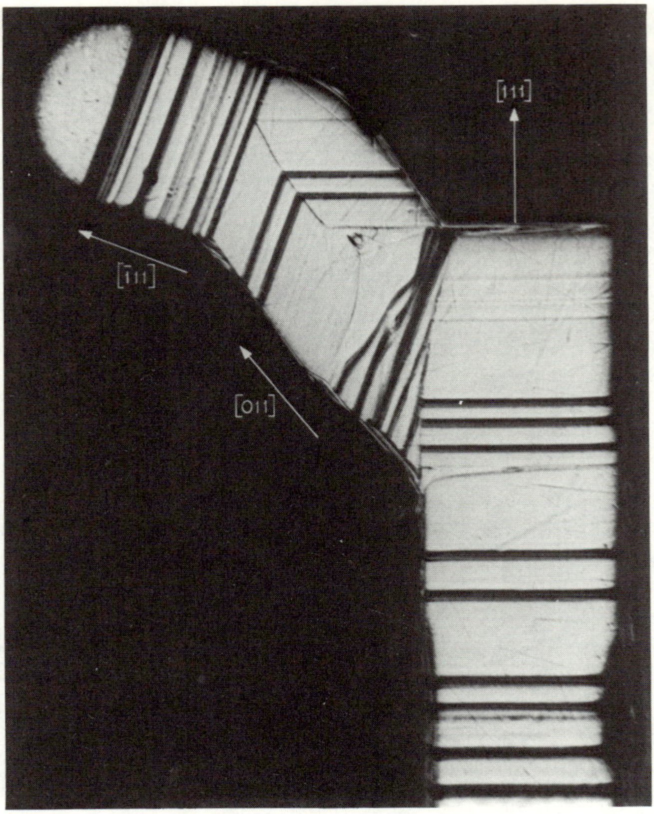

Figure 3.42 Etched section of a silicon crystal kinked by a sudden increase in temperature, doped with phosphorus (mag. approx. 100×).

and gradually changed to the stable [111] direction. No dislocation etch pits could be found on this sample.

Six crystals that branched or kinked by the discussed procedure were examined by X-ray topography. Entrapments of Au-Si alloy were found in four samples. Such a topograph and the actual sample are shown in Figure 3.43. The hemispherical tips, composed of Au-Si alloy, were removed from the crystals by etching. A significant aspect is that the crystallography of the branch and kink crystals is the same as that of the main crystal. Some of the fine crystals which can be seen on the actual sample could not be reproduced properly from the X-ray topograph. An island of Au-Si eutectic is indicated by an arrow. As usual, the inclusion gives rise to formation of defects.

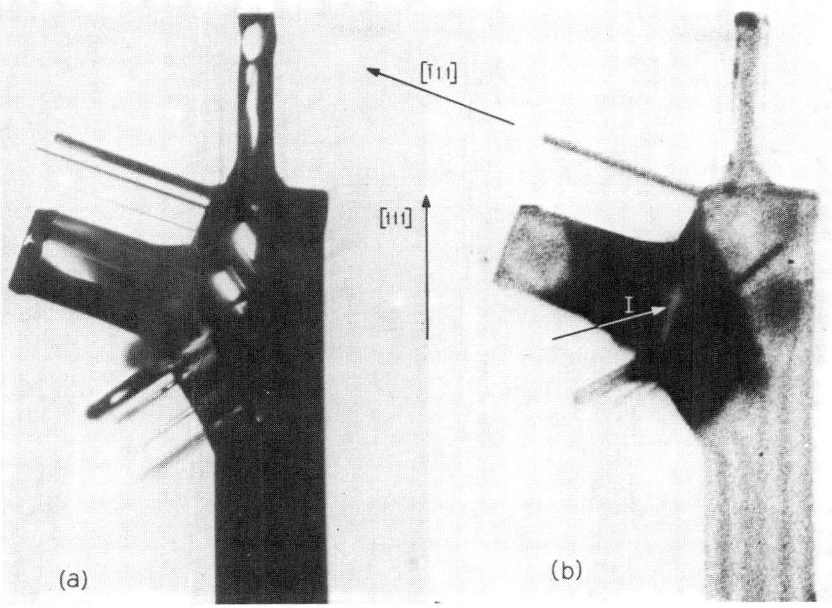

Figure 3.43 (a) Photograph of a branched crystal with alloy tips removed by etching. (b) (111) X-ray topograph showing Au-Si entrapment (mag. approx. 50×).

Sudden Decrease in Temperature

This mechanism is basically similar to the one discussed in the preceding section. The volume of the liquid droplet decreases if the temperature is lowered during VLS growth. The excess of silicon freezes out of solution, by normal alloy regrowth, in order to establish the equilibrium composition. The solid-liquid contact angle, however, remains essentially constant during the change in temperature. This implies that a reduction in SL interfacial area is required to satisfy the same contact angle configuration. Therefore, a sudden decrease in temperature causes a sudden increase in growth rate because of alloy regrowth and a reduction of the diameter of the growing crystal. Small droplets frequently separate from the meniscus region of the droplet during the transition in volume. These droplets then give rise to formation of branches during subsequent VLS growth. The growth of branch crystals by this process is illustrated by the time-lapse sequence in Figure 3.44. The crystal is shown in Figure 3.44a during growth at about 1050°C. The temperature was subsequently decreased by 100°C in about 30 sec.

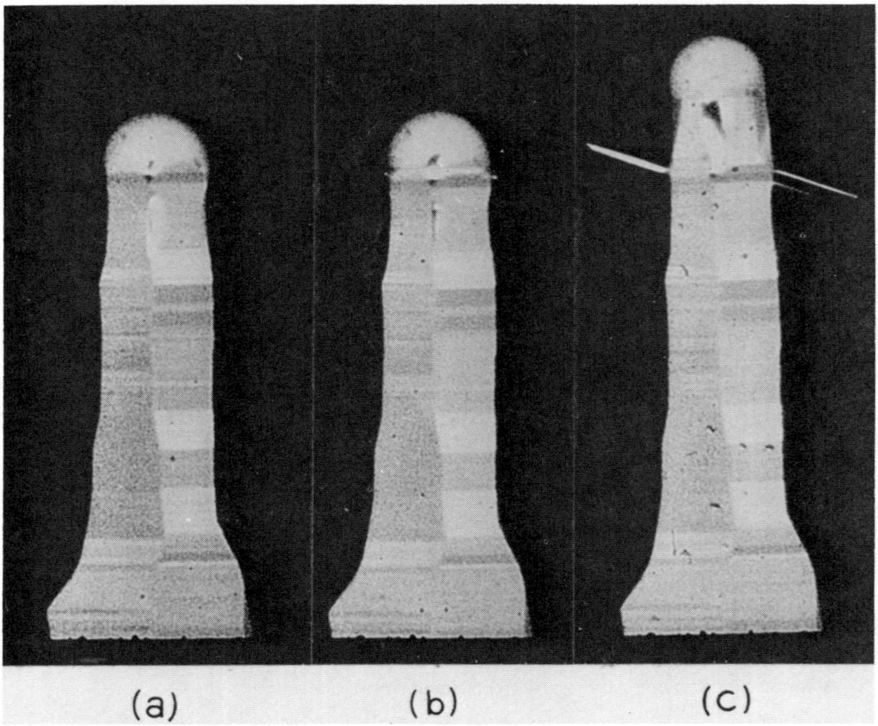

Figure 3.44 Time-lapse sequence of branching caused by sudden decrease in temperature (mag. approx. 45X).

Three small droplets were "pinched off" the main droplet during the transition (Figure 3.44b). The last figure shows the crystal after continued VLS growth. The three branches grew in the same $\langle 111 \rangle$ direction, although inclined to the $[111]$ direction of the main crystal.

It has been found in many experiments that this process of branch formation can be initiated easily by a sudden drop in deposition temperature. However, there is little control over the number of branches formed. The change in temperature must be rapid enough to cause separation of small droplets from the main body of the liquid alloy. A gradual decrease in temperature only causes a reduction in diameter of the growing crystal. The deposition temperature can be gradually raised after completion of branching. However, a sudden large reduction in temperature must be avoided because this may result in nucleation of silicon crystals in the liquid alloy.

The most common direction of the side branches is $\langle 111 \rangle$. The $\langle 110 \rangle$ direction has been observed in a few cases. The diameters of the branches are usually small compared to the diameter of the main crystal. X-ray topography and etching studies indicate that the majority of the crystals are dislocation free.

Co-Deposition of Liquid-Forming Impurities

The most likely process of branch and kink formation during vapor-phase growth is due to co-deposition of impurities. Impurities that satisfy the criteria for VLS growth may condense on an already grown whisker crystal. The transport can occur by a chemical reaction, such as disproportionation, or simply by an evaporation and condensation process. Such an impurity transport frequently has been observed during whisker growth of silicon or germanium using either a closed or dynamic system. Each alloy droplet formed by this means may give rise to VLS growth of a branch crystal. The original crystal serves as substrate material. It should be emphasized that only extremely small amounts of impurities are required for this process. For example, a gold-silicon alloy droplet of about 100 Å diameter can result in the growth of a silicon whisker a few millimeters long. The final morphology will be altered drastically in most cases because of VS deposition during and after VLS growth.

Many branched and kinked whisker crystals of silicon and of germanium were grown using co-deposition of impurities. The formation of branch crystals is initiated by a sudden increase in supersaturation of liquid-forming impurities in the growth zone, during or after completion of VLS growth. Numerous new alloy droplets form and branch growth occurs. It is obvious that this process may not be used for controlled growth of branched and kinked crystals.

The mechanism discussed involving the co-deposition of impurities probably is the most common cause of uncontrolled VLS growth. Note that careful experimentation with co-deposition of impurities originally led to the discovery of the VLS mechanism [6].

Summary and Discussion

The growth direction of silicon crystals prepared by the VLS technique is determined by both the shape of the solid-liquid interface and the contact angle configuration of the alloy droplet. Unidirectional growth can occur only if these two parameters are kept constant during growth. A necessary condition for unidirectional growth is therefore an isothermal environment. Most crystals grown under this condition grow with a single $\{111\}$ interface in a $\langle 111 \rangle$ direction. However, the macroscopic growth direction is not necessarily perpendicular to the growth interface. Specific interface configurations have been found where the interface consists of more than one {111} plane of equal area. These interfaces also form a stable contact angle configuration of the alloy droplet. The resulting growth direction is the vector sum of the individual interface growth directions.

A sufficiently large lateral temperature gradient changes the equilibrium interface, which invariably leads to a change in growth direction. A new stable interface will be formed during growth which is as parallel as possible to the isotherm. While a curved crystal is grown during the interface transition, subsequent growth is again unidirectional. Branches are formed if the alloy droplet ruptures during the kinking sequence. The likelihood of kinking, for a given lateral gradient, depends on the diameter of the growing crystal. It is the absolute temperature difference across the interface that causes interface transition.

Either a sudden increase or decrease in deposition temperature may introduce instability of the alloy droplet. A sudden rise in temperature causes an unstable contact angle configuration. The droplet may wet the side faces of the crystal to increase the interfacial area. The new droplet configuration results in formation of branches and/or kinks. Similarly, a sudden decrease in temperature may cause pinching off of small droplets from the main droplet. These two processes of branching and kinking are more or less independent of the diameter of the growing crystal.

The most common cause of uncontrolled VLS growth of silicon or germanium was found to be the result of co-deposition of liquid-forming impurities. The impurities can be transported to the reaction zone by a chemical reaction or by evaporation. Branches are formed when liquid-forming impurities condense on an already grown whisker crystal.

SUMMARY OF MATERIALS GROWN BY THE VLS TECHNIQUE

α-Al_2O_3. Barber [72] grew α-Al_2O_3 whiskers by condensation and oxidation of aluminum on an alumina substrate and examined the crystals by transmission electron microscopy and diffraction. He found that the whisker crystals invariably terminate in a small globule of aluminum in a "drumstick" form. The experimental observations suggest that the α-Al_2O_3 crystals grow by VLS with an excess of aluminum acting as the agent.

Boron. Sitarik [73] prepared α-rhombohedral boron crystals using the hydrogen reduction of BCl_3 as transport reaction and platinum as the agent. Both straight and kinked whisker crystals were produced with $\langle\bar{4}401\rangle$ growth direction.

GaAs, GaP, *and* Ga(As$_{1-x}$P$_x$). There is a series of papers [75–78, 80] describing VLS growth of GaAs, GaP, and Ga(As$_{1-x}$P$_x$) crystals. Barns [74] and Frosch [79] were the first to use gallium as the agent for the growth of GaAs or GaP respectively. This procedure has the advantage of avoiding the introduction of impurities into the reaction chamber.

Germanium. The author [44] studied VLS growth of germanium using two different transport reactions. It was found that the hydrogen reduction of the halides, $GeBr_4$ and $GeCl_4$, could not be used for VLS growth of germanium. Very likely the rate of these reactions is controlled by diffusion of the reagents and reaction products to and from the crystal surface. However, the disproportionation reaction of the halides, GeI_2, $GeBr_2$, and $GeCl_2$, can be conveniently used for VLS growth in the temperature range of 400 to 500°C using gold as the agent. The morphology, defect structure, branching, and kinking of VLS-grown germanium crystals were found to be almost identical to the silicon crystal previously studied.

MgO. Heuer [81] observed VLS growth of this material inside the electron microscope during grain growth studies. The impurity agent could not be identified. It is speculated that a molten phase containing Al, Ca, and Si could probably meet the requirements of VLS growth.

$NiBr_2$. Sickafus [82] found that the mobile particles that are seen forming on the surface of $NiBr_2$ crystals during their growth, by sublimation, at nearly 500°C below the melting point of $NiBr_2$, are liquid-phase droplets. The droplets are nucleated by accumulation of impurities that exist in the source material in relatively low concentrations. When these droplets become attached to growth steps, an accelerated growth, taking the form of surface spikes, is observed. The process can best be described as a two-dimensional mode of VLS phase reaction.

NiO. The growth of NiO whiskers from molten nickel was described by Ahmad [83]. It apparently involves initial vaporization of nickel onto the surface of the boat in the form of tiny globules that become supersaturated with dissolved oxygen, with subsequent precipitation of the oxide nucleus at the solid-liquid interface. This nucleus grows as more oxygen is taken up by the liquid globule.

Selenium. Keezer [84] prepared selenium crystals by an evaporation-condensation reaction with thallium as the agent. The growth direction of the whisker crystal is perpendicular to the *c* axis and the crystals were found to be dislocation free.

Silicon. In addition to the author's own work, there are a number of papers describing various aspects of VLS growth of silicon. Filby [85] investigated the possibility of large-area epitaxial VLS growth with thin liquid alloy zones. Thornton [86] studied VLS growth of silicon from a gold-silicon alloy. The crystals were examined by scanning electron microscopy. The growth obtained various forms, from epitaxial layers at high temperature to fine whiskers at low temperature. The morphology of the whiskers was found to vary greatly as the experimental conditions were changed. The experiments confirm the important role played by the surface tension of the molten alloy zone. In a similar study, James [87] found that epitaxial layers and oriented whiskers are obtained using gold, while the growth from nickel appears random. Various growth forms are explained in terms of the temperature variation of the surface tension of the liquid alloy. The work emphasizes the value of the VLS process and has shown the importance of temperature and temperature gradients in its control. Filby [88] could not grow silicon whiskers with gold as agent, using the sublimation of silicon as transport process. No difficulties were encountered when the decomposition of $SiHCl_3$ was used. However, Ishiwatari [89] prepared silicon whiskers by sublimation of silicon, with gold as agent, by carefully controlling the supersaturation of the vapor phase. Komatsu [90] prepared a *p-n* junction in a silicon whisker crystal grown by VLS. The junction was identified by etching and scanning electron microscopy.

SiC. Ryan [91] grew beta silicon carbide by the hydrogen reduction of methyltrichlorosilane (CH_3SiCl_3) onto carbon substrates at 1500°C. He has shown that alpha inclusions present are the rare 2H (Wurtzite) modification of silicon carbide and that their presence resulted from a VLS process which was dominated by impurities in the substrate. By carefully cleaning the substrate and purifying the CH_3SiCl_3, the alpha inclusions were eliminated. The 2H alpha crystals were then deliberately grown by introducing selected impurities locally on the substrate. Beta crystals were also intentionally grown by VLS by using appropriate impurities.

CONCLUDING REMARKS

Needless to say, the author is unable to describe in detail how the VLS concept should be applied for the growth of whisker crystals of specific materials useful for technological application. Specific difficulties must be expected for each individual material, be it an element or compound. Of paramount importance is the proper selection of the transport process and the liquid forming agent. In addition, the transport process and the agent must be compatible for useful application to VLS. It is very likely that the low-yield transport processes, such as condensation or sputtering, are not economical for mass production of whisker crystals. Much more promising is the use of chemical transport reactions, for example, the decomposition or disproportionation reaction. Fortunately, the practical importance of crystal growth from the vapor has greatly stimulated the compilation of thermodynamic data for chemical transport reactions. The usefulness of such data in planning processes for growing a desired material is well recognized. The selection of a proper agent might even be more difficult than finding a suitable transport process. Very little is known about the solid solubility of liquid-forming impurities, for example, in refractory materials. For most materials, there is a complete lack of data on the wetting characteristic of the liquid alloy on the crystalline material to be grown by VLS.

The work reported in this chapter has clearly outlined the basic principles of the VLS process. The author is convinced that these principles can be applied to grow whisker crystals of almost any material. The successful application, however, still depends to some extent on the originality of the individual investigator. I hope that my own contribution and that of my collaborators will help in making the growth of whisker crystals more a science than an art.

REFERENCES

[1] F. C. Frank, *Discussions Faraday Soc.*, **5,** 48 (1949).

[2] W. K. Burton, N. Cabrera, and F. C. Frank, *Phil. Trans. Roy. Soc.* (*London*), **243A,** 299 (1951).

[3] G. W. Sears, *Acta Met.*, **1,** 457 (1953); **3,** 361 (1955).

[4] P. B. Price, D. A. Vermilyea, and M. B. Webb, *Acta Met.*, **6,** 524 (1958).

[5] W. W. Webb, *Growth and Perfection of Crystals*, John Wiley and Sons, New York, 1958, 230.

[6] R. S. Wagner, W. C. Ellis, K. A. Jackson, and S. M. Arnold, *J. Appl. Phys.*, **35,** 2993 (1964).

[7] E. S. Greiner, J. A. Gutowski, and W. C. Ellis, *J. Appl. Phys.*, **32,** 2489 (1961).

[8] W. J. McAleer, H. R. Barkemeyer, and P. I. Pollak, *J. Electrochem. Soc.*, **108,** 1168 (1961).

[9] R. S. Wagner and W. C. Ellis, *J. Metals*, **15,** 76 (1963).

[10] A. V. Sandulova, P. S. Bogojavlensky, and M. I. Dronjuk, *Fiz. Tverd. Tela* **5,** 2580 (1963).

[11] H. Schäfer and B. Morcher, *Z. Anorg. Allgem. Chem.*, **290,** 279 (1957).

[12] E. S. Wadja and R. Glang in "Metallurgy of Elemental and Compound Semiconductors," *Met. Soc. Conf.*, **12,** 229 (1961).

[13] R. V. Coleman and G. W. Sears, *Acta Met.*, **5,** 131 (1957).

[14] W. W. Webb and E. F. Riebling, *J. Chem. Phys.*, **28,** 1242 (1958).

[15] S. S. Brenner, *Acta Met.*, **7,** 519 (1959).

[16] C. R. Morelock and G. W. Sears, *J. Chem. Phys.*, **31,** 926 (1959).

[17] J. A. Gutowski and W. D. Beaver, private communication.

[18] R. G. Treuting and S. M. Arnold, *Acta Met.*, **5,** 598 (1957).

[19] R. S. Wagner and R. G. Treuting, *J. Appl. Phys.*, **32,** 2490 (1961).

[20] R. G. Treuting, R. S. Wagner, and W. C. Ellis, *Trans. Met. Soc. AIME*, **236,** 994 (1966).

[21] J. E. Gordon, *Nature*, **179,** 1270 (1957); in *Growth and Perfection of Crystals*, John Wiley and Sons, New York, 9 (1958).

[22] S. S. Brenner in *The Art and Science of Growing Crystals*, John Wiley and Sons, New York, 30 (1963).

[23] J. D. Eshelby, *J. Appl. Phys.*, **24,** 176 (1953); *Phil. Mag.*, **3,** 440 (1958).

[24] K. A. Jackson and R. S. Wagner, *J. Appl. Phys.*, **36,** 2132, (1965).

[25] R. S. Wagner, *Acta Met.*, **8,** 57 (1960).

[26] R. S. Wagner and W. C. Ellis, *Appl. Phys. Letters*, **4,** 89 (1964).

[27] R. S. Wagner and W. C. Ellis, *Trans. Met. Soc. AIME*, **233,** 1053 (1965).

[28] G. G. Lemmlein, E. D. Dukova, and A. A. Chernov, *Kristallografiya*, **2,** 428 (1957); translation: *Soviet Phys. Cryst.*, **2,** 426 (1957).

[29] E. Yoda, *J. Phys. Soc. Japan*, **15,** 821 (1960).

[30] R. O. Grisdale, *The Art and Science of Growing Crystals*, John Wiley and Sons, 1963, 163.

[31] L. Kowarski, *J. Chim. Phys.*, **32,** 303 (1935).

[32] W. J. Allan and W. W. Webb, *Acta Met.*, **7,** 646 (1959).

[33] W. W. Webb and W. D. Forgeng, *J. Appl. Phys.*, **28,** 1449 (1957).

[34] D. J. Barber, *Phil. Mag.*, **10,** 75 (1964).

[35] K. M. Merz, *Silicon Carbide—A High Temperature Semiconductor*, Pergamon Press, New York, 1960, 73.

[36] P. L. Edwards and R. J. Happel, Jr., *J. Appl. Phys.*, **33,** 943 (1962).

[37] H. C. Theuerer, *J. Electrochem. Soc.* **108,** 649 (1961).

[38] J. P. Hirth and G. M. Pound, *J. Phys. Chem.*, **64,** 619 (1960); *Progr. Mater. Sci.*, **11,** 107 (1963).

[39] K. A. Jackson, *Growth and Perfection of Crystals*, John Wiley and Sons, New York, 1958, 319.

[40] W. G. Pfann, *Zone Melting*, John Wiley and Sons, New York, 1960, 254.

[41] W. Steinmaier, *Philips Res. Rept.*, **18,** 75 (1963).

[42] W. C. Dash, *Growth and Perfection of Crystals*, John Wiley and Sons, New York, 1958, 361.

[43] R. S. Wagner and C. J. Doherty, *J. Electrochem. Soc.*, **113,** 1300 (1966).

[44] R. S. Wagner, unpublished work.

[45] D. Turnbull, *J. Appl. Phys.*, **21,** 1022 (1950).

[46] W. H. Shepherd, *J. Electrochem. Soc.*, **112,** 988 (1965).

[47] R. S. Wagner, *J. Appl. Phys.*, **38,** 1554 (1967).

[48] A. J. Goss, K. E. Benson, and W. G. Pfann, *Acta Met.*, **4,** 332 (1956).

[49] W. W. Mullins and R. F. Sekerka, *J. Appl. Phys.*, **34,** 323 (1963).

[50] J. W. Cahn, *Crystal Growth*, Pergamon Press, Oxford and New York, 1967, 681.

[51] A. R. Lang, *J. Appl. Phys.*, **29,** 597 (1958); *J. Appl. Phys.*, **30,** 1748 (1959).

[52] W. C. Dash, *J. Appl. Phys.*, **27,** 1193 (1956); *J. Appl. Phys.*, **30,** 459 (1959).

[53] E. E. Jenkinson and A. R. Lang, *Direct Observation of Imperfections in Crystals*, Interscience Publishers, John Wiley and Sons, New York, 1962, 471.

[54] H. J. Queisser, *J. Appl. Phys.*, **33,** 1536 (1962).

[55] R. A. Logan and W. L. Bond, *J. Appl. Phys.*, **30,** 322 (1959).

[56] R. H. Finch, H. J. Queisser, G. Thomas, and J. Washburn, *J. Electrochem. Soc.*, **34,** 406 (1963).

[57] T. B. Light, "Metallurgy of Semiconductor Materials," *AIME Met. Soc. Conf.*, **15,** 137 (1962).

[58] A. G. Revesz and R. J. Evans, *Trans. Met. Soc. AIME*, **230,** 581 (1964).

[59] G. K. Gaule and J. R. Pastore, "Metallurgy of Elemental and Compound Semiconductors," *Trans. AIME*, **12,** 201 (1961).

[60] S. K. Tung, *J. Electrochem. Soc.*, **112,** 436 (1965).

[61] R. S. Wagner, "Conference on Advanced Electronic Materials," *Trans. AIME*, **19,** 239 (1963).

[62] F. A. Nichols and W. W. Mullins, *Trans. Met. Soc. AIME*, **253,** 1840 (1965).
[63] S. R. Coriell and R. L. Parker, *J. Appl. Phys.*, **37,** 1548 (1966).
[64] W. W. Webb, *J. Appl. Phys.*, **36,** 214 (1965).
[65] J. P. Hirth and F. C. Frank, *Phil. Mag.*, **3,** 1110 (1958).
[66] S. Amelinckx, *Phil. Mag.*, **3,** 425 (1958).
[67] F. R. N. Nabarro and P. J. Jackson, *Growth and Perfection of Crystals*, John Wiley and Sons, New York, 1958, 81.
[68] S. Amelinckx, *ibid.*, p. 144.
[69] R. S. Wagner and C. J. Doherty, *J. Electrochem. Soc.* 115, 93 (1968).
[70] R. S. Wagner, C. J. Doherty, and S. M. Arnold, *J. Metals*, **17,** 1031 (1965).
[71] R. S. Wagner, *Crystal Growth*, Pergamon Press, Oxford and New York, 1967, 347.
[72] D. J. Barber, *Phil. Mag.*, **10,** 75 (1964).
[73] J. P. Sitarik and W. C. Ellis, *J. Appl. Phys.*, **37,** 2399 (1966).
[74] R. L. Barns and W. C. Ellis, *J. Appl. Phys.*, **36,** 2296 (1965).
[75] P. Knoll and R. Zuleeg, *J. Appl. Phys.*, **37,** 5006 (1966).
[76] C. M. Wolfe, C. J. Nuese, and N. Holonyak, Jr., *J. Appl. Phys.*, **36,** 3790 (1965).
[77] S. Iida and Y. Sugita, *Appl. Phys. Letters*, **8,** 77 (1966).
[78] B. D. Joyce and J. B. Mullin, *Solid State Comm.*, **4,** 463 (1966).
[79] C. J. Frosch, *Crystal Growth*, Pergamon Press, Oxford, 1967, 305.
[80] J. J. Grossman and M. W. Wegner, *Crystal Growth*, Pergamon Press, Oxford, 1967, 89.
[81] A. H. Heuer and P. Burnett, *J. Am. Ceram. Soc.*, **50,** 627 (1967).
[82] E. N. Sickafus and D. B. Barker, *J. Crystal Growth*, **1,** 93 (1967).
[83] I. Ahmad and G. P. Capsimalis, *Crystal Growth*, Pergamon Press, Oxford, 1967, 325.
[84] R. C. Keezer and C. Wood, *Appl. Phys. Letters*, **8,** 139 (1966); R. C. Keezer, C. Wood, and J. W. Moody, *Crystal Growth*, Pergamon Press, Oxford, 1967, 119.
[85] J. D. Filby and S. Nielsen, *Brit. J. Appl. Phys.*, **17,** 81 (1966); *J. Electrochem. Soc.*, **113,** 1091 (1966).
[86] P. R. Thornton, D. W. F. James, C. Lewis, and A. Bradford, *Phil. Mag.*, **14,** 165 (1966).
[87] D. W. F. James and C. Lewis, *Brit. J. Appl. Phys.*, **16,** 1089 (1965).
[88] J. D. Filby, S. Nielsen, G. J. Rich, G. R. Booker, and J. M. Larcher, *Phil. Mag.*, **16,** 565 (1967).
[89] K. Ishiwatari, T. Oka, and K. Akiyama, *J. Appl. Phys.* (*Japan*), **6,** 1170 (1967).
[90] E. Komatsu, Y. Higuchi, and T. Niima, *Appl. Phys. Letters*, **10,** 42 (1967).
[91] C. E. Ryan, I. Berman, R. C. Marshall, D. P. Considine, and J. J. Hawley, Physical Sciences Research Papers, No. 266, AFCRL-66-641, September 1966.

A. I. Mlavsky and H. E. LaBelle, Jr.

4 Growth of Continuous Whiskers from the Melt

ABSTRACT

Sapphire (α-Al_2O_3) whiskers have been produced by a melt growth technique. These whiskers range from a few mils to 20 mils in diameter. Continuous single-crystal lengths up to 100 ft have been produced. A maximum tensile strength of 625,000 psi has been observed for the smaller diameter samples. A brief description of the process and the mechanical properties of the whiskers is given. The process is not limited to the growth of sapphire nor to the growth of strictly whisker or filament shapes. Spinel filaments and sapphire multibore tubing, for example, have been also produced.

INTRODUCTION

The growth of continuous sapphire whiskers and filaments from the melt [1] was undertaken following the demonstration that small sapphire whiskers produced in a batch process can exhibit extremely interesting mechanical properties and that these properties can be exploited in composites. It was expected that a continuous preparation process would lead to ease of handling, improved composite fabrication, and economical cost of production of the filament.

As a candidate for use in a composite, sapphire has very attractive physical properties including a high modulus of elasticity, low density, and potentially high tensile strength at both room and elevated temperatures. Its chemical inertness suggests that it would be promising in high-temperature metallic composites, and its high dielectric strength makes it an excellent choice for epoxy composites in applications where electromagnetic radiation transparency is essential, for example, in radomes.

Single-crystal sapphire (α-Al_2O_3) has previously been available in three distinctly different physical forms: bulk crystals grown by melt [2, 3], vapor [4], and hydrothermal [5] techniques; platelets grown by flux/solvent [6] techniques; and whiskers grown by vapor techniques [7]. Except for vapor growth, these techniques, as previously practiced, have not yielded a continuous process.

A good deal of information has been generated on the mechanical properties of both large melt-grown materials [8] and micron-sized vapor-grown whiskers [9]. It is obvious that whiskers can, in fact, approach the theoretical strength of sapphire, while melt-grown material has generally been considerably weaker, typically 20 to 30 times lower in tensile strength. (A mean strength of 1.6×10^6 psi has been reported [9] for whiskers that have cross-sectional areas in the $0.1\,\mu^2$ to $1.0\,\mu^2$ range.) The most generally accepted reason for the high strength of whiskers is that they have fewer defects, based simply on the fact that the probability of a surface flaw is considerably less for the micron-sized whisker. Investigations have shown, however, that even submicron-sized whiskers have surface steps or flaws. In recent work by F. P. Mallinder and B. A. Proctor [10], centerless-ground sapphire rods were carefully fire-polished before testing, yielding maximum bending strengths close to 1 million psi. This, of course, is strong evidence that the apparent weakness of bulk sapphire is mainly due to its surface quality. As Mallinder and Proctor state, this evidence suggests that large sapphire crystals could be produced having strengths comparable to those of whiskers. In fact, measurements on sapphire filaments grown from the melt by the process to be described below tend to corroborate this hypothesis.

HISTORICAL SURVEY OF THE MELT GROWTH OF SHAPED CRYSTALS

For a number of years scientists have concerned themselves with the problem of producing objects from the melt in a directly usable form, that is, in contrast to melting the raw material, solidifying it in one form, and then forging and working it into another form. It is obvious, for instance, that any continuous production of sapphire whiskers should avoid mechanical working of the surface.

Experiments along these lines were reported as early as 1857. Thus, according to Stepnov [11], the English metallurgist G. Bessemer experimented in forming metal articles directly from the melt, and this work was later extended by Stepnov and co-workers.

In 1922 von Gomperz [12] reported the use of a floating mica disk with a hole at its center to control the shape of metal crystals grown from the melt. In recent years this type of approach was introduced into the semiconductor industry for the production of device quality ribbons of germanium and silicon continuously from the melt. The advantages of such a process are obvious if we consider the economics involved in slicing and polishing device material from a large boule compared with transversely cutting the ends of a ribbon to the desired length.

The interest in the growth of semiconductor ribbon can be appreciated from the large volume of scientific literature which has been published on the topic. There were at least two basic approaches used for ribbon growth. Initially, the majority of investigators used dendritic growth [13–18] with specially twinned seed crystals. In dendritic growth, the growth interface advances into a supercooled area of the melt. Since the latent heat of solidification can readily be transferred from the growth interface to the cooler surrounding liquid, growth usually takes place at a rapid rate (e.g., 7 in./min for sapphire filaments). In the special case of dendritic ribbon growth of germanium, a multi-twinned seed crystal [13] has been found necessary for continuous propagation.

Simultaneous work initiated on the ribbon growth of germanium by the Czochralski technique through melt guidance [19] seems to have been neglected later, compared with the investigations on dendritic growth. Briefly, the Czochralski technique employs a seed crystal which is in contact with a melt of the same material, the melt being contained in a nonreactive crucible. As the seed is withdrawn, molten material solidifies on to it, producing a crystal at a relatively slow rate (e.g., 6 in./hr for silicon). In this process, the solid-liquid interface is maintained close to the equilibrium crystallization temperature.

Recently, the observed nondendritic growth of both silicon [20] and germanium [21] has been published. In comparing this with the dendritic technique, it would appear that dendritic growth should offer the greater growth speeds (due to the supercooled dendritic-growth mechanism), while the nondendritic technique should offer the greatest versatility (a single crystal is grown instead of a specially twinned morphology).

EXPERIMENTAL APPROACH

In an attempt to grow sapphire whiskers from the melt, both dendritic and nondendritic techniques were considered. A prerequisite to continuous dendritic growth experiments was determination of the proper seed to employ (with respect to orientation) and twinning. Sapphire dendrites were

Figure 4.1 Sapphire dendrite grown from the melt (mag. 42X).

produced spontaneously to obtain this information. Dendrites were produced in a standard Czochralski puller using a molybdenum crucible with a carbon susceptor. A sapphire seed of arbitrary orientation or a tungsten rod was introduced into a crucible containing molten alumina, and the power to the furnace was shut off. Dendritic growth proceeded radially from the seed along the top surface of the melt; the growing mass was then rapidly removed before complete solidification of the melt could occur. Dendrites such as the one shown in Figure 4.1 were broken from the mass. X-ray examination of some dendrites produced in this manner showed the c-axis to be at least one of the major axes for dendritic growth. Therefore, we chose to attempt filament growth using a c-axis (0001) seed.

The apparatus used was essentially a modified crystal pulling furnace. A major modification was the use of a hydraulic-pneumatic piston that was controlled by a needle valve to produce uniform linear motion of the seed shaft over a speed range of several inches per hour to several inches per minute. A schematic of the crucible, susceptor, and so on, is shown in Figure 4.2. A feature of this part of the apparatus is the use of carbon cloth for radiation shielding: three turns of cloth held in place by carbon yarn increased the temperature by 200 to 500°C for a given power input. Another feature is the use of water cooling between the two concentrically positioned quartz tubes. Since a majority of the infrared is absorbed in the water, a microscope for visual observation can be positioned close to the *r-f* coil without excessive heating. Helium or argon was used as the atmosphere.

Early attempts to grow sapphire filament employed the setup shown in Figure 4.2, with various additional radiation shielding on top of the crucible. Since induction heating was used, the temperature distribution could be changed by variations in the radiation shield design and by deliberately supplying more heat to either the bottom or top of the crucible through changes in its position relative to the work coil.

A typical experiment involved (1) melting the sapphire, (2) inserting the *c*-axis seed into the melt, (3) supercooling the melt, and (4) withdrawing the seed. Under some conditions of thermal distribution, we observed that dendrites up to several millimeters in length propagated vertically downwards into the melt.

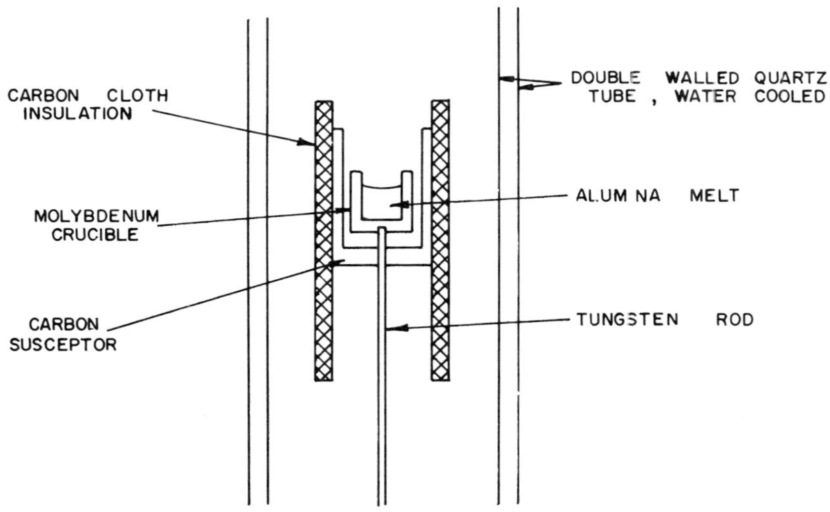

Figure 4.2 Apparatus used for melt growth.

The experimental evidence implied that controlled vertical growth required the use of radiation shielding such that annular rings having a small inside diameter (on the order of $\frac{1}{32}$ in.) should be placed on top of the crucible. This assumes that heat loss would be greatest in a small region at the center of the melt surface.

An alternate technique that suggested itself was to place a shield directly on the surface. If such a shield were rigidly held by mechanical means, however, a continuous adjustment would be required as a consequence of the liquid level changing as material was removed. A more practical solution was found by using a shield that would float and would therefore always remain on the melt surface regardless of its position relative to the crucible. This type of arrangement is shown in Figure 4.3. A molybdenum annulus with a $\frac{1}{16}$ in. diameter orifice at its center was placed directly on the melt surface. Since molybdenum is heavier than molten alumina, the floating orifice was fashioned into a boat, although it might be expected from the high surface tension of the molten material (690 dynes/cm) that a thin sheet of molybdenum placed on the surface would float.

Using a floating shield and orifice in the apparatus shown in Figure 4.2, we were able to grow filaments of sapphire simply by seeding in the orifice, supercooling the melt, and subsequently retracting the seed at rates up to several inches per minute. Accurate temperature measurements are difficult at these temperatures (the melting point of sapphire has been reported by several investigators, the most generally accepted being 2072°C). However, evidence that molten alumina could be supercooled up to at least 100°C at fast cooling rates was obtained by placing a standard tungsten-rhenium

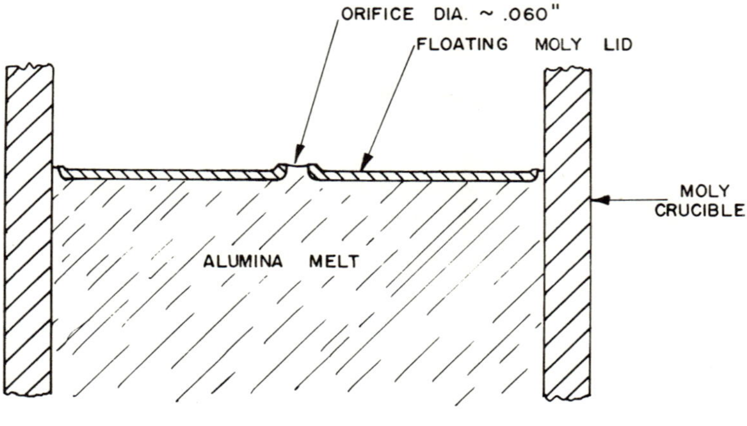

Figure 4.3 Floating orifice used for melt growth of whiskers.

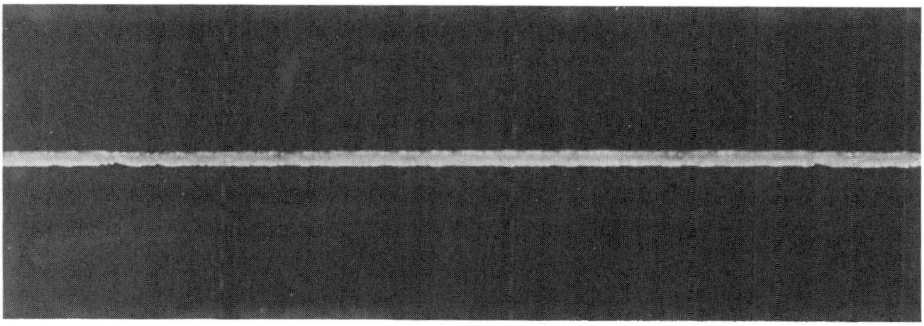

Figure 4.4 Sapphire filament grown in the *c*-direction (mag. 4×).

thermocouple in the melt and recording its output as the material was cooled and solidified. The latent heat of solidification released is substantial and can be seen as a bright flash as the material solidifies. This, of course, is further evidence that solidification can take place at a very rapid rate.

A *c*-axis filament grown at 6 in./min is shown in Figure 4.4. It should be noted that the filament was between .004 and .020 in. in diameter (tapered along its length) and was grown from a $\frac{1}{16}$ in. diameter orifice. This filament is typical of our early results, although in some cases much more complicated habits were formed. Figure 4.5 shows four typical types of habits formed.

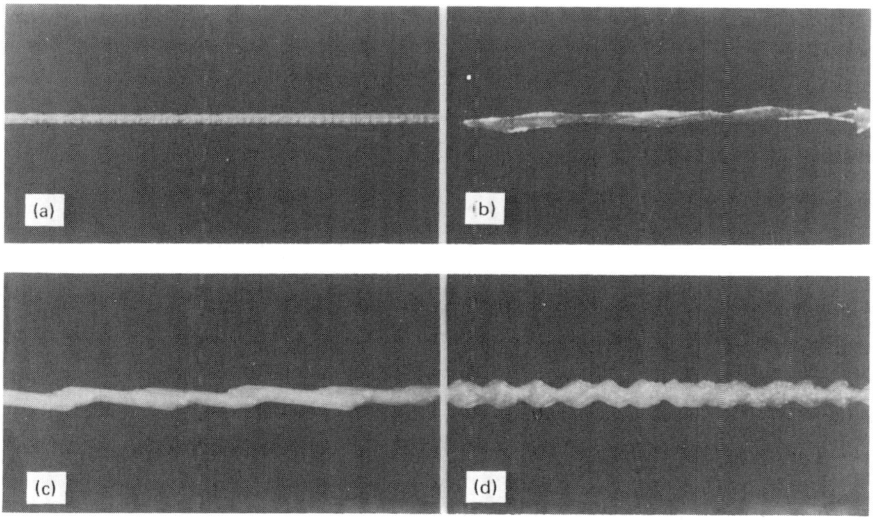

Figure 4.5 Sapphire filaments exhibiting different habits (mag. 5×).

Figure 4.6 Subgrains revealed by etching (mag. 190×).

All of these crystals had faceting on their surface, with type A being most nearly circular. A detailed analysis of these different filaments was not made; however, Laue-back reflection photographs showed type A to be grown in the *c*-direction, with subgrain boundaries being present.

Evidence of subgrain boundaries was also obtained by polishing the cross sections and etching (Fig. 4.6) in boiling phosphoric acid according to the procedure outlined by R. Scheuplan and P. Gibbs [23]. The presence of subgrains was found to be dependent on growth speed and temperature.

Examination of growth tips pulled free from the melt as the crystal was growing showed that, in the case of the nearly circular or polygonal cross sections (generally type A of Fig. 4.5), the filament was growing as one or more highly faceted fingers that appeared as rhombohedra stacked on top

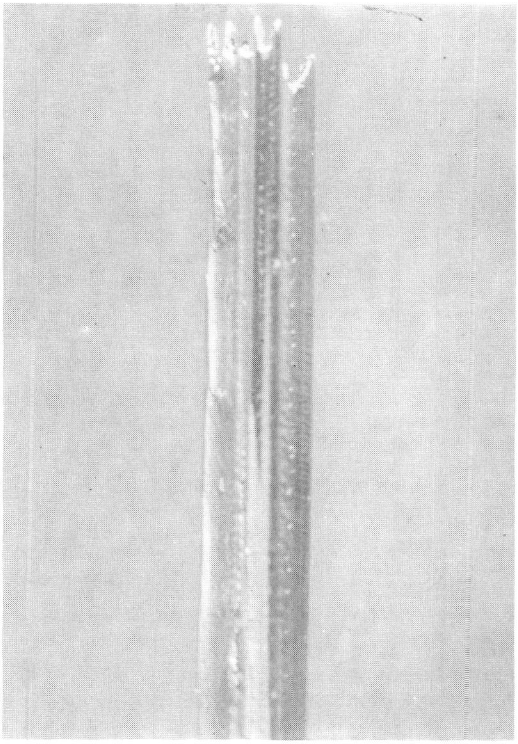

Figure 4.7 Several c-axis dendrites growing approximately parallel to form a rod (mag. 5×).

of one another. It appears that when subgrains are present, they represent a series of these individual fingers that are slightly misoriented with respect to each other, each one being a single crystal. An extreme case of these dendrites growing together is shown in Figure 4.7.

The tensile strength of these faceted filaments was approximately 125×10^3 psi as measured in tension. In these cases, the filaments were growing dendritically from an orifice $\frac{1}{32}$ to $\frac{1}{16}$ in. in diameter, at rates up to several inches per minute. It should be noted that although a floating orifice was used, it was employed not to guide the melt, but to achieve the proper thermal distribution. This is attested to by the fact that the filaments are not necessarily the shape or size of the orifice.

Recent advances in technique have enabled the production of single-crystal sapphire filaments as small as 1.5 mils in diameter, with nonfaceted surfaces (Fig. 4.8) and free from easily detectable subgrain boundaries. Preliminary evidence indicates that these are possibly grown by a cellular-growth mechanism.

Figure 4.8 Sapphire filament having smooth surfaces (mag. 25×).

MECHANICAL PROPERTIES

As mentioned earlier, the tensile strength of the highly faceted filaments appeared to be low. It should be noted, however, that cross-sectional area measurements were difficult in many cases. The smooth filaments recently produced yielded an average room-temperature tensile strength greater than 370,000 lb/in.^{2}, with maximum values in excess of 500,000 lb/in.^{2} in the 4 to 6 mil diameter range. A maximum value of 625,000 psi has been obtained on a 1.5-mil-diameter filament. All of these strengths were measured in tension. Modulus of elasticity measurements on long filaments yield a value of 67 million psi in tension.

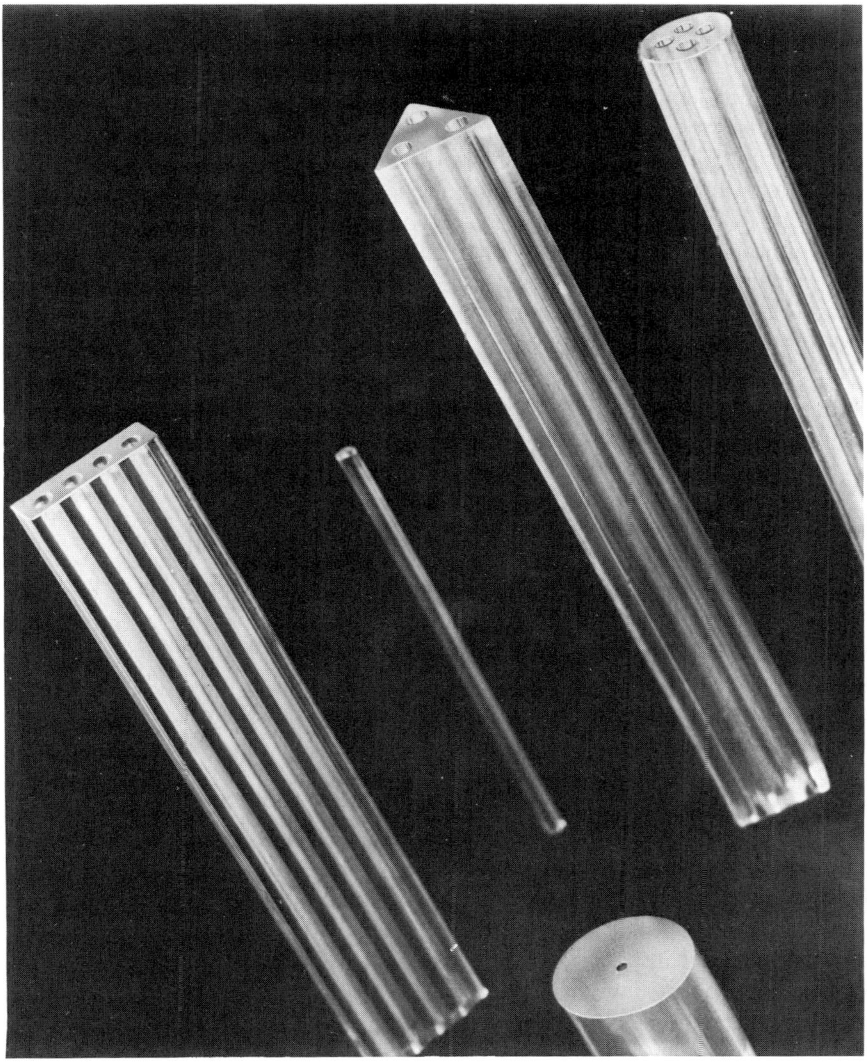

Figure 4.9 Sapphire shapes produced from the melt (mag. 3.25×).

In comparing the strength of these filaments with other forms of sapphire, the material is much stronger than the bulk sapphire previously tested by Watchman and Maxwell [9] where bending strengths of 102,000 psi were observed on centerless-ground, fire-polished *c*-axis sapphire.

The results of Mallinder and Proctor [10] showed bending strengths ranging from 280,000 psi to nearly 1 million psi for 1-mm diameter, centerless-ground, fire-polished sapphire. In the latter case extreme care was taken during the fire-polishing technique; the rods were carefully selected for perfection; the apparent gauge length was 3 mm. In contrast, the melt-grown sapphire filaments were tested in tension with 25 to 50 mm gauge lengths and without critical inspection or handling. As the bending strengths generally give greater values than the tensile strength, the filaments on the average should be considerably stronger than the 1 mm rods tested. The average strengths are weaker than vapor-grown whiskers; however, the larger whiskers that have diameters in the range of the filaments have strengths equal to or lower than the filaments [9]. This is more significant since the gauge length for the whiskers is considerably smaller.

SUMMARY

The process described can successfully produce single-crystal sapphire filament of 1 mil in diameter or greater. The strength of these filaments is essentially the same as that of other high-modulus filamentary materials commercially available, such as boron and silicon carbide. The feasibility of producing many filaments continuously from a single piece of equipment has been demonstrated. The successful use of this material now depends on its compatibility in composites and, ultimately, on the economics of the process.

It should be noted that the process has been applied to the growth of sapphire with essentially any cross section (or very broad limits). Sapphire filaments, rods, ribbons, and tubes of various shapes have been grown (Fig. 4.9). It is expected that some shapes other than filament can be useful in reinforcement applications. It is also of interest that the process is not limited to sapphire and has already been extended to other congruently melting materials, such as spinel.

ACKNOWLEDGMENTS

We would like to acknowledge the valuable assistance of our many colleagues who have contributed to this work, specifically Dr. G. Hurley and W. Sherman who are responsible for the mechanical measurements; B. diBenedetto, W. Clemments, and C. Cronan who grew many of the sapphire shapes as per Figure 4.9; S. Mermelstein, L. Bellomo, and H. Binney who designed and constructed equipment for continuous filament growth; Dr. G. A. Wolff and Dr. B. Das who assisted in crystallographic and X-ray analysis; and J. Bailey and W. Little who assisted in various phases of crystal growth.

This work was largely supported by the Air Force Materials Laboratory, Wright-Patterson Air Force Base, Ohio.

REFERENCES

[1] H. E. LaBelle, Jr., and A. I. Mlavsky, *Nature*, **216,** 5115 (1967), 574.
[2] A. Verneuil, *Ann. Chim. Phys.*, **3** (1904), 20.
[3] A. E. Paladino and B. O. Roiter, *J. Am. Ceram. Soc.*, **47** (1967), 456.
[4] P. S. Schaffer, *J. Am. Ceram. Soc.*, **48** (1965), 508.
[5] R. A. Laudise and A. A. Ballman, *J. Am. Chem. Soc.*, **80** (1958), 2655.
[6] E. A. D. White, *Nature*, **191,** 4791 (1961), 901.
[7] W. W. Webb and W. D. Forgeng, *J. Appl. Phys.*, **28** (1957), 1449.
[8] J. B. Watchman and L. H. Maxwell, *J. Am. Ceram. Soc.*, **42,** 432 (1959).
[9] R. L. Mehan and E. Feingold, Chapter 15 in *Strengthening Mechanisms, Metal and Ceramics*, J. J. Burke, N. L. Reed, and V. Weiss, Eds., Syracuse University Press, 1966.
[10] F. P. Mallinder and B. A. Proctor, *Phil. Mag.*, **13** (1966), 121.
[11] A. V. Stepanov, *Soviet Phys. Tech. Phys.*, **4** (1959), 339.
[12] E. von Gomperz, *Z. Physik.*, **8,** 184 (1922).
[13] G. F. Bolling and W. A. Tiller, in *Metallurgy of Elemental and Compound Semiconductors*, Interscience Publishers, New York and London, 1960, 97.
[14] H. F. John and J. W. Faust, Jr., *ibid.*, p. 127.
[15] S. O'Hara, *ibid.*, p. 148.
[16] P. J. Holmes, *ibid.*, p. 161.
[17] R. S. Wagner, *ibid.*, p. 177.
[18] D. C. Jillson and R. E. Hysell, *ibid.*, p. 187.
[19] G. K. Gaule and J. R. Pastore, *ibid.*, p. 201.
[20] J. C. Boatman and P. C. Goundry, *J. Elec. Tech.*, **5** (1967), 98.
[21] D. E. Swets, *J. Elec. Tech.*, **5** (1967), 385.
[22] S. J. Schneider, U.S. Nat. Bur. Std., Monograph 68 (1963).
[23] R. Scheuplan and P. Gibbs, *J. Am. Ceram. Soc.*, **9** (1960), 458.

Robert H. Kelsey

5 Mechanical Testing of Whiskers

INTRODUCTION

Before we can sensibly consider the uses of a particular kind of whisker in a composite structure, we must first evaluate the whisker in its role as a material, and in its role as a structural member. The material properties—chemical, thermal, and electrical—are usually derived from the corresponding properties of the bulk substance. The elastic moduli are also generally considered to be characteristic of a given material. There is not yet a body of knowledge about the general properties of whiskers which will justify conclusions as to which are "material" properties and which are properties of the whisker form only. Soltis [1] found unusually high values for Young's modulus in alumina whiskers, and also found some evidence of plastic flow at high loadings. To date, these values have not been verified by other investigators. In the absence of contrary evidence, such properties as thermal expansion, density, chemical reactivity, dielectric constant, electrical resistivity, and the like are assumed to be the same in whiskers as in large single crystals of the same material.

The mechanical properties that will determine the ultimate properties of the composite, however, are the result of the geometric form and size of the whisker, and must be evaluated by measurements made on whiskers. The ultimate tensile strength, yield point, or proportional limit and the ultimate extension are peculiar to the whisker form.

These tests are made at the present time on individual whiskers, often in some form of tensile-testing device that will yield a stress-strain curve for the specimen. It is hoped that means will eventually be developed for estimating these properties by incorporating the whiskers in matrix materials of known or standard properties. Such a technique would yield average or effective values directly which could be used as engineering data or in the evaluation of new combinations of materials. In spite of many excellent studies, both theoretical and applied, relating to the effect of fibers in continuous matrices, it is still not possible to prepare specimens in which the properties of unselected whiskers may be deduced from the composite properties.

Thus, we are left with the problem of measuring the tensile properties of large numbers of individual whiskers, and from these data estimating their average values.

TENSILE TESTS

The ultimate tensile strength of whiskers and the breaking extension have been estimated by three methods: direct application of a longitudinal stress, bending, and stress applied by the use of an electrostatic field. Most often, perhaps, some form of tensile machine that applies longitudinal stress to the whisker has been used. Some of these devices are described later. Gordon [2] and Campbell [3], among others [4, 5], have employed simple bending tests to evaluate the strength and extension at rupture.

A whisker having thickness t and Young's modulus E is bent into a loop of radius r. The strain ε at the surface is given by

$$\varepsilon = \frac{t}{2r} \tag{1}$$

and the stress σ at the surface by

$$\sigma = \varepsilon E = \frac{Et}{2r} \tag{2}$$

The radius r is reduced until fracture occurs; σ then represents the ultimate tensile strength and ε the ultimate extension. The test may be made under a microscope by embedding the whisker in a film of grease on a slide and manipulating it with a dissecting needle [2]. Photographs may be made at intervals as the whisker is progressively bent. Campbell [3] and Gyulai [4] attached one end of the whisker to a rod and expressed the radius in terms of the deflection of the free end. Gyulai's technique, which he used on NaCl whiskers, is illustrated in Figure 5.1. The thickness t (see Fig. 5.1) is estimated by the eyepiece micrometer or by a reticle before bending is begun. If the thickness t is much less than 2 μ, the error in its measurement by visible light becomes excessively large and propagates a correspondingly large error into

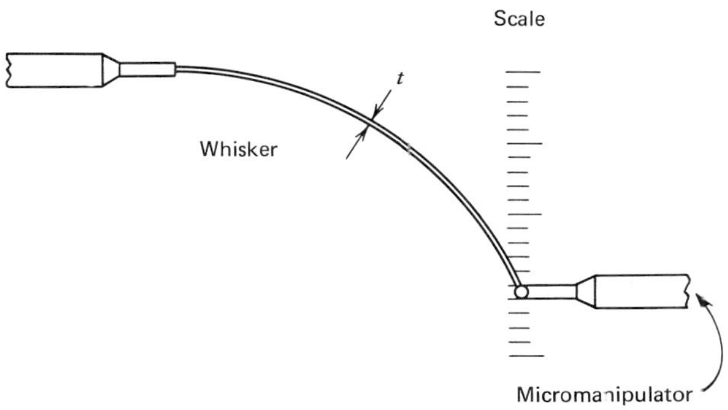

Figure 5.1 Bending test for whiskers (after Gyulai [4]).

the strength and extension data. The limitations inherent in measurement of small dimensions by light microscopes or other light optical devices are discussed in the section, "Gauge Length and Cross-Sectional Area."

The method fails if plastic flow occurs before fracture. Its use is therefore limited to those ceramic whiskers in which plastic flow at ordinary temperatures is negligible. This type of bending test does not conveniently measure the stress-strain relation for whiskers because of the difficulty of evaluating the force applied to form the loop. Young's modulus must be known or measured independently. Many kinds of whiskers are anisotropic with respect to the elastic modulus, and few have convenient rectangular cross sections. The elastic modulus to be applied in Eq. 2 must be appropriate to the plane in which bending occurs. Equation 2 is valid for cross-sectional areas of any shape if the thickness t is the maximum thickness within the plane of bending. The bending test is, however, simple to perform and requires no equipment beyond a microscope with measuring eyepiece and a camera.

Although it is not generally applicable as a testing technique for whiskers that have been harvested, Gomer's method [6] is perhaps unique in that it measures the strength of whiskers without removing them from the sites in which they are grown. Working with mercury whiskers grown in a field-emission tube, he was able to estimate the strength of his whiskers from the electrostatic field required to shatter them.

Mercury whiskers were grown on the cathode of the field-emission tube, which was the polished end of a tungsten rod 1 mm in radius. It was maintained at a temperature of $-78°C$. An attached reservoir containing the mercury supply could be cooled to control the vapor pressure of mercury in the field-emission tube. Whiskers grew on the tungsten substrate at mercury vapor pressures of 10^{-6} to 10^{-5} torr.

An electric field up to 3×10^7 volts/cm was generated at the whisker tips by applying a potential difference between the fluorescent anode and the tungsten substrate. This field, which was used to produce an image of the whisker ends, also applied an axial stress to the whiskers. The stress is related to the field F by

$$\sigma = \left(\frac{F}{300}\right)^2 \quad \text{dynes/cm}^2 \tag{3}$$

Whiskers were broken rather than separated from the substrate. Stresses of the order of 10^{10} dynes/cm^2 (1.45×10^5 psi) were observed.

TENSILE MACHINES

The devices most often used are designed to stress the whisker in axial tension and to measure the elongation as a function of the load, or conversely, the load as a function of the elongation. This distinction is real and calls attention to the two general types of tensile-testing machine: the so-called "hard" and "soft" machines. For a "hard" machine, one end of the specimen is attached to an essentially rigid load-measuring device, the other end attached to a member that moves away at a constant rate. If the movement of the load-measuring device is small compared with the change in length of the specimen, the extension rate is approximately constant.

Most of the large commercially available tensile machines used for general laboratory testing are hard machines. They are advantageous in the testing of ductile metals or high-elastic materials such as rubber and plastics, because they record the specimen loading even though it may diminish at some point in the test. For brittle, high-modulus materials, such as ceramic whiskers, hard machines are less effective.

Because the rate of extension is considered the independent variable in hard tensile machines, the rate at which the load is applied to brittle materials may become excessively high. Testing short whiskers requires extremely slow, precisely controlled drive rates to avoid shock loading. For example, consider a whisker 1-mm-long with an ultimate extension of 2.5%. The total elongation will be 25 μ, or about 0.001 in. Any inaccuracy in the drive mechanism of the testing machine which will cause a movement of 0.001 in./sec will break the whisker at once. Although the mechanical or hydraulic drive trains in hard machines may be built to yield average rates that are low (0.002 in./min or less), it is extremely difficult to guarantee uniformity of motion when the total movement is only a few thousandths of an inch at most.

"Soft" tensile machines are those in which the load on the specimen is the independent variable, with the elongation at any load assuming a value appropriate to that load. For ductile materials that may severely reduce cross-sectional area at some stress, soft machines are at a disadvantage: they do not reveal diminished loading, but instead induce immediate failure in the specimen. However, for short, brittle materials, soft machines (including most of the devices that have been used in the testing of whiskers) permit loads to be applied as slowly as required. In soft tensile machines, loads are typically applied by springs, weights, electromechanical devices, or other means by which the force may be evaluated as it is applied. One end of the specimen is usually held fixed in position, while a known or measurable force is applied to the other end. The extension of the specimen is then dependent upon the force.

To return to the example given for hard machines, suppose the whisker to have a cross-sectional area of 10 μ^2 (10^{-7} cm^2) and an ultimate tensile strength of 10^6 psi, or 7.03×10^7 g/cm^2. To achieve a strain rate equivalent to that given for the hard machine, we would need to increase the force at a rate of 7.03 g/sec. It is not difficult to control the application of forces three orders of magnitude smaller than this. Ordinary laboratory balances routinely deal with forces four orders smaller.

FASTENING AND MEASURING PROBLEMS

Regardless of the form of testing device used in evaluating whisker mechanical properties, it is necessary for the experimenter to select the specimen to be tested from a mass of whiskers, and then to attach it to some holding device through which a stress can be applied. Provision must be made at this time for estimating a gauge length, and usually for estimating the cross-sectional dimensions.

The small size of the specimens which must be handled leads to formidable experimental problems. Although no completely general statement can be made with respect to whisker dimensions, they are typically 0.5 to 10 mm in length and may have cross-sectional areas of the order of 0.01 to 100 μ^2. All manipulation must be done under the microscope, using specially ground forceps or micromanipulators. In this connection it must be pointed out that some damage in handling is inevitable. This may lead to premature breakage in whiskers that are initially weak, and may thus introduce into the data a bias favoring higher strength whiskers. Usually no attempt is made to estimate the effect of this bias; indeed, some workers have deliberately selected whiskers by bending, rejecting at the outset those that were least likely to survive the mounting procedure.

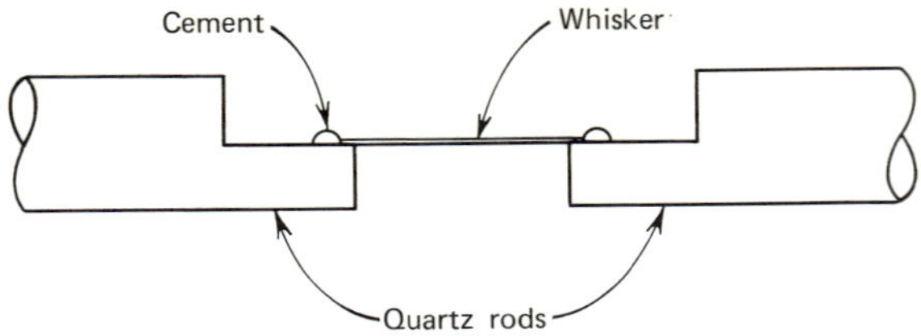

Figure 5.2 Method of cementing whisker (after Marsh [7]).

The problem of attaching a whisker to some type of holding or gripping device in order to apply a stress is a difficult one, for which many solutions have been tried, with varying success. Conventional mechanical grips are totally useless with small whiskers; highly specialized grips or adhesive must always be used. The most popular of the adhesives have been epoxy cements [1], diphenyl carbazide [2, 7], collodion cements, and various combinations of wax and shellac. All introduce some error into the measurement of grip movement because they yield under stress more readily than the whisker does. The writer has had good success with gelatin or flake glue. In dilute solution it flows readily to form a thin bond, providing a good continuous film around the whisker.

Attachment devices take various forms. Marsh [7] mounted the whisker between rods of quartz, as shown in Figure 5.2. Soltis [1] used a similar scheme with metal rods, each having a small groove in the direction of the whisker axis to aid in aligning the whisker. Campbell [3] used hypodermic tubing to help alignment. Shlichta [8] melted small spheres of diphenyl carbazide on the whisker ends, which were then inserted in double metal hooks as shown in Figure 5.3. This method allows some self-alignment of the whisker. Emrick and Gegel [9] and Herzog [10] cemented whiskers to wire loops as shown in Figure 5.4. These also assist in aligning the specimen, and provide a large area of contact between whisker and grips. Grips have

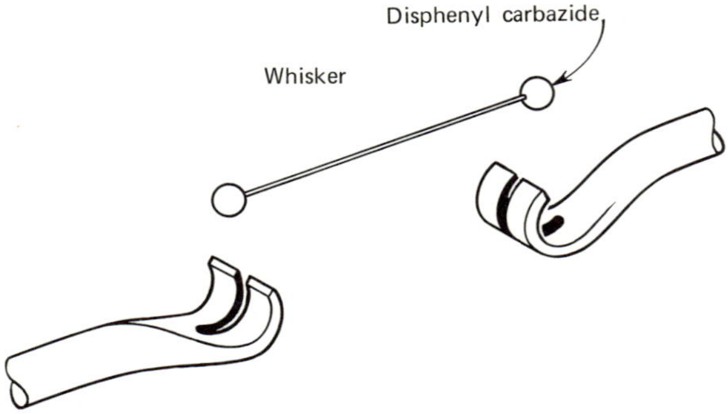

Figure 5.3 Method of gripping whisker (after Shlichta [8]).

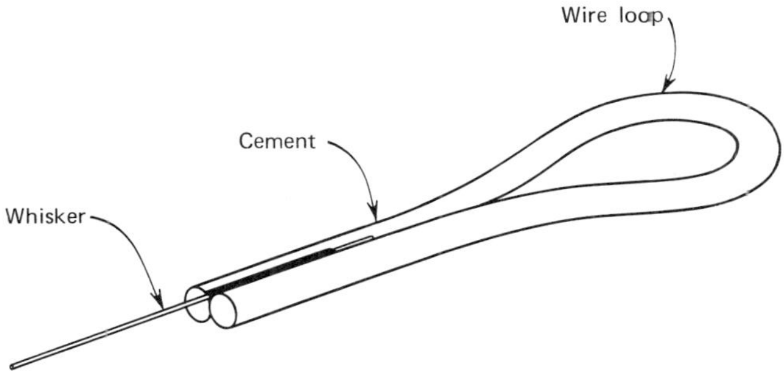

Figure 5.4 Wire loop holder for whisker (after Emrick and Gegel [9] and herzog [10]).

been made either to provide some degree of self-alignment [8], or have been attached by some device to permit realignment as it is needed [3, 7]. Alignment of the tensile grips and of the whisker in the grips must be made with care to avoid the introduction of bending moments in the whisker. Inadvertent bending during the test can introduce stresses of considerable magnitude [3] at the whisker surface, and thus can yield fictitiously low values for the measured tensile strength.

GAUGE LENGTH AND CROSS-SECTIONAL AREA

Perhaps the most critical problem in evaluating the fracture stress of individual whiskers is that of estimating the cross-sectional area. Obviously, errors in the measurement of the cross-sectional dimensions will be reflected as corresponding errors in the derived values for the elastic moduli and the tensile strength.

In measuring the cross-sectional area prior to testing, it is usual to ignore surface features and to assume that the whisker is smooth (in the absence of gross defects). It is also customary to select whiskers that have no obvious taper. These judgments of uniformity and surface perfection are probably poorly founded in most cases.

The usual practice in evaluating cross-sectional areas is to examine the whisker after mounting, using a light microscope of the highest available resolution. For whiskers of large size, this procedure is useful; for small whiskers, the data may be essentially meaningless. The bulk of the tensile data on whiskers has been based upon measurements with the light microscope and filar eyepiece. It seems important to remind the reader of the limitations of the instrument. Generally, the limit of resolution of a microscope [11] is given by

$$d = \frac{0.61\lambda}{N} \tag{4}$$

where d is the minimum distance between two resolved points on the object, λ is the wavelength of the light used, and N is the numerical aperture of the microscope objective. With an objective having $N = 0.95$ and light of $\lambda = 5500$ Å ($0.55\ \mu$), d is about $0.35\ \mu$. The normal eye is most sensitive to yellow-green light, with a wavelength about 5500 Å [9].

In practical terms, there will be an uncertainty of $0.35\ \mu$ in each separate measurement required to estimate the whisker area. Kelsey and Krock [12] found the standard deviation of repeated measurements of the dimensions of small whiskers to be 0.20 to $0.28\ \mu$.

Calow [13] and his co-workers used an interference microscope to measure the thickness of silicon nitride and alumina whiskers, and an image-shearing eyepiece to measure their width. They reported errors of $\pm 10\%$ on small (about $0.55\ \mu^2$ cross-sectional area) whiskers, and $\pm 0.5\%$ on larger whiskers (about $50\ \mu^2$ cross-sectional area).

In estimating the ultimate tensile strength of alumina whiskers, Mehan, Feingold, and Gatti [14, 15] employed a direct measurement of the cross-sectional areas and an indirect method involving the apparent elastic modulus and the strain at failure. Reasonable agreement was found between the two methods.

In a tensile test, if the grip displacement is measured as a function of the force applied, and the force is increased until the whisker breaks, one can, with the aid of an independently measured value of Young's modulus, deduce the cross-sectional area and the tensile strength of the specimen. Thus:

$$A = \frac{Fl_o}{E(l - l_o)} = \frac{F}{E\varepsilon} \tag{5}$$

where A is the cross-sectional area, F is the force, l is the attained length, and l_o is the initial gauge length of the whisker; the elongation is called ε. It is usually best to plot ε against F, averaging the data to a straight line of slope AE.

The simple analysis above ignores plastic or elastic deformation of the cement, and assumes that the whisker has a cross-sectional area that is constant throughout the gauge length. Mehan, Feingold, and Gatti [15] have analyzed the effect of deformation of the cement, using a two-spring model. If E is the "true" elastic modulus of the whisker and E^* is the modulus evaluated from the force-extension data, it is shown that

$$\frac{1}{E^*} = \frac{AK}{l_o} + \frac{1}{E} \tag{6}$$

where K is a constant depending on the diameter of the whisker, type of cement used, and perhaps other factors. Once K has been experimentally determined, the tensile strength and area may be deduced.

If the whisker to be tested is not uniform in cross-sectional area, a correction is needed to yield an "effective" area. An appropriate correction for uniformly tapered whiskers has been given by Calow, Boyer, and McCarthy [13]. Using a two-dimensional model, they represent the "effective" area A_{eff} by

$$A_{\text{eff}} = \frac{\alpha L}{\log (1 - \alpha L / A_o)} \tag{7}$$

where α is a constant defining the taper, L is the specimen length, and A_o is the minimum cross-sectional area. Note that the area at any point is given by

$$A = A_o + \alpha L \tag{8}$$

The extension problem may be eliminated by the use of fiducial marks on the whisker. Herzog [10] used whisker fragments cemented on the whisker surface to represent the gauge length. This is the method of choice, although the experimental difficulties inherent in the technique probably rule it out for mass or quality-control testing.

TESTING DEVICES AND TECHNIQUES

Most of the devices which have been built for testing whiskers are the result of a particular need that could not immediately be met by commercially available equipment. The instruments have generally been "soft" devices, as discussed earlier. They have been designed to meet the design criteria adopted by particular workers in the field. Various inventors have chosen purely mechanical systems to be free of electronic instabilities, or electronic systems that yield high sensitivity without mechanical complexity. Some have chosen springs as loading devices, others balances or dead weights. Most of the devices have some merit which makes them worthy of examination.

It has seemed to the writer that such devices ought to satisfy a few basic requirements, whatever form they take in the operating state. First, the art of handling whiskers is a difficult one at best; the devices associated with it should offer the maximum possible convenience in mounting and examining the specimen before testing. Second, the system should permit a reasonably precise alignment of the specimen in its holding devices, or should incorporate some form of self-alignment. Third, the measuring system should be as insensitive as possible to random vibrations. In practice, this means that the system must have a low natural frequency; below one cycle per second if

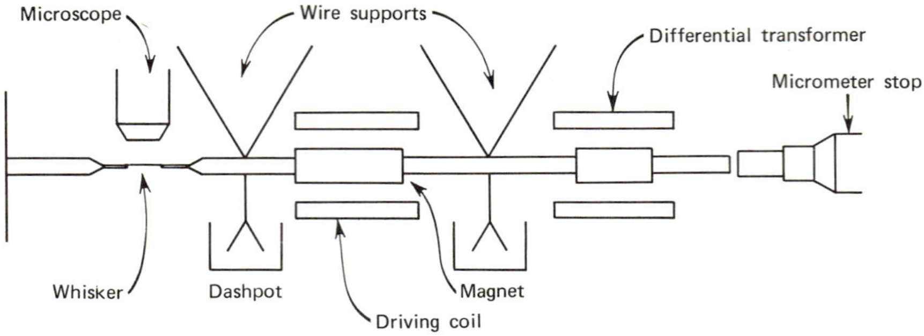

Figure 5.5 Electromechanical whisker tensile device (after Brenner [16]).

possible. Finally, it should be free from arbitrary calibrations. These are essentially for convenience and reliability in operations and are additional to the usual requirements of precision and stability.

Although it could not strictly be called a "machine," the technique of Gyulai [4] which he applied to whiskers of sodium chloride represents the class of bending tests. Its simplicity commends it to our attention.

Brenner [16] used a much more sophisticated device, shown schematically in Figure 5.5. The device was essentially friction-free, its moving element suspended from fine wires. The force was applied to the whisker by a solenoid operating upon a magnetic armature. The current through the solenoid measured the force. The extension was determined by a differential transformer attached to the moving rod. An important feature of the device was the micrometer aligned on the rod axis, which permitted the rod to be stopped when plastic flow began. Measurement could then be resumed at reduced loads. This feature permitted the examination of the plastic-flow region in metal whiskers. Dashpots prevented sudden movement of the moving element.

A somewhat similar device was used by Cabrera and Price [17]. This instrument was designed to be used under the microscope, permitting the whisker to be observed at a magnification of 1000 × during the test. It is shown in Figure 5.6. A pair of leaf springs supported the moving element. The force on the specimen was measured by the current in the solenoid. The displacement of the moving rod was measured by an electro-optical scheme. The shadow of a fine wire attached to the rod was magnified 50 ×, and centered on a pair of cadmium selenide photocells. These were connected in a bridge configuration in which the unbalance was proportional to the wire displacement. The unit is simple and advantageous in that it permits direct observation of the whisker being stressed.

An ingenious device was used by Lemke and Kraft [18] which offered a solution to the alignment problem. Figure 5.7 shows its elements. One end

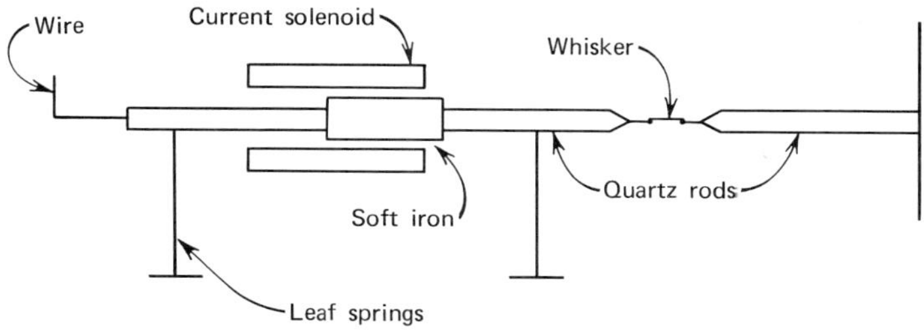

Figure 5.6 Whisker tensile machine (after Cabrera and Price [17]).

of the whisker was attached to a 1-mil tungsten wire with diphenyl carbazide. The other end of the wire was looped for attachment to a light jeweler's chain. The free end of the whisker was cemented to a 1-mil tungsten wire held by a two-circle goniometer. The whisker could then be maneuvered into alignment before the microscope objective. The chain was carried on a drum with a spiral groove cut into it, so that many turns could be stored.

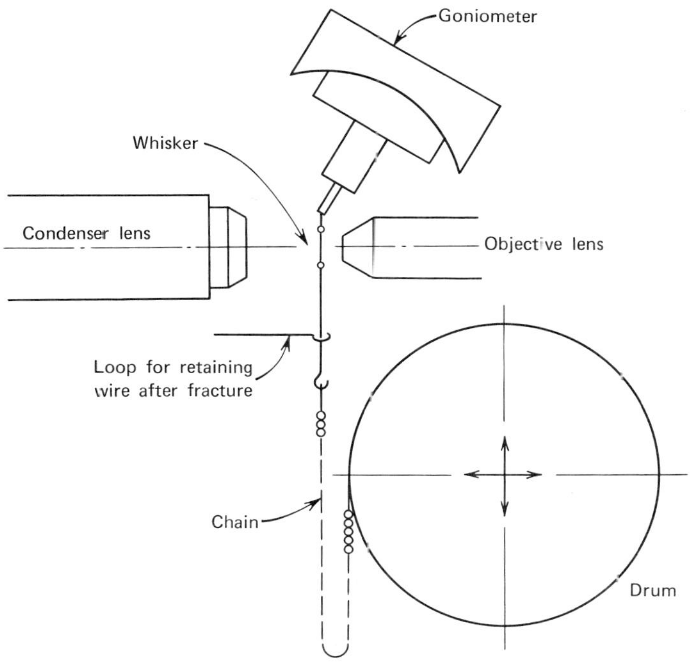

Figure 5.7 Whisker tensile machine (after Lemke and Kraft [18]).

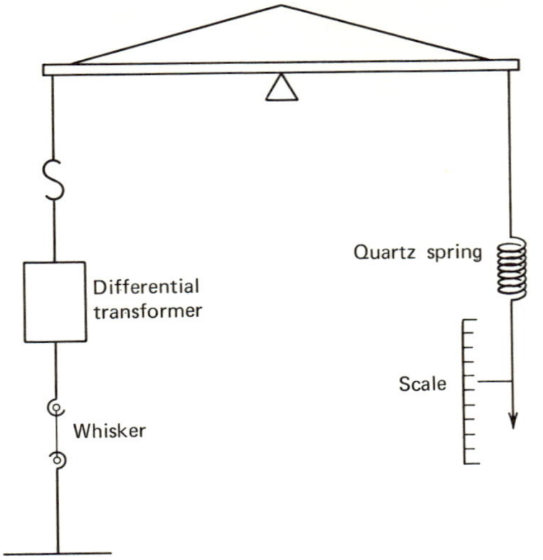

Figure 5.8 Whisker tensile device from analytical balance (after Shlichta [8]).

The drum was arranged to advance axially as the chain was unwound, a feature which released the chain at the same point in space as the load was applied. The drum mechanism was mounted on an *XYZ* micromanipulator, to allow the chain to be placed at the same level and directly adjacent to the bottom loop. The drum was calibrated for several chain sizes. The whisker extension was measured by taking photomicrographs, at intervals as the load was applied, of diphenyl carbazide beads on the whisker.

Another solution to the problem of alignment was used by Shlichta [8], mentioned earlier, with the apparatus shown in Figure 5.8. A quartz spring was used to apply the load to the specimen through a modified analytical balance. A differential transformer measured the elongation of the whisker.

Kelsey and Krock [12, 19] used a variant of the analytical balance, together with a capacitive extension-sensing device, shown in Figures 5.9 and 5.10. This system permits mounting and observation of the whisker in a separate unit (Fig. 5.9) which is then inserted in the balance for measurement. The grips, which are modified hypodermic needles, are held in rigid alignment by a chuck during the manipulation of the whisker. The chuck is released after the auxiliary unit is inserted in the balance, allowing friction-free application of the load. Forces up to 10 g are applied with the calibrated chain; if more than 10 g are required, weights may be added to the balance pan.

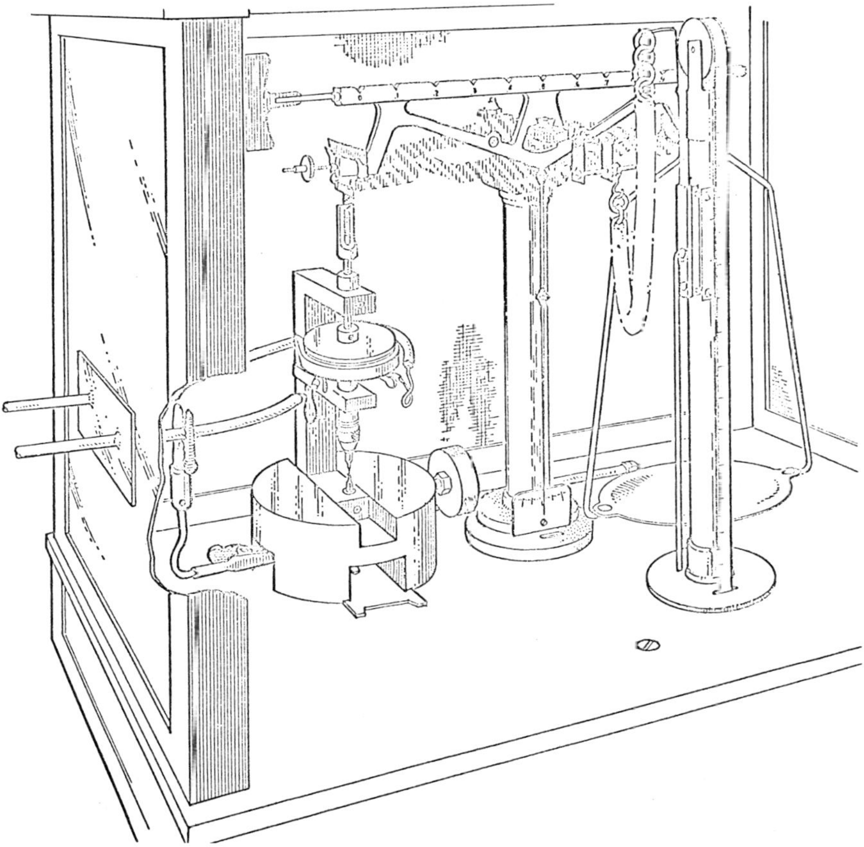

Figure 5.9 Whisker tensile device with capacitance measurement of elongation (after Kelsey and Krock [12, 19]).

Soltis [1] and Herzog [20] have used micrometer-loaded calibrated springs to apply tensile forces directly to the whisker, along with optical determination of the extension. Herzog used a device that incorporated a double helical spring, wound out-of-phase to avoid twisting, and an automatic follower to determine the separation of fiducial marks on the specimen. This permitted direct plotting of the load-elongation curve on an x-y plotter.

Campbell [3] devised a motor-driven machine for the Army Materials and Mechanics Research Center. This machine, shown schematically as Figure 5.11, employed a crosshead that could be driven at speeds from 0.05 to 1.0 in./min. As the load was applied to the whisker, the stainless steel beam

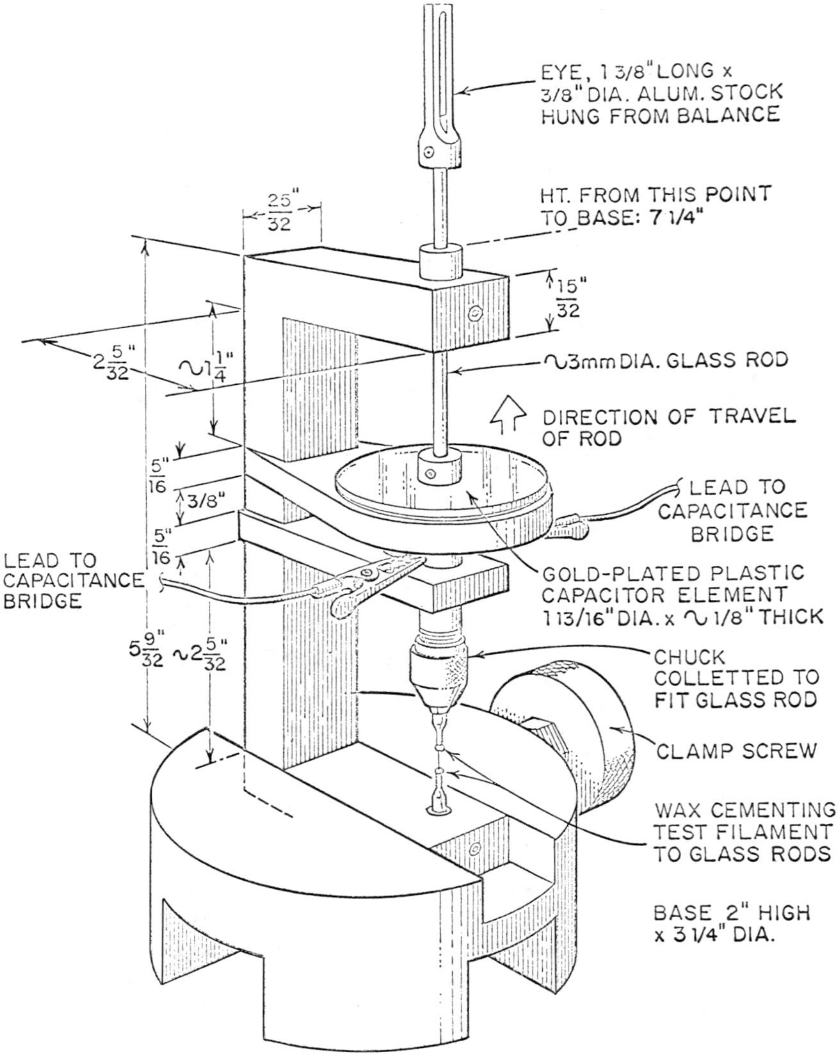

Figure 5.10 Detail of capacitance elongation unit (after Kelsey and Krock [12, 19]).

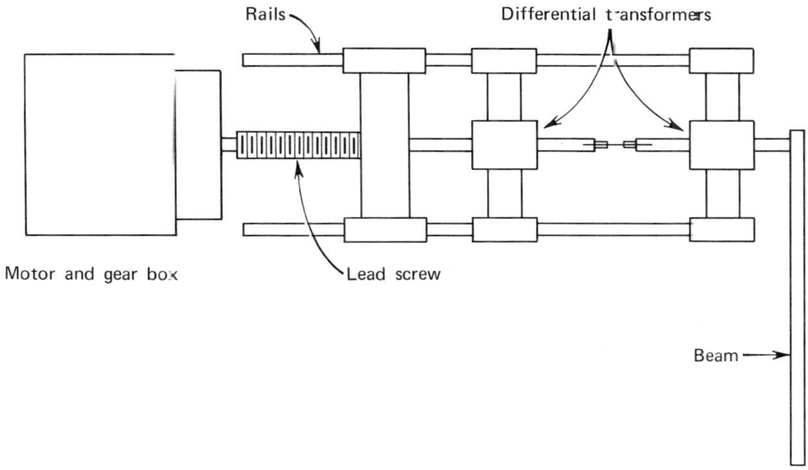

Figure 5.11 AMMRC whisker tensile machine (after Campbell [3]).

was bent. The deflection of the beam was measured by a differential transformer. A second differential transformer was attached to the crosshead. The load signal was plotted automatically against the difference signal between the two differential transformers. Thus an xy plot was automatically generated, representing the load-elongation relation for the specimen.

Horizons Inc. [21] uses a somewhat similar device developed by the writer. A differential screw arrangement driving the crosshead provides extremely low rates of travel (5 to 50 μ/min). A massive measuring head mounted on flat cantilever springs gives the necessary sensitivity, while making the measuring system essentially immune to vibrations in the sonic range.

A device developed by Marsh [7], and offered for commercial sale under the name "Tecam", is completely mechanical in its operation to avoid the instabilities that sometimes occur in electrical or electronic sensing devices. This instrument, shown schematically in Figure 5.12, uses a torsion balance to apply stress to the whisker. Once the whisker is mounted in the grips, a force is applied to it by adjustment of the lever system with a micrometer head. The torsion spring is restored to its original position by rotating the torsion head. An optical lever system permits very precise restoration. The force is read from the torsion head indicator, and the elongation is given by the change in reading of the micrometer.

Most mechanical testing of whiskers has been done at room temperature. Perhaps the most detailed study of the elevated-temperature mechanical properties was described by Brenner [27]. The whisker (sapphire) was surrounded in the testing device by a coil of tungsten or platinum-rhodium wire.

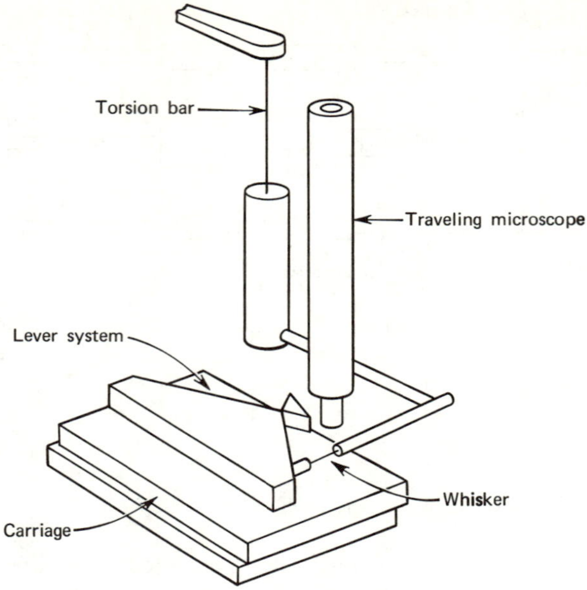

Figure 5.12 "Tecam" whisker tensile machine (after Marsh [7]).

A glass jacket surrounding the specimen and the heating coil contained a protective atmosphere. The maximum fracture stress and the time to fracture at constant stresses were measured as functions of temperature.

Mehan and Feingold [28] used a bending test to estimate the degradation in the strength of coated (titanium/nickel, 1 μ thick) Al_2O_3 whiskers resulting from elevated-temperature testing temperatures (650 to 1800°C). The whiskers were tested in a conventional laboratory tensile-testing machine (Instron), using a three-point bending technique. The furnace, resistance wire wrapped on a small ceramic tube, surrounded the whisker and its supporting structure.

Only large whiskers (above 10 μ diameter) could be tested in this way. Further, the authors felt that anisotropy and varying cross-sectional areas would make absolute measurements of the tensile strength by this method unreliable. The technique used was to test a long whisker at room temperature, then to test the broken remnants at high temperature. Comparison could then be made between the elevated- and room-temperature data for the same whisker. The data presented are normalized to 100% at room temperature.

Many, perhaps most, workers have developed individual machines and techniques to be used with them. The catalog given here is intended to be representative rather than exhaustive. In addition to these machines, which are intended to evaluate the tensile strength and ultimate elongation, or to provide data for a stress-strain curve, there are also a number of rather clever devices for measuring independently the elastic moduli of whiskers.

MEASUREMENT OF ELASTIC MODULI

For measuring the shear stress in iron and silicon whiskers, Eisner [22] used the device shown in Figure 5.13. The torque was applied to the whisker by a magnet in the field of a pair of coils in which electric current flowed. The magnet was attached to a float in a tube of water, to avoid applying any longitudinal stress to the whisker. A second pair of coils, at right angles to the measuring coils, was used to compensate for the effect of the earth's magnetic field. Because the torque diminishes as the fourth power of the

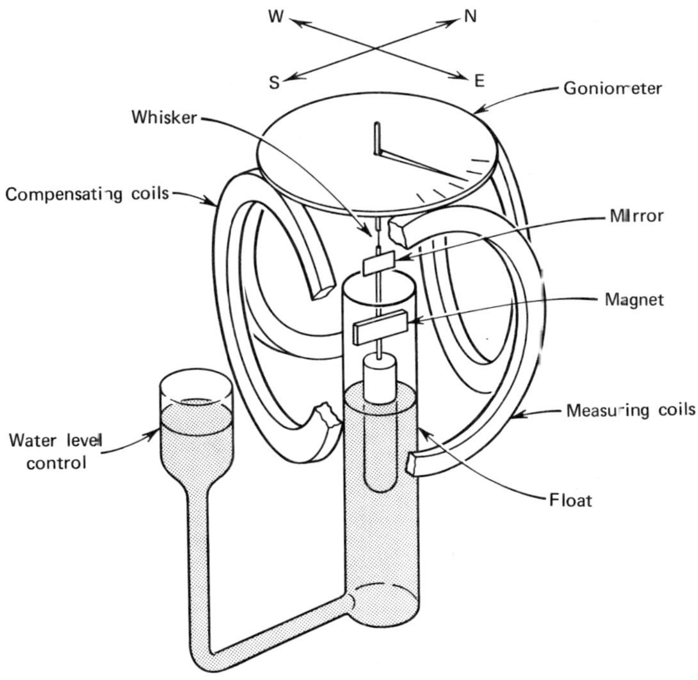

Figure 5.13 Torsion device for measuring shear stress in whiskers (after Eisner [22]).

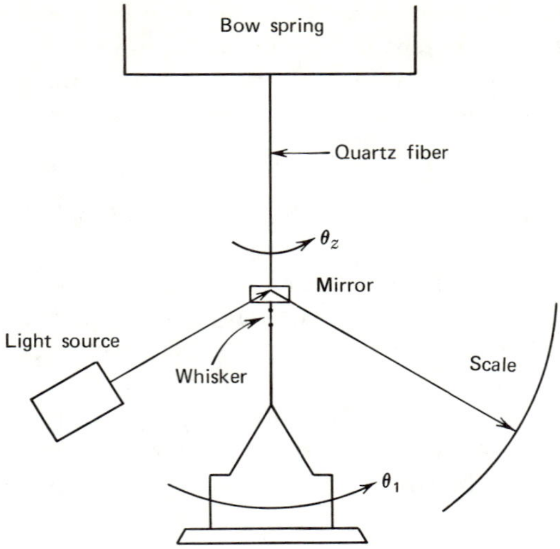

Figure 5.14 Method of measuring whisker shear modulus (after Brenner [23]).

diameter of the whisker, the tests are difficult for whiskers much below 3 μ in diameter.

Brenner [23] described the device shown in Figure 5.14, which he used to determine the shear moduli of copper and iron whiskers. The whisker was twisted through an angle indicated by the goniometer. This twist is transmitted to a quartz fiber. The rotation of the quartz fiber is indicated by a mirror-and-scale arrangement. Equations 9 and 10 are appropriate to the arrangement.

$$M_t = \frac{\theta \pi G_q d_q^4}{32 L_q} \tag{9}$$

where M_t = twisting moment;
L = length;
G = shear modulus;
d = diameter;
subscript q = quartz.

$$Y = \frac{d_w(\theta_1 - \theta_2)}{2L_w} \tag{10}$$

where Y is the shear strain at the surface of the whisker, and $(\theta_1 - \theta_2)$ is the angular displacement between the ends of the whisker. Both the quartz fiber and the whisker are assumed to have a uniform circular cross section.

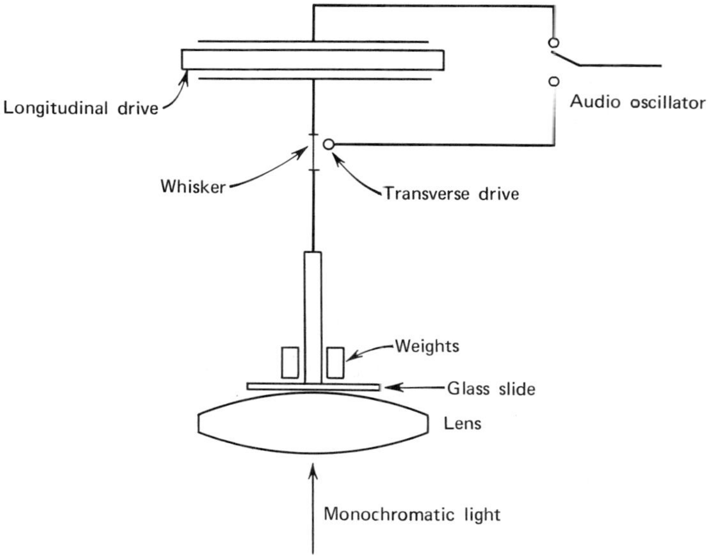

Figure 5.15 Method of measuring Young's modulus of whiskers (after Bacon [24]).

A very sensitive method for measuring Young's modulus in whiskers was adapted by Bacon [24] from similar techniques that have been used in the textile industry for evaluating single filaments of polymeric materials. Gonsalves [25] and Montgomery and Milloway [23] reported employing the method as early as 1947. The apparatus is shown in Figure 5.15. The whisker may be vibrated in resonance longitudinally by an audio transducer, or transversely by electrostatic attraction. A very long focus lens was used against a flat glass plate to generate Newton's rings. Resonance in either mode is indicated by the blurring or disappearance of the rings. Young's modulus may be expressed in terms of the resonant frequency by

(a) longitudinal vibration

$$F_L = \frac{1}{2\pi}\sqrt{\frac{EA}{lm}} \tag{11}$$

(b) transverse vibration

$$F_t = \frac{1}{2l}\sqrt{\frac{mg}{PA}} \tag{12}$$

where F = resonant frequency;
A = area of the whisker;
l = whisker length;
m = load;
g = acceleration of gravity;
ρ = whisker density.

CONCLUSION

Much clever work has been done in developing devices and techniques for estimating the mechanical properties of whiskers. The testing of whiskers one at a time remains, however, a tedious and delicate process. As the technology of preparing composites from whiskers advances, we may expect that detailed examination of individual whiskers, and the averaging processes now required, will be replaced by the evaluation of whiskers from the mechanical properties of standard composites.

REFERENCES

[1] P. J. Soltis, Report No. NAEC-AML-1831, Naval Engineering Center, Philadelphia, Pa.
[2] J. E. Gordon, *Endeavour*, **23**, 8–12 (January 1964).
[3] W. B. Campbell, W. J. Mallio, and R. C. Folweiler, Fifth Quarterly Report, Contract DA-19-020-AMC-0068 (X) (15 August, 1964).
[4] Z. Gyulai, *Z. Physik*, **138**, 317 (1954).
[5] C. K. Herring and J. K. Galt, *Phys. Rev.*, **85**, 1062–1063 (1952).
[6] R. J. Gomer, *J. Chem. Phys.*, **28**, 457 (1958).
[7] D. M. Marsh, *J. Sci. Instr.*, **38**, 229 (1961).
[8] P. J. Shlichta, *Growth and Perfection of Crystals*, Edited by R. H. Doremus, B. W. Roberts and D. Turnbull, John Wiley and Sons, New York, 1958, p. 217.
[9] J. E. Emrick and H. L. Gegel, ASD-TR-61-168 (1961).
[10] J. A. Herzog, *Metallo*, **17**, 1 (1963).
[11] F. W. Sears, *Principles of Physics, III*, Addison-Wesley, Cambridge 1947, p. 259.
[12] R. H. Kelsey and R. H. Krock, *J. Materials*, **2**, 146 (1967).
[13] C. A. Calow, P. D. Boyer, and G. P. R. McCarthy, *Brit. J. Appl. Phys.*, **18**, 1715 (1967).
[14] R. L. Mehan and E. Feingold, *J. Materials*, **2**, 146 (1967).
[15] R. L. Mehan, E. Feingold, and A. Gatti, Technical Report, Air Force Materials Laboratory, AFML-TR-65-275 (August 1965).
[16] S. S. Brenner, *J. Appl. Phys.*, **28**, 1023 (1957).
[17] N. Cabrera and P. B. Price, *Growth and Perfection of Crystals*, Edited by R. H. Doremus, B. W. Roberts, and D. Turnbull, John Wiley and Sons, New York, 1958, pp. 206, 210.
[18] F. D. Lemke and R. W. Kraft, *Rev. Sci. Instr.*, **33**, 8, 846 (1962).
[19] R. H. Kelsey and R. H. Krock, *Rev. Sci. Instr.*, **36**, 7, 1031 (1965).
[20] R. L. Mehan, W. H. Sutton, and J. A. Herzog, *Amer. Inst. Aero. Astro. J.* (November 1966), p. 1889.

[21] Private communication.
[22] R. S. Eisner, *Acta Met.*, **3,** 419 (1955).
[23] S. S. Brenner, *Growth and Perfection of Crystals*, Edited by R. H. Doremus, B. W. Roberts, and D. Turnbull. John Wiley and Sons, New York, 1958, p. 164.
[24] R. Bacon, *ibid.*, p. 201.
[25] V. E. Gonsalves, *Textile Res. J.*, **17,** 369 (1947).
[26] D. J. Montgomery and W. T. Milloway, *Textile Res. J.*, **22,** 729 (1952).
[27] S. S. Brenner, *J. Appl. Phys.*, **33,** 1 (January 1962), p. 33.
[28] R. L. Mehan and E. Feingold, *J. Mater.*, **2,** 239 (1967).

R. L. Mehan and J. A. Herzog

6 Mechanical Properties of Whiskers

INTRODUCTION

The use of whiskers to reinforce various matrices, as discussed in Chapter 11, can result in composites of impressive strength. Although the mechanical properties of whisker composites depend on a variety of factors, the properties of the whiskers themselves are the most important. Indeed, the whole impetus behind current research on whisker composites lies in the fact that whiskers are the strongest form of solids yet discovered. Their strength approaches the theoretical cohesive strength of solid matter, which lies in the range of 0.03 E to 0.17 E [1], where E is the modulus of elasticity.

The technological use of whiskers, as opposed to their purely scientific interest, lies in incorporating them into composites that have more attractive properties than conventional materials. As such, the properties of the whiskers in the composite itself are of prime concern. It is useless, in this sense, to be concerned with whiskers that have attractive properties measured in the as-grown state if, by the act of incorporating them into a composite, these properties (particularly strength) are badly degraded. For example, most whiskers of pure metals, such as copper and iron, nucleate slip bands at local regions either under direct loading or by handling, and the strength of the whisker drops drastically [1]. This behavior in a copper whisker is illustrated in Figure 6.1. Such whiskers are not generally considered for use in composite applications.

Although the whisker properties "in the composite" are of prime importance, nevertheless it is necessary that accurate and fairly extensive data be available for the whiskers in the as-grown condition. The need for such data

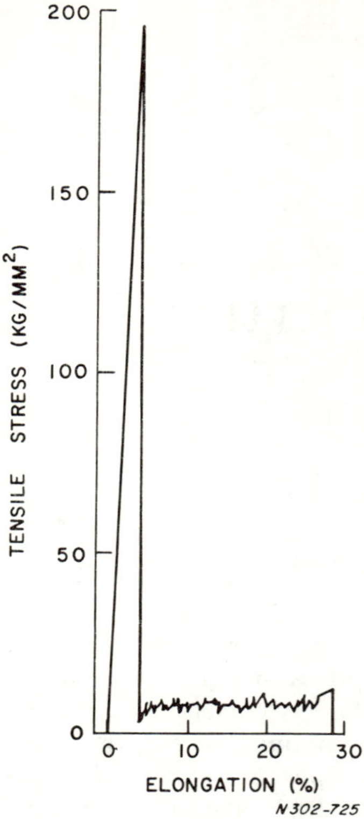

Figure 6.1 Stress-strain curve of a copper whisker whose growth direction is ⟨100⟩ [36].

is twofold. First, some basis is needed in order to choose the particular type of whisker to reinforce a specific matrix (in addition to factors such as chemical inertness, size, etc.). Second, it is important to know the fraction of the whisker property that is being utilized in the composite; that is, how good is the composite fabrication technique. In assessing this factor, it is necessary also to take account of the strength dispersion when calculating a theoretical composite strength. This factor, together with the necessary theory for calculating composite properties, is discussed in Chapter 7.

In the present state of whisker technology, strength is the property that has been studied to the greatest extent. This is understandable because it is the strength of whiskers which makes them unique. However, other properties of perhaps equal interest have not been studied to any great extent and data are unfortunately scarce. Such properties include elevated-temperature strength, elastic constants, creep behavior, and fatigue.

It is the purpose of this chapter to review the properties of whiskers that are of prime technological importance. These whiskers are almost exclusively ceramic-type whiskers, such as α-Al_2O_3, SiC, B_4C, and Si_3N_4. These whiskers have been studied because most of the effort in the whisker-composite field has been directed toward reinforcing relatively ductile metals such as aluminum, nickel, and copper. Metallic whiskers undergo metallurgical reactions (interdiffusion) with the matrix during fabrication and at elevated temperatures, where the greatest potential benefits are desired. Also, as mentioned earlier, dislocations are easily introduced into most metallic whiskers, leading to a large loss in strength. Furthermore, most metallic whiskers have a low modulus and high density relative to ceramic whiskers, which make them less efficient from a modulus/density and strength/density viewpoint. Metallic whiskers may eventually be useful in reinforcing ceramics [2] or polymers, but little work has yet been done with such systems.

The properties of whiskers have been reviewed several times in the past [1, 3, 4], most recently (1964) by Brenner [1]. This chapter emphasizes work that has been published since 1964 but, for completeness, information which has previously appeared in review-type articles is also presented.

WHISKER PROPERTIES

Table 6.1, reproduced from Brenner [1], gives the available data for most whiskers produced to date. It may be seen that the theoretical strength

TABLE 6.1 MATERIALS PRODUCED IN THE FORM OF FILAMENTARY CRYSTALS [1][a]

Material	Maximum Measured Strength, % E	Preparation
Fe, Cu, Ag, Co, Ni, Pd, Mn, Hg, Zn, Cd, Sn, Cr	4.9 tension	Halide reduction, vapor condensation, eutectic solidification, or recrystallization
Si, Ge	3.6 tension	Halide reduction
Mn_5Si_3 (italic), Cu_6Sn_5, $NiAl_3$	2.7 bending	Eutectic solidification
NaCl, KCl, LiF	2.6 tension	Precipitation, cleavage, or vapor
Al_2O_3, *SiO_2*, ZnO, MoO_3, MgO	5.2 tension	Vapor reaction condensation
Mg_2SiO_4, Cr_3O_4, BeO, Si_3N_4, SiC, Fe_3C, ZnS	11.0 bending	Cleavage or eutectic solidification
C, Hydroquinone, *Pthalocyanine*, Trioxane	2.1 tension	Precipitation of vapor condensation

[a] For details of data, see reference 1. Values of strengths refer to materials in italics.

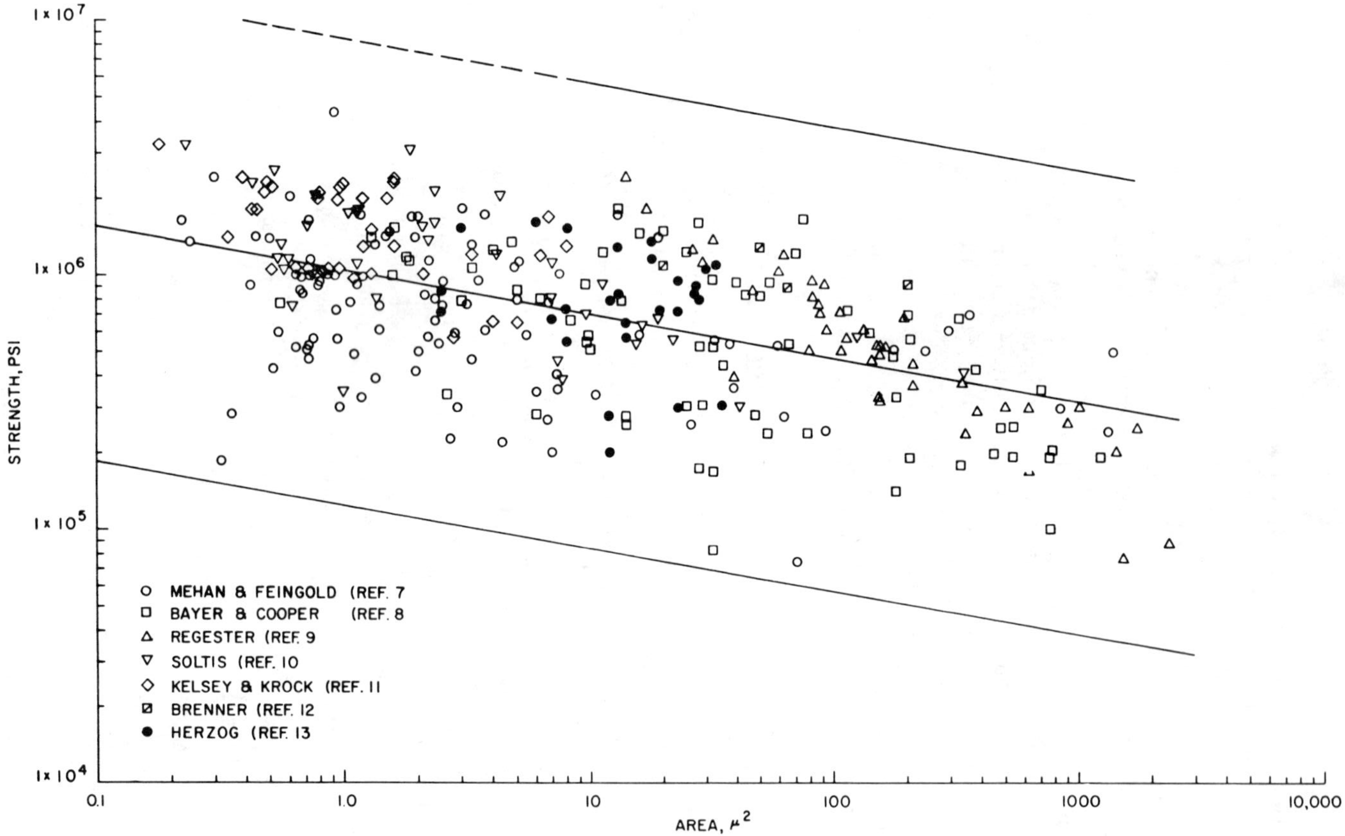

Figure 6.2 Tensile strength of α-Al_2O_3 whiskers.

($0.03\,E < \sigma < 0.17\,E$) [1] was achieved for many whiskers. The variety of preparation techniques used, which are discussed in more detail in Chapters 2, 3, and 4, should also be noted.

In the following section, the properties of ceramic whiskers are presented, and in the subsequent section metallic whiskers are discussed. With few exceptions, ceramic whiskers are brittle at room temperature and fracture without plastic deformation. A notable exception is magnesium oxide whiskers, which have been found to exhibit plastic strains of the order of 10% in bending [27].

The following discussions of whisker strength are concerned primarily with data obtained from tension tests. Direct-loaded bend tests, another common method of determining the strength of brittle materials, are generally unreliable because of the extremely low loads involved, the nonlinearity at large deflection [5, 6], and the difficulty in determining the correct moment of inertia to use for the common case of noncircular whiskers. In the so-called "loop" tests, the fiber is drawn or pushed into a loop and the strength is calculated from the radius of curvature at fracture, the fiber diameter, and the modulus of elasticity. A noncircular cross section can also lead to large errors in strength. In all bend tests, the volume of material under stress is far less than in a tensile test, so that unless the specimen is flaw-free, a bend test will generally overestimate the strength of the whisker.

One point should be kept in mind when examining strength data on whiskers. Almost all investigators test only those whiskers that appear to be perfect and free from surface defects, at least in the size range where defects can be detected. Because whiskers in a composite cannot be subjected to such scrutiny,* the average whisker strength in a composite will be lower than that predicted from published strength data. To date, no one seems to have performed the formidable task of randomly testing a large whisker population to obtain a true measure of strength and strength dispersion.

Properties of Ceramic Whiskers

Aluminum Oxide Whiskers. The properties of α-Al_2O_3 whiskers have received the most attention in the open literature. More than 300 individual α-Al_2O_3 whiskers have been tested in tension by a number of investigators [7–13] using at least five different sources of whiskers. In addition, the elastic modulus has been studied in some detail [7, 8, 10, 13], the effect of temperature on whisker strength has been determined [7, 12], and delayed failure phenomena have been investigated [12].

Figure 6.2 shows the tensile-strength data of α-Al_2O_3 whiskers of all orientations as a function of the cross-sectional area. A least-squares line,

* In an $\frac{1}{8}$-in. diameter composite 2 in. long, containing 20 vol% fine whiskers (area $\cong 4.0\ \mu^2$), there are about 10^6 whiskers.

together with the 95% confidence limits, has been fitted to the data. A decrease in strength with increasing size is evident. Note also the large amount of scatter of strength values. Part of this scatter is due to experimental error in determining the strength and cross-sectional area, part is due to the intrinsic scatter of brittle materials, and some may be due to the fact that different sources of whiskers were used.

The strength data in Figure 6.3 (and in subsequent strength presentations) are shown as a function of cross-sectional area primarily because the available data can conveniently be treated in this manner. However, detailed work by Register [9] and Bayer and Cooper [8] has indicated that, at least for the case of the most common "A-type" sapphire whiskers,* the strength can be expressed by the relation:

$$\sigma_f = KA_s^{-\alpha} \tag{1}$$

where A_s is the surface area and α and K are constants. Register finds a value of α of 0.54, and Bayer and Cooper $0.39 < \alpha < 0.56$. Both conclude that the strength of A-type sapphire whiskers is dependent on the surface area, which suggests that surface flaws govern the strength. These flaws may be of various types, and a particular type is shown in Figure 6.3. This defect is termed an overgrowth.

Bayer and Cooper have also studied the strength dependence of "C-type" whiskers and found that they obey the relation:

$$\sigma_f = Kd^{-0.64} \tag{2}$$

where d is the whisker diameter. This type of whisker shows no dependence of strength on length, and Bayer and Cooper suggest that a dislocation pile-up or interaction process causes fracture. However, no direct evidence of this mechanism has been found [8].

The determination of the elastic modulus of whiskers, as discussed in Chapter 5, is a difficult task.† For sapphire whiskers, some controversy exists as to the value of the elastic modulus. Moduli as high as 330×10^6 psi for whiskers with an "A-type" orientation have been reported [10]. However, other investigators [7, 13, 14] have not reported such high values. It is the authors' opinion that the high values reported are due to experimental error, perhaps in measuring the extension or testing badly tapered whiskers. The question, however, must remain open until further work of greater

* "A-type" whiskers have axial orientations of $\langle 210 \rangle$ or $\langle 100 \rangle$, "C-type" of $\langle 001 \rangle$, and "A-C type" of $\langle 111 \rangle$ [7].

† Many whiskers have a complex crystal structure and consequently more independent elastic constants than those for isotropic materials. The elastic modulus referred to here and subsequently will be that particular elastic constant measured along the length of the whisker under consideration.

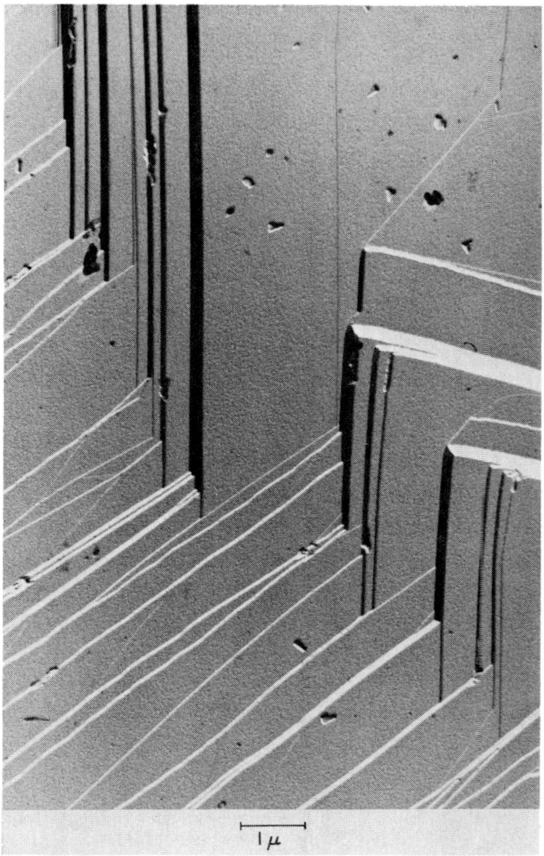

Figure 6.3 Overgrowths in a sapphire whisker.

precision is performed. The possibility of materials with exceedingly high elastic moduli is an interesting theoretical and practical prospect.

Figure 6.4 shows the available data [13, 15] (excluding those of [10] for the reasons previously discussed) for sapphire whiskers. The bulk single-crystal data are due to Wachtman and others [16]. Although there is considerable scatter in the small size range, it may be seen that the values in general fall near those expected for bulk sapphire.

Recent work by Calow, Bayer, and MacCarthy [14] has also confirmed the similarity between the elastic modulus of sapphire whiskers and bulk single-crystal sapphire. Their data are shown in Figures 6.5 and 6.6. Because the data were obtained on a Tecam machine (see Chapter 5), the measured

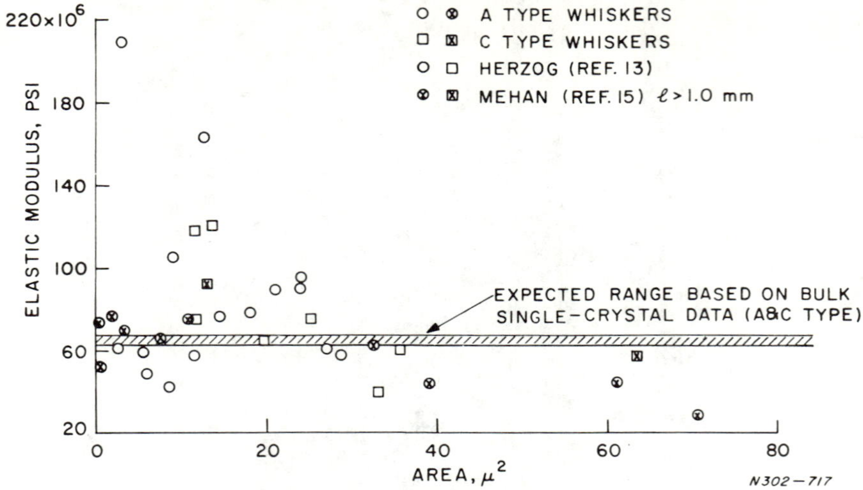

Figure 6.4 Elastic modulus of α-Al_2O_3 whiskers.

extension of short whiskers is in error because of the contribution of extension from the testing machine. However, it is possible to correct for this error [7, 14] and obtain a relationship between the apparent elastic modulus, E^*, and the reciprocal of the gauge length, $1/l_o$. The solid curves in Figures 6.5 and 6.6 were obtained using such a relation and assuming bulk values of the elastic constants of sapphire [16].

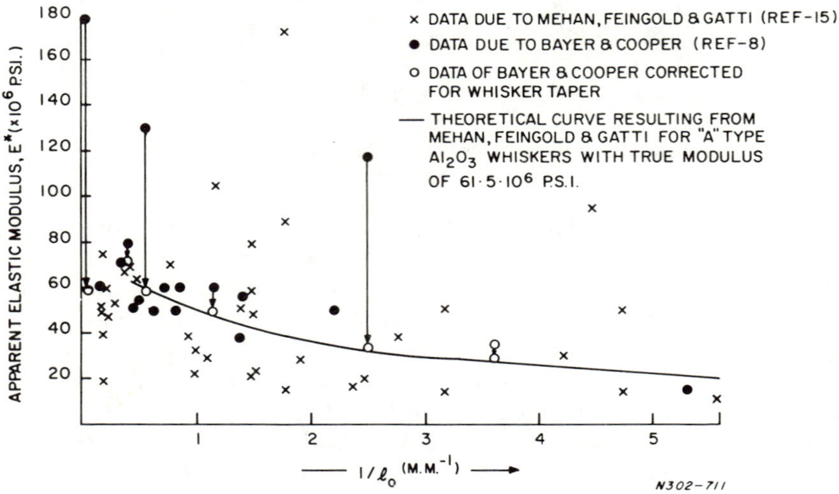

Figure 6.5 Apparent elastic modulus of A-type α-Al_2O_3 whiskers as a function of reciprocal gauge length [14].

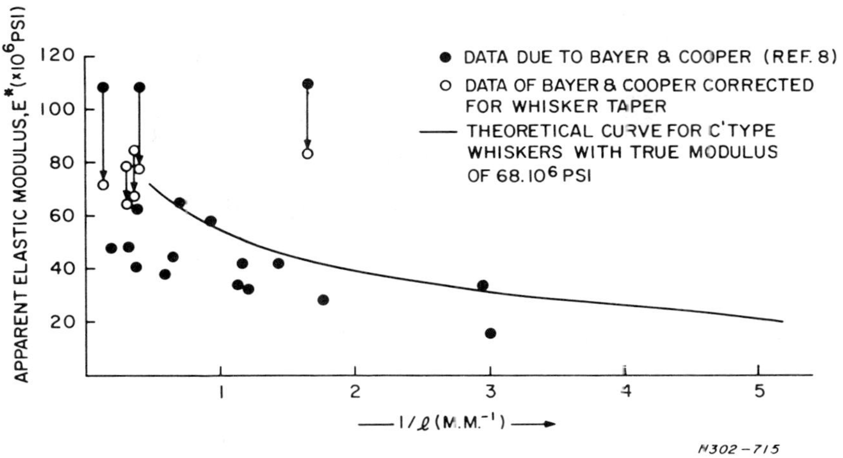

Figure 6.6 Apparent elastic modulus of C-type α-Al_2O_3 whiskers as a function of reciprocal gauge length [14].

It is immediately evident from Figures 6.5 and 6.6 that the careful work of Calow, *et al.* yields results that are very close to those obtained by assuming that the elastic constants of sapphire whiskers are identical to those of bulk sapphire single crystals. By using interferometry techniques, Calow, *et al.* were able to keep the error in area measurement quite low, and hence their scatter is very much less than that reported by Mehan and Feingold for very fine sapphire whiskers [7]. Of equal interest are the corrections made by Calow, *et al.* for whisker taper. They computed an effective area given by:

$$A_{\text{eff}} = \frac{\alpha L}{\log\,(1 + \alpha L/A_o)} \tag{3}$$

where L is the gauge length, A_o the minimum cross-sectional area, and α a constant defining the taper. Using this value of A_{eff}, very high moduli values were reduced to those expected from theory. It is apparent that the use of tapered whiskers can lead to large errors in moduli values, depending on where the area is measured.

The effect of temperature on the strength of sapphire whiskers has been studied by Brenner [12] and Mehan and Feingold [7]. Brenner used a direct tension method, first thinning the gauge section of the whisker by vaporization at 1800 to 1900°C in hydrogen in order to prevent grip failures. Mehan and Feingold used a bending method, valid to about 1200°C, at which point sapphire deforms plastically and bend strength data are unreliable. Mehan and Feingold also utilized whiskers that had been coated (by

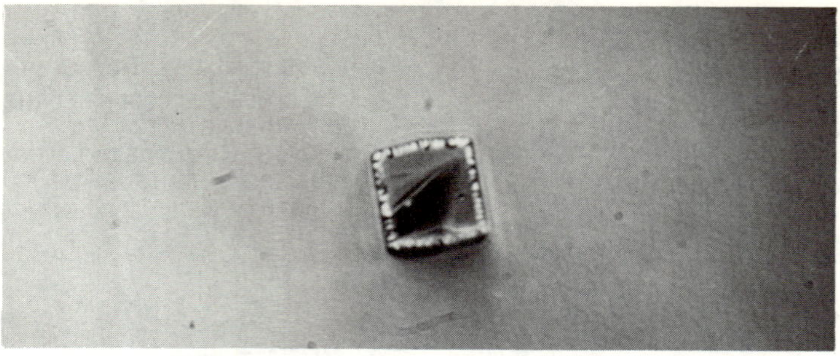

Figure 6.7 Ti-Ni coated whisker (General Electric Company) (mag. 1070X). Photographed by Mrs. M. Schueller, AFML.

a sputtering process) with a thin film (about $\frac{1}{2}\,\mu$ thick) of titanium and nickel. The cross section of a whisker with such a coating is shown in Figure 6.7.

The results of these high-temperature tests are shown in Figure 6.8. The two methods yield similar results, indicating that the elevated-temperature strength of sapphire is not greatly impaired by a metallic coating. The data show that sapphire whiskers must be heated to 1200 to 1300°C before 50% of the room-temperature strength is lost.

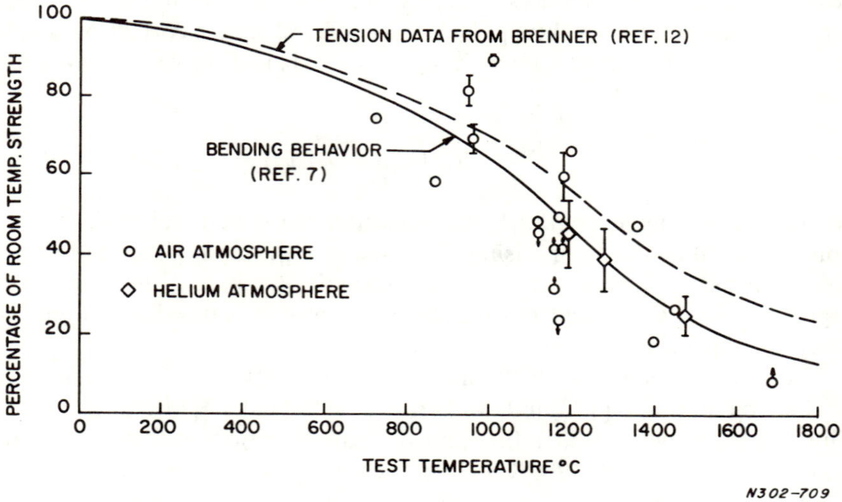

Figure 6.8 Bend strength of titanium-nickel coated alumina whiskers as a function of temperature [7].

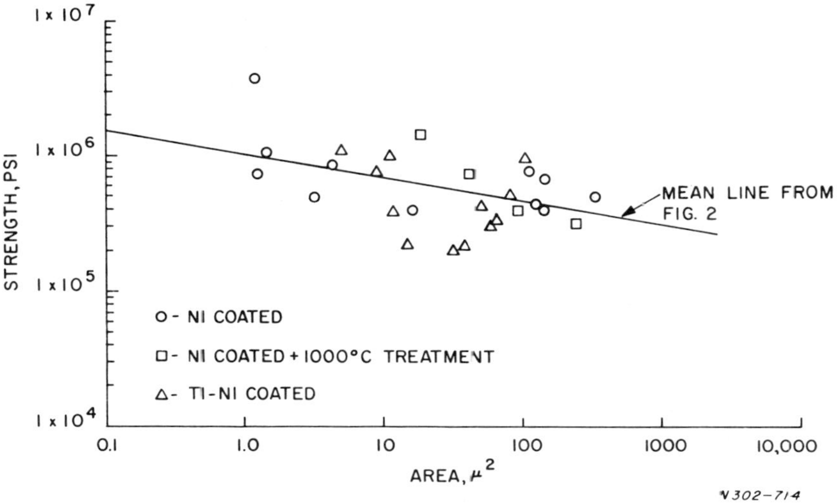

Figure 6.9 Effect of coating on strength of α-Al_2O_3 whiskers [15].

The effect of thin metallic coatings on sapphire whiskers, or on any ceramic whisker, is a matter of considerable importance. As discussed in Chapter 8, such a coating is necessary in order to promote wetting by most metallic matrices, and it is quite important to have assurance that the whisker strength is not degraded by application of the coating. In addition to the high-temperature tests discussed above, Mehan and Feingold [7] and Herzog [13] have investigated the room-temperature tensile strength of whiskers with both Ni and Ti/Ni coatings. The results are shown in Figure 6.9, together with the least-squares line of Figure 6.2. Within the experimental error, it appears that the coating does not decrease the strength of the whiskers.

Delayed failure of sapphire whiskers was studied by Brenner [12] at 630, 1060, and 1550°C. The data are shown in Figure 6.10. Brenner found that the time for failure at the three temperatures studied was related by the expression:

$$\bar{\sigma} = \bar{\sigma}_1 - 0.1 \log t \qquad (4)$$

where $\bar{\sigma}$ is the average stress that could be supported for t seconds, and $\bar{\sigma}_1$ is the average 1-sec strength. No direct evidence of plastic deformation prior to failure was observed, although Brenner pointed out that the test sections were too short for quantitative measurements.

Recently, another type of sapphire filament has become available [17]. These filaments are continuous, as contrasted to the discontinuous vapor-grown whiskers, but are true single crystals. They are grown from the melt

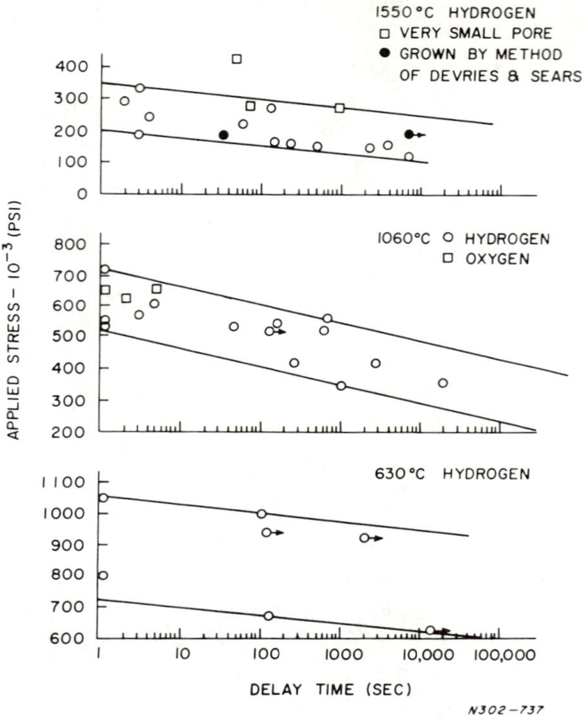

Figure 6.10 Delay time for fracture of α-Al_2O_3 whiskers as a function of stress and temperature [12].

(see Chapter 4) with diameters in the range of 0.05 to 0.50 mm and at rates up to 150 mm/min. Property measurement have yielded average tensile strengths of 300,000 psi, with high values of the order of 500,000 psi. Moduli values fall into the range of 40 to 68 $\times$ 10^6 psi [17].

The development of such continuous single-crystal sapphire filaments with properties approaching those of whiskers is of considerable importance, both from a scientific and technological viewpoint. If the desirable properties of sapphire whiskers (high strength, high modulus, low density, chemical inertness, and good elevated-temperature properties) could be combined with those inherent advantages of continuous filaments, truely impressive high-strength, high-temperature composite structures could perhaps be attained.

Silicon Carbide Whiskers. Although the available data for SiC whiskers is not as voluminous as that for α-Al_2O_3, a considerable body of information does exist. The strength of SiC whiskers has been studied both in tension [13, 18] and in bending [19], and the elastic modulus has been well characterized [13, 18].

SiC occurs in many different crystallographic modifications that differ only in the stacking arrangement of the atom planes. The sequence formed by stacking the Si-C double-layer in the sequence *ABCABC* ... results in the structure of cubic β-SiC (zincblende structure). All other sequences are referred to as α-SiC [18]. Noone [18] has pointed out that most SiC whiskers are disordered and contain both α- and β-SiC, with the $\langle 111 \rangle$ axis and the *C*-axis coincident with each other and the whisker axis. In general, therefore, the distinction between α- and β-SiC whiskers is somewhat academic. In a similar vein, Webb, Batha, and Shaffer concluded that when α- and β-SiC whisker crystals were identified and tested, no significant strength difference was found between these two structures [11].

Figure 6.11 shows the available room-temperature tensile strength of SiC whiskers as a function of cross-sectional area. The whiskers studied by Noone [18], which he grew using a gaseous pyrolysis method, seem to have somewhat higher strengths than those studied by Herzog [13], who used whiskers supplied by Thermokinetic Fibers, Inc. Considering both sets of data, the scatter is remarkably similar to that of α-Al_2O_3 whiskers. Because

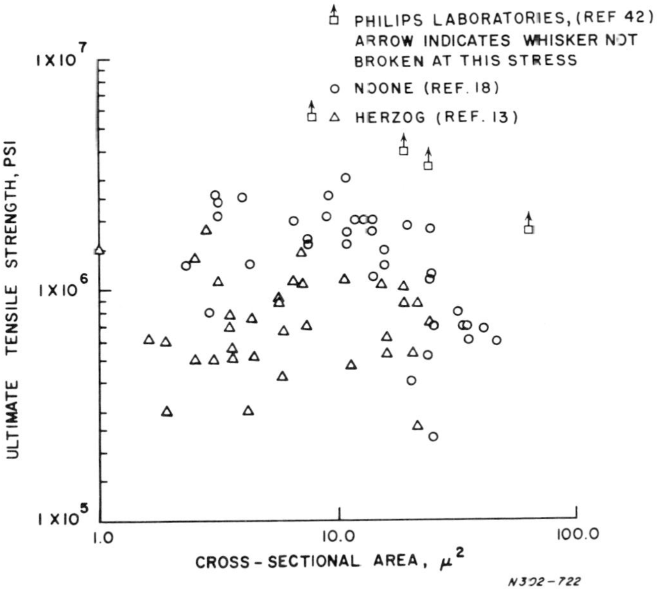

Figure 6.11 Tensile strength of SiC whiskers.

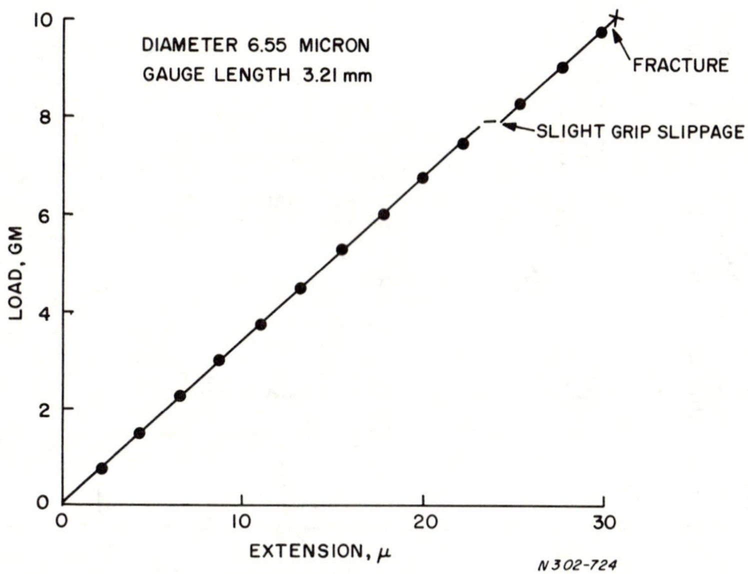

Figure 6.12 Load-extension curve for a typical SiC whisker [18].

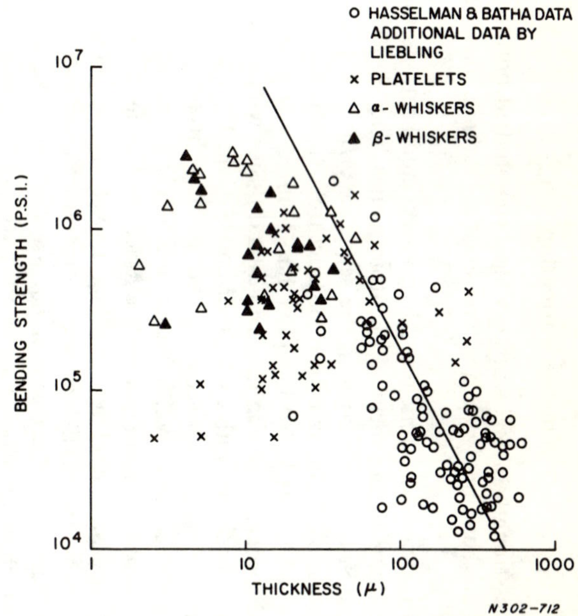

Figure 6.13 Room-temperature bend strength of SiC whiskers [21].

only a relatively small size range was studied, it is not possible to draw firm conclusions regarding the dependence of strength on whisker size. It is probable that a similar relation exists as that found for sapphire whiskers.

Some recent data of Philips Laboratories, Eindhoven are also shown [42]. These strengths are significantly higher than what has been previously observed.

As in the case of sapphire whiskers, no evidence of plastic flow is found at room temperature. A typical load-extension curve, obtained on a Tecam machine, is shown in Figure 6.12. This purely elastic behavior is typical of almost all ceramic-type whiskers at low temperatures.

Tensile strength at room temperature of SiC whiskers has also been determined by Evans, Gordon, and Marsh [19] and Kirchner and Knoll [20]. The former authors reported a strength of 2×10^6 psi, and the latter a range between 1.0×10^5 to 1.65×10^6. These values are consistent with those shown in Figure 6.11.

The bend strength of SiC whiskers and platelets has been reported by Webb, Batha, and Shaffer [21]. The tests were conducted in three-point bending at room temperature and at 1750°C, and the data are shown in Figures 6.13 and 6.14. Strengths in excess of 10^6 psi at room temperature

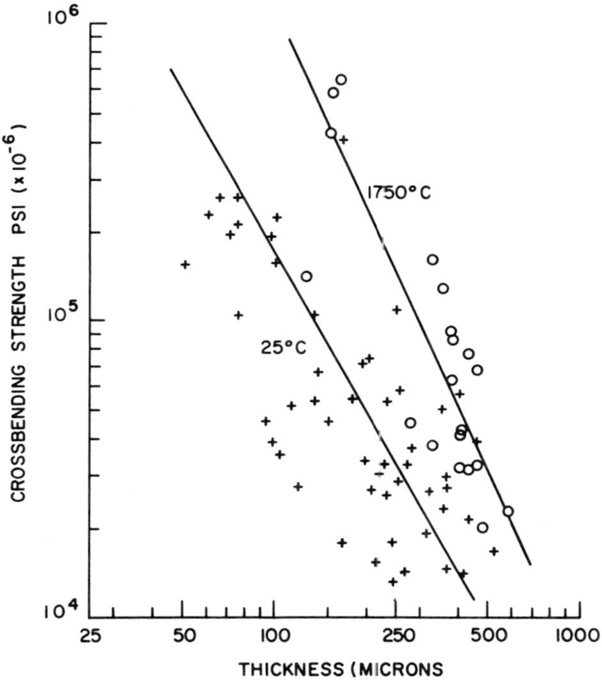

Figure 6.14 Strength of hexagonal SiC crystals at 25 and 1750°C versus crystal thickness [21].

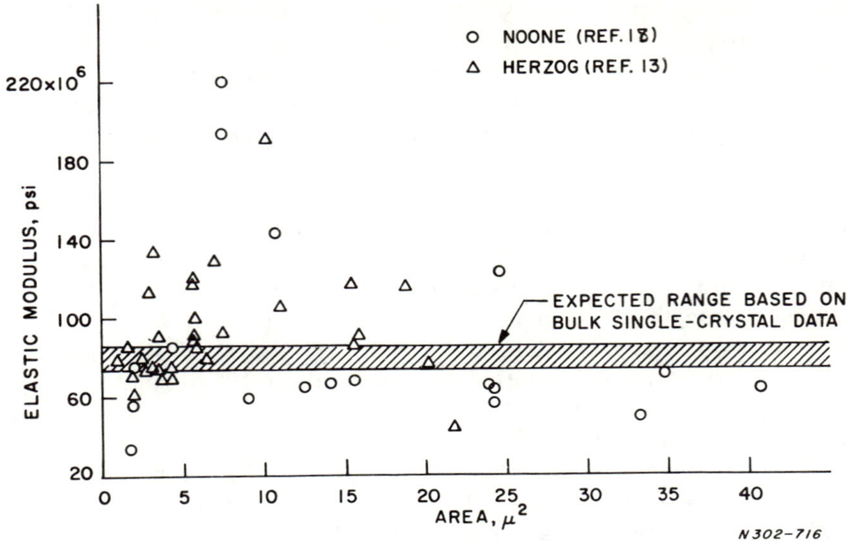

Figure 6.15 Elastic modulus of SiC whiskers.

were obtained. Direct comparison in equivalent size ranges between the tensile values shown in Figure 6.11 and the three-point bend tests are not possible because of different ways of reporting size but, in general, three-point bend strength is higher than tensile strength [22].

The increase in strength at 1750°C (Fig. 6.14) is puzzling. The most immediate explanation is that SiC undergoes plastic deformation at this temperature, which would increase the bend strength as computed from the $\sigma = Mc/I$ relation. However, Webb, et al. [21] do not find evidence of macroscopic plasticity below 1800°C. The reason for the increase in strength at 1750°C, which is contrary to the results obtained for sapphire whiskers (see Fig. 6.8), is not known.

The elastic modulus for SiC whiskers has been determined by Noone [18] and Herzog [13]. The data are shown in Figure 6.15 and are very similar to those obtained for sapphire whiskers (Fig. 6.4). The values fall about those expected for bulk properties. The occasional high values obtained are attributed by Noone [18] to experimental error in cross-sectional area measurement of fine whiskers.

Boron Carbide Whiskers. The tensile strength of B_4C whiskers has been determined by Gatti, Mancuso, Feingold, and Mehan [23]. The data are shown in Figure 6.16. These whiskers do not show the usual decrease in

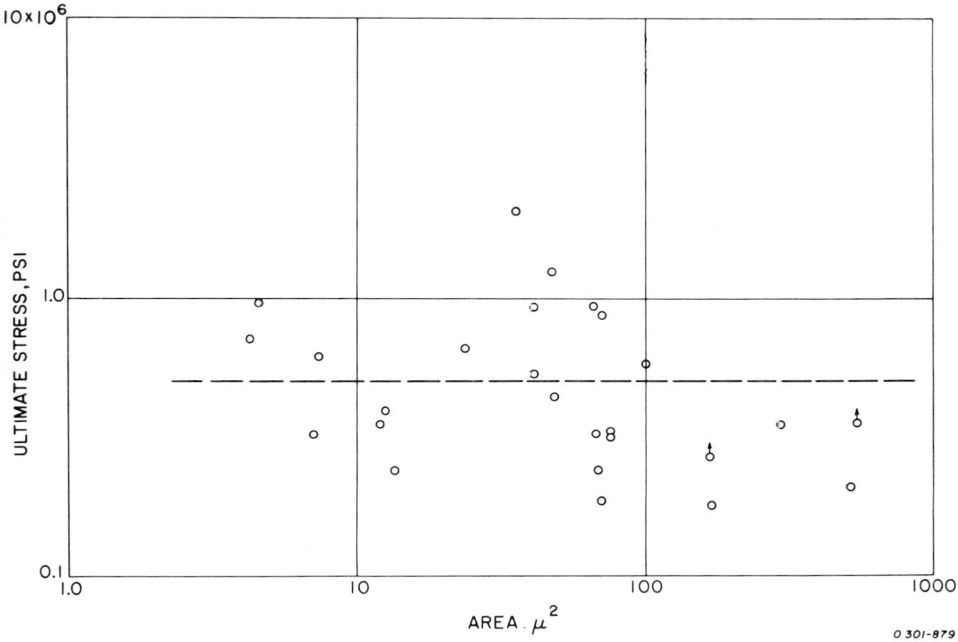

Figure 6.16 Strength of B_4C whiskers [23].

strength with increasing size, and the data are best fitted by a horizontal line giving an average whisker strength of about 5.0×10^5 psi irrespective of size. This behavior (discussed in more detail in the section, Failure and Deformation Mechanisms) is attributed to the nature of the growth defects.

The elevated temperature strength of B_4C whiskers has also been determined [23]. A bending technique, identical to that used for sapphire whiskers [7], was employed. The results, shown in Figure 6.17, indicate that B_4C whiskers are quite susceptible to oxidation attack at temperatures as low as 600°C. Even a low concentration of oxygen (the tests were performed in a nominal helium atmosphere) is detrimental at temperatures above 1000°C.

Silicon Nitride Whiskers. The strength of Si_3N_4 whiskers has been determined by Bayer and Cooper [8]. The data, shown in Figure 6.18, show a pronounced size dependency, with the average strength falling from about 1.2×10^6 psi at $1.0\ \mu^2$ to about 1×10^5 psi at $100\ \mu^2$. Bayer and Cooper found that the size-strength relation could be expressed as

$$\sigma_f = KA_s^{-\alpha} \tag{5}$$

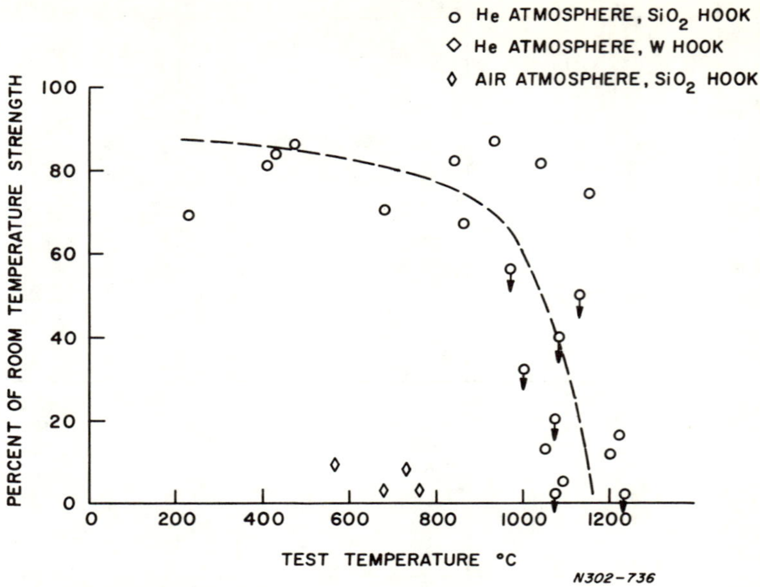

Figure 6.17 Hot bend strength of B_4C whiskers [23].

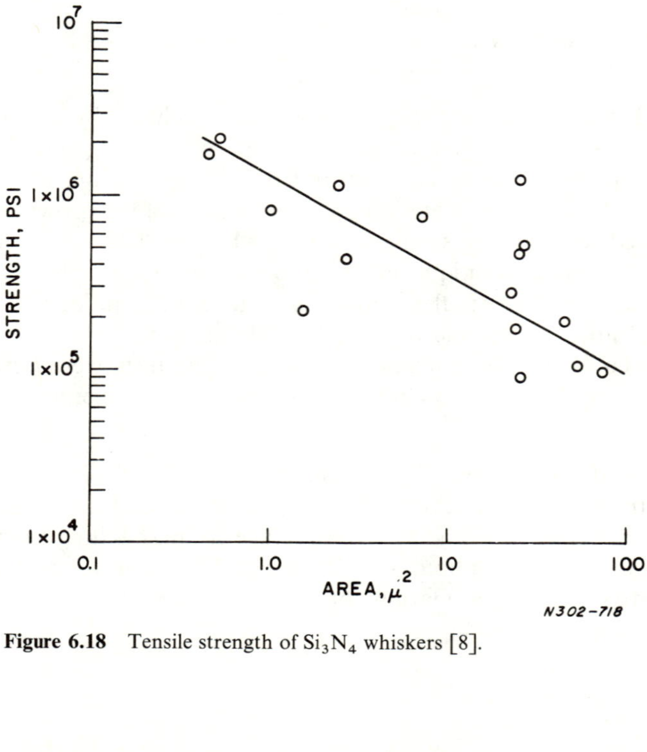

Figure 6.18 Tensile strength of Si_3N_4 whiskers [8].

where $0.81 \lesssim \alpha \lesssim 1.11$. This indicates, as with the case of "A-type" sapphire whiskers, that the strength of Si_3N_4 whiskers is dependent on the surface area.

Aluminum Nitride Whiskers. The strength of aluminum nitride whiskers has been studied by Davis and Evans [24]. Both bend tests and tensile tests were conducted. The bend tests were of the "loop" variety, where the fiber was pushed into a loop and the stress calculated using the elementary beam-bending equation. The data are shown in Figure 6.19. The size-strength dependency is flat, and the bend strength and tensile strengths are very similar. This indicates that either very few flaws are present, or they are spaced so close together that bend and tensile tests yield equivalent strength values, or the flaws are such that the strength is independent of size. Whiskers with hexagonal cross sections, as can be seen, yielded appreciably lower strength than whiskers with square cross sections. The reason for this behavior is not known.

Graphite Whiskers. The strength and elastic modulus of graphite whiskers has been determined by Bacon [25], who found a maximum room-temperature strength of 2.85×10^6 psi and an elastic modulus greater than 100×10^6 psi. These properties, together with a low density and good

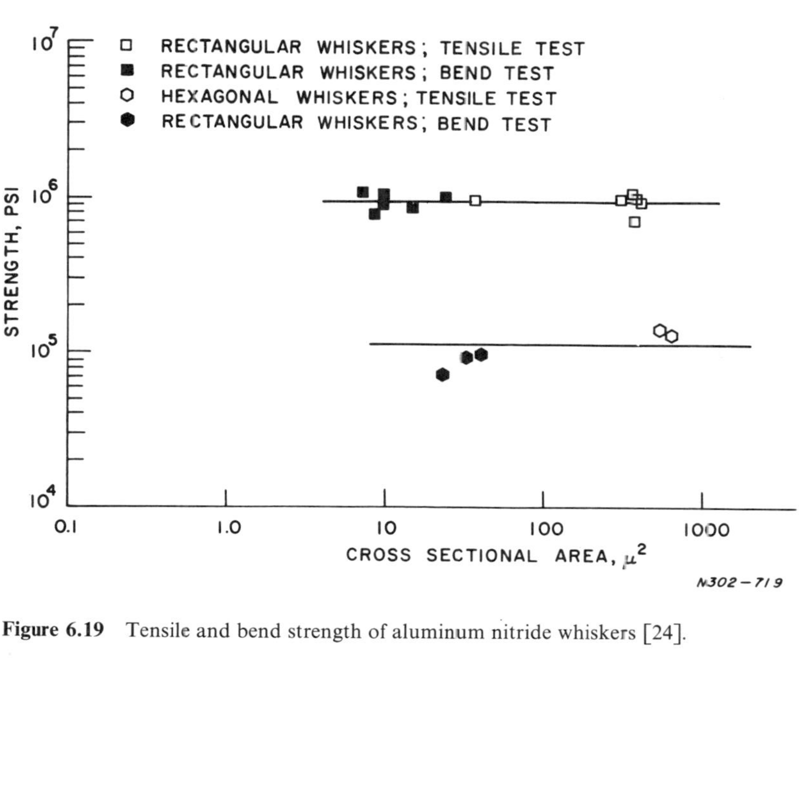

Figure 6.19 Tensile and bend strength of aluminum nitride whiskers [24].

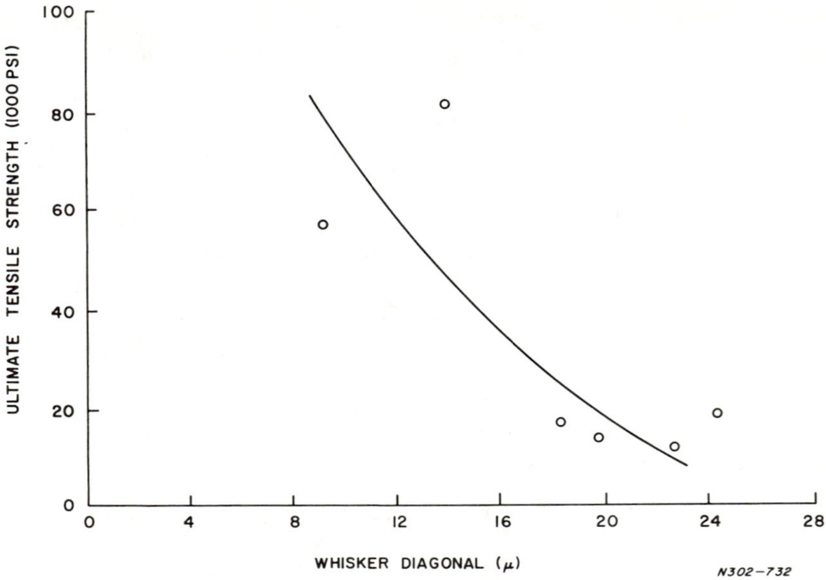

Figure 6.20 Ultimate tensile strength of MgO whiskers versus diagonal of the square cross section [27].

thermal and chemical properties, give these whiskers a very attractive strength-to-density ratio. Not a great deal of additional work has been performed with graphite whiskers, and current emphasis has been placed on continuous filaments prepared by controlled thermal degradation of synthetic organic fibers.

Magnesium Oxide Whiskers. MgO whiskers have been studied by Wolf and Coskren [27], Hulse [28], and Brubaker [26]. Hulse, using whiskers formed by fracture of MgO single crystals, found strengths up to 2.5×10^6 psi in the 1–3 μ diameter range as measured under pure bending. Wolf and Coskren found somewhat lower strengths for larger sizes as shown in Figure 6.20. They suggested that the lower strength was due to the fact that no precautions were taken to avoid testing whiskers with growth steps, and felt that in the size range of 1–3 μ strengths of $\geq$400,000 psi are possible, analogous to those found by Hulse [28]. Brubaker [26] did not measure strengths but found that about 10% of his whiskers could be plastically deformed of the order of 10%.

Properties of Metallic Whiskers

Metallic whiskers have been prepared from a number of elements, as shown in Table 6.1. Historically, much of the early work on whisker properties was performed on metallic whiskers, and it was the original discovery of Herring and Galt [29] of the remarkable strength of tin whiskers that led to the current interest in this field.

Metallic whiskers have been of great interest scientifically because of the information they have provided on the structure of solids and the mechanism of crystal growth. However, they have not been used to any great extent as a reinforcing fiber in composites because of the ease with which dislocations are introduced. Tensile tests on iron whiskers, for example, where considerable efforts were made in growth procedures to reduce the impurities and dislocation defects, showed strength decreases from 1.5×10^6 psi to below 0.01×10^6 psi with increasing impurity content [31]. Plastic deformation occurs early in the deformation process for pure metal whiskers such as iron, copper, and many others. In these cases, the yielding of a specimen with a smooth surface is indicated by slip lines at an angle to the specimen

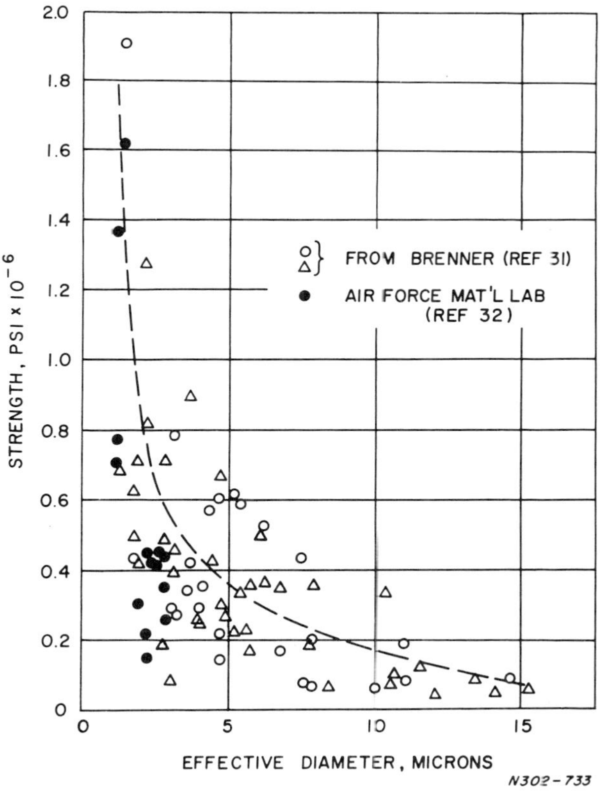

Figure 6.21 Strength of iron whiskers at room temperature as a function of size [39].

axis. In addition, pure metal whiskers do not have the high elastic modulus and high strength required for successful applications in composites, and also they are handicapped by fairly high densities and the probability of interdiffusion with metallic matrices. As a reinforcing fiber, therefore, their properties are somewhat academic and will be discussed only briefly. For further information, several of the earlier reviews of whisker properties may be consulted [1, 3, 4].

The strength of iron whiskers has been studied extensively by Brenner [30] and Herzog [31]. The data are shown in Figure 6.21. The degree of scatter is quite similar to that observed for ceramic whiskers. In spite of this large scatter, Brenner [1] has shown that the average strength and both the minimum and maximum strength follow a $1/d$ relation, similar to that found by Bayer and Cooper for "C-type" α-Al_2O_3 whiskers.

The elastic modulus of iron whiskers has been determined by Herzog [31], and the data are presented in Figure 6.22. The apparent size dependency seems to result from an oxide coating on the whisker surface. Effects such as these are not uncommon and point to the necessity for very careful interpretation of experimental results obtained from measurements of whisker properties.

The strength of tungsten whiskers, grown by the thermal decomposition of the oxides, has been studied by Aleksandrov and Kogan [32]. The data

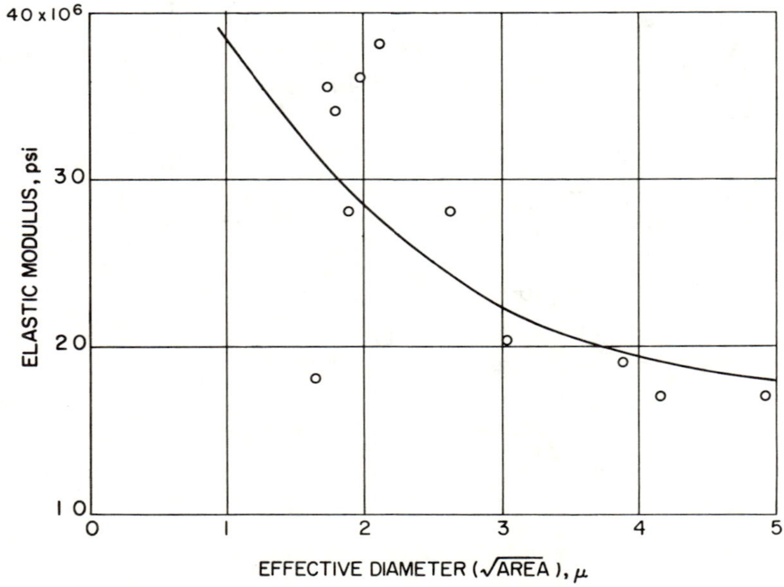

Figure 6.22 Elastic modulus or iron whiskers as a function of size [39].

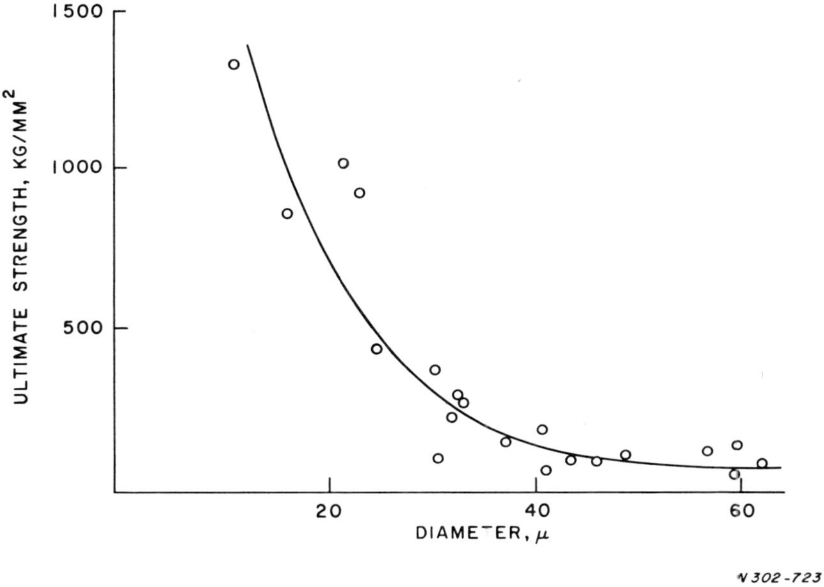

Figure 6.23 Strength of tungsten whiskers as a function of size [32].

are shown in Figure 6.23. The highest strength obtained, 1.87×10^6 psi, is 37% of the theoretical strength, as calculated by the $E/10$ relation and using a modulus of 50×10^6 psi. Tungsten wire of the same diameter, obtained by drawing, did not exceed 7.1×10^5 psi [32].

Chromium whiskers have been studied by Amonenko and Vasyutinskiy [33]. The data (Fig. 6.24) show a very abrupt drop in strength with increasing size. The maximum strength, obtained on a whisker with a cross-sectional area of 20 μ^2, was 3.15×10^5 psi. This is quite low, suggesting the presence of growth defects on the whisker surfaces or internal imperfections.

Copper whiskers have been studied by Brenner [1] and Kostyuk, *et al.* [34] and the data have been discussed by Brenner [1]. Additional results have been obtained by Bokshtein, *et al.* [35], who also studied the strength-size dependency of nickel and cobalt whiskers. These data are shown in Figures 6.25, 6.26, and 6.27, together with single-crystal data where available. Crystal orientation of three types was studied (⟨111⟩, ⟨100⟩, and ⟨110⟩), and it is evident that no effect of orientation on strength was found. The strength of the whiskers drops sharply, and at 15–20 μ diameter have strengths essentially similar to bulk single-crystal properties. The average strength obeys the $1/d$ relation found by Brenner for iron whiskers, as shown in Figure 6.28. It may be noted that Bokshtein, *et al.* carefully selected their whiskers

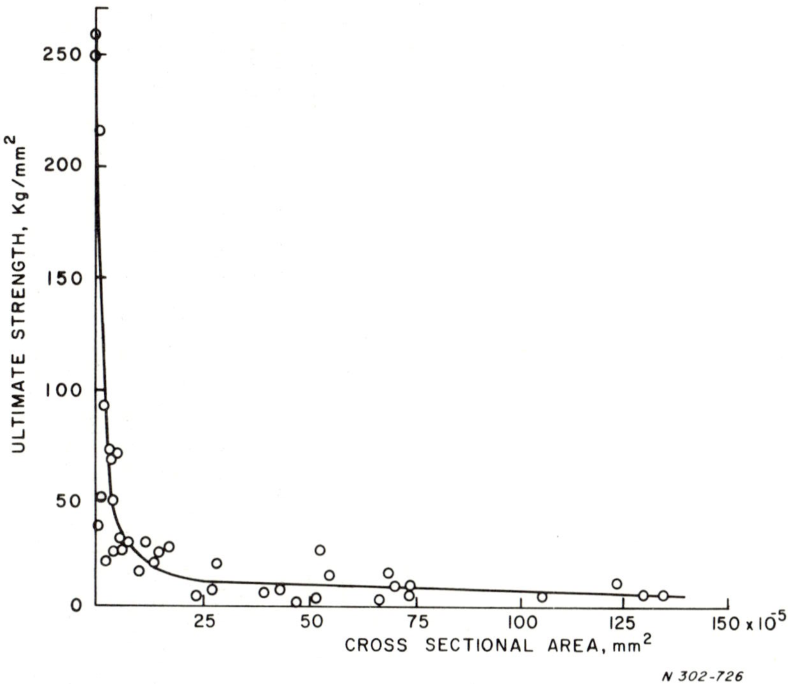

Figure 6.24 Strength of chromium whiskers as a function of size [33].

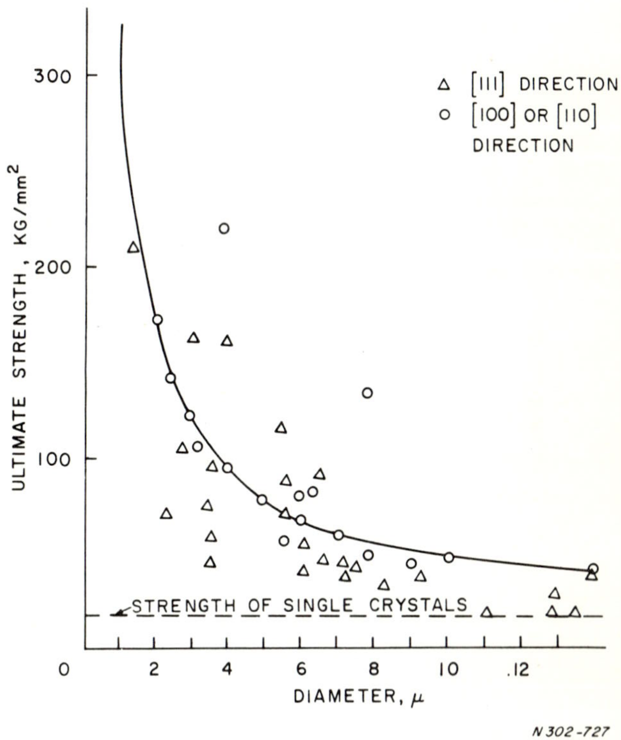

Figure 6.25 Strength of copper whiskers as a function of size [35].

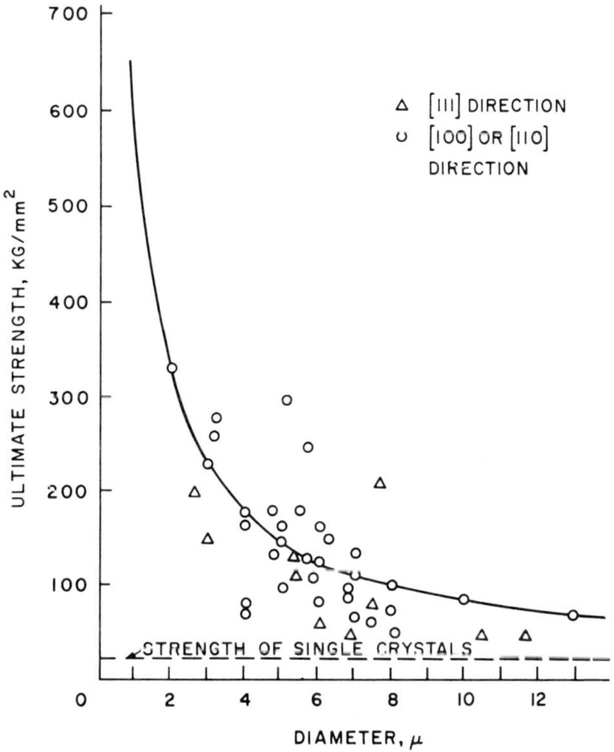

Figure 6.26 Strength of nickel whiskers as a function of size [35].

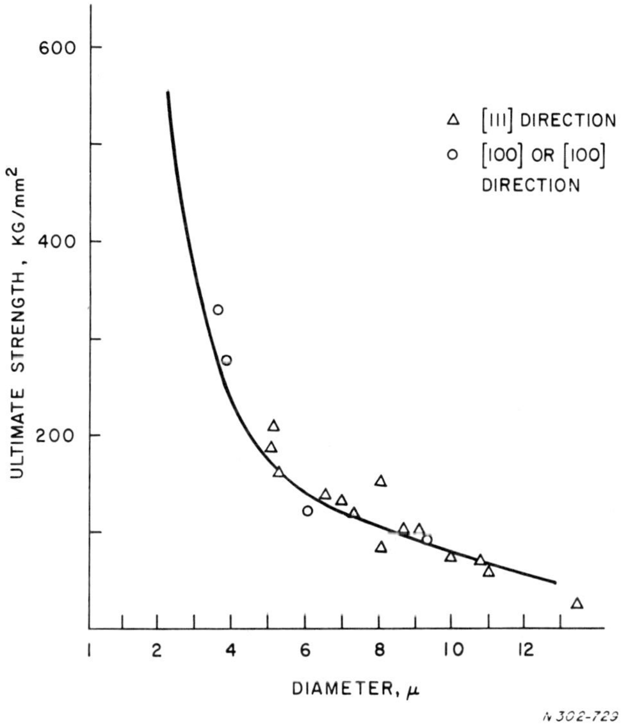

Figure 6.27 Strength of cobalt whiskers as a function of size [35].

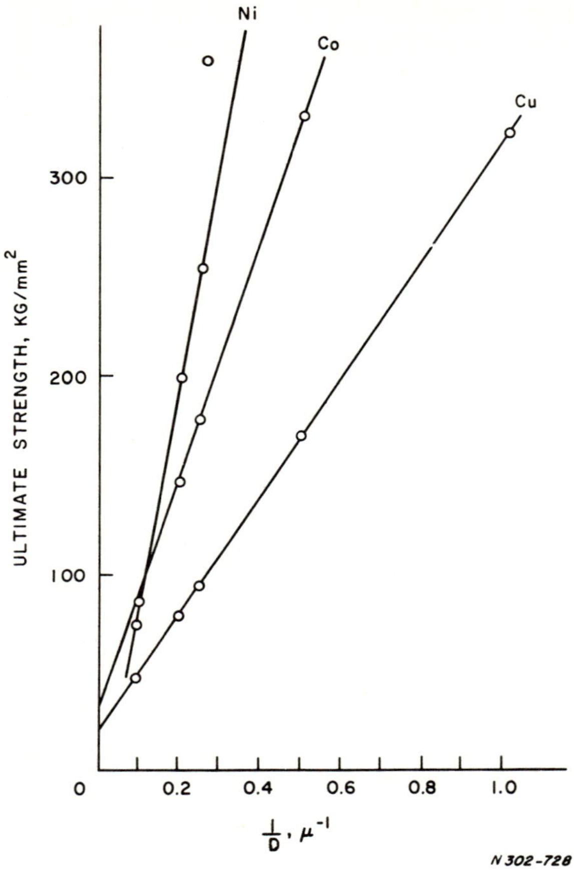

Figure 6.28 Strength of nickel, cobalt, and copper whisker as a function of reciprocal diameter [35].

before testing, and estimate that only 1% of the whisker population was strong.

FAILURE AND DEFORMATION MECHANISMS

The failure of tensile specimens usually occurs at the location of the smallest cross-sectional area. In carefully prepared specimens with an accurately maintained diameter over the whole gauge length, the failure location is determined by the pattern of either surface or subsurface imperfections or dislocations in the gauge length. This generally holds for both polycrystalline and single-crystal specimens. Whiskers, however, are generally "unsuitable" tensile specimens since they cannot be machined to desired shapes and dimensions. They must be tested in the as-grown condition and this often requires that the tensile test be conducted on a specimen with changing cross-sectional area and containing possible growth flaws in the gauge length. Therefore, the study of whisker morphology is of great importance. It is known that most whisker specimens have either the tapers or step growths of α-Al_2O_3 whiskers, the subboundaries of iron whiskers, the sectional growths (commonly described as bamboo structure) of SiC whiskers, or any combination of these morphological types. Some whisker types grow in sections rotated with respect to each other. Certain growth conditions produce this result on silicon carbide whiskers of triangular cross section. The rotation angle around the axis of one section with respect to the adjacent section is 60°. The conditions leading to this peculiarity are presently not thoroughly understood but may be due to twinning. The areas of adjacent sections may also differ from each other, and this results in a complex stress pattern in the vicinity of the change of section. Complex stress patterns may also exist between adjacent sections in the bamboo structure of silicon carbide whiskers. It is natural that these morphologies strongly influence the stress distribution and that consequently the measurement of the ultimate stress of a number of whiskers will reflect this in the observed scatter.

The compilation of data on whisker properties presented in the previous section showed that in general the strength of whiskers decreased with increasing size. It is also expected that the strength would decrease with increasing whisker length, although little systematic work in this area has been done. Geminov and Kop'yev [37] have studied the strength of copper whiskers at a fixed length and varying diameter, and also at a fixed diameter and varying length. These data are shown in Figure 6.29 and are considered to be typical of the behavior of most whiskers. Evidently defects are present in most whiskers and their severity increases as the whisker size increases. Longer whiskers have a greater probability of having a strength-reducing flaw in the gauge length.

An attempt to quantitatively account for decreasing whisker strength with increasing size has been made by Marsh [38]. This analysis is applicable to whiskers having step growths, such as α-Al_2O_3 whiskers or B_4C whiskers. Using photoelastic techniques, Marsh experimentally determined the stress-concentration factor for sharp steps typical of those encountered on whisker surfaces. The results (see Fig. 6.30) show that steps can produce stress concentrations of the same order of magnitude as cracks. Marsh also found that the stress distribution around cracks and steps is quite similar.

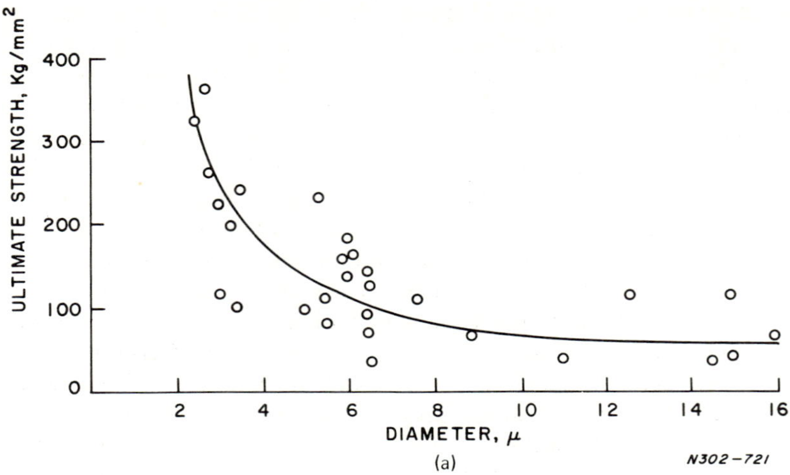

Figure 6.29a Strength of copper whiskers as a function of diameter at a gauge length of 2.0 mm [27].

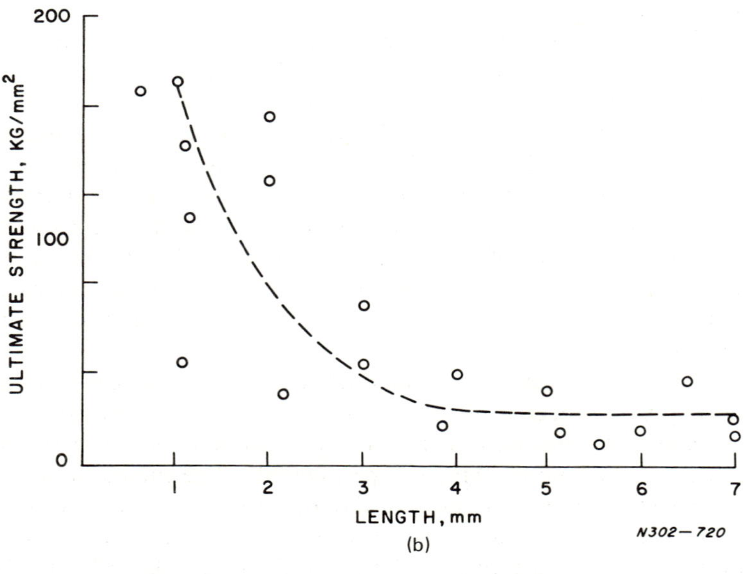

Figure 6.29b Strength of copper whiskers as a function of length at a diameter of 6.5 μ [37].

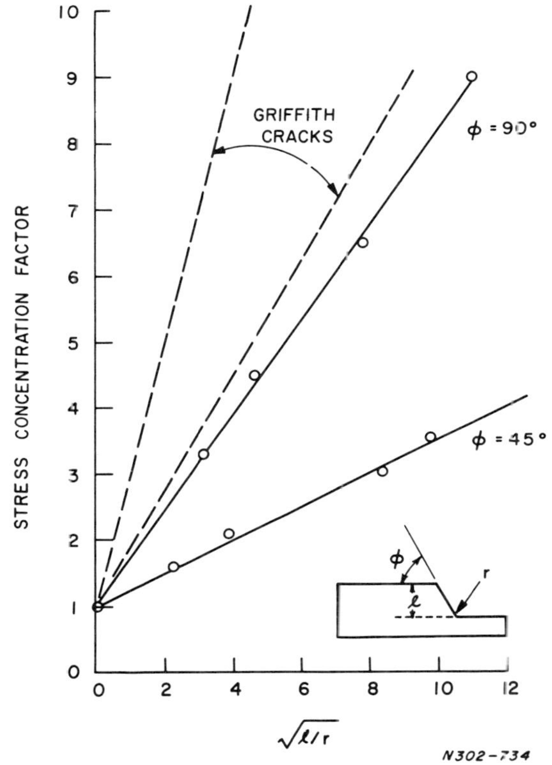

Figure 6.30 Stress concentration at steps of various dimensions compared with equivalent crack values [38].

Marsh applied these results to test data of silicon whiskers. By determining the root radius, r, from electron micrographs (found to be 30 Å) and using a step height of $\frac{1}{40}$ of the whisker diameter, obtained by optical microscopy, a stress concentration factor could be obtained which increased as the whisker size increased. Using these values and a theoretical strength of 0.05 E, Marsh was able to predict a size-strength relation for silicon whiskers. The data are shown in Figure 6.31. It may be seen that the theoretical curve is a very good upper bound for the experimental values. It is clear, as Marsh points out, that the stress concentration due to a step provides a reasonable quantitative explanation of the size-strength variation for brittle whiskers.

The effect of surface steps in reducing the strength of sapphire whiskers has been treated in a semiquantitative fashion by Mehan, et al. [39]. Figure 6.32 shows the surface of two separate whiskers, one of high strength (Fig. 6.32b) and the other of low strength (Fig. 6.32a). The differences between the

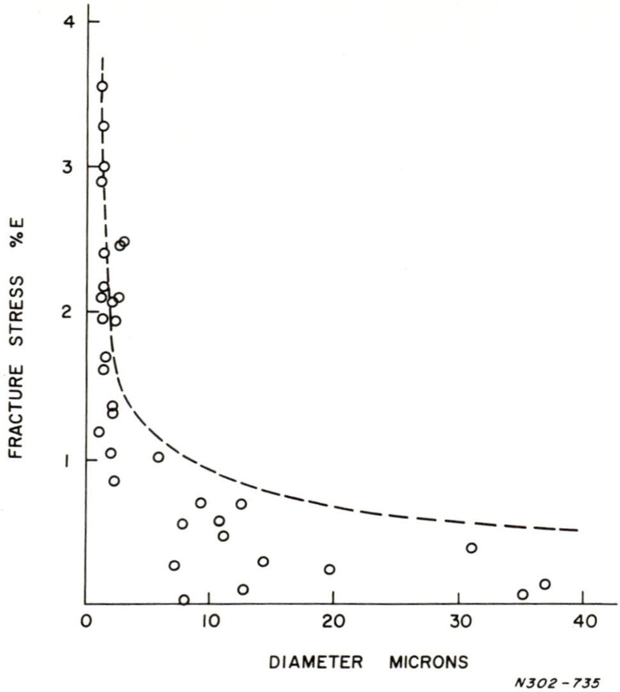

Figure 6.31 The strength of silicon whiskers as a function of their diameter. The theoretically predicted strength/diameter curve is superposed [38].

surfaces are evident. It is possible, on the basis of Marsh's analysis, to assign a value to the stress-concentration factor associated with surface steps such as those depicted in Figure 6.32a. Using a value of $r = 30$ Å (based on measurements by Marsh on silicon whiskers) and a value of $l = 2.5 \times 10^{-5}$ mm, a stress concentration factor of 2.8 is obtained from Figure 6.30. Although exact agreement between the two specimens is not obtained (1.95×10^6 psi predicted versus 1.27×10^6 psi measured), the order of magnitude is correct.

The Marsh analysis can also be used to explain the size-strength independency of B_4C whiskers, which were discussed in the section on whisker properties (Fig. 6.16). A typical B_4C whisker is shown in Figure 6.33, and Gatti, et al. [23] found that the step height, *l*, was random and did not vary with whisker size. They concluded that the growth steps in B_4C whiskers lead to a stress-concentration factor somewhere between 5 and 10, irrespective of the whisker size. This postulated effect is in accord with the data shown in Figure 6.16.

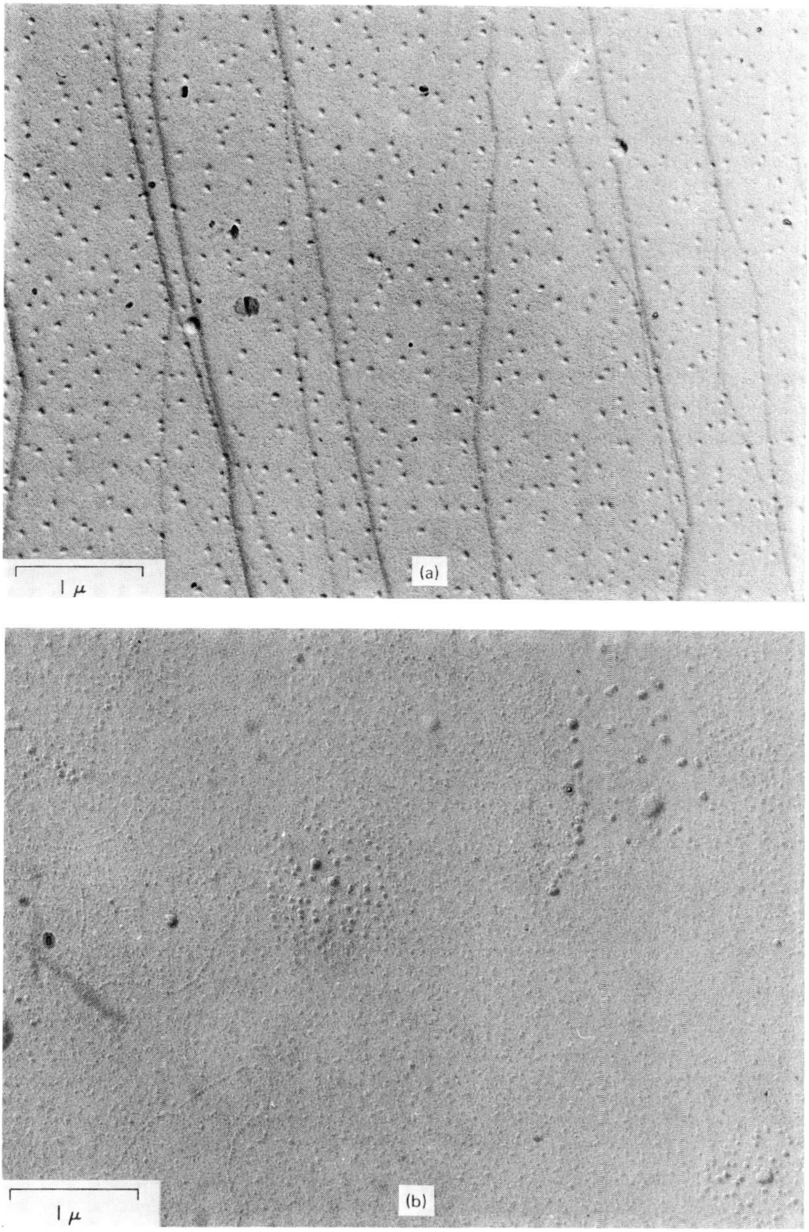

Figure 6.32 Electron photomicrograph of the replicated surface of two α-Al_2O_3 whiskers of different strengths [39]. (a) Tensile strength, 698,000 psi; (b) tensile strength, 1,275,000 psi.

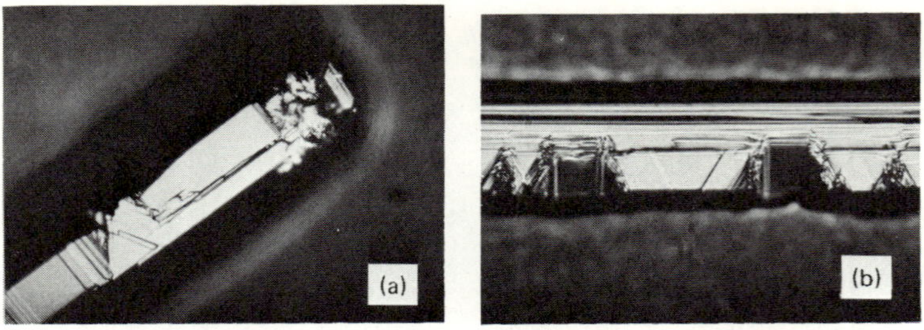

Figure 6.33 Photograph of B_4C whiskers showing step defects [23].

Another example illustrating the potent effect of surface steps in reducing the strength of whiskers may be cited, again using the work of Gatti, et al. [23]. One B_4C whisker had a strength of 2.0×10^6 psi. For the Marsh analysis to be applicable, surface steps on this particular whisker must be absent, be quite small, or be oriented parallel to the whisker axis. Figure 6.34 shows this whisker and it may be seen that the growth step is parallel to the whisker axis and hence not effective in reducing its strength.

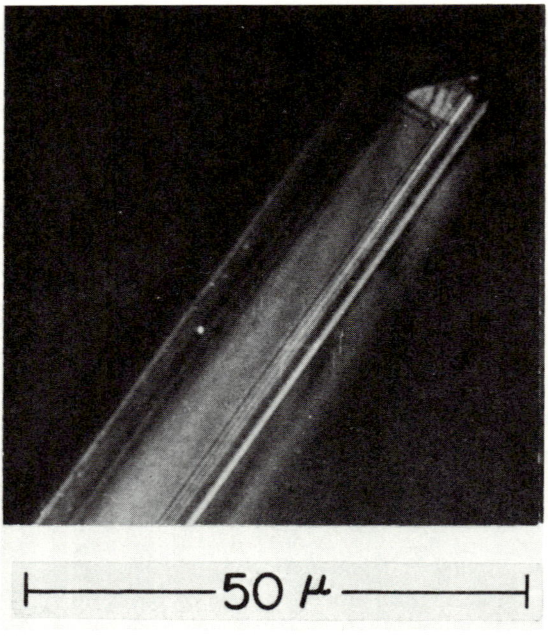

Figure 6.34 The surface of a strong B_4C whisker having a strength of 2×10^6 (1200X) [23].

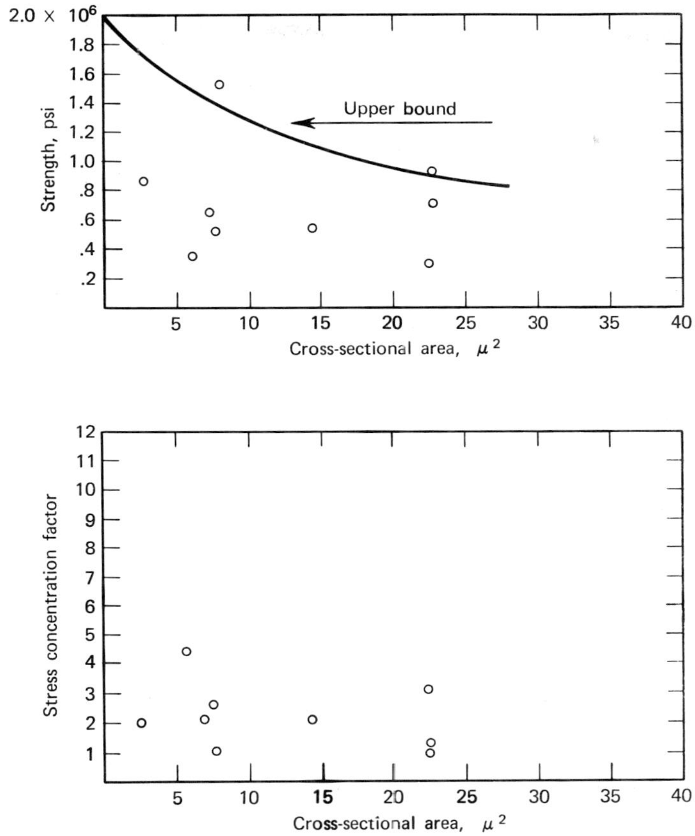

Figure 6.35 Strength and deduced stress concentration factor for α-Al_2O_3 whiskers [41].

A somewhat related approach of the effect of surface steps or irregularities on whisker strength has been taken by Herzog [41]. By assigning an upper bound to the experimentally determined strength values as a function of cross-sectional area, it was possible to deduce what the stress concentration factor would have to be in order to reduce the measured strength of those whiskers falling below the upper bound. These data, for Al_2O_3 and SiC whiskers, are shown in Figures 6.35 and 6.36. Stress concentration factors as high as 6 were obtained by this method, and the numerical value decreased somewhat with increasing cross-sectional area. The magnitude of the deduced stress-concentration factors are of the same order as those measured by Marsh.

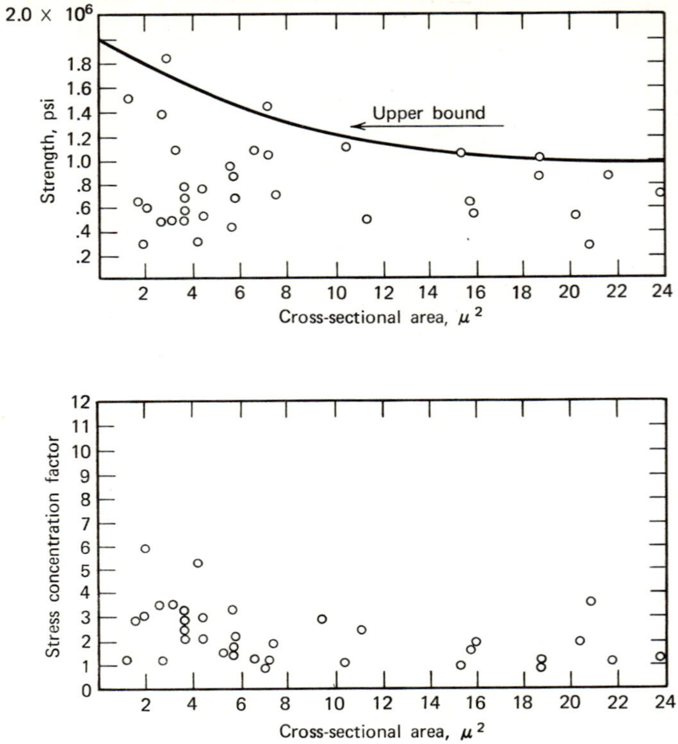

Figure 6.36 Strength and deduced stress concentration factor for SiC whiskers [41].

Some whiskers do not display the sharp growth steps shown by α-Al_2O_3 but display about the same strength. Evidently, then, other defects may be responsible for reducing whisker strength. These may include a dislocation substructure, dislocation pile-ups, microcracks, and thermally induced stresses. In general, the number and severity of these defects will increase as the whisker size increases. Whatever the type, if the whisker strength is dependent on surface area, the defects will be primarily on the whisker surface. If not, the defects must be subsurface; the strength may be governed by a dislocation substructure; or as in the case of B_4C whiskers, the severity of flaws may be independent of whisker size.

The above discussion applies to single-phase crystal whiskers. However, it is known that such crystals can also grow with surface films of crystalline or amorphous nature. On iron whiskers, iron oxide shells have been observed [31], and these shells affect the whisker modulus as discussed previously

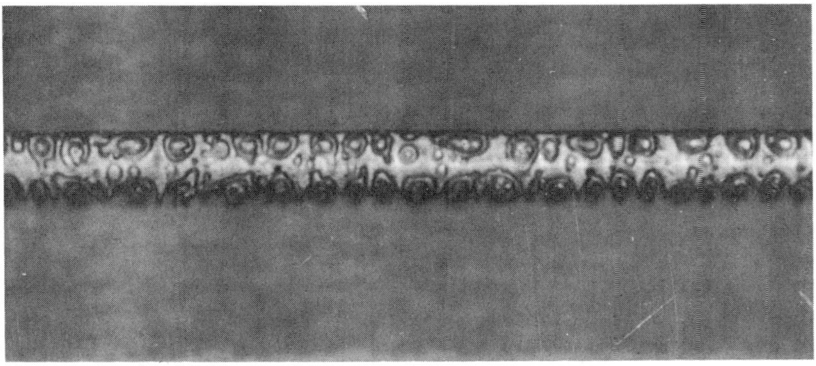

Figure 6.37 As received SiC whisker from Thermokinetic Fibers, Inc. showing glassy phase (mag. 1070X). Photographed by Mrs. M. Schueller, AFML.

(Fig. 6.22). Silicon carbide whiskers may be covered with what appears to be a glassy material which can be removed by etching with hydrofluoric acid, as shown in Figures 6.37 and 6.38. This shell and core combination represents a "composite fiber" and may produce a built-in residual stress system that could affect the measured ultimate tensile stress. This situation is possible if the shell and the core consist of two widely different materials such as a metal and an oxide. It develops fairly easily on several other ceramic whiskers, since glass formation is certainly a possibility under certain growth conditions. This area of growth is presently not very thoroughly investigated, but can be of importance in future developments. Since most whiskers will need an interface material when embedded in a metal matrix, the reinforcing fiber in this case will also be a "composite" with an enclosed stress system. Naturally, such coated whiskers and the relation of their strength to that of

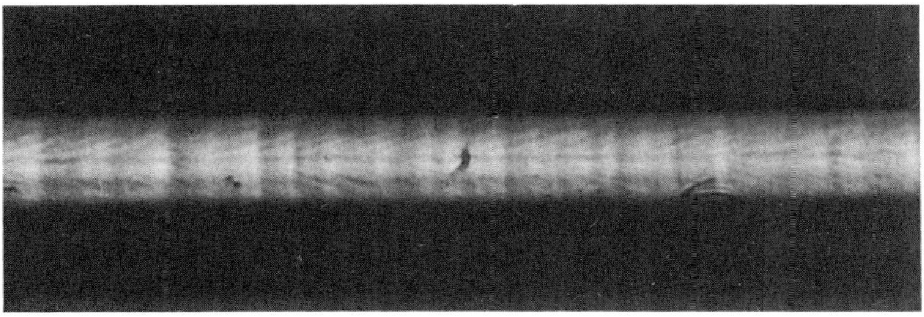

Figure 6.38 Same whisker as Figure 6.37 after etching in hydrofluoric acid (mag. 1070X). Photographed by Mrs. M. Schueller, AFML.

the "as-grown" material is of considerable interest to the manufacturer of whisker composites. It has already been shown, for example, that α-Al_2O_3 whiskers do not have their strength impaired by a thin metallic film (see Fig. 6.9).

Another factor that can affect the properties of whiskers is a taper along the whisker axis, which is common in many whiskers such as α-Al_2O_3, SiC, and B_4C. In this case, the failure will occur close to the grip holding the smaller cross section, provided no more severe irregularity is present in the gauge length. If the area is evaluated at the correct point, the stress will be correct, provided that no unsymmetrical stress distribution due to eccentric gripping causes early failure. However, the elastic modulus will be affected, as discussed earlier (Calow, et al. [14]). An analysis of the effect of a tapered whisker on modulus has also been performed by Gatti, et al. [40], and an illustration of the effect which can be obtained is shown in Figure 6.39. Depending on where the area is evaluated, the error can be large or small. Normally, the area will be evaluated at the minimum section where the fiber breaks, and for a uniform taper (either cylindrical or prismatic) the error can be quite large. This was confirmed by the work of Calow, et al.,

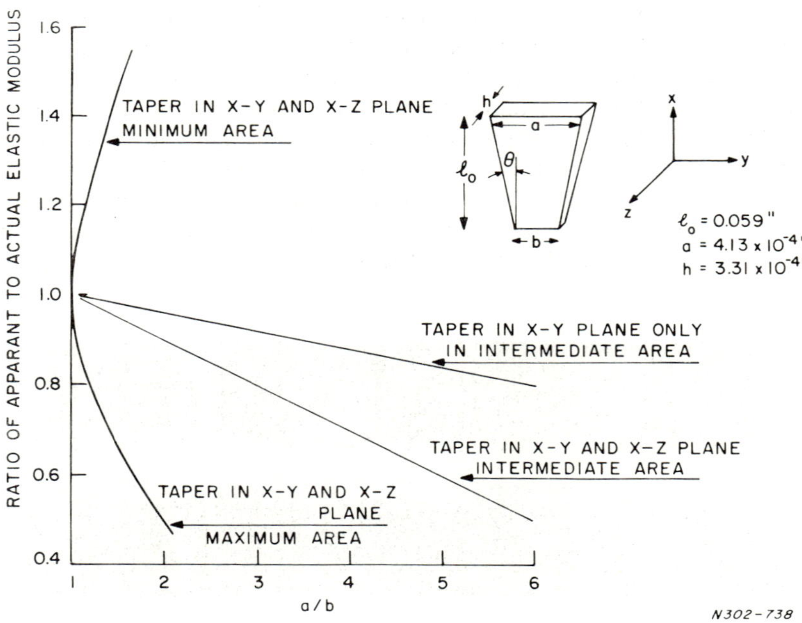

Figure 6.39 Effect of whisker taper on the apparent elastic modulus [40].

and may account for the occasional high moduli values reported in the literature. Herzog [41] has also considered the effect of surface steps or irregularities on the elastic modulus of Al_2O_3 and SiC whiskers. He concluded that the presence of such irregularities will, in general, result in an elastic modulus that is somewhat higher than that which would have been obtained on a smooth specimen of an equivalent diameter.

PREVIEW OF FUTURE DEVELOPMENT

Despite the fact that whiskers can be produced today and successfully incorporated into composites, improvements in certain areas are desirable. Although it may not be possible to immediately achieve the full whisker strength in composites, an increase in strength properties of whisker composites by a factor of two to three would represent valuable progress for design purposes.

Some features remain to be improved such as (*a*) increasing the smoothness of whisker surfaces to increase strength and reduce the strength scatter; (*b*) an increase in length of some crystal types, especially those obtained from less costly production methods; (*c*) the preparation of their surfaces with respect to wetting without the penalty of strength losses, and (*d*) the development of easily and reasonably fabricated coatings leading to stable interfaces at elevated temperatures. Data are also needed on randomly chosen whiskers rather than the carefully selected ones that have been used in the past, as it is not possible to adequately predict properties without a knowledge of the actual spread in strength data. Finally, more data are needed on elevated-temperature properties, delayed failure phenomena, creep behavior, and fatigue characteristics.

ACKNOWLEDGMENTS

The authors would like to express their appreciation to H. W. Rauch, General Electric Company for his aid in the literature review of whisker properties and to M. J. Noone, General Electric Company, and Mrs. K. Branch, Air Force Materials Laboratory, for reviewing the manuscript.

REFERENCES

[1] S. S. Brenner, "Factors Influencing the Strength of Whiskers," in *Fiber Composite Materials*, American Society for Metals, 1965, p. 11.

[2] J. R. Tinklepaugh, "Ceramic-Metal Fiber Composite Systems," in J. J. Burke et al., Ed., *Strengthening Mechanisms, Metals and Ceramics*, Syracuse University Press, New York, 1966, p. 579.

[3] R. H. Doremus, B. W. Roberts, D. Turnbull, Ed., *Growth and Perfection of Crystals*, John Wiley and Sons, New York, 1958.

[4] R. V. Coleman, *Met. Rev.*, **9**, 261 (1964).

[5] R. L. Mehan and J. M. Berry, *J. Am. Ceram. Soc.*, **49**, 169 (1966).

[6] R. D. Schile and R. L. Sierakowski, *Intern. J. Non-Linear Mech.*, **2**, 61 (1967).

[7] R. L. Mehan and E. Feingold, *J. Materials*, **2**, 239 (1967).

[8] P. D. Bayer, R. E. Cooper, *J. Materials Sciences*, **2**, 333 (1967).

[9] R. F. Regester, P. D. Gorsuch, L. Girifalco, *Mater. Res. Std.*, **7**, 203 (1967).

[10] P. J. Soltis, *J. Materials*, **2**, 300 (1967).

[11] R. H. Kelsey and R. H. Krock, *J. Materials*, **2**, 146 (1967).

[12] S. S. Brenner, *J. Appl. Phys.*, **33**, 33 (1962).

[13] J. A. Herzog, unpublished data.

[14] C. A. Calow, P. D. Bayer, G. P. R. MacCarthy, *Brit. J. Appl. Phys.*, **18**, 1715 (1967).

[15] R. L. Mehan, E. Feingold, A. Gatti, Technical Report AFML-TR-65-275, August 1965.

[16] J. B. Wachtman, W. G. Tefft, D. G. Lam, R. P. Stinchfield, *J. Res. Nat. Bur. Std.*, 64A, (1960).

[17] H. E. LaBelle and A. I. Mlavsky, *Nature*, **216**, 575 (1967).

[18] M. J. Noone, "The Preparation, Structure, and Mechanical Properties of Filamentary Forms of Silicon Carbide," Ph.D. Thesis, University of Leeds, 1967.

[19] C. C. Evans, J. E. Gordon, D. M. Marsh, Tube Investments Technical Report No. 150, 1962 (Tube Investments Research Labs, Cambridge, England): through M. J. Noone, Ph.D. Thesis, University of Leeds, 1967.

[20] H. P. Kirchner and P. Knoll, *J. Am. Ceram. Soc.*, **46**, 299 (1963).

[21] W. W. Webb, H. D. Batha, P. T. B. Shaffer, "Strength Characteristics of Whisker Crystals, Macrocrystals, and Microcrystals," in J. J. Burke et al., Ed., *Strengthening Mechanisms, Metals, and Ceramics*, Syracuse University Press, New York, 1966, p. 239.

[22] P. I. Vincent, *Brit. J. Appl. Phys.*, **13**, 578 (1962).

[23] A. Gatti, C. Mancuso, E. Feingold, R. Mehan, "The Growth and Characterization of Boron Carbide Whiskers," in *Crystal Growth*, Pergamon Press, Oxford and New York, 1967, p. 317.

[24] T. J. Davies and P. E. Evans, *Nature*, **207**, 254 (1965).

[25] R. Bacon, *J. Appl. Phys.*, **31**, 283 (1960).

[26] B. D. Brubaker, *J. Am. Ceram. Soc.*, **50**, 624 (1967).

[27] E. G. Wolff and T. D. Coskren, *J. Am. Ceram. Soc.*, **48**, 279 (1965).

[28] C. O. Hulse, *J. Am. Ceram. Soc.*, **44**, 572 (1961).

[29] C. Herring and J. K. Galt, *Phys. Rev.*, **85**, 1060 (1952).

[30] S. S. Brenner, *Science*, **128**, 569 (1958).

[31] J. A. Herzog, *Metall.*, **17**, 7 (1963).

[32] L. N. Alexandrov and A. N. Kogan, *Soviet Phys. -Solid State*, **6**, 246 (1964).

[33] V. M. Amonenko and B. M. Vasyutinskiy, AN SSSR. Institut Metallurgi Issledovaniya P Vysokoprochuym Splavan I Nitevidnym Kristallam. Doklady Na Nauchnoy Sessii 1962, volume unknown, 34 (1963), through Translation FTD-65-339/1 + 2 + 4, AD618630 (1965).

[34] V. G. Kostyuk, K. K. Zilng, A. V. Serebyakov, *Soviet Phys. -Solid State*, **5**, 2241 (1964).

[35] S. Z. Bokshtein, S. T. Kishkin, I. L. Svetlov, *Soviet Phys. -Solid State*, **4**, 1272 (1963).

[36] K. Yoshida, Y. Goto, M. Yamamoto, *J. Phys. Soc. Japan*, **21**, 825 (1966).

[37] V. N. Geminov, I. M. Kop'yev, AN SSR Institut Metallurgii Imevi A. A. Baikova Trudy, No. 10, 202 (1962): through translation FTD-65-14-/1 + 2, AD 615 875 (1965).

[38] D. M. Marsh, "Stress Concentrations at Steps on Crystal Surfaces and Their Role in Fracture," in D. G. Drucker and J. J. Gilman, Ed., *Fracture of Solids*, Interscience, New York-London, 1963, p. 119.

[39] R. L. Mehan, W. H. Sutton, and J. A. Herzog, *AIAA J.*, **4,** 1889 (1966).

[40] A. Gatti, R. Cree, E. Feingold, R. L. Mehan, "The Synthesis of Boron Carbide Filaments," Final Report, NASA Contract NASw-670, July 10, 1964.

[41] J. A. Herzog, Entwicklungen Und Zukunft in der Whiskerforschung (Development and Potential of Whiskers for Composites Presentation) at the Wissenschaftliche Gesellschaft Fur Luftfahrt (WGLR) and Deutsche Gesellschaft Fur Raketen- und Raumfahrt (DGRR), published in *Jahrbuch der Wissenschaftlichen Gesellschaft Fur Luftfahrt* (*WGLR*), 1967.

[42] Data of Philips Laboratories, Eindhoven. Courtesy of Norton Research Corporation.

7 Mechanics of Whisker-Reinforced Composites

INTRODUCTION

The extremely high strength and stiffness of the whiskers discussed in this book are most attractive for potential structural applications. It has been demonstrated that some composites that have high strength and stiffness can be fabricated from these whiskers. The problems of doing this in a consistent and economical fashion have also been emphasized. In this chapter, we treat the potential performance of whisker composites by analysis of the mechanical response of such materials to a range of loading conditions. These analyses have two major goals. The first is to establish the capability to predict the response of a given whisker composite to specified loads or displacements so that these materials can be incorporated into the design of structures. The second goal is to develop an understanding of the influence of constituent properties upon composite performance so that suitable materials can be designed for particular desired properties. This capability to "tailor" a material to its design environment is the hopeful outcome of the combined studies of both the mechanics of materials and materials engineering. The mechanics analyses are intended to supplement and to simplify the experimental studies that must be conducted, not to replace them.

This chapter is intended to convey an outline of rational methods of approach to the analysis of whisker-composite mechanical properties. Attention is focused on the basic problems of composite stiffness and strength under static loads. A complete review of the literature has not been attempted, but rather effort has been directed toward defining the nature of the important properties and approaches toward their solutions. This work draws heavily upon the authors' previous research and that of their laboratory colleagues, and reproduces portions of earlier publications. This research is presented as a review article to make this book a more complete reference for workers in the field of whisker composites. Mathematical derivations of important theoretical results are included for completeness. The principal results, however, are available in the discussion sections alone.

The major areas treated in this chapter are elastic moduli of discontinuous fiber and whisker composites, internal stress distributions under simple extensional loads, and failure models for tensile strength. Encouraging relationships between composite properties and the reinforcement properties are presented. The prognosis for composites containing discontinuous fibers appears to be favorable.

DEFINITION OF PROPERTIES

Structural applications for whisker-reinforced composites will not become practical until sufficient knowledge of the response of such materials to various loading conditions is established reliably. The structural designer requires information in terms of average stress and average strain relations. One of the principal aims of the study of the mechanics of composites is to provide these relationships between the average response of a composite and the properties, both mechanical and geometrical, of the constituents. As an example, consider the simple problem of a rod subjected to an axial tensile load when the rod is made of a whisker-composite material. The details of the stress field are extremely complex, particularly in the vicinity of the whisker ends (see Chapter 11). For design purposes, however, it is only necessary to specify such overall measures of this stress field as the relative average displacement of the rod ends and the average stress at failure. Indeed, from the designer's viewpoint this complex nonhomogeneous composite may be replaced by a homogeneous (usually anisotropic) material that has the same displacement-load relationship and the same failure loads, and so forth. We may consider this aspect of the micromechanics studies to be directed to the definition of the characteristics of this effective homogeneous material. There is a second goal of these studies, which is equally important; namely, the definition of desired changes in constituent geometry or mechanical properties to yield an improved composite material. We will direct our attention toward this phase of the work subsequently.

Returning now to the example of the composite tension rod, we can illustrate some basic concepts by considering definitions. Thus the load, P, is related to the axial stress, σ_z, by

$$P = \int_A \sigma_z \, dA \tag{1}$$

where A is the area of any cross-section perpendicular to the load direction.

If we assume that the failure surface is a plane cross-section, the failure load, $P_{\max}$, is similarly related to the axial stress at failure, σ_{zm}:

$$P_{\max} = \int_A \sigma_{zm} \, dA \tag{2}$$

Conventionally the strength of this rod in tension, σ_t, is defined by

$$\sigma_t = \frac{P_{\max}}{A} \tag{3}$$

From Eqs. 2 and 3:

$$\sigma_t = \frac{1}{A} \int_A \sigma_{zm} \, dA \tag{4}$$

Since the failure plane is generally partly in each of the two phases, we may write

$$\sigma_t = \frac{1}{A} \int_{A_f} \sigma_{zm} \, dA + \frac{1}{A} \int_{A_b} \sigma_{zm} \, dA \tag{5}$$

where A_f is the fiber or whisker area and A_b is the binder or matrix area on the failure plane.

Equation 5 may be rewritten as

$$\sigma_t = \frac{A_f}{A} \bar{\sigma}_{zm}^{(f)} + \frac{A_b}{A} \bar{\sigma}_{zm}^{(b)} \tag{6}$$

where $\bar{\sigma}_{zm}^{(f)}$ and $\bar{\sigma}_{zm}^{(b)}$ are the average axial stresses in the fiber and binder respectively at the point of composite failure.

It is evident that if all cross-sections are assumed to have the same area fraction of fibers, then Eq. 6 may be written as:

$$\sigma_t = v_f \bar{\sigma}_{zm}^{(f)} + v_b \bar{\sigma}_{zm}^{(b)} \tag{7}$$

where v_f and v_b are the fiber and binder volume fractions in the composite.

Equation 7 now begins to resemble the popular "rule of mixtures" procedure for computing composite properties. Indeed, one aim of this discussion is to place this "rule" in proper perspective. Thus, if we consider the average density, $\bar{\rho}$, of a composite, it is most reasonable to apply the rule:

$$\bar{\rho} = v_f \rho_f + v_b \rho_b \tag{8}$$

as can be verified by taking a volume integral of the weight (for a small displacement theory.) However, when this principle is applied to strength, it is generally assumed that the average fiber stress at composite failure, $\bar{\sigma}_{zm}^{(f)}$, is equal to the average strength of a large number of similar fibers individually tested to failure. It will become abundantly clear that this is an unreasonable assumption for brittle fibers, which includes whiskers, when the statistical nature of the strength of brittle fibers is discussed subsequently. Equation 7 is limited in validity by assumptions such as the one regarding the shape of the failure surface, but it is useful when the average constituent stresses at failure are known. The problem of relating these quantities to the constituent properties is treated in the section on tensile strength.

The rule of mixtures is therefore useful for certain composite properties which are simply the weighted average (weighted on a volume basis) of the constituent properties (density being the notable example). Its use is incorrect for other properties, such as strength, although the fallacy in its application is the lack of knowledge of the individual phase averages and not the overall averaging process.

The problem of determining composite stress-strain response is simplified by certain relations between body averages and surface values. For example, let us consider a composite body subjected to specified surface tractions and body forces—the tensile rod of the previous example can be treated as such a body—and compute the average stresses, $\bar{\sigma}_{ij}$, defined by

$$\bar{\sigma}_{ij} = \frac{1}{V} \int_V \sigma_{ij} \, dV \tag{9}$$

It can be shown that this yields

$$\bar{\sigma}_{ij} = \frac{1}{V} \int_S T_i x_j \, dS + \frac{1}{V} \int_V F_i x_j \, dV \tag{10}$$

where T_i = the components of the surface tractions;
F_i = the components of the body forces;
x_i = the coordinates;
V = the volume;
S = the surface area.

Notice that Eq. 10 defines the average stress components in terms of the specified surface tractions, body forces, and surface geometry without regard to the internal properties. For the simple case of no body forces and surface tractions derived from constant states of stress:

$$F_i = 0$$

$$T_i = \sigma_{ij}^o n_j$$

it follows that

$$\bar{\sigma}_{ij} = \sigma_{ij}^o \tag{11}$$

Thus, for the rod subjected to an axial stress σ_0, the volume average of the axial stress, σ_z, in the rod is equal to σ_0, regardless of the phase geometry or properties.

Similarly, we may treat the average strain components, $\bar{\epsilon}_{ij}$:

$$\bar{\epsilon}_{ij} = \frac{1}{V}\int_V \epsilon_{ij}\, dV \tag{12}$$

and show that

$$\bar{\epsilon}_{ij} = \frac{1}{V}\int_S u_i n_j\, dS \tag{13}$$

where u_i = the displacement components:
n_i = the components of the unit surface normal.

Equation 13 relates the average strain components to the surface displacements and geometry only. When the surface displacements are linear functions of the coordinates, that is,

$$u_i(S) = \epsilon^o_{ij} x_j \tag{14}$$

then it follows from Eq. 13 that

$$\bar{\epsilon}_{ij} = \epsilon^o_{ij} \tag{15}$$

This discussion demonstrates that volume averages of the stresses and strains are functions of surface values (and body forces) only. Thus when solving a boundary value problem, it is reasonable to replace the composite by a homogeneous material having the same average stress-strain response. This is discussed in the following section.

ELASTIC PROPERTIES

The elastic moduli are among the basic properties used in structural applications. Their evaluation has received much attention in the recent literature on composites. For continuous fiber composites there exist valid treatments of all the independent elastic moduli. The state of stress in an oriented whisker composite differs from that of the similar continuous fiber composites primarily in the vicinity of the fiber ends. Therefore, the elastic moduli of discontinuous fiber composites can be computed approximately by the methods developed for continuous fibers. These methods are described first. Then a treatment of the stresses in the vicinity of the fiber ends is used to determine the degree of approximation involved in using the continuous fiber results.

Methods of Evaluating Elastic Moduli

A variety of methods for computation of composite properties when the constituent properties are known have appeared in the literature. A review of these methods has been presented by Hashin [1]. The following discussion indicates the available methods and presents a recommended theory.

The direct approach to the evaluation of elastic constants is to determine the internal response in the inhomogeneous composite medium subjected to various simple boundary conditions. By obtaining appropriate body averages, the moduli are evaluated by treating the composite as an equivalent homogeneous material and relating average stresses to average strains. These boundary value problems require the complete specification of the phase geometry (e.g., the relative location of fibers over the cross-sectional plane normal to the fiber axes) and the extensive application of numerical methods. This approach has recently been utilized by Pickett [2] for a hexagonal array and by Adams and Doner [3] for a square array of circular fibers. These methods produce the desired results for a known geometry. However, they require extensive use of high-speed computers and they leave unanswered the question of the magnitude of possible variations in moduli associated with the variation in cross-sectional location of fibers.

A second approach to this problem utilizes known and regular phase geometries and gross approximations to the nature of the stress field. Thus, the materials are usually represented as various combinations of simple elements in series and in parallel with one another (e.g., Sonneborn [4] and Shaffer [5]). Discussions of this type range from simple approximations of the stiffness parallel to the fibers to transverse Young's and shear moduli obtained by assuming uniform strain or by assuming that the material can be represented by a truss or a laminate. These methods can be justified as providing working estimates only if no better methods are available. Thus at the present state of development they are not of major importance.

The third approach to the elastic constants problem is the use of variational principles to obtain bounds on the desired moduli. In this method the effective elastic moduli are expressed in terms of the strain energy, and bounds on the strain energy are found for simple, applied average stress and strain fields, thus also bounding the elastic moduli. This definition of moduli in terms of strain energy is directly equivalent to the definition in terms of average stress and strain. Hashin and Rosen [6] used strain energy methods to evaluate elastic constants as functions of constituent moduli and volume fractions. Uniaxially oriented circular fibers randomly distributed over a cross-sectional plane were considered. Such a material is transversely isotropic and is defined by five independent elastic constants. These constants can be bounded for several different types of geometries.

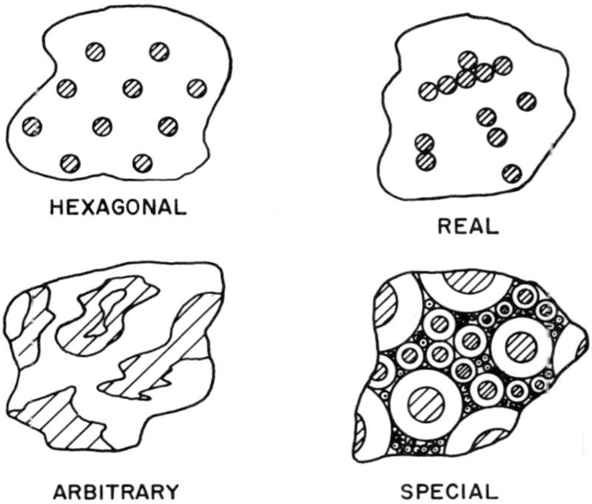

Figure 7.1 Cross-sectional geometries of various types of fibrous composites.

Several geometries of interest are shown in Figure 7.1. The hexagonal array is typical of the ordered arrays that are susceptible to rigorous bounding procedures. The contrast between such orderly arrays and the arrays that result from common manufacturing procedures is indicated by comparison of the upper two sketches in Figure 7.1. Indeed, the contiguity between the fibers shown in cross sections of real fibrous composites has motivated some interest in the possible effect of continuous or semicontinuous load transmission paths through the fiber material. The extreme of such geometric considerations is represented by the arbitrary phase geometry in which both phases are cylindrical, but a cross section perpendicular to the generator of the phase geometry interfaces is completely arbitrary and a distinction between the two materials enabling one to identify a fiber and a binder is no longer possible. Rigorous bounds have been obtained for this arbitrary geometry material by Hashin [7] and Hill [8]. The problem, however, is that the distance between such bounds is usually quite large and the use of the upper bound for application to the case of rigid fibers in relatively less rigid binder materials is questionable. That is, it appears intuitively that a continuous binder phase will result in modulus values closer to the lower bound of the arbitrary phase geometry material.

Another set of bounds exists, however, which gives extremely useful results. These bounds are presented by Hashin and Rosen [6] for the special material also shown in Figure 7.1. This material is composed of concentric circular cylinders of fiber and binder, with the fiber material as the core. The volume fractions of fiber and binder in this typical composite cylinder are equal to the respective volume fractions of the two materials in the bulk composite. These volume fractions are maintained in a set of cylinders having various diameters which are used to completely fill out the entire volume by going to an infinitesimally small fiber diameter. The bounds for this material coincide for four of the five independent elastic constants. These are the so-called random array results. For the real material, it is always possible to surround each circular fiber with a cylinder of binder material such that the fiber binder composite has the proper volume fractions. However, the outer surface of the binder will not be a circular cylinder. The results of the random array can be used as an approximation to this real material by assuming that the strain energy in this irregularly shaped composite cylinder can be approximated by the strain energy in a concentric circular cylinder having the same areas of fiber and binder. This assumption makes the random array results directly applicable to the real material behavior. However, there is some uncertainty associated with this assumption and alternate assumptions have been proposed in the form of interpolating factors between the upper and lower bounds of the arbitrary phase geometry results. It should be noted that for those cases where the bounds of the random array moduli are coincident, they also coincide with the arbitrary phase geometry bounds when the binder and fiber materials are alternately used as the core of the typical composite cylinder. This information is of value in that it establishes for those cases the fact that the arbitrary phase geometry bounds are the best possible bounds.

Various interpolation schemes have been proposed to yield a set of values for elastic constants which can be used in structural analysis. Tsai [9] has proposed that a linear interpolation between the upper and lower bounds is a proper approach. The contiguity factor represents the interpolation between the bounds. Wu [10] has proposed an interpolation factor based on physical reasoning which has been shown by Hashin [11] to be equivalent to a linear interpolation factor between the bounds. Such approaches assume that a single interpolation between the bounds will be valid for all volume fractions when dealing with a pair of constituents in a fibrous composite. Furthermore, their value is quite limited unless the interpolation factor is also insensitive to the ratio of the elastic constants of the two phases. These questions can only be answered by comparison with experimental data. Such a comparison by Rosen [12] emphasized the merits of the Hashin/Rosen composite cylinder assemblage results as a rational, convenient, and reliable analysis of elastic moduli of fibrous composites.

Results for Elastic Moduli of Uniaxial Continuous Fiber Composites

As indicated in the preceding discussion, the "composite cylinder assemblage" model of Hashin and Rosen [6] yields results that appear to be appropriate for determining design values of composite elastic moduli. The model incorporates randomness of structure and permits the derivation of simple closed-form expressions for the effective elastic moduli, as functions of the elastic properties of the constituents and of their volume fractions. These expressions are obtained by the application of variational principles which yield upper and lower bounds on the value of each elastic constant. For four of the five independent constants, the bounds coincide to yield the simple results given by Eqs. 19 through 22. For the fifth constant, the transverse shear modulus, the bounds do not coincide and the upper bound has been found to be in good agreement with experiment. The expression for this upper bound, which has recently been put into simpler form than that originally given in [6], is presented in Eq. 24.

Use of the "composite cylinder assemblage" model also permits the evaluation of stresses in the individual composite cylinder elements. These stresses appear to be no less reliable than those obtained by tedious numerical calculations.

The analysis of Ref. 6 treats hollow fiber composites. In the following discussion, only the special case of solid or zero hollowness fibers will be presented. Considering a Cartesian coordinate system with axes 1, 2, 3, the stress-strain relations for a transversely isotropic material may be written as (with 1 as the fiber direction and the 2-3 plane as the transverse plane):

$$\begin{aligned}
\bar{\sigma}_{11} &= c_{11}\bar{\epsilon}_{11} + c_{12}\bar{\epsilon}_{22} + c_{12}\bar{\epsilon}_{33} \\
\bar{\sigma}_{22} &= c_{12}\bar{\epsilon}_{11} + c_{22}\bar{\epsilon}_{22} + c_{23}\bar{\epsilon}_{33} \\
\bar{\sigma}_{33} &= c_{12}\bar{\epsilon}_{11} + c_{23}\bar{\epsilon}_{22} + c_{22}\bar{\epsilon}_{33} \\
\bar{\sigma}_{12} &= 2c_{44}\bar{\epsilon}_{12} \\
\bar{\sigma}_{13} &= 2c_{44}\bar{\epsilon}_{13} \\
\bar{\sigma}_{23} &= (c_{22} - c_{23})\bar{\epsilon}_{23}
\end{aligned} \tag{16}$$

The moduli which were evaluated may be expressed as

$$\begin{aligned}
k_{23} &= \tfrac{1}{2}(c_{22} + c_{23}) \\
G_{23} &= \tfrac{1}{2}(c_{22} - c_{23}) \\
G_{12} &= c_{44} \\
E_1 &= c_{11} - \frac{2c_{12}^2}{c_{22} + c_{23}}
\end{aligned} \tag{17}$$

and c_{11}.

From these we may obtain:

$$\begin{aligned}
\nu_{21} &= \tfrac{1}{2}\left(\frac{c_{11} - E_1}{k_{23}}\right)^{\frac{1}{2}} \\
E_2 &= \frac{4G_{23}k_{23}}{k_{23} + \psi G_{23}} \\
\nu_{23} &= \frac{k_{23} - \psi G_{23}}{k_{23} + \psi G_{23}}
\end{aligned} \tag{18}$$

where $\psi = 1 + \dfrac{4k_{23}\nu_{21}^2}{E_1}$

The equations for five independent moduli (based on the results of Refs. 6 and 8) are as follows:

Plane-strain bulk modulus, k_{23}:

$$k_{23} = \frac{k_f k_b + G_b(v_f k_f + v_b k_b)}{G_b + v_b k_f + v_f k_b} \tag{19}$$

where k_b and k_f are the plane-strain bulk moduli of the binder and fiber respectively; thus:

$$k = K + \tfrac{1}{3}G$$

In-plane shear modulus, G_{12}:

$$G_{12} = G_b \frac{G_f(1 + v_f) + G_b v_b}{G_f v_b + G_b(1 + v_f)} \tag{20}$$

Longitudinal Young's modulus, E_1:

$$E_1 = v_f E_f + v_b E_b + \frac{4v_f v_b(\nu_b - \nu_f)^2}{(v_f/k_b) + (v_b/k_f) + (1/G_b)} \tag{21}$$

For most practical purposes the last term is negligible.

Longitudinal Poisson's ratio, ν_{21}:

$$\nu_{21} = v_f \nu_f + v_b \nu_b + \frac{v_f v_b(\nu_f - \nu_b)[(1/k_b) - (1/k_f)]}{(v_f/k_b) + (v_b/k_f) + (1/G_b)} \tag{22}$$

Transverse plane shear modulus, G_{23}:
This is the only modulus for which the upper and lower bounds are not coincident.

Lower bound (from Ref. 7):

$$G_{23L} = G_b + \frac{v_f}{1/(G_f - G_b) + (k_b + 2G_b)v_b/[2G_b(k_b + G_b)]} \tag{23}$$

Upper bound (equivalent to result of Ref. 6)

$$G_{23U} = G_b \frac{(\alpha + \beta_b v_f)(1 + \rho v_f^3) - 3v_f v_b^2 \beta_b^2}{(\alpha - v_f)(1 + \rho v_f^3) - 3v_f v_b^2 \beta_b^2} \tag{24}$$

where

$$\alpha = \frac{\gamma + \beta_b}{\gamma - 1} \qquad \rho = \frac{\beta_b - \gamma\beta_f}{1 + \gamma\beta_f}$$

$$\gamma = \frac{G_f}{G_b} \qquad \beta_b = \frac{1}{3 - 4\nu_b} \qquad \beta_f = \frac{1}{3 - 4\nu_f}$$

Typical results of this analysis are shown in Figures 7.2 to 7.5.

The curves for the effective transverse Young's modulus, E_2, of glass fiber reinforced epoxy are shown in Figure 7.2. This indicates the distance between the bounds and the fact that the transverse modulus is low compared to the fiber modulus except for very high, and therefore generally unobtainable, fiber volume fractions. The fact that moduli other than the simple Young's modulus in the fiber direction are strongly influenced by the matrix modulus is shown more clearly in the curves of Figures 7.3 to 7.5 wherein the three principal moduli for a two-dimensional sheet of a unidirectional fibrous composite are plotted. Thus, while the longitudinal modulus, E_1, shown in Figure 7.3 rises linearly with the fiber modulus, both the transverse modulus, E_2 (Fig. 7.4) and the in-plane shear modulus, G_{12} (Fig. 7.5) increase with increasing fiber stiffness but become asymptotic to the finite value associated with infinitely rigid fibers.

As a result of these differing behaviors of the various elastic moduli, it is clear that the ratios of any two moduli can depart significantly from the similar ratio for an isotropic material. Thus the ratio of axial Young's modulus, E_1, to axial shear modulus, G_{12}, for the material of Figures 7.3 through 7.5, with a fiber modulus of 50×10^6 psi, is approximately 20. For the constituents separately, the modulus ratio is 2.6. This extreme change in relative properties presents both a problem and a challenge to the structural designer. A frequent outcome is the use of multidirectional fiber orientation in the composite material.

Elastic Constants for Multidirectional Stiffening

For the practical composite material, with whiskers oriented in more than one direction, as in a laminate, or for composites whose constituent whiskers are not all uniaxially aligned, the results of the previous section can be modified provided the whiskers are straight and have some known distribution of orientations. To do this, whisker alignment is idealized to the extent that all whiskers are considered parallel to a given plane. The composite can then be treated by the methods developed for anisotropic laminates, in which each layer contains a uniform array of parallel fibers oriented in a prescribed direction. The elastic behavior of such laminates has been extensively studied (e.g., see Dong, et al. [13] and Reissner and Stavsky

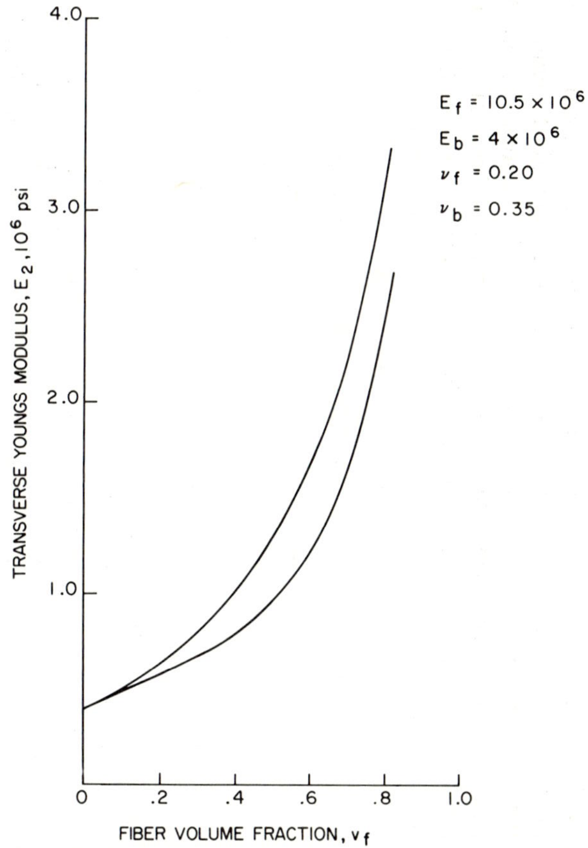

Figure 7.2 Bounds on the effective transverse Young's modulus of unidirectional glass-reinforced plastic composites.

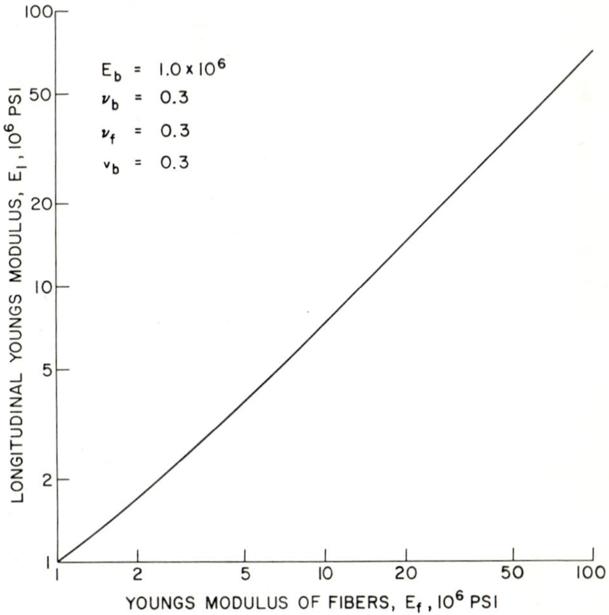

Figure 7.3 Effective longitudinal Young's modulus of a fibrous composite as a function of the Young's modulus of the fibers.

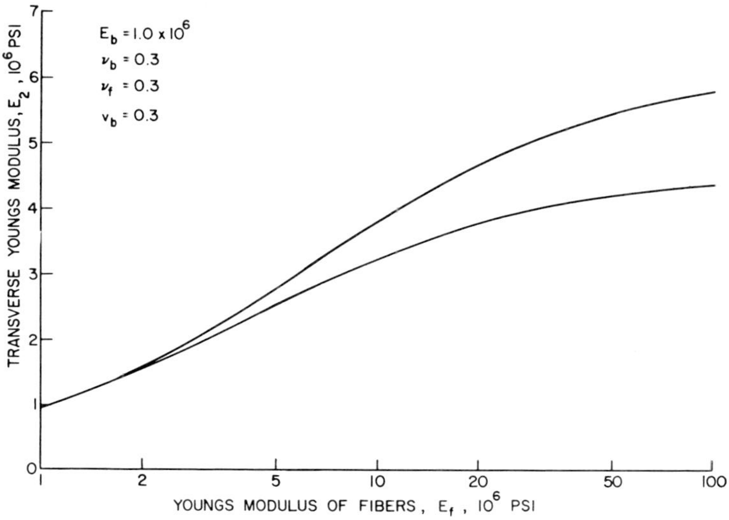

Figure 7.4 Bounds on the effective transverse Young's modulus of a fibrous composite as a function of the Young's modulus of the fibers.

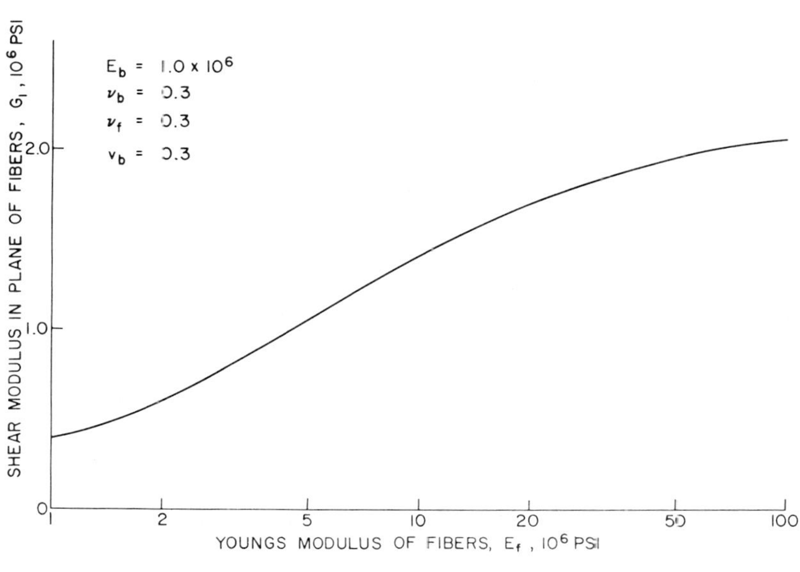

Figure 7.5 Effective shear modulus of a fibrous composite in the plane of the fibers as a function of the Young's modulus of the fibers.

[14]). The following analysis is similar to that of [13], but it includes, and emphasizes, the computation of the laminate elastic moduli.

For a thin plate or shell the stress-strain analysis will be a two-dimensional one. Thus each layer of the laminate will be described by the following stress-strain relations:

$$\begin{Bmatrix} \sigma_1 \\ \sigma_2 \\ \sigma_3 \end{Bmatrix} = \begin{bmatrix} C_{11} & C_{12} & C_{13} \\ C_{21} & C_{22} & C_{23} \\ C_{31} & C_{32} & C_{33} \end{bmatrix} \begin{Bmatrix} \epsilon_1 \\ \epsilon_2 \\ \epsilon_3 \end{Bmatrix} \tag{25}$$

In terms of the previous notation for orientation of principal elastic axes of the uniaxial fiber composite:

$$\begin{aligned} \sigma_1 &= \sigma_{11} & \epsilon_1 &= \epsilon_{11} \\ \sigma_2 &= \sigma_{22} & \epsilon_2 &= \epsilon_{22} \\ \sigma_3 &= \sigma_{12} & \epsilon_3 &= \epsilon_{12} = \tfrac{1}{2}\gamma_{12} \end{aligned} \tag{26}$$

also:

$$\begin{aligned} C_{11} &= \frac{E_1}{1 - \nu_{21}\nu_{12}} \\ C_{22} &= \frac{E_2}{1 - \nu_{21}\nu_{12}} \\ C_{33} &= 2G_{12} \\ C_{12} &= C_{21} = \frac{\nu_{21}E_2}{1 - \nu_{12}\nu_{21}} \\ C_{13} &= C_{23} = C_{31} = C_{32} = 0 \end{aligned} \tag{27}$$

and

$$\frac{\nu_{21}}{E_1} = \frac{\nu_{12}}{E_2} \tag{28}$$

Note that the tensor components of both stress and strain are being used. The transformation to components with respect to Cartesian axes, x_i, rotated at angle θ, in the plane of the plate, with respect to initial axes, $\bar{x}_i$, is thus

$$\begin{aligned} \sigma_i &= T_{ij}\bar{\sigma}_j \\ \epsilon_i &= T_{ij}\bar{\epsilon}_j \end{aligned} \tag{29}$$

where

$$[T_{ij}] = [T] = \begin{bmatrix} \cos^2\theta & \sin^2\theta & 2\sin\theta\cos\theta \\ \sin^2\theta & \cos^2\theta & -2\sin\theta\cos\theta \\ -\sin\theta\cos\theta & \sin\theta\cos\theta & \cos^2\theta - \sin^2\theta \end{bmatrix}$$

If the stress-strain relations with respect to the $\bar{x}_i$ coordinates are written as

$$\bar{\sigma}_i = \bar{C}_{ij}\bar{\epsilon}_j \tag{30}$$

Then, from Eqs. 29, 30, and 25 rewritten as

$$\sigma_i = C_{ij}\epsilon_j \tag{31}$$

we obtain

$$\bar{C}_{ij} = L_{im}C_{mn}T_{nj} \tag{32}$$

where

$$[L] = [T]^{-1} \tag{33}$$

is the inverse of the matrix $[T]$.

For simplicity we consider a laminate composed of a large number of orthotropic layers. For each fiber direction we consider principal axes x_1 and x_2 with the former oriented at an angle θ_k with respect to the laminate longitudinal axis x_L. The composite has elastic properties, C_{ij}, with respect to the lamina principal axes as computed in the previous section. The fraction of material with this orientation is denoted by t_k. The stress and moment resultants may now be defined in terms of the strains and elastic constants. Thus

$$N_i = \int_{-t/2}^{t/2} \bar{\sigma}_i \, dz \tag{34}$$

where the barred quantities denote components with respect to the laminate axes, x_L and x_T. Therefore,

$$\begin{aligned} N_i &= \int_{-t/2}^{t/2} \bar{C}_{ij}\bar{\epsilon}_j \, dz \\ N_i &= a_{ij}\epsilon_j \end{aligned} \tag{35}$$

where $a_{ij} = \int_{-t/2}^{t/2} \bar{C}_{ij} \, dz$

$$\text{or} \qquad a_{ij} = \sum_{k=1}^{n} \bar{C}_{ij}^{(k)}(h_k - h_{k-1}) \tag{36}$$

$$\text{with} \qquad h_k = -\frac{t}{2} + \sum_{i=1}^{k} t_i \qquad k = 1, 2 \ldots n \tag{37}$$

$$\text{with} \qquad h_o \equiv -\frac{t}{2}$$

The desired laminate stress-strain relations, given by Eqs. 35 and 36 may be written in matrix form as:

$$\{N\} = [a]\{\epsilon\} \tag{38}$$

The matrix $[a]$ is symmetric since the elastic constants C_{ij} are symmetric. Thus, these equations may be inverted to yield:

$$\{\epsilon\} = [A]\{N\} \tag{39}$$

where $[A] = [a]^{-1}$. The notation $[a^{-1}]$ indicates the inverse of the matrix $[a]$.

Elastic constants may be defined as the ratio of stress to strain (e.g., Young's modulus), or of strain to strain (e.g., Poisson's ratio) when a single non-zero stress resultant is applied. Thus, it follows that we may write:

$$\begin{aligned} E_L &= \frac{1}{tA_{11}} \\ E_T &= \frac{1}{tA_{22}} \\ \nu_{TL} &= -\frac{A_{12}}{A_{11}} \\ \nu_{LT} &= -\frac{A_{12}}{A_{22}} \\ G_{LT} &= \frac{1}{2tA_{33}} \end{aligned} \tag{40}$$

This procedure is particularly applicable to composite tapes in which the fibers are all essentially parallel to the same plane.

For three-dimensional reinforced composites, an approximate method by Dow [15] appears suitable. This approach relates to the one followed by Dow, et al. [16] to determine the properties of integrally stiffened plates. Therein the reinforcement provided by integral stiffening is evaluated as fully effective in the direction of the stiffening but reduced in stretching effectiveness transverse to the stiffening by a factor β. Similarly the transverse shearing effectiveness is evaluated as reduced by another factor β'. With the longitudinal and transverse effectivenesses established, the remainder of the analysis is a straightforward elasticity problem of trigonometric resolution and summation to yield the desired elastic constants.

The results are illustrated by the following typical equation:

$$\begin{aligned} E_{T_o} = E_b\Bigg[& \frac{1}{1-\nu^2} - \beta_{-o}(1 - v_b) + \beta_o(1 - v_b)\frac{E_f}{E_b} \\ & - \frac{[\nu/(1-\nu^2)]^2}{1/(1-\nu^2) - (1 - v_b) + (1 - v_b)(E_f/E_b)}\Bigg] \end{aligned} \tag{41}$$

where for simplicity the Poisson's ratios of filaments and binder have been assumed equal.

In these equations:

E_{T_0} = stiffness transverse to circular fibers in binder
E_b = Young's modulus of binder material
E_f = Young's modulus of fiber material
ν = Poisson's ratio
β_{-0} = transverse effectiveness lost by making holes in binder
β_o = transverse effectiveness of the filament
v_b = volume fraction of binder material

Extension of this type of analysis to three dimensions and its application to filaments and binders of different Poisson's ratios is discussed by Dow [15]. This approach is an alternate to the treatment of a general fibrous composite as a three-dimensional array of composite elements which can be approximated by circular cylinders. In the latter treatment the results of [6] could be used to estimate the strain energy in each element and, therefore, to obtain bounds on the composite moduli. This work is presently in a developmental state and a perusal of [15] is recommended.

Effect of Finite Length Fibers

The stress fields used in the preceding analyses of elastic moduli are valid when the reinforcing fibers are continuous. For discontinuous fiber composites, they are not applicable in the vicinity of the fiber ends. Thus, for application to whisker composites, it is desired to obtain an approximate solution for the end stress field and to determine the influence of this perturbed field upon the effective properties.

The method of approach is to consider the model of Figure 7.6 with a uniform stress applied to the end faces, and to determine the average longitudinal strain. The problem cannot be solved exactly and a shear lag type analysis is utilized herein. The method is not aimed at the determination of an exact stress distribution, but rather at an approximation of the contribution of the internal shear stresses to the longitudinal extension. The need for refinement of some of the initial assumptions will be largely determined by the results obtained with this model.

Load is applied parallel to the fiber direction. The fiber is assumed to be of circular cross section and to carry only extensional stresses, while the matrix transmits only shear stresses. The material surrounding this typical fiber-matrix element is assumed to have uniform characteristics which are those to be evaluated. This "average" material will have the effective properties of the composite. No stress is transmitted axially from

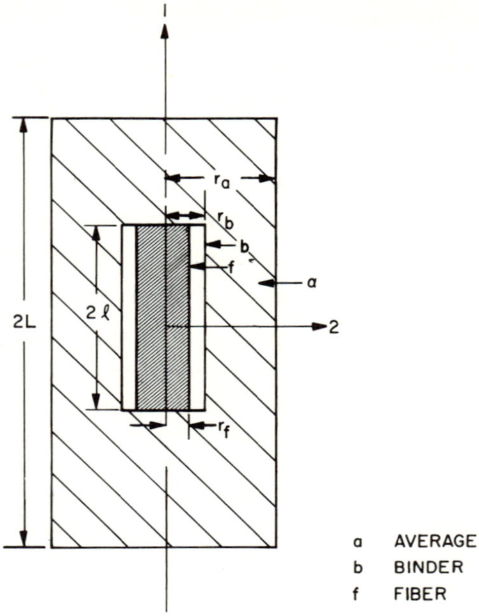

Figure 7.6 Model for evaluation of stresses in the vicinity of a whisker end.

the end fiber to the average material. Shear stresses in the average material are considered to decay in a negligible distance from the inclusion interface.

For equilibrium of a fiber element in the axial direction:

$$\tau + \frac{r_f}{2}\frac{d\sigma_f}{dz} = 0 \tag{42}$$

where τ = shear stress in matrix material

σ_f = axial stress in fiber

For equilibrium of the composite in the axial direction:

$$\left(\frac{r_f^2}{r_a^2}\right)\sigma_f + \left(\frac{r_a^2 - r_b^2}{r_a^2}\right)\sigma_a = \bar{\sigma} \tag{43}$$

where σ_a = axial stress in average material

$\bar{\sigma}$ = applied axial stress

The axial displacements in the fiber, u_f, and in the average material, u_a, define the binder shear strain, γ, as follows:

$$u_a - u_f = (r_b - r_f)\gamma \tag{44}$$

Differentiating Eq. 44 with respect to z, and using Hooke's law, yields

$$\frac{\sigma_a}{E_a} - \frac{\sigma_f}{E_f} = \left(\frac{r_b - r_f}{G_b}\right)\frac{d\tau}{dz}$$

Differentiating again, yields

$$\frac{1}{E_a}\frac{d\sigma_a}{dz} - \frac{1}{E_f}\frac{d\sigma_f}{dz} = \left(\frac{r_b - r_f}{G_b}\right)\frac{d^2\tau}{dz^2} \tag{45}$$

where E_a = effective Young's modulus of the composite
E_f = Young's modulus of the fiber
G_b = shear modulus of the binder

Differentiating Eq. 43 and substituting the result and Eq. 42 into Eq. 45 yields

$$\frac{d^2\tau}{dz^2} - \beta^2\tau = 0 \tag{46}$$

where

$$\beta^2 = \frac{2G_b}{E_f(r_b - r_f)r_f}\left[1 + \frac{E_f}{E_a}\frac{r_f^2}{r_a^2 - r_b^2}\right]$$

The solution to Eq. 46 is of the form:

$$\tau = A \sinh \beta z + B \cosh \beta z$$

The boundary conditions are

$$\tau(0) = 0$$

$$\sigma_f(l) = 0$$

Hence,

$$B = 0$$

$$A = \frac{G_b\bar{\sigma}r_a^2}{\beta E_a(r_b - r_f)(r_a^2 - r_b^2)\cosh \beta l}$$

and

$$\tau = \frac{G_b\bar{\sigma}r_a^2 \sinh \beta z}{\beta E_a(r_b - r_f)(r_a^2 - r_b^2)\cosh \beta l} \tag{47}$$

From Eqs. 42 and 47:

$$\sigma_f = -\frac{\bar{\sigma}r_a^2 E_f}{[E_a(r_a^2 - r_b^2) + E_f r_f^2]}\left(\frac{\cosh \beta z}{\cosh \beta l} - 1\right) \tag{48}$$

From Eqs. 43 and 48:

$$\sigma_a = \frac{\bar{\sigma}r_a^2}{r_a^2 - r_b^2}\left\{1 + \frac{(\cosh \beta z/\cosh \beta l - 1)}{(E_a/E_f)[(r_a^2 - r_b^2)/(r_f^2)] + 1}\right\} \tag{49}$$

With the stress distribution established, the displacements can be determined and the average strain will be defined. From Hookes' Law, and the strain-displacement relation for uniaxial stress:

$$\frac{du_a}{dz} = \frac{\sigma_a}{E_a} \tag{50}$$

The average strain, $\bar{\epsilon}$, need only be considered over the length of the fiber since deviations from average strain are considered small over the end region; thus,

$$\bar{\epsilon} = \frac{1}{l}\,u_a(l) \tag{51}$$

The effective modulus, E_a, is given by

$$E_a = \frac{\bar{\sigma}}{\bar{\epsilon}} \tag{52}$$

From Eqs. 49 through 52 the following equation for the effective modulus is obtained:

$$v_i\left[1 + \left(\frac{1 - v_i}{v_i}\right)\frac{E_a}{v_f E_f}\right] + \frac{\tanh \beta l}{\beta l} = 1 \tag{53}$$

$$\text{where } (\beta l)^2 = \frac{2G_b}{E_f}\left(\frac{l}{r_f}\right)^2\left(\frac{v_f^{\frac{1}{2}}}{1 - v_f^{\frac{1}{2}}}\right)\left[1 + \frac{E_f}{E_a}\left(\frac{v_f v_i}{1 - v_i}\right)\right] \tag{54}$$

v_i = volume fraction of inclusion relative to model $= (r_b/r_a)^2$

v_f = volume fraction of fiber relative to inclusion $= (r_f/r_b)^2$

For a meaningful model the volume fraction of the inclusion is small. Thus, the modulus is found by setting $v_i = 0$ in Eqs. 53 and 54:

$$\frac{1}{v_f}\left(\frac{E_a}{E_f}\right) + \frac{\tanh \beta l}{\beta l} = 1 \tag{55}$$

or

$$E_a = v_f E_f\left[1 - \frac{\tanh \beta l}{\beta l}\right]$$

$$\text{where } (\beta l)^2 = \frac{2G_b}{E_f}\left(\frac{l}{r_f}\right)^2\left(\frac{v_f^{\frac{1}{2}}}{1 - v_f^{\frac{1}{2}}}\right) \tag{56}$$

The modulus, E_a, is plotted in Figure 7.7 for two values of l/r_f and for a fiber-to-binder modulus ratio corresponding to those of glass filament-epoxy composites. The weighted average of the constituent moduli is used as the normalizing factor, so that the reduction of modulus due to shear deformations, and thus due to fiber discontinuities, is indicated by a modulus ratio

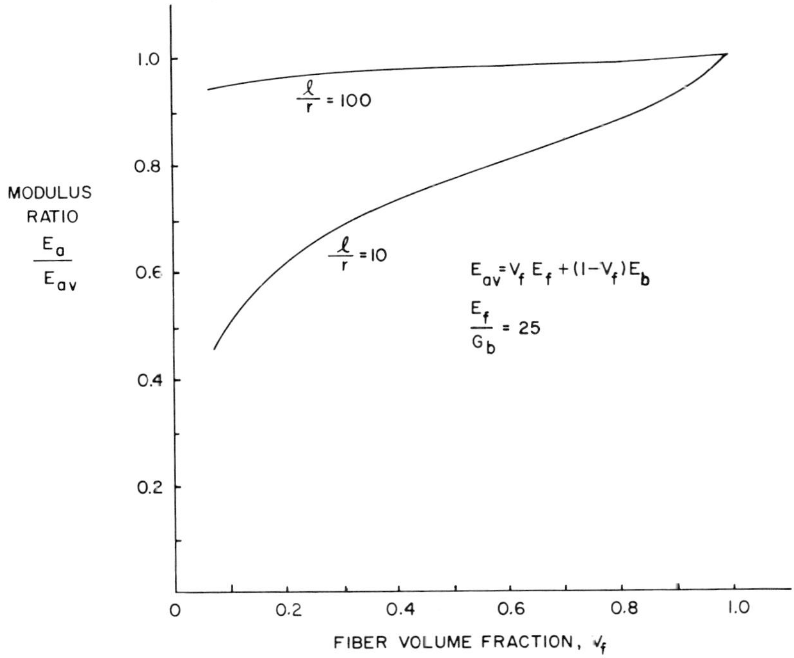

Figure 7.7 Effect of finite length on axial Young's modulus of a unidirectional whisker composite.

less than unity. (In this case, neglect of extensional stresses in the binder reduces the weighted average, E_{av}, to $E_{av} = v_f E_f$.) The modulus ratio is seen to be very close to unity for even a very small aspect ratio of $l/r_f = 100$. This corresponds to a length of only 0.02 in. for conventional glass filaments used in filament wound construction. Typical whisker aspect ratios are an order of magnitude greater than this value. The curves for $l/r_f = 10$ are shown to indicate the aspect ratios necessary to provide a significant deviation from the modulus of a continuous filament composite.

INTERNAL STRESSES

The evaluation of the details of the stress distribution in a fibrous composite can be important for the study of failure mechanisms. Overall inelastic response or failure may be studied without knowledge of the local stresses by the use of limit analysis theorems (for example, Rosen and Shu [17] and

Shu [18]). However, models such as that of the statistical tensile failure study of Rosen [19] do serve to emphasize the usefulness of a knowledge of the stress distribution, particularly in the vicinity of initial small internal fractures and, for discontinuous fiber composites, in the vicinity of fiber ends. This attempt to relate failure processes to constituent characteristics is based on the hypothesis that, given a complete set of constitutive relations and failure levels for a uniaxial fibrous composite, it will be possible to determine the behavior of a laminate of layers of uniaxial composites and thereby to have relations defining material behavior which are suitable for complete structural analysis. In this section, the approaches used and results obtained from various theoretical studies of the internal stress distribution of the heterogeneous material will be discussed, and comparisons made with the results of photoelastic analyses. Photoelastic results are also shown to be useful with respect to the determination of various effects which have not been studied analytically. The application of these studies to the definition of failure mechanisms and the establishment of rational failure criteria will be treated subsequently.

Theoretical Treatments

Here, as in the case of the elastic constants, a complete solution of the stress problem is a complex undertaking; hence, the literature consists of a limited number of exact and approximate solutions based on idealized constituent geometry. Numerical methods for determining the elastic stress distribution when the fiber array is a regular one have been discussed in the previous section. Further work is required to obtain the influence of non-uniform fiber spacing upon the stress distribution. These results would be of principal value for consideration of the failure mechanisms associated with loads applied in a plane transverse to the fiber axes. For loads applied parallel to the fiber axes, on the other hand, failure is generally considered to result from either an accumulation of fiber fractures, or excessive matrix or interface stresses. For discontinuous fiber composites, stress concentrations are built up in the vicinity of fiber ends, as well as around fiber breaks, during loading. Thus, the stress analysis of interest for the discussion of failure mechanisms for loading in the fiber direction is the one associated with the perturbed stress field in the vicinity of a fiber end or break. Knowledge of this stress field also sheds considerable light on the nature and effectiveness of load transmission between fiber and matrix in these regions. The stresses in the vicinity of such a discontinuity do not lend themselves readily to analysis. This has resulted in a variety of approximate treatments.

A number of idealized models postulate that the internal stress field may be simulated by considering a concentric cylinder model in which the core represents a single fiber. The works of Cox [20] and Dow [21] appear to be the initial treatment of this problem. In both of these works, the fiber is surrounded by the matrix material and is assumed to undergo negligible shear strains compared to those in the matrix. Cox [20] calculates interfacial shear stresses on the basis of the stress being proportional to the difference in displacements calculated at the interface and those calculated at the same point in the absence of the fiber. Dow [21] utilizes the shear lag approximation that extensional stresses in the binder are negligible with respect to those in the fiber. The shear strains in the matrix are also taken to vary in a prescribed way. In the previous section this work was extended to a three-concentric cylinder model (see Rosen [19]) where the third cylinder attempts to represent an average effect of the surrounding material. These models generally retain some realism with regard to the constituent geometry but involve gross approximations of the stress field.

An alternative approach utilized by Sadowsky [22] considers an exact elasticity analysis for fiber geometries that approximate the true configuration. Thus, fibers having ends of a paraboloidal shape which are perfectly bonded to the surrounding matrix material have been studied. The results of these studies, however, indicate that a large portion of the load carried in the fiber is transmitted to the matrix by stress components normal to the interface and localized at the end of the fiber. Thus, these results are perhaps most appropriate for discontinuous fibers prior to an interface failure. The influence of interface shear stresses appears to be better represented in the shear lag type models. Sadowsky has also presented a two-dimensional analysis for rigid fibers with alternate fibers discontinuous [23]. A two-dimensional model has also been utilized by Hedgepeth [24] for extensional fibers using the shear lag analysis assumptions to approximate the stress field in the matrix. In a general way, all of the studies described above support the conclusion that the load is transferred around a discontinuity in a fiber, in a very short distance by relatively high stresses. However, the details of the load transmission, which is the important consideration in a failure mechanism analysis, can be quite different, as will be shown by a comparison of the results of References 19, 21, and 24.

As a basis for comparison, expressions are given for the extensional stress in the fiber and the shear stress at the fiber-matrix interface as functions of the distance from the fiber midpoint. For the sake of consistency and ease of comparison, the nomenclature used in each of these papers has been altered somewhat. For all three treatments, the constituent materials are assumed to be elastic. The fiber stress distributions, assumed uniform across any given cross section, are all of the form:

$$\sigma_f(z) = \sigma_{fo}\left[1 - \frac{\cosh K(z/d_f)}{\cosh K(l/d_f)}\right] \tag{57}$$

where σ_{fo} = stress in infinitely long fiber (for fibers of sufficiently large aspect ratio the stress near the center of the fiber approaches this value)

z = distance measured from center of fiber

d_f = diameter of fiber

l = one half of fiber length

For equilibrium of a fiber element of infinitesimal length and circular cross section, the interface shear stress is found to be related to the fiber

TABLE 7.1 WHISKER END STRESSES

Author	K^2
Dow [21]	$\dfrac{8\sqrt{2}(G_b/E_f)\{1 + [v_f/(1 - v_f)](E_f/E_b)\}}{(\sqrt{2} - 1)(G_b/G_f) + [(1 - v_f)/v_f]^{\frac{1}{2}} - \sqrt{2}}$
Hedgepeth [24]	$\pi^2 \dfrac{G_b}{E_f} \dfrac{v_f}{1 - v_f}$
Rosen [19]	$8 \dfrac{G_b}{E_f} \dfrac{v_f^{\frac{1}{2}}}{1 - v_f^{\frac{1}{2}}}$

where G = shear modulus

E = Young's modulus

v_f = volume fraction of fiber relative to composite

b, f = subscripts referring to binder and fiber, respectively.

stress by

$$\tau = -\frac{d_f}{4}\frac{d\sigma_f}{dz} \tag{58}$$

Substituting Eq. 57 into Eq. 58:

$$\tau(z) = \tfrac{1}{4}\sigma_{fo}K\frac{\sinh K(z/d_f)}{\cosh K(l/d_f)} \tag{59}$$

It is thus seen that as the fiber stress is relaxed near the fiber end, the mechanism of stress transfer from fiber to matrix results in a buildup of the shear stress.

Expressions for K^2, which can be considered as the square of four times the maximum normalized shear stress [i.e., $K = 4(\tau_{\max}/\sigma_{fo})$], are given in Table 7.1 for the three cases considered. These treatments have been chosen for comparison in order to demonstrate the frequent disparity of results

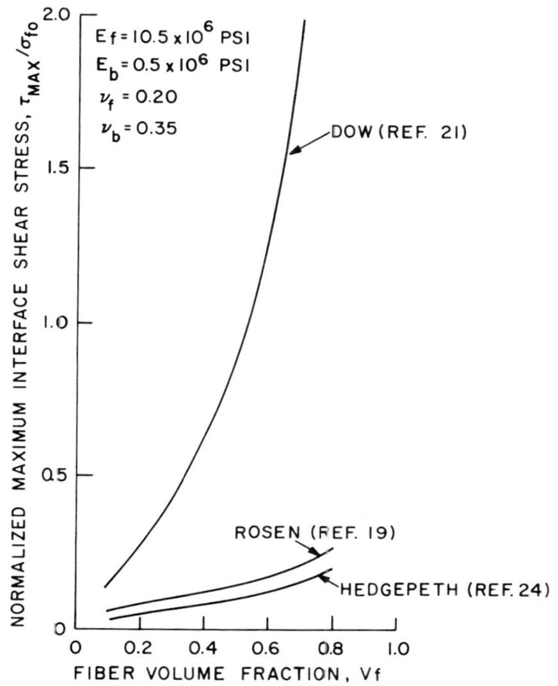

Figure 7.8 Comparison of analyses of influence of whisker volume fraction upon interface shear stress at fiber end.

obtained from theoretical models which are, to some degree, quite similar. For each of these cases, for example, both fiber and matrix are taken to be elastic, and a shear lag type analysis is used to simulate the stress transfer mechanism.

For illustrative purposes, the variation of the maximum interface shear stress—an excessive value of which may be a dominant failure mode—with fiber volume fraction and with the elastic constants of the constituent materials are plotted in Figures 7.8 and 7.9, respectively. It is seen from both these figures that the results obtained from the models of Rosen [19] and Hedgepeth [24] are quite similar, both in trends and magnitudes, even though the former model is circularly symmetric and the latter essentially two-dimensional (planar). Referring now to the results of Dow [21], in which an axisymmetric configuration is also used, the magnitudes of the maximum shear stresses are seen to be substantially in excess of those of the other two treatments, for reasonable volume fractions and constituent

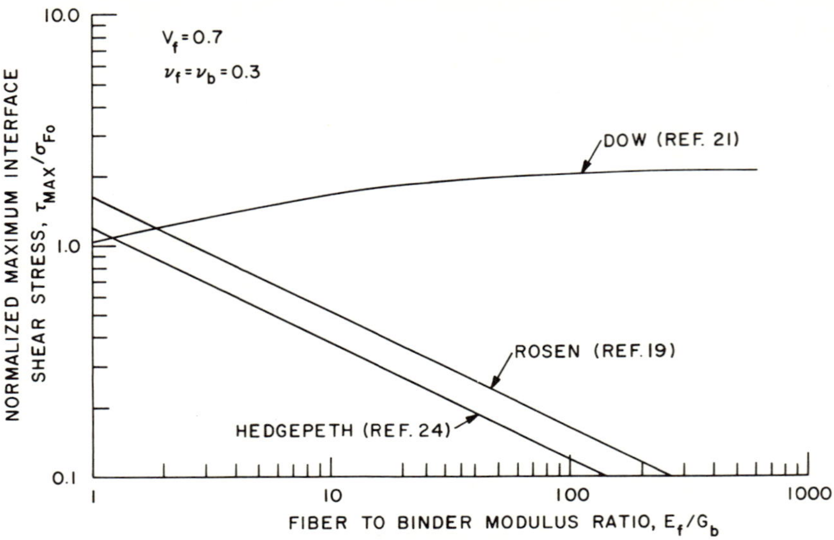

Figure 7.9 Comparison of analyses of influence of whisker and binder stiffness upon interface shear stress at fiber end.

moduli. The variations of shear stress with increasing fiber-to-matrix modulus ratio for a fixed fiber volume fraction are even of different sign.

The dilemma is thus apparent with respect to the choice of an existing theoretical model—or, perhaps, with respect to whether or not a new model should be formulated—to adequately simulate the perturbed stress field in the vicinity of a fiber discontinuity. Any tensile failure mechanism for a uniaxial composite loaded in a direction parallel to the fibers must account in some manner for this phenomenon. To weigh the merits of the various theoretical models that have been proposed, it appears that a well-conceived set of experiments would be a desirable, if not a necessary, contribution.

Prior to a discussion of experimental treatments of the end stress problem, it is of interest to note the effects of plastic flow in the matrix on the end stresses. This is of special importance when considering metal matrix composites in which potentially high interface and matrix shear stresses near a fiber discontinuity may result in yielding.

Kelly and Tyson [25] considered fiber stresses near a fiber end for a composite subject to average stresses of sufficient magnitude to result in the entire matrix being in a plastic state of stress. Thus, the flow stress of the matrix in shear (or the fiber-matrix interface, if its limiting shear stress is smaller) cannot be exceeded at any surface of the interface. As expected, these results show that the existence of a limiting value of shear stress gives rise to the decrease of the fiber stress level, from its maximum value at the middle of the fiber to zero at the end, occurring over a longer distance than for a corresponding composite with an elastic matrix. The quantity l_c, which is the perturbed fiber length, is given in [25] as

$$\frac{l_c}{d} = \frac{\sigma_f}{2\tau} \tag{60}$$

where d is the fiber diameter, σ_f the unperturbed fiber stress, and τ the matrix or interface yield stress, whichever is less. Rosen [26] performed a similar analysis for a matrix behaving as an elastic, perfectly plastic solid, with a yield stress equal to 5% of the fiber tensile strength. This analysis also demonstrated that the limiting matrix stress results in a less effective stress transfer mechanism at the fiber end.

Thus for a brittle fiber-metal matrix composite, in which the mechanism of crack propagation may be arrested due to the ductility of the matrix, a statistical mode of failure may become dominant. The magnitude of the region in which the stress field is perturbed then becomes an important parameter, as will be shown in a subsequent section. Increased ductility of the binder is therefore advantageous in one way and detrimental in another. This is another example of the necessity of obtaining a reliable estimate of the stress field around a fiber discontinuity when examining failure characteristics of the composite.

Experimental Treatments

Theoretical predictions of stresses in the fiber-matrix microstructure are based on gross idealizations of one form or another. This results in the obvious necessity of requiring some degree of experimental verification. The macroscopic, or average, properties of uniaxial composites or composite laminates are determined experimentally by use of tests that relate gross, or average, response to the overall, uniform loading applied to the specimen. When considering the microstructure, however, knowledge of overall response is of little value, and an experimental procedure capable of determining the internal stresses is required. For discontinuous fiber composites, in particular, a determination of stress concentrations induced in the vicinity of fiber ends is especially important. The techniques of photoelasticity are extremely useful in this regard. They not only lend themselves to correlations with the results of the approximate analytical treatments, but also serve to measure the effects of certain properties, which are not readily subject to analysis by theoretical means, on the internal stress distribution. Among these properties are the shape of the fiber cross section, the nature of the fiber end condition, and the type of discontinuity existing (i.e., whether a fiber end or a fiber break). Thus, in the study of whisker-reinforced composites, supporting photoelastic studies can be of marked value. (See also Chapter 11 for some results of photoelastic studies of stresses around fiber ends.)

Photoelastic analyses of stresses around fiber discontinuities have been carried out by using test specimens, each of which consists of a single opaque fiber embedded in a birefringent matrix. Although the effects of mutual interaction between adjacent fibers in a real composite material cannot be studied by use of a model of this sort (see Schuster and Scala [27] and Chapter 11 for a discussion of interaction effects), a reasonable assessment of the nature of stress transfer through shear in the matrix around a fiber end or break—of prime importance for the study of failure mechanisms—can be expected.

Photoelastic studies are generally carried out either on specimens that can be considered three-dimensional, in which the fiber is embedded in, and completely surrounded by, a mass of birefringent material, or on two-dimensional models, in which the fiber (or the material simulating the fiber) is embedded in a planar slab of the matrix material. A distinct advantage of the former relates to its capability of using the actual fibers or whiskers that are of interest, and surrounding them by the matrix, thus well simulating both the fiber-matrix configuration in an actual composite, and the configurations used in the theoretical models cited in, References 19, 21, and 24, for example.

Correlations between the results of photoelastic analyses and those of the theoretical model of Dow [21] have been made in several cases. Schuster and Scala [28], for example, used a three-dimensional model in which a single sapphire (α-Al_2O_3) whisker of hexagonal cross section was embedded in plastic. They found that the shear stress distribution at the fiber-matrix interface in the vicinity of the whisker tips was in good agreement with that predicted by the Dow theory. High shear stresses were observed in the tip region for a length of about one-half the fiber diameter.

Stress transfer at the end face of a fiber has been investigated theoretically only for idealized—and often not very realistic—tip geometries (see, for example, Sadowsky [22] where the paraboloid-shaped fiber tip eliminates geometric discontinuities). Photoelastic studies of the effects of the end condition, however, have led to some noteworthy conclusions. In the test setup used in [28], it was possible to analyze matrix stresses as a function of the end condition only, and independent of any other geometric factors. This was done by breaking one end of a whisker so that that end would be squared, while the other end remained in its original as-formed tapered condition. It was found that the stress concentrations at the tapered end were of significantly lesser magnitude than at the square end. This phenomenon was also observed by Hill [29] who embedded a single glass filament in epoxy resin, and photoelastically observed residual stresses around the fiber due to shrinkage during curing. For this test configuration, which is also three-dimensional, the highest matrix shear stresses were observed in the vicinity of the fiber end for square-ended and round-ended fibers. These stresses were about 30% greater than the corresponding ones observed for pointed-ended fibers.

Hill [29] also made some interesting observations regarding fracture in the matrix due to high shear stress, for both the square-ended and pointed-ended fibers. For the square-ended fiber, a rather flat, saucer-shaped fracture was formed in the binder at the fiber end. With regard to the pointed-end fiber, on the other hand, fracture was initiated at the tip and propagated at about a 45° angle around the fiber so as to follow the path of the threads of a screw. No preferential direction of the screw path (i.e., left-handed or right-handed) was observed, thus indicating a brittle tensile fracture in the matrix along a principal stress direction.

Theoretical results have been correlated with photoelastic results for two-dimensional models as well. Tyson and Davies [30] tested a specimen consisted of a Dural (aluminum alloy) bar cemented in a rectangular slot cut from a sheet of resin. They found that high shear stresses resulted in the immediate vicinity of the fiber tip. It is seen in Figure 7.10 that these shear stresses are somewhat higher than those predicted by both Cox [20] and Dow [21], while at distances greater than two diameters from the fiber end, the shear stresses are in good agreement with those predicted by both

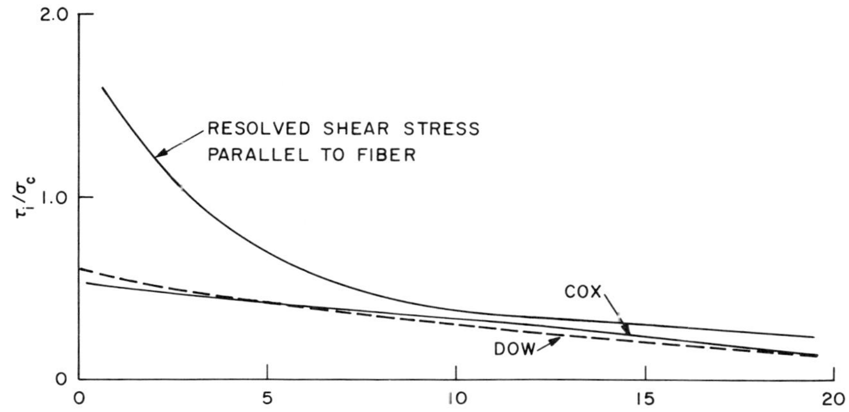

Figure 7.10 Comparison between theory and experiment for interface shear stress near whisker end.

theoretical treatments. It is interesting to note that the stress concentration factor (ratio of maximum shear stress to stress applied to composite specimen) for the 4-mm-wide, square-ended Dural bar [30] is virtually the same as that for the whisker (also square-ended) of [28], whose diameter is on the order of 0.002 in., or one-eightieth that of the bar.

Two-dimensional photoelastic models have been especially useful in the study of the effects of fiber end conditions. MacLaughlin [31] investigated end effects by embedding steel bars in epoxy. He found that, for this test model, interfacial shear stresses were somewhat greater at the ends of tapered fibers than at the ends of square-ended fibers, and the smallest interfacial shear stresses were found for round-ended fibers. This appears to be at odds with the conclusions reached in References 28 and 30.

The effect of end bonding of square-ended fibers was considered by Tyson and Davies [30] and MacLaughlin [31], who both indicated that the interfacial shear stresses in the vicinity of the fiber end are greater for fibers whose ends are intimately bonded to the matrix, than for fibers unbonded at the ends. This, of course, is to be expected since tensile load is transferred across the end surface if adequate bonding exists. Otherwise, stress transfer can only be accomplished by increased shear across the interface.

One question which is of special importance when considering failure of discontinuous fiber composites concerns the difference between the perturbed stress field at a fiber end and that at a fiber break. The various failure models which require some knowledge of the stress field in the microstructure (e.g., Friedman [32] and Kelly and Tyson [25]) make no distinction between these stress fields. Schuster and Scala [28] studied the stresses around a whisker fractured prior to final curing and observed that fiber fracture gave rise to a transverse crack in the matrix. In the vicinity of this crack, the matrix stresses near the fiber interface were seen to be greater than those stresses in the vicinity of the fiber end. The square-ended tip geometry was used for the comparison, since it should be similar to the fracture surface geometry of the break. This effect was also measured by MacLaughlin [31], wherein two steel bars were embedded end-to-end in the binder. It was found that stress concentrations are greater in the vicinity of the fiber discontinuity for this configuration than for that of a single fiber. The greater the spacing between the fibers, the less the interfacial stresses and, as the end-to-end spacing gets quite large, the shear stress distribution in the matrix approaches the distribution around the end of a single fiber.

TENSILE STRENGTH

Many of the practical fibrous composites contain brittle high-strength fibers. For these materials, the fiber strength is very nonuniform, and fiber fractures are distributed throughout the material prior to composite failure. The whisker composite fits easily into this model when it is assumed that a fiber discontinuity is equivalent to a fiber fracture occurring at an arbitrarily low stress level.

In a practical fibrous composite, loads are resisted primarily by the high-strength filamentary materials, while the matrix serves both to protect the filaments and to distribute load between and among the filaments. In the simplest of such materials, all of the fibers are parallel to a given axis, and the load is a tensile load also applied parallel to that same axis. Considerable attention has been devoted to relating the tensile strength of the composite to the properties of the constituents for this simplest case. A comprehensive review of much of this literature has been presented recently by Kelly and Davies [33]. Principal attention in that review was directed toward composites in which both phases were metals.

It has been shown (e.g., McDanels, et al. [34]) that, for ductile constituent phases, composite strength can be related to fiber strength by the so-called "rule of mixtures." This is illustrated in Figure 7.11 wherein the fiber strength, σ_f, considered to be uniform for all fibers, is equivalent to composite strength for a unit fiber volume fraction. At lower (and possible) fiber contents the strength is assumed to be

$$\sigma_t = v_f \sigma_f + v_m \sigma_m \tag{61}$$

where σ_t = composite tensile strength

σ_f = fiber strength

σ_m = matrix strength

v_f = fiber volume fraction

v_m = matrix volume fraction

A refinement of this equation is shown as the solid curve in Figure 7.11. This refinement uses the matrix stress at the fiber failure strain in place of the ultimate matrix strength, σ_m. At high volume fractions note that inasmuch as $\sigma_f \gg \sigma_m$, the rule of mixtures is adequately represented by $\sigma_t = v_f \sigma_f$ as shown by the dotted line of Figure 7.11. A modification of the rule of mixtures equation to account for the presence of discontinuous fibers as the reinforcing material will be discussed presently.

The high-strength and stiffness filaments, which make the potential of composites seem high, usually brittle failure characteristics. These fibers, of which whiskers are included, are subject to the distribution of flaws and imperfections inherent in brittle materials, and their strength can generally only be characterized statistically. A bell-shaped strength distribution curve is representative of a brittle material strength; it is indeed the type of result obtained by testing glass or boron filaments of a given gauge length or

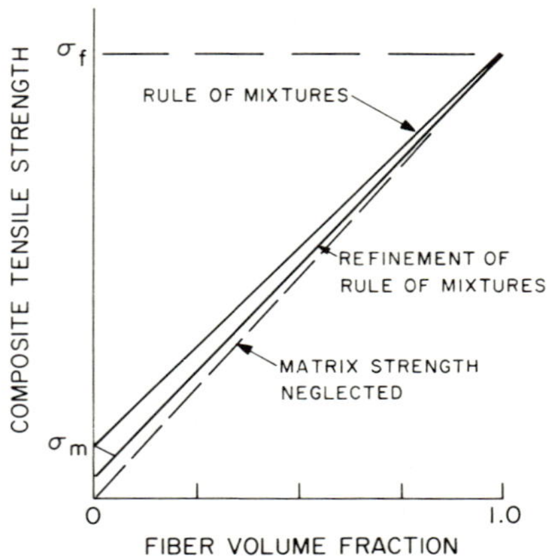

Figure 7.11 Rule of mixtures results for strength of unidirectional fibrous composites.

alumina whiskers of similar gauge lengths and cross-sectional areas. The extreme case is represented by the mathematical idealization—the Dirac delta function—for the case in which all fibers have the same strength.

Because of the statistical nature of the strength of brittle fibers, there is a dilemma as to the definition of the appropriate fiber strength and the resulting composite strength. For example, if we consider a set of brittle filaments embedded in a homogeneous matrix, and assume that at some relatively low stress level the weakest member of the fiber population fractures, conceivably for certain types of constituents this initial fracture could be followed by a crack propagation which would cause complete failure of the specimen in question. If this were so, the appropriate fiber strength to use in the rule of mixtures equation would be the strength of the weakest element in the population. Another significant possibility is that the fracture of a fiber would produce a crack which would propagate along the interface between the fiber and the matrix for the full length of the specimen. Thus the composite would then behave as if it were a bundle of unbonded fibers, and the appropriate fiber strength would be the mean strength of a bundle of fibers which is, of course, not the same as the mean strength of the individual fibers. A fracture propagation mode which accounts for stress concentrations in fibers adjacent to a broken one has recently been discussed by Zweben [35] for continuous fiber composites. Results were obtained which were reported to be in good agreement with experimental data. It is also possible that as the stress level increases, potential propagation of cracks due to the initial breaks will be inhibited by the relatively soft matrix. This would result in an accumulation of fiber fractures throughout the specimen which would eventually lead to the weakening of one cross section and the failure of the composite.

Other failure modes may also be considered, but these few examples illustrate the dilemma one faces in defining an appropriate characteristic fiber stress level which can be used to define composite strength. In view of this dilemma, the statement that the composite failure stress is governed by the rule of mixtures appears to be merely a subterfuge which changes the problem to that of defining a governing fiber stress at composite failure. This stress level is, in general, unrelated to fiber strength when considering brittle fibers.

With the aim of presenting an integrated view of the tensile failure mechanisms that can occur in fibrous composites having differing constituent properties, the following review considers fibers which are characterized by a statistical frequency distribution function of strength, and matrix materials which are homogeneous and may have inelastic properties. Accordingly, a statistical analysis of the failure process is required which is in no way affected by inelastic matrix behavior. As has been shown, plastic flow in the matrix increases the ineffective, or characteristic link, length, which is merely a parameter used in the model (see Rosen [19] and Friedman [32]). This will be discussed subsequently when the statistical failure model is set down in detail. It will be shown, however, that for a certain limiting case the statistics will become the secondary aspect of the problem and that certain failure modes which appear in the literature will be naturally derived.

Single Fiber Strength

When fibers of a given gauge length, tested in large quantities, yield a distribution curve, of even moderate dispersion, an essential consequence of this result is that different strength distributions would be obtained at different gauge lengths of the same filament. This is due to an assumed random distribution of flaws on the surface of the fiber. When the fiber is loaded, stress concentrations result in the vicinity of the flaws. It is these regions of high stress which give rise to brittle fiber fracture. The concentration of flaws per fiber is, therefore, related to the surface area of that fiber. Thus, if longer test gauge lengths are used, the probability of encountering weaker elements of the fiber is enhanced and, as a result, the mean strength of the longer filament can be expected to decrease. For whiskers, in which cross-sectional geometry varies, strength distributions would vary with cross-sectional area as well. This general effect will be illustrated by first considering a particular type of strength distribution function for filaments of varying length but of fixed diameter. This characterization will then be extended for fibers, or whiskers, of varying cross-sectional area, as well as length.

Various analytical forms for the frequency distribution function have been proposed. Most prominent has been the Weibull distribution function, which was employed by Rosen [19]. A Weibull distribution may be considered in the form:

$$f(\sigma) = \alpha L \beta \sigma^{\beta - 1} \exp(-\alpha L \sigma^{\beta}) \tag{62}$$

where α, β = two parameters characterizing the distribution function

L = fiber length

σ = fiber stress

For fibers characterized by this function, the mean strength at any given length is found from

$$\bar{\sigma}_L = \int_o^{\infty} \sigma[f(\sigma)]\, d\sigma \tag{63}$$

Substituting Eq. 62 into Eq. 63 we obtain

$$\bar{\sigma}_L = \alpha^{-1/\beta} L^{-1/\beta} \Gamma\left(1 + \frac{1}{\beta}\right) \tag{64}$$

The quantity $\alpha^{-1/\beta}$ may be considered as a reference stress parameter, though its units are not those of stress. $\Gamma(x)$ is the gamma function defined by

$$\Gamma(x) = \int_o^{\infty} z^{x-1} e^{-z}\, dz$$

In logarithmic form, Eq. 64 may be written as

$$\log \bar{\sigma}_L = -\frac{1}{\beta} \log L + \log \left[\alpha^{-1/\beta} \Gamma\left(1 + \frac{1}{\beta}\right)\right] \tag{65}$$

Thus, the Weibull statistical distribution function for fiber strength is such that a logarithmic plot of mean strength versus length yields a straight line whose slope is determined by β. Conversely, a straight-line variation of log of mean strength versus log of gauge length implies a unique pair of parameters for a Weibull distribution function. Experimental data for glass filaments and for boron filaments have shown this linear dependence (e.g., see Schmitz, et al. [36]).

Similar relationships are found to apply when relating the strength of Al_2O_3 whiskers to their circumference (see, for example, Brenner [37]). Postulating the strength of a given fiber to be a function of its surface area, the Weibull distribution function of Eq. 62, valid for a population of fibers of equal cross-section, may be generalized such that

$$f(\sigma) = \alpha L d \beta \sigma^{\beta - 1} \exp(-\alpha L d \sigma^{\beta}) \tag{66}$$

The mean strength at any given length and diameter is now

$$\bar{\sigma}_{L,d} = \alpha^{-1/\beta}(Ld)^{-1/\beta}\Gamma\left(1 + \frac{1}{\beta}\right) \tag{67}$$

and the logarithmic relationship between mean strength and surface area can be expressed as

$$\log \bar{\sigma}_{L,d} = -\frac{1}{\beta}\log(Ld) + \log\left[\alpha^{-1/\beta}\Gamma\left(1 + \frac{1}{\beta}\right)\right] \tag{68}$$

That is, a straight-line variation is said to exist between the log of the mean strength and the log of the fiber surface area.

Composite Failure Model

When the individual fiber strength-distribution function has a non-zero coefficient of variation (i.e., a failure stress level not defined by a unique value), the first fiber breaks will undoubtedly occur at relatively low stress levels, but then it will be possible to continue to increase the applied load and thereby to obtain breaks at other points of weakness on the fibers. This mechanism has been studied by Rosen [19] for continuous fiber composites with a model which shows an increasing statistical accumulation of fiber fractures with increasing load until a sufficient number of fractures occur at some cross-sectional region of the composite to reduce the strength at that cross-section by an amount sufficient to result in failure. Such an approach was suggested by Parratt [38] who proposed that composite failure occurs when the accumulation of fiber fractures shortens the fiber lengths to the point that further increases in load cannot be transmitted to the fibers because the matrix shear strength is exceeded. The fiber fractures are thus eventually followed by a shear failure of the matrix producing composite failure. In both Parratt's and Rosen's models, the local perturbations of the stress field which occur in the vicinity of the end of a broken fiber have a major influence upon the composite failure. At such a fiber break the axial stress in the fiber vanishes and gradually builds back up to its undisturbed stress value due to shear stresses being transferred across the fiber-matrix interface. The general form of the fiber stress is disturbed for only a short distance, with the result that very high interface shear stresses are produced. This mechanism of stress transfer was dealt with in the internal stress discussion. These shear stresses are those that Parratt proposed as being the limiting mechanism in tensile failure. In the Rosen model the characteristic dimension of the perturbed stress field is utilized to define artificially a dimension of the basic structural element. The composite is then represented as being composed of a series of bundles of such elements. These elements, or links, or ineffective lengths, are the basic load-carrying members, so that a fracture of one link eliminates that element from its participation in the load-carrying function. When the fibrous composite is reduced to this type of model, it is similar to the model constructed by Gucer and Gurland [39] for particulate reinforced materials, and the analytical approach of that study utilizing the statistics of chains of bundles can be followed. It remains to define the link length, the link strength distribution function, and then to make the model applicable to discontinuous fiber composites, in which the population of fibers may have arbitrary distributions of length L, and diameter d. For a fiber of noncircular cross-section, d is a characteristic dimension proportional to the fiber circumference.

Considering the composite to be a series of m layers, each layer being a bundle of N fiber elements of length l as shown in Figure 7.12, the distribution of strength values for a single fiber element, or link, must first be determined. From this function, the distribution of bundle strength values will be calculated which, in turn, leads to the distribution of composite strength values.

Consider the distribution of fiber element strength values, denoted by $f_l(\sigma, L, d)$, and the corresponding cumulative distribution function, denoted by $F_l(\sigma, L, d)$, where

$$F_l(\sigma, L, d) = \int_o^\sigma f_l(\sigma, L, d)\, d\sigma \tag{69}$$

A fiber of length L can be thought of as a chain consisting of n links, each link being a fiber element of length l ($n = L/l$). Failure of the fiber is then

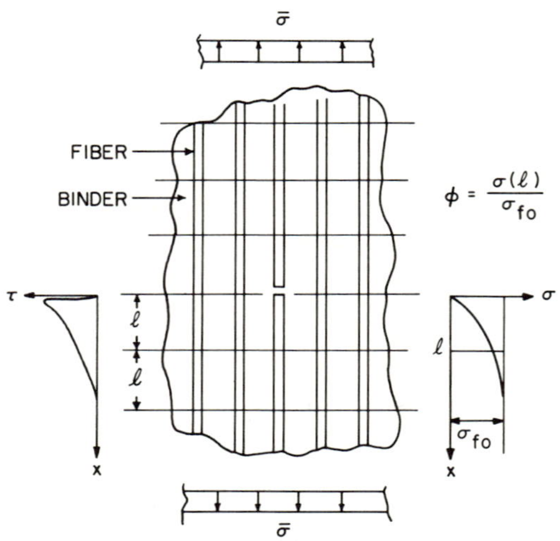

Figure 7.12 Tensile failure model for discontinuous fiber composites.

postulated as resulting from failure of its weakest link (see Epstein [40] for a discussion of weakest link concepts). Denoting the distribution of fiber strength values by $f_L(\sigma, L, d)$ and the cumulative function by $F_L(\sigma, L, d)$, it has been shown by Epstein [40] that for all links having the same failure distribution function, the distribution and cumulative distribution functions for the fiber element (link) in terms of those of the fiber (chain)—the latter being prescribed functions—are, respectively:

$$f_l = \frac{1}{n} f_L (1 - F_L)^{-(n-1)/n} \tag{70}$$

$$F_l = 1 - (1 - F_L)^{1/n} \tag{71}$$

The probability, p, that a fiber end (initial discontinuity) exists in a given fiber element, which is equivalent to the probability of fiber failure within the element occurring at an arbitrarily low stress level, is given by

$$p = l/L \tag{72}$$

Therefore, the distribution of fiber element strength values including the effect of initial fiber discontinuities is given by

$$f(\sigma, L, d) = p\,\delta(\sigma - 0) + (1 - p) f_l(\sigma, L, d) \tag{73}$$

δ is the Dirac-delta function.

With the total load on any cross-section or in any bundle of fiber elements distributed uniformly among the unbroken fibers in that bundle, the total load, P, acting on the bundle, is given by

$$P = \sigma_B A_T = \sigma A \tag{74}$$

where σ_B is the average bundle stress acting on all N fiber elements in the bundle, and A_T is the total cross-sectional area of the fiber elements. A, on the other hand, is the area of those fibers that remain unbroken at the actual fiber stress level σ.

Assuming all bundles to be statistically identical, and defining $g(L)$ and $h(d)$ as the distributions of fiber length and diameter, respectively, in the composite, the following expression for average bundle stress is obtained (Friedman [32]):

$$\sigma_B = \left\{ 1 - \frac{\int_d \int_L d^2 F(\sigma, L, d) h(d) g(L)\, dL\, dd}{\int_d d^2 h(d)\, dd} \right\} \sigma \tag{75}$$

$\int_d$ and $\int_L$ are integrals over the entire range of d and L prescribed by the nature of the distribution functions $h(d)$ and $g(L)$, respectively, while $F(\sigma, L, d)$ is the cumulative distribution function associated with $f(\sigma, L, d)$. The second term within the brackets of Eq. 75 can be defined as a weighted average of probability of failure at fiber stress σ. Letting this term be represented by $\bar{F}(\sigma)$,

$$\sigma_B = [1 - \bar{F}(\sigma)]\sigma \tag{76}$$

With the average probability of fiber failure thus defined, the statistical bundle strength theory developed by Daniels [41] can be used to formulate a distribution function for bundle failure. This theory shows rigorously that the distribution of bundle strength values for large N approaches the Gaussian, or normal, form with mean value $\bar{\sigma}_B$ being the maximum value of σ_B obtained from Eq. 76, and standard deviation given by:

$$\psi_B = \bar{\sigma}_B \left\{ \frac{\bar{F}(\sigma^*)}{N[1 - \bar{F}(\sigma^*)]} \right\}^{\frac{1}{2}} \tag{77}$$

σ^* is the value of fiber stress corresponding to maximum bundle stress. Hence,

$$\bar{\sigma}_B = [1 - \bar{F}(\sigma^*)]\sigma^* \tag{78}$$

where

$$\left. \frac{d\sigma_B}{d\sigma} \right|_{\sigma = \sigma^*} = 0 \tag{79}$$

is the equation whose solution gives σ^*.

The distribution of bundle strength values and the associated cumulative distribution function are, respectively, given by

$$\omega(\sigma_B) = \frac{1}{\sqrt{2}\,\psi_B} \frac{1}{\sqrt{\pi}} \exp\left[-\left(\frac{\sigma_B - \bar{\sigma}_B}{\sqrt{2}\,\psi_B} \right)^2 \right] \tag{80}$$

and

$$\Omega(\sigma_B) = 1 - \tfrac{1}{2} erfc\left(\frac{\sigma_B - \bar{\sigma}_B}{\sqrt{2}\,\psi_B} \right) \tag{81}$$

where $erfc(x)$ is the complementary error function defined by

$$erfc(x) = \frac{2}{\sqrt{\pi}} \int_x^{\infty} e^{-z^2}\, dz$$

Considering now the composite itself to be a chain of links (Fig. 7.12), the strength of each link characterized by the above functions, the application of the weakest link concept results in composite failure occurring when its weakest link (or weakest bundle of fiber elements) fails. Defining composite failure stress (stress acting on the fibers) as the most probable value of failure stress of the chain (i.e., the statistical mode of the chain's failure distribution function), it was shown by Rosen [19] that, for (*a*) a large number of fiber elements in a bundle and (*b*) a large number of bundle links in the composite chain, the failure stress level approaches the mean value of the Gaussian bundle strength distribution $\bar{\sigma}_B$. In terms of stress acting on the composite, rather than just on the fibers, composite strength is thus given by

$$\sigma_t = v_f \bar{\sigma}_B \tag{82}$$

where v_f is the volume fraction of fibers in the composite. The tensile load transmitted by the matrix is neglected in this equation. Though Eq. 82 is of the same form as that of the rule of mixtures for ductile fibers, $\bar{\sigma}_B$ is in this case the mean value of the bundle strength distribution, rather than a unique value of the fiber strength. The two are identical, of course, when the fiber strength is defined by a unique stress level, in which case the fiber strength distribution is described by the Dirac-delta function.

Implications of the Failure Model

Based on the use of the Weibull-type statistical distribution function characterizing fiber strength in the analytical failure model in which the composite failure mechanism is that due to an accumulation of fiber fractures, certain results are obtained which are, in general, applicable toward continuous or discontinuous fiber reinforced composites.

In those cases in which the fibers are whiskers, considerable deviations of fiber geometry within the composite can be expected. It is, therefore, important to determine the effects of the geometric characteristics of the fibers on composite strength.

Consider first the effects of mean fiber diameter and deviations from the mean on composite strength. Plots of composite strength against mean diameter are given in Figure 7.13 for the full range of fiber coefficients of variation, which is a measure of the dispersion of fiber strength. Both the composite strength and the mean diameter are normalized, so that a composite with mean diameter equal to the reference value has a strength equal to the reference strength. It is seen that the effect of mean diameter on composite strength is more pronounced for higher coefficients of variation. This is to be expected since the decrease of strength of individual fibers with increased diameter is more pronounced for higher fiber strength dispersion. For alumina whiskers, in which the coefficient of variation has been calculated to be on the order of 0.2 based on the available strength data of Mehan, et al. [42], an order of magnitude increase of mean fiber diameter results in a greater than 50% decrease in composite strength. It therefore is advantageous to utilize reinforcing whiskers with cross sections as small as is practicable in a composite, while maintaining the overall fiber volume fraction. It was

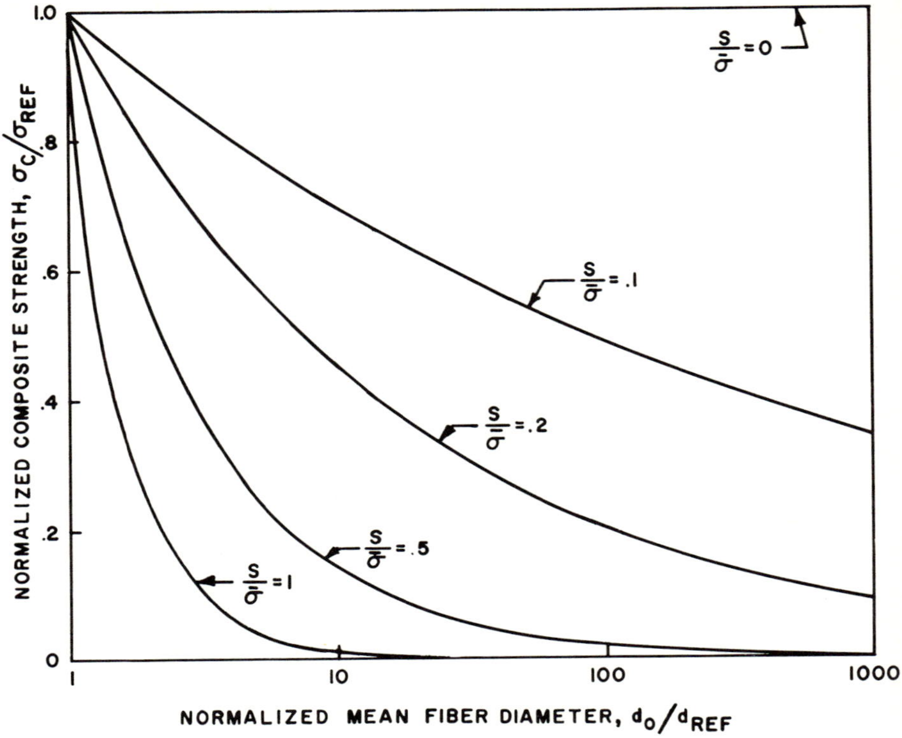

Figure 7.13 Variation of composite strength with mean fiber diameter.

found by Friedman [32] that deviations from the mean diameter have little effect on the strength characteristics.

In investigating the strength of discontinuous fiber composites, it is obviously important to determine the nature of the decrease in composite strength relative to the strength of the corresponding continuous filament composite whose other constituent properties are the same. The fiber discontinuities are characterized by a statistical distribution of fiber length values (for continuous filament composites, fiber length would be taken as infinite). For illustrative purposes, consider a distribution of fiber lengths that is uniform over a specified range. Normalizing this distribution function with respect to the mean fiber length, plots of composite strength against mean fiber length are presented in Figure 7.14 for the entire range of values of the parameter $\bar{y}$, which is a measure of the dispersion of fiber length within the composite. For these curves, composite strength is normalized with respect to the strength of the corresponding continuous filament

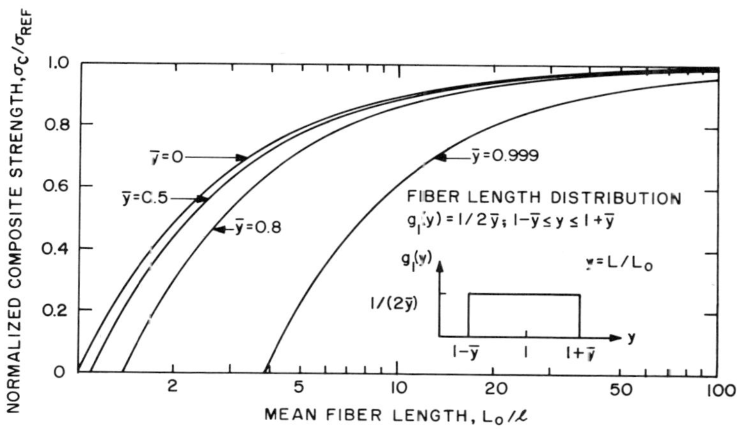

Figure 7.14 Variation of composite strength with mean fiber length and deviations from the mean.

composite, while mean fiber length is normalized with respect to the ineffective length of a fiber of diameter equal to the mean. The ineffective length is generally an order of magnitude greater than the fiber diameter. The normalized mean fiber length, then, is an order of magnitude less than the mean fiber aspect ratio. It is seen that, for normalized mean fiber lengths greater than 15, composite strength is at least 90% of the strength of the corresponding continuous filament composite for fibers with length variation of up to 80% of the mean. Even for greater variations in length, a desirable composite strength is attained at slightly higher values of mean fiber length, or aspect ratio. A significant decrease in composite strength is evident only when a combination of small mean fiber length and increased length variation results in a large percentage of the fibers being too short to effectively transmit stress. The primary length criterion, then, for an efficient discontinuous fiber composite is a sufficiently large mean fiber length, or mean fiber aspect ratio, rather than the degree of dispersion from the mean.

It is of interest to compare these results with those obtained using the modified rule of mixtures equation given by Kelly and Tyson [25] for discontinuous fiber composites with ductile constituent phases, and fibers which are all of equal length and cross section. Neglecting the contribution of the matrix in transmitting tensile load, this equation is as follows:

$$\sigma_t = v_f \sigma_f \left[1 - (1 - \beta) \frac{l_c}{L} \right] \tag{83}$$

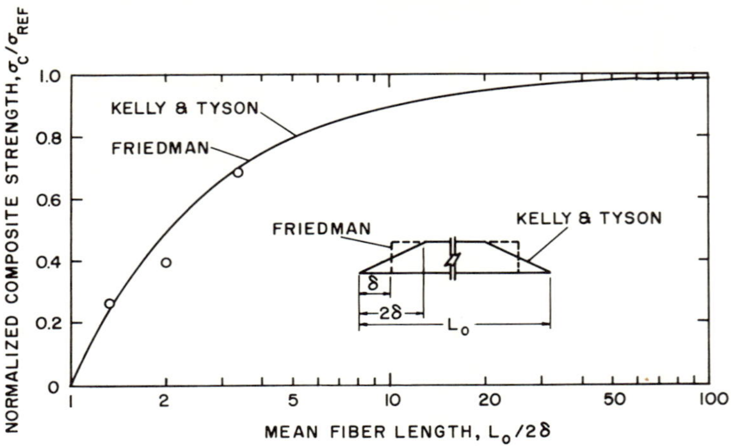

Figure 7.15 Comparison between theory and experimental for normalized tensile strength of whisker composites.

where $l_c/2$ is defined as the fiber transfer length and is related to the characteristic dimension of the perturbed stress field around a fiber discontinuity. β is the ratio of the average stress in a fiber within a distance $l_c/2$ from the fiber end to the uniform fiber stress outside this perturbed region. For the case in which the fiber stress is built up linearly from the ends, which is the case considered here (see Fig. 7.15), $\beta = 1/2$.

A plot of composite strength, normalized with respect to the strength of the corresponding continuous filament composite, against the dimensionless fiber length, is presented in Figure 7.15, together with the curve $\bar{y} = 0$ of Figure 7.14. In order to do this, it is necessary to adjust for the difference in definition of ineffective, or transfer, length in References 25 and 32 (as shown in Fig. 7.15.) When this is done, the two theoretical curves are identical. This results because, for the statistical failure model, it was shown in [32] that the statistical strength parameters and the distribution of fiber diameter values—properties that determine the strengths of both continuous and discontinuous fiber composites—do not contribute at all to the *relative* reduction in composite strength resulting from finite-length fiber reinforcement. Thus, for both theoretical models, the reduction in strength due to the reinforcing fibers being discontinuous depends only on the fiber transfer length and the mean fiber length.

These results can be compared with the experimental data for discontinuous fibrous composites obtained by Kelly and Tyson [25]. Figure 7.16 shows those data for composites having different initial length to diameter

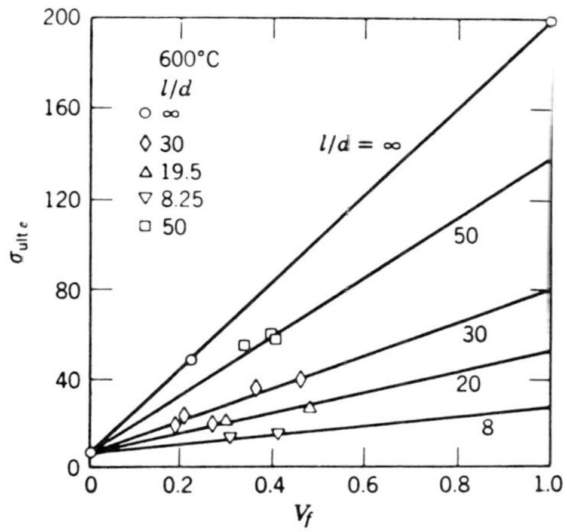

Figure 7.16 Experimentally measured values of strengths of discontinuous fiber-reinforced composites (from Kelly and Davies [33]).

ratios with best-fit lines drawn through the data. The ratios of the strength of these discontinuous fiber composites to the strength of the continuous fiber composite made from the same constituents is also plotted in Figure 7.15. These are experimental points, both in terms of ineffective length and strength level. Again, it appears that the statistical model offers a means to assess the behavior of the composite.

It is seen from Figure 7.14 that l, the mean ineffective length, is a factor that greatly influences the attainment of a high tensile strength in discontinuous fiber composites. Since ineffective length is also an inverse measure of the efficiency with which stress is transferred by the matrix around a fiber discontinuity, whether the discontinuity is an end or a break, the influence of ineffective length on composite strength is twofold. The latter effect results in a decrease in composite strength due to increased ineffective length which is shown in Figure 7.17 for the entire range of fiber coefficients of variation. For an alumina whisker-epoxy resin composite, an order of magnitude increase in ineffective length results in a 30% decrease in composite strength.

It was shown in the models of Dow [21], Rosen [19], and Hedgepeth [24] that the fiber-matrix stress transfer mechanism, which dictates the magnitude of the ineffective length, depends on the ratio of fiber modulus to matrix modulus, and the fiber volume fraction v_f. Using the three-concentric cylinder model of [19], postulating that the average axial stress along the entire length of fiber for the approximate stress distribution be made equal

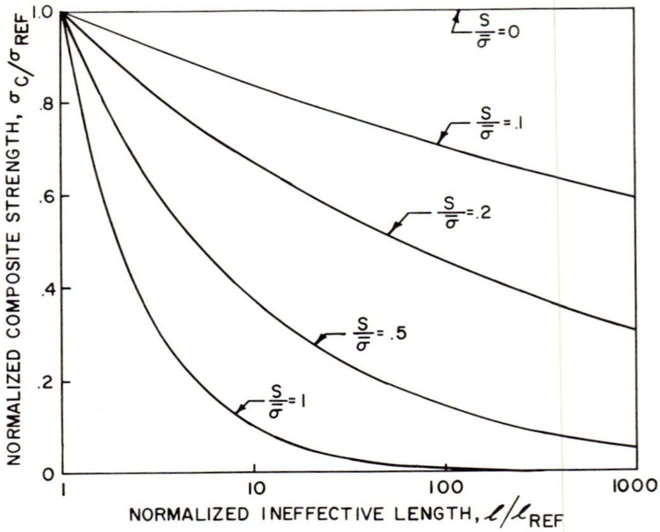

Figure 7.17 Variation of composite strength with ineffective length.

to that of the actual distribution,* and assuming that both fiber and matrix behave elastically, the ineffective length is found to be given by

$$l/d_o = \left[\frac{1}{2}\frac{E_f}{G_b}\frac{1 - v_f^{\frac{1}{2}}}{v_f^{\frac{1}{2}}}\right]^{\frac{1}{2}} \tag{84}$$

Plots of ineffective length against modulus ratio for several values of fiber volume fraction are presented in Figure 7.18. It is seen that a low volume fraction and an insufficiently stiff matrix are both detrimental to the efficient transfer of stress around a fiber discontinuity.

Though a stiffer matrix will tend to result in a stronger composite due to the improved efficiency of this stress transfer mechanism, it will also give rise to higher matrix and fiber-matrix interface shear stresses (see Fig. 7.9), thus increasing the probability of composite failure due primarily to either of these phenomena. A less ductile matrix may also increase the probability of a crack initiated at a fiber break, propagating through the matrix to adjacent fibers, thus causing premature composite failure. As pointed out previously, investigations into the nature of these other potential failure mechanisms,

* The "approximate" stress distribution corresponds to the idealization wherein the fiber stress is completely unperturbed except in the length of fiber which is considered ineffective. Within this ineffective region, the fiber is said to transmit no stress. The "actual" stress distribution is that obtained from the model (see Eq. 57).

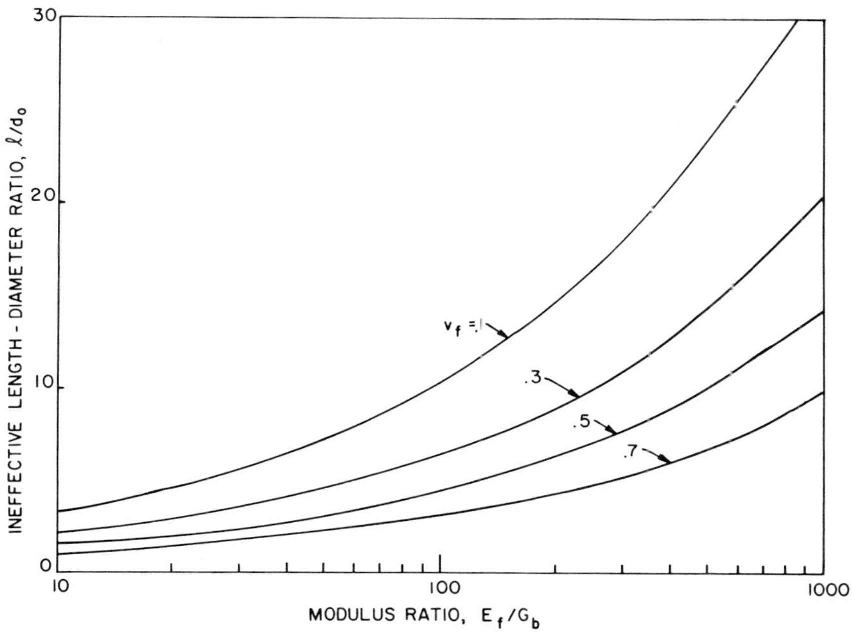

Figure 7.18 Influence of moduli upon elastic ineffective length.

and the factors that make any one of them dominant or contributory with respect to ultimate failure, are therefore required before making definitive statements concerning the optimum constituent properties of a composite under a given set of conditions.

CONCLUDING REMARKS

Although the computation of properties of whisker composites can be expected to be more complex than for the equivalent fiber composite, the results presented have demonstrated that certain whisker composite properties are well understood and subject to rational analysis. In the case of elastic moduli, it was demonstrated that the available continuous fiber results are directly applicable to whisker composites except for the impractical extreme of short fibers having a length of less than 10 diameters. Thus, all the procedures previously established for evaluating elastic stiffnesses of continuous fiber composites, including laminated composites, are immediately available for application to similar discontinuous fiber composites.

Review of the methods of analysis of internal stresses in the vicinity of a whisker end showed that this problem is not yet solved precisely, but that the general nature of the local stress fields is well understood. Improving experimental techniques generally show reasonable agreement with the approximate theories except at very short distances from the discontinuity. This, unfortunately, leaves an area of uncertainty in the subsequent failure analyses. When one is concerned with a stress average over some moderate dimension, as in the cumulative weakening tensile-failure model, the results are probably reliable. However, when one is concerned with a single local stress value, as in a crack-propagation tensile-failure model, then there is uncertainty in the use of the available analyses of stresses in the vicinity of a fiber end.

The statistical tensile-failure model described herein provides a basis for evaluating the effects of fiber geometry. Measurable, but not catastrophic, decreases in the tensile strength of the composite are associated with the finite length-to-diameter ratio. Effects of fiber distribution are shown to be important and almost unavoidable. The understanding of failure is, however, incomplete as the stress concentration effect has not as yet been fully evaluated. Extensive experimental mechanics studies are required.

This chapter demonstrates that the effect of constituent properties upon the mechanical strength and stiffness of nearly uniaxial whisker composites can be determined analytically. Stiffnesses close to those of continuous fiber composites can be achieved and, although further analysis of stress concentration effects is indicated, high-strength composites appear to be attainable.

REFERENCES

[1] Z. Hashin, "Theory of Composite Materials," presented at the GE-ONR International Conference on the Mechanics of Composite Materials, Phila., Pa., May 1967, Pergamon Press, 1970.

[2] G. Pickett, "Analytical Procedures for Predicting the Mechanical Properties of Fiber Reinforced Composites," AFML TR-65-220, June 1965.

[3] D. F. Adams and D. R. Doner, "Longitudinal Shear Loading of a Unidirectional Composite," *J. Composite Materials*, **1** (January 1967).

[4] R. H. Sonneborn, *Fiberglass Reinforced Plastics*, Reinhold, New York, 1954.

[5] B. W. Shaffer, "Stress-Strain Relations of Reinforced Plastics Parallel and Normal to Their Internal Filaments," *AIAA J.* (February 1954).

[6] Z. Hashin and B. W. Rosen, "Elastic Moduli of Fiber Reinforced Materials," *J. Appl. Mech.* (June 1964).

[7] Z. Hashin, "On Elastic Behavior of Fiber Reinforced Materials of Arbitrary Transverse Phase Geometry," *J. Mech. Phys. Solids*, **13,** 119 (1965).

[8] R. Hill, "Theory of Mechanical Properties of Fiber-Strengthened Materials: I. Elastic Behavior," *J. Mech. Phys. Solids*, **12,** 199 (1964).

[9] S. W. Tsai, "Structural Behavior of Composite Materials," NASA CR-71, July 1964.

[10] T. T. Wu, "On the Parametrization of the Elastic Moduli of Two Phase Materials," *J. Appl. Mech.*, **32,** 211 (1965).

[11] Z. Hashin, "Discussion of: On the Parametrization of the Elastic Moduli of Two Phase Materials," *J. Appl. Mech.*, 953 (December 1965).

[12] B. W. Rosen, "Strength and Stiffness of Fibrous Composites" in L. J. Broutman and R. H. Krock, Ed., *Modern Composite Materials*, Addison-Wesley, Reading, Mass., 1967.

[13] S. B. Dong, et al., "Analysis of Structural Laminates," ARL Report #76, USAF, September 1961.

[14] E. Reissner and Y. Stavsky, "Bending and Stretching of Certain Types of Heterogeneous Aeolotropic Elastic Plastics," *J. Appl. Mech.* (September 1961).

[15] N. F. Dow, B. W. Rosen, L. S. Shu, and C. H. Zweben, "Design Criteria and Concepts for Fibrous Composite Structures," NASA Final Report, NASw-1377, July 1967.

[16] N. F. Dow, C. Libove, and R. E. Hubke, "Formulas for the Elastic Constants of Plates with Integral, Waffle-Like Stiffening," NACA TR 1195, 1954.

[17] L. S. Shu and B. W. Rosen, "Strength of Fiber-Reinforced Composites by Limit Analysis Methods," *J. Composite Materials*, **1,** 4 (1967), 366.

[18] L. S. Shu, "The Evaluation of the Strength of Fiber-Reinforced Composites," presented at the GE-ONR International Conference on the Mechanics of Composite Materials, Phila., Pa., May 1967, Pergamon Press, 1970.

[19] B. W. Rosen, "Tensile Failure of Fibrous Composites," *AIAA J.*, **2,** 11 (November 1964).

[20] H. L. Cox, "The Elasticity and Strength of Paper and Other Fibrous Materials," *Brit. J. Appl. Phys.* (August 1951).

[21] N. F. Dow, "Study of Stresses Near a Discontinuity in a Filament-Reinforced Composite Material," General Electric Company Space Sciences Laboratory, Space Mechanics Memo #102, January 1961.

[22] M. A. Sadowsky, "Transfer of Force by High Strength Flakes in a Composite Material," Watervliet Arsenal, WVT-RR-6105-R, June 1961.

[23] M. A. Sadowsky, "Effect of Poisson's Ratio of an Elastic Filler on Force Transfer between Embedded Microfibers," Watervliet Arsenal, WVT-RR-6108-R, September 1961.

[24] J. M. Hedgepeth, "Stress Concentrations in Filamentary Structures," NASA TN D-882, May 1961.

[25] A. Kelly and W. R. Tyson, "Fiber-Strengthened Materials" in V. F. Zackay, Ed., *High Strength Materials*, John Wiley and Sons, New York, 1965.

[26] B. W. Rosen, "Mechanics of Composite Strengthening," in *Fiber Composite Materials*, American Society for Metals, Metals Park, Ohio, 1965.

[27] D. M. Schuster and E. Scala, "Single and Multifiber Interactions in Discontinuously Reinforced Composites," *AIAA J.*, **6,** 3 (March 1968), 527.

[28] D. M. Schuster and E. Scala, "The Mechanical Interaction of Sapphire Whiskers with a Birefringent Matrix," *Trans. Met. Soc. AIME*, **230,** 1635 (1964).

[29] J. E. Hill, "An Experimental Three-Dimensional Analysis of Residual Stresses Around Fiber Inclusions in Epoxy," Report #726, Materials Science Center, Cornell University, 1967.

[30] W. R. Tyson and G. J. Davies, "A Photoelastic Study of the Shear Stresses Associated with the Transfer of Stress During Fibre Reinforcement," *Brit. J. Appl. Phys.*, **16,** 199 (1965).

[31] T. F. MacLaughlin, "Effect of Fiber Geometry on Stress in Fiber-Reinforced Composite Materials," *Exp. Mech.* **6,** 10 (October 1966), 481.

[32] E. Friedman, "A Tensile Failure Mechanism for Whisker Reinforced Composites," SPI-RPD 22nd Annual Conference, Paper 4a, February 1967.

[33] A. Kelly and G. J. Davies, "The Principles of the Fiber Reinforcement of Metals," *Met. Rev.*, **10,** 37 (1965).

[34] D. L. McDanels, et al., *Metal. Progr.*, (1960), 118.

[35] C. Zweben, "Tensile Failure Analysis of Fibrous Composites," AIAA Paper No. 67–173, 1968.

[36] G. K. Schmitz and A. G. Metcalfe, "Effect of Length on the Strength of Glass Fibers," ASTM 1964, Preprint No. 87, June 1964.

[37] S. S. Brenner, "Factors Influencing the Strength of Whiskers," in *Fiber Composite Materials*, American Society for Metals, Metals Park, Ohio, 1965.

[38] N. J. Parratt, "Defects in Glass Fibers and Their Effect on the Strength of Plastic Mouldings," *Rubber and Plastics Age* (March 1960), 263.

[39] D. E. Gucer and J. Gurland, "Comparison of the Statistics of Two Fracture Modes," *J. Mech. Phys. Solids* (1962), 365–373.

[40] B. Epstein, "Statistical Aspects of Fracture Problems," *J. Appl. Phys.* **19,** 140 (1948).

[41] H. E. Daniels, "The Statistical Theory of the Strength of Bundles of Threads," *Proc. Roy. Soc.*, **183A,** 405 (1945).

[42] R. L. Mehan, E. Feingold, and A. Gatti, "Evaluation of Sapphire Wool and Its Incorporation into Composites of High Strength," General Electric Co., TIS R65SD36, August 1965.

8 Whisker-Matrix Bonding and Interaction

INTRODUCTION

Whisker-reinforced composites have great potential for applications where very high-strength, high-modulus, and high-temperature stability is required. Their successful utilization is critically dependent on the chemical, mechanical, and physical stability of the whisker-matrix system. Chemical stability is needed to minimize deleterious reactions during fabrication and use at elevated temperatures; mechanical stability is essential for control of whisker and/or interfacial failure under applied loads. This chapter discusses interfacial stability considerations in the fabrication and use of composites.

Because little information is available on whisker-matrix interfacial stability, data reported for some bulk materials will be presented. If extensive reactions occur between a bulk material and a particular matrix, it is probable that similar reactions will occur between this matrix and whiskers of the same composition as the bulk material, although correlating the stability of a material in bulk and whisker form can only be qualitative. For example, bulk and whisker diffusivities may differ if defect structures and/or impurity levels vary. Surface defects of a size acceptable in bulk materials may drastically reduce whisker strengths. Fabrication of such whisker-matrix composites might better be deferred until the nature and kinetics of the interfacial reactions have been determined.

DEFINITION AND FUNCTIONS OF THE INTERFACIAL ZONE

Because of the small diameter (generally less than 5 μ) of whiskers, whisker-matrix composites are characterized by very large interfacial areas per unit volume. Interfacial zones are required to provide the transition

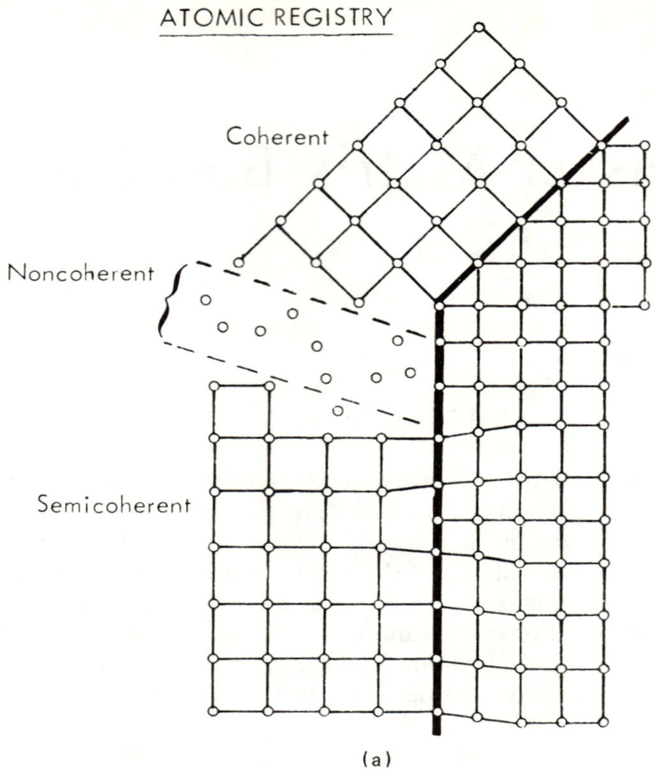

Figure 8.1a Degrees of coherency at an interface [1]. Courtesy of Society of Aerospace Materials and Process Engineers.

between the fiber and the matrix since these differ in structure and bond types. The interfacial zone ranges in thickness from several angstroms to several thousand angstoms in whisker-metal composites and thus is smaller than that generally encountered in other ceramic metal composites (for example, cermets and ceramic metal seals). Differing from both the fiber and matrix, the interfacial zone represents a chemical, physical, and mechanical discontinuity in the composite material which may thicken or change in nature as chemical reactions and/or diffusion proceed.

The changes occurring in the interfacial zone (as defined above) are both physical and chemical in nature. Possible physical changes include the state of aggregation (e.g., crystalline or amorphous), the crystal structure, and the orientation relationship and degree of coherency between the fiber and the matrix. Coherency occurs if the atoms at the interface are common

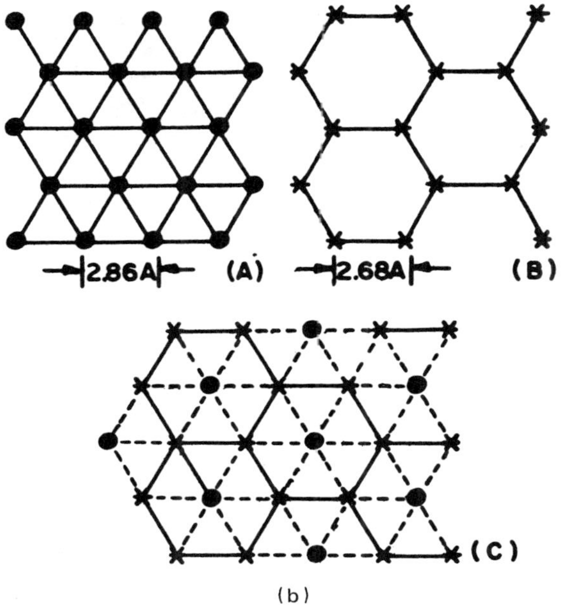

Figure 8.1b Possible coherent interface between the (111) plane of aluminum (A), the (0001) plane of Al_2O_3 (B), and superposition of these two planes (C) [2]. Courtesy of American Society for Metals.

to the crystal lattices of the mating phases (Fig. 8.1a). Semicoherent and noncoherent interfaces have progressively less registry between adjacent phases. Although geometric matching between the (111) plane of aluminum and the (0001) plane of Al_2O_3 is nearly perfect (Fig. 8.1b), this system appears to be noncoherent, indicating other factors must be considered.

A chemical change will be reflected in variations in composition and in bonding across the interfacial zone. The desirability of designing composites so that an interfacial phase of controlled strength is formed should be considered and is discussed with respect to spinels in a later section.

If a noncoherent interface is formed, chemical bonding through the interfacial zone (or across the interface, if a discrete surface) may contribute less to interfacial strength than mechanical gripping between the two mating phases. The aluminum-alumina interface appears to be noncoherent, although Pask* suggests that some chemical bonding occurs. At present, data distinguishing the degree of chemical bonding from mechanical

* J. Pask, University of California, Department of Mineral Technology, Berkeley, California, private communication, 1968.

TABLE 8.1 FUNCTIONS OF INTERFACE IN FIBER-REINFORCED COMPOSITES [3]

Functions of Interface	Consideration In		
	Fabrication	Use	Type Stability
1. Determines Wetting & Bonding	X		Chemical, Mechanical
2. Couples Fibers With Matrix		X	Chemical, Mechanical
3. Transfers Stresses From Matrix To Fibers		X	Chemical, Mechanical
4. Controls Fiber Spacing & Volume Fraction	X		Mechanical
5. Prevents Direct Fiber-Fiber Contact	X	X	Mechanical
6. Aids in Fiber Alignment (i.e., Magnetic)	X		
7. Protects Fiber From Mechanical Damage	X		Chemical, Mechanical
8. Serves As A Diffusion Barrier	X	X	Chemical, Mechanical

gripping have not been reported; such data are needed because mechanical gripping may not operate at elevated temperatures (due to differential thermal expansion), where whisker-reinforced composites appear most attractive.

The functions of the whisker-metal matrix interface are summarized in Table 8.1 [3]. These involve wetting, bonding, chemical stability (liquid and solid phase), and mechanical stability in the composite and will be discussed in turn.

WETTING

Fabrication of void-free composites by methods involving liquid-phase infiltration of a whisker mat requires that the liquid wet the whiskers or their coating. Wetting occurs when the equilibrium contact angle (θ) between the liquid matrix and the solid whisker (or coating on the whisker) is less than 90°, as determined by vectorial summation of the surface tensions of the liquid and solid in equilibrium with the vapor phase (γ_{LV} and γ_{SV}), and the interfacial tension between the two phases (γ_{SL}) (Fig. 8.2):

$$\gamma_{LV} \cos \theta = \gamma_{SV} - \gamma_{SL} \tag{1}$$

Wetting occurs only if $\gamma_{SL} < \gamma_{SV}$, and reducing γ_{SL} with respect to γ_{SV} improves wetting (or reduces the degree of nonwetting, where $\theta > 90°$). The γ_{SL} will generally be determined by the degree of compatibility between the

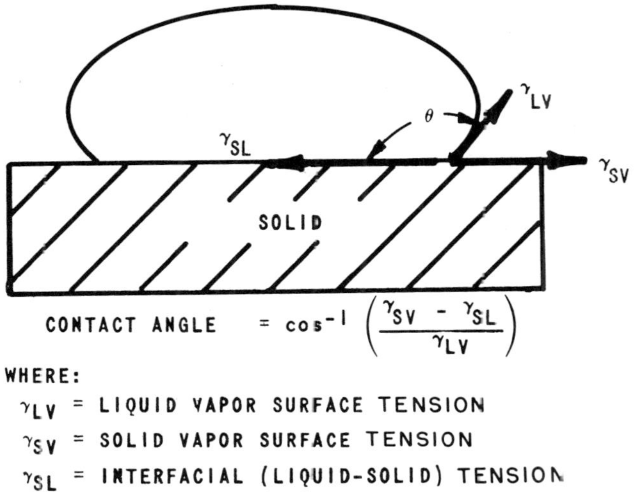

Figure 8.2 Solid, liquid, vapor under equilibrium conditions.

liquid and solid phases and can be reduced by the migration of surface-active elements to the interface, where they are either adsorbed or absorbed to minimize the free energy of the system. Absorption causes a chemical alteration of the substrate, and changes in the interfacial energy depend on the specific reactions occurring. Adsorption effectively coats the substrate with the surface-active element, and the change in interfacial energy proceeds according to the Gibbs adsorption isotherm [5]:

$$d\gamma = RTd\ln a \tag{2}$$

where Γ is the excess concentration at the interface, a is the activity of the adsorbed species, R is the universal gas constant, and T is the absolute temperature.

Although wetting between whiskers or whisker mats and possible matrices is not easily studied because of the small size and different crystallographic surfaces of whiskers and the small capillaries in whisker mats through which the molten matrix must flow, wetting is readily investigated by substituting bulk materials for whiskers. Available data indicate that plastics will generally wet both metals and ceramics, that glasses will wet metals, and that metals and crystalline ceramics may or may not wet other metals and ceramics. Wetting of ceramics by metals will be explored further because of

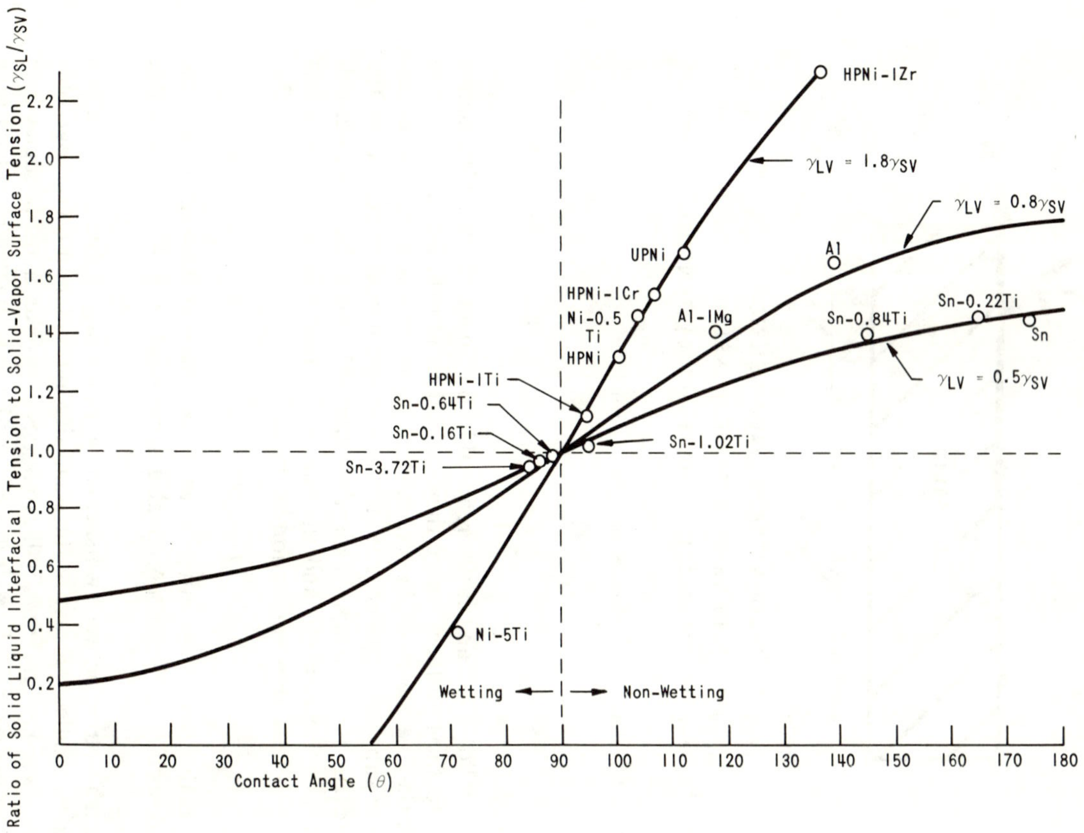

Figure 8.3 Relation between contact angle and surface tensions in Young's equation $[\cos\theta = (1 - \gamma_{SL}/\gamma_{SV})/(\gamma_{LV}/\gamma_{SV})]$ for several metals on alumina.

the great potential of ceramic whisker-metal matrix composites* for extending the high-temperature limits of monolithic materials, the attractiveness of fabricating them by liquid-phase infiltration or sintering, and the relative superiority of alumina and silicon carbide whiskers.

Wetting of ceramics by metals cannot be obtained solely by selecting $\gamma_{LV} < \gamma_{SV}$ as shown in Figure 8.3 for wetting of alumina by tin, aluminum, and nickel, for which $\gamma_{LV}:\gamma_{SV} = 0.5$, 0.8, and 1.8, respectively. The most striking difference between wetting behavior of these metals is that as $\gamma_{LV}:\gamma_{SV}$ decreases, the degree of wetting or nonwetting increases more rapidly as $\gamma_{SL}:\gamma_{SV}$ deviates from unity. Data from Allen and Kingery [6] for the wetting between tin-titanium alloys and polycrystalline aluminum oxide show that wetting can be obtained if sufficient titanium is added to the tin. Adsorption of titanium in a monolayer on the alumina according to the packing of titanium atoms in the (111) plane of titanium monoxide was the postulated mechanism for improving wetting.

Results of sessile drop tests between aluminum alloys and sapphire ($\gamma_{LV}:\gamma_{SV} = 0.8$) suggest that wetting in this system may not be strongly dependent on the crystallographic surface of sapphire exposed to the metal, since the contact angles on the basal plane and a random plane at 60° to the *c*-axis were identical [4]. These contact angles were not affected by holding times up to 1 hr after equilibrium was established (which occurred in less than 10 min). Wetting behavior was not sensitive to decreasing residual impurity levels in the base aluminum from 0.9% by weight (primarily 0.5% Fe, 0.3% Si, and 0.1% Mn) to <42 ppm (<10Cu, <10Fe, <10Mg, ~10Mn, and ~2Si). Selected alloying additions did affect contact angles (Table 8.2), the minimum angle of 94° being observed with an Al-0.94 Mg (at.%) alloy.

Increasing the temperature of formation of the specimen from 1600 to 2100°F reduced the observed contact angle from 120° to 70° (Fig. 8.4). Above 1900°F ($\theta \sim 90°$), the molten drop partially dissolved the sapphire plaque, and at 2100°F, the depth of attack exceeded 15 μ. Therefore, whiskers with diameters up to 30 μ would have been completely consumed if infiltrated with molten aluminum at 2100°F. An investigation of wetting of silicon carbide by aluminum alloys shows similar results to those obtained with sapphire.†

The results of an investigation of wetting behavior between nickel and nickel-titanium alloys and polycrystalline aluminum oxide [6] are shown on

* Whiskers should be strong, rigid, refractory, low in density, and chemically and mechanically stable; matrices should have sufficient shear strength to utilize the whisker strength by load transfer, enough ductility to allow some plastic formability of the composite, and a large temperature range where they are chemically stable but mechanically weak unless reinforced.

†A. P. Levitt, S. M. Wolf, and E. Harvey, *Wetting and Bonding Between Aluminum Alloys and Silicon Carbide*, Army Materials and Mechanics Research Center, to be published.

TABLE 8.2 EFFECTS OF ALLOYING ELEMENTS ON ALUMINUM-SAPPHIRE CONTACT ANGLE

Effect on Contact Angle	Alloying Element (atomic percent)	Category	Contact Angle (deg)	Apparent Shear Strength (ksi)	
				As Cast	Heat Treated
	Pure Al	-	139 ± 5	6.3	6.9
Decreased	0.23 Cu	I	119 ± 5	3.2	3.5
	0.94 Mg	I, II	118 ± 5	5.2	5.2
	0.13 Zr	II, III	118 ± 5	1.5-4.2	0.7-1.0
	0.94 Mg*	I, II	94 ± 5	7.3	7.0
	7075 Alloy†	-	115 ± 5	4.8	11.7
Not Affected	0.019 Pb	I, IV	139 ± 5	4.6	6.9
	0.26 V	III	139 ± 5	5.7	1.0
	0.53 Mn	I, III	137 ± 5	7.8	2.2
	0.14 Cr	I, III	134 ± 5	3.6	2.4
	0.12 Bi	IV	134 ± 5	12.0	-
	2024 Alloy‡	-	139 ± 5	1.6	0.4
Increased	0.014 Se	III	155 ± 13	11.5	13.0
	0.017 Te	III	152 ± 6	3.0	-
	~0.02 Ge	IV	146 ± 9	9.1	-
	1100 Alloy**	-	144 ± 5	10.2	12.6

*Melted at 1300 F, remelted at 1600 F in vacuum of 5 x 10^{-5} torr
†Al-5.6Zn-3.0Mg-0.2.Cr (weight Percent)
‡Al-4.0Cu-0.6Mn-1.5Mg (weight percent)
**Al-1.0(Fe+Si) (weight percent)

the curve with the liquid-to-solid surface tension ratio equal to 1.8 (Fig. 8.3). Increasing additions of titanium to nickel also resulted in wetting, with the postulated adsorption of titanium on the aluminum oxide surface corresponding to its position in the (111) plane of titanium oxide. It was suggested that in commercial nickel-base alloys which contain approximately the same amount of titanium as the above alloys but which show a higher contact angle, that is, a lesser tendency towards wetting, the attraction of titanium to the nickel-aluminum oxide interface is retarded by other elements in the alloys.

The effect of small additions of reactive metals to nickel on its wetting and bonding behavior with single-crystal sapphire plaques has been carefully studied with sessile drop specimens by Sutton and co-workers [7, 8] at the General Electric Company (Table 8.3). Chromium, titanium, and zirconium were preferentially attracted to the interface and formed new phases within 2 min after melting (as discussed in the next section). Titanium additions lowered the contact angle, interfacial tension, and bonding strength (the stress required to shear the metal drop off the ceramic plaque), and raised the work of adhesion. Chromium additions increased the contact angle, bonding strength, and interfacial tension, and lowered the liquid surface energy and work of adhesion. Zirconium additions increased the interfacial

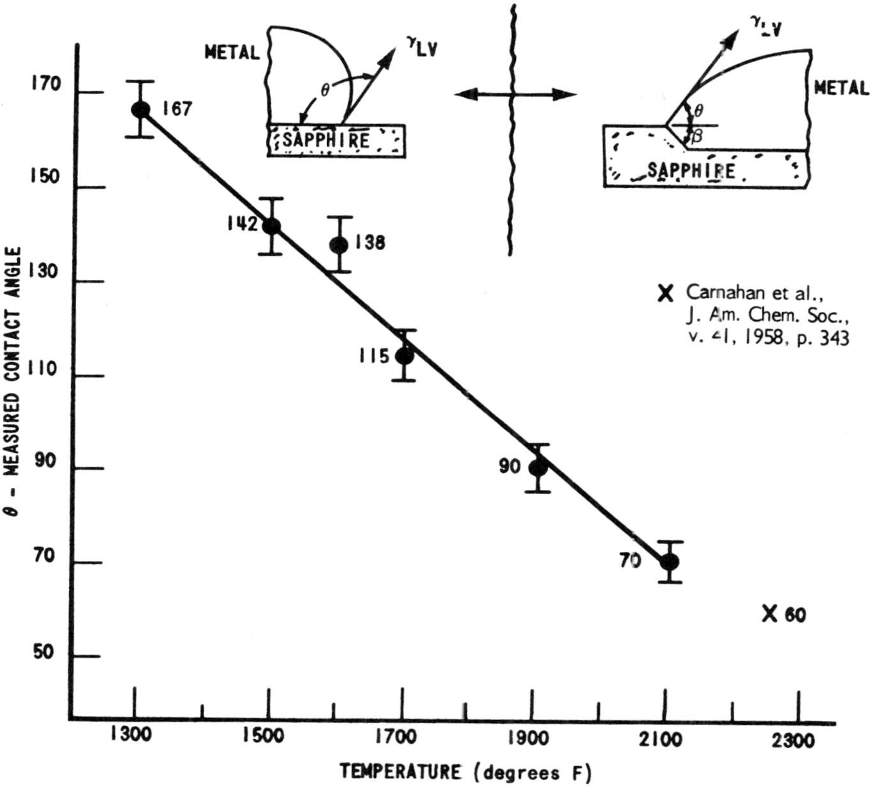

Figure 8.4 Temperature dependence of aluminum-alumina contact angle [4].

tension and contact angle and markedly reduced the bonding strength and work of adhesion.

An important question raised by Sutton's work is the effect of minute impurity levels in wetting studies. For example, results obtained with an ultra-pure nickel containing <15 ppm impurities differed from those obtained with a high-purity nickel containing <31 ppm impurities. As can be seen in Table 8.3, the effect of decreasing the impurity levels from 31 ppm to 15 ppm (primarily in the iron, carbon, and oxygen levels) caused a larger increase in the contact angle than a 1 at.% addition of chromium to high-purity nickel.

The dramatic effects of low impurity levels have also been demonstrated in the copper-tungsten system, where wetting was obtained with spectrographically pure copper (<1 ppm oxygen) between 1095 and 1225°C, but

TABLE 8.3 DATA ON TEST CONDITIONS, CONTACT ANGLES, AND SURFACE ENERGIES FOR NICKEL/NICKEL ALLOYS ON SAPPHIRE SUBSTRATES* [8]

Specimen	Test Temp. (deg C)	Pressure (mm Hg)	Contact Angle θ	γ_{LV}† (dynes/cm)	γ_{SL}† (ergs/cm²)	AD‡ (ergs/cm²)	Interfacial Thickness and Composition
UPNi**	1470	5×10^{-6}	112.5	1740	1610	1080	Same as rest of drop
HPNi††	1490	4	100.7	1770	1290	1390	Same as UPNi
HPNi +1.00Zr	1470	2	136.2	1770	2220	500	ZrO_2∿30μ into sapphire and ∿15μ into metal; Zr diffusion ∿100μ into sapphire
HPNI +0.99Ti	1520	6	94.5	1770	1090	1640	Ti(or Ti_3O_5?)∿5μ into sapphire; Ti diffusion ∿100μ into sapphire
HPNi +1.35Cr	1520	2	107.7	1670	1470	1170	α-Cr, β-Cr, and γ-Cr in metal within ∿75μ of interface; diffusion of Cr∿10μ into sapphire; diffusion of Ni∿4μ into sapphire
HPNi +0.71Cu	1520	3	100.4	1720	1270	1410	Same as UP Ni
HPNi +0.82Al	1520	7	100.8	1770	1290	1460	Al at L-V interface
HPNi +0.76 In	1550	5	106.8	1575	1430	1130	In at L-V interface

*Impurities (ppm) in Sapphire (α - Al_2O_3): 20 Mg, 80 Si (several others not detected)

†Corrected to 1500 C (where $\gamma_{SV} \approx$ 965 erg/cm²)

‡W_{AD} - Work of Adhesion (1500 C)

**Impurities (ppm) in UP (ultrapure) Ni: 5 Si, 2 Fe, 2 Cu, 2 Mg, 1 Al, 1 Ca, 1 Na, <1 Mn, <1 Ag; Total <15 ppm (C and O among others not detected)

††Impurities (ppm) in HP (high purity) Ni: 5 Si, 10 Fe, <15 C, <0.5$_2$, 0.3 N_2; Total <31 ppm

wetting between high conductivity copper (251 ppm oxygen) was obtained only above 1125°C, and as little as 41 ppm oxygen in copper inhibited wetting for several minutes at the lower temperatures [9]. It was concluded that oxygen combines with tungsten at the interface to form its oxide and inhibit wetting. Although additions of other metals to copper also inhibited wetting, even with oxygen-free copper, it would be useful to know whether oxygen getters such as titanium would improve wetting with high oxygen concentrations.

Because metals do not generally wet ceramic whiskers, metallic coatings may be required. Of the feasible coating techniques (Table 8.4), sputtering [10]* and organometallic vapor reaction (from carbonyls) [11] have been

* When a metal coating is sputtered onto a clean ceramic surface, an adherent coating is obtained despite the fact that the molten metal will not wet the ceramic. The precise reason for this is not known although it is postulated that the metal adheres to the ceramic surface because its impact velocity in the vapor phase is high enough to bring it sufficiently close to the ceramic surface for Van der Waal's forces or chemical bonding to be operative. This illustrates the point that wetting is not required for bonding.

successfully used on a laboratory scale. Coating stability cannot be overemphasized because dewetting will occur if the coating is dissolved in the molten metal. In such a case, although a completely dense composite is obtained for a few seconds after the molten matrix infiltrates coated whisker mats, dewetting and void formation occur before solidification is complete. The importance of thermal contraction that occurs concurrently with

TABLE 8.4 POSSIBLE TECHNIQUES FOR METALLIZING Al_2O_3 WHISKERS [11]

Technique	Advantages	Disadvantages
Organometallic Fired After Application	Simplicity in technique	Trace organics in coating, coatings not adherent; process not reproducible
Powder Metallurgical (Active Metal)	Commonly used ceramic-to-metal seal process	Non-uniform film on microscopic-scale, not uniform coverage
Pyrolytic Deposition	Continuous process	Too many process variables
Vacuum Evaporation	High vacuum (purity)	Shadowing effects and possible discontinuous coatings
Cathode Sputtering	Most adherent	Process requires reduced pressure
Electroless Plating	Thick coatings	Voids and Impurities in coating
Electroplating	Thick coatings	Requires prior metallizing for non-conducting surfaces

solidification in promoting dewetting or void formation has not been determined. The coatings are very specific in nature; for example, molten aluminum dissolved a nickel coating on sapphire whiskers in just a few seconds, whereas a nickel coating was stable for several minutes in an aluminum-20% nickel alloy [12]. Tungsten coatings dissolve in just a few seconds when infiltrated by molten nickel. Duplex tungsten over titanium coatings have been partially successful in fabricating nichrome (75% Ni–25% Cr) reinforced with sapphire whiskers (Fig. 8.5). It seems probable that multilayer coatings can be obtained to suit each whisker-metal system. Also, the effect of these coatings during long-term elevated temperature exposure has not yet been determined.

Coatings that are mechanically adherent on the solids may also enhance spreading of liquids on their substrates through the mechanism called "tunneling" [13]. The coating is selected with $\gamma_{CV} \gg \gamma_{CL}$ so that

$$\gamma_{CV} + \gamma_{SV} > \gamma_{LS} + \gamma_{CL} \tag{3}$$

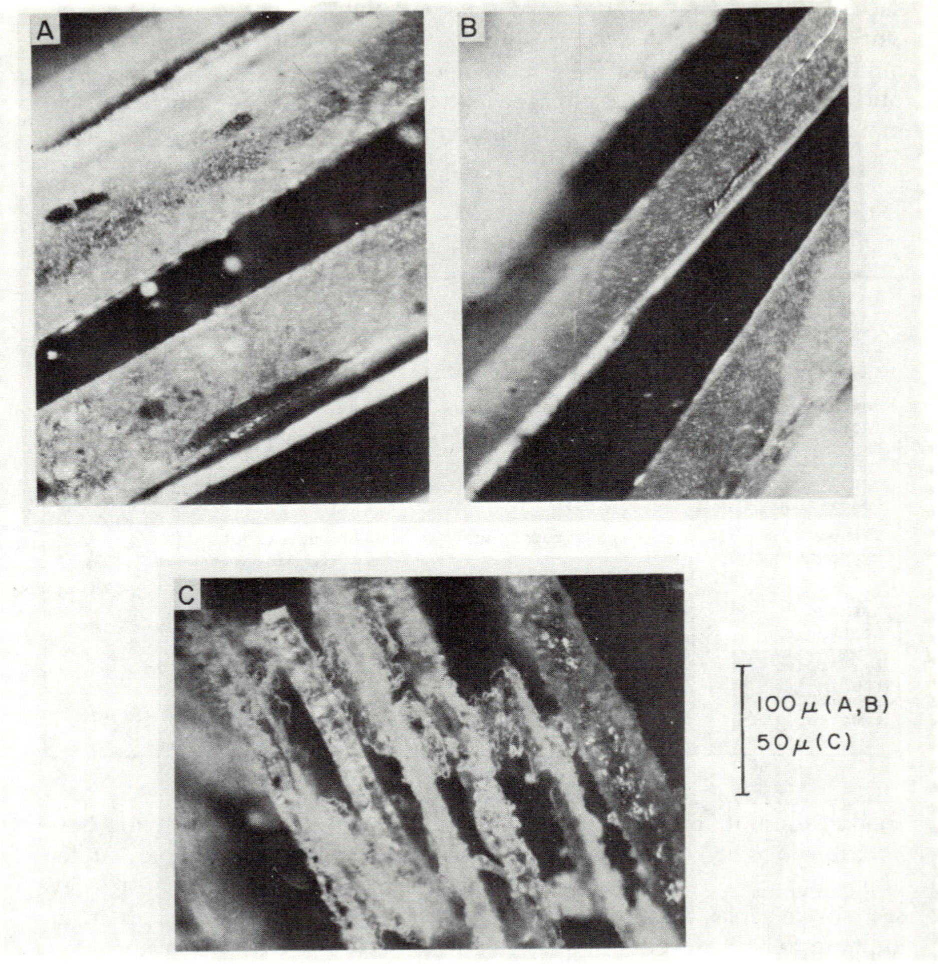

Figure 8.5 Microscopic observations of the interior portion of Nichrome-Al_2O_3 whisker composites [12]. A and B, W-Ti coated sapphire whiskers in nichrome matrix (225X); C, W-coated sapphire whiskers in nichrome matrix (450X).

even if $\gamma_{SV} < \gamma_{LS}$ (i.e., the liquid does not wet the solid). The subscripts C, V, L, and S refer to the coating, vapor, liquid, and solid, respectively.

The inequality of Eq. 3 provides a driving force for the liquid to penetrate through the coating at selected sites (e.g., pores and micro-cracks). Because of the coating interfacial energy inequality, the liquid wets the mechanically adherent coating at the coating-solid interface and is "pulled" over the solid surface. It should be noted that the liquid lifts the coating away from the interface so it cannot act as a reaction barrier between the solid and the liquid. The ability of coatings to promote this type of spreading is dependent on coating thickness and perfection, liquid temperatures, and reaction between the coating and the liquid, as well as interfacial energy relationships. Titanium, zirconium, and chromium coatings have induced spreading of Al-5Si on polycrystalline alumina [13], and titanium coatings have induced wetting of the (100) planes of magnesia by silver [14].

BONDING

Correlating the wetting and bonding behavior in a sessile drop experiment is very difficult. Wettability is a measure of the extent to which a liquid will spread on a solid. The greater the wettability, the lower the contact angle and the greater the work required to separate the two phases. This work is known as the work of adhesion W_{AD} and is given by:

$$W_{AD} = \gamma_{LV} + \gamma_{SV} - \gamma_{SL} = \gamma_{LV}(1 + \cos\theta) \tag{4}$$

Good wetting, because it implies intimate interfacial contact between liquid and solid, promotes good bonding. However, the interfacial bond strength will be determined by the extent to which primary or secondary atomic bonding between the two phases is achieved [1]. As with homogeneous materials, interfacial bond strengths will not equal theoretical bond strengths —on the order of 10^8 psi for primary (atomic) bonds and 10^5 psi for secondary (molecular or Van der Waals) bonds—because of local stress concentrations.

An empirical relationship between interfacial bonding strength and the contact angle may be obtained if such a relationship is based on physical or chemical characteristics of the interface such as surface roughness or surface cleanliness [15, 16]. However, such relationships may not apply to whisker-matrix interfacial bonding because defects which are too small (e.g., <50 Å) to affect the strength of polished bulk materials (e.g., a 1-μin. finish corresponds to a surface roughness of 250 Å) markedly reduce whisker strengths.

Ceramic-metal bonding has been qualitatively studied by shearing sessile drop specimens. Although the stress state set up is complex and not equivalent to that at a whisker-matrix interface, the value of such shearing tests should not be overlooked because of the ease of fabricating sessile drop specimens and the absence of reproducible composites. For example, the effect of chemical reaction on bonding strength with active metal additions to nickel-base alloys when melted on sapphire plaques was related to the degree of reaction and interfacial zone thickness [7, 8] (Fig. 8.6). The high purity

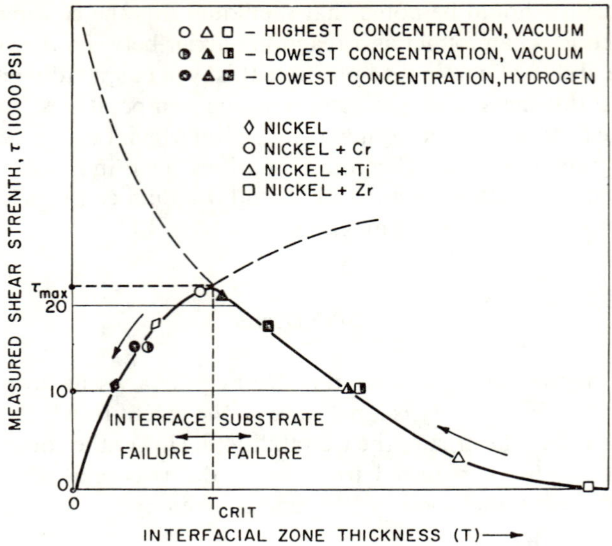

Figure 8.6 Measured shear strength between nickel, dilute nickel alloys, and sapphire as a function of the thickness of the interfacial zone [8].

nickel showed a bonding strength of about 18,000 psi and the smallest reaction zone ($\sim 0.01\ \mu$ zone thickness). Additions of 1 at.% chromium resulted in the highest bonding strength with an interfacial thickness of about 5 μ. Titanium additions reduced strength to about 5000 psi and additions of zirconium to nickel resulted in interfacial separation upon cooling to room temperature. The drastic reductions in strength with the titanium and zirconium alloys were attributed to the formation of brittle interfacial phases (growing into the substrate) in which failure was initiated. It is important to note that the interfacial zones formed by chromium, titanium, or zirconium additions would have consumed most whiskers. (Whiskers are purposely grown with small diameters ($< 5\ \mu$) because their strength decreases with increasing size.) It is encouraging that with an extremely small reaction zone the high purity nickel-sapphire bonding strength was 90% of the maximum strength obtained. Thus, adhesion between relatively unreactive whisker-metal combinations can develop significant bonding strengths required for load transfer in a whisker-metal composite. That little or no reaction occurred between nickel and sapphire is further indicated by the constant contact angle maintained by the sessile drop and the continuous reduction of the drop size and interfacial area as evaporation occurred, causing a continuous reduction of the contact angle. With active metal

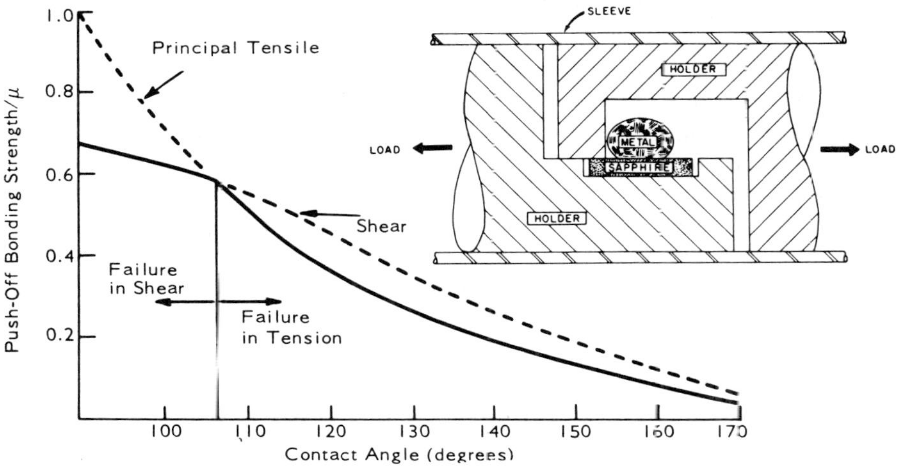

Figure 8.7 Calculated variation of push-off bonding strength with U, the tensile strength of the interface, and the contact angle for failure by shear and tension [17].

additions, the interfacial area remained constant for a period, even though the surface area to volume ratio was increasing above equilibrium because of evaporation, until the drop suddenly contracted and reestablished the initial contact angle to minimize the free energy of the system.

The relationship of bonding strengths obtained in the sessile drop test to those encountered between whisker and matrix in an actual composite is not clear for several reasons. First, the stress system in the sessile drop test can range from shear to tension depending on the contact angle (Fig. 8.7) [17]. Second, the stresses induced in the sessile drop system by differential thermal contraction are very complex and differ significantly from those in an actual whisker-matrix interface because of different geometries. Third, since mechanical bonding may predominate in such systems, the bonding strengths measured in sessile drop shear tests may be much different from those in an actual whisker composite. This is because whiskers have inherently smoother surfaces than polished plaques of the same material.

CHEMICAL STABILITY

The chemical stability of the initial components of a composite can be estimated from reactions between bulk phases as well as from thermodynamic and kinetic data, if available. In general, neither metals nor ceramics are chemically degraded by plastics, but both metals and ceramics react to varying degrees with other metals, glasses, and ceramics. Reported data on titanium-silicon carbide [18]; nickel, cobalt, and silver-silicon nitride [19]; alloying additions to copper-tungsten [20], nickel-sapphire [7], and aluminum-sapphire [4]; and nickel-silicon carbide [18] indicate that rapid and severe whisker degradation occurs if extensive chemical reactions occur between bulk phases of the whisker and matrix, so sessile drop specimens can provide preliminary data concerning liquid and solid phase stability.

Segregation in nickel alloy-sapphire sessile drops was previously mentioned in connection with wetting and bonding. Interfacial phases and/or reaction products were determined to be metallic in nature with chromium additions and ceramic in nature with titanium and zirconium additions (Table 8.2). Each reactive metal produced characteristic interactions, and the severity of these interactions was dependent on the concentration of the reactive metal added. Although maximum reaction and diffusion occurred with zirconium additions, the interaction thickness with each addition exceeded the diameter of most alumina whiskers.

Investigating interfacial stability after exposure of composites to elevated temperatures can indicate their temperature limitations. Although whisker recrystallization is a possibility, the small quantity of reported data suggests that chemical reactions are more important. One reported exception is the recrystallization of silicon nitride whiskers in nickel in less than half the time required for chemical dissolution in nickel [19]. Unidirectionally solidified composites exhibit excellent stability, as would be expected since the systems can form low-energy interfaces upon solidification. For instance, no recrystallization of the nickel aluminide whiskers in an aluminum matrix was observed even after 500 h exposure at 610°C which is only 30°C below the melting point of the system (640°C) [21]. After 264 h at 1370°C, no microstructural changes were observed in the columbium-columbium carbide and tantalum-tantalum carbide systems [22]. It is likely that the system "selects" those specific interfaces to maximize interfacial stability. For example, electron diffraction showed that both blade- and rodlike carbide whiskers grew preferentially in the $\{10\bar{1}0\}$ direction (indices based on hexagonal crystallography). The broadest faces of most tantalum carbide rodlike whiskers were parallel to the type 2 prismatic plane $(1\bar{2}10)$, and some faces were parallel to the basal plane (0001). The planes of the matrix that mated with the whiskers at the interface were not reported; however, electron diffraction studies did show that columbium grew primarily in a $\langle 110 \rangle$ direction.

Alumina whiskers showed no microstructural change in a nickel matrix [12] (electroplated over a sputtered nickel or nickel titanium coating) after 100 h at 1000°C in a static argon atmosphere (10^{-3} mm Hg); after 1000 h at

1000°C the microstructural examination was not possible because of matrix embrittlement. Loss of strength of similar nickel-alumina composites occurred rapidly when heated above 1200°C, but whiskers extracted from a composite held at 1175°C for 22 h showed tensile strengths (in bending) of 900 ± 250 ksi and were structurally unchanged. Although alumina whiskers are thermodynamically stable with respect to nickel, if nickel oxide is formed with diffusion of oxygen into the nickel, the stability criterion is the formation of a spinel between nickel oxide and aluminum oxide, as shown below:

$$NiO + Al_2O_3 \rightleftharpoons NiAl_2O_4 \quad (5)$$

It may be possible for this reaction to occur in nickel if continuous dissolution of oxygen in the matrix compensates for depletion of oxygen combining with nickel to form the spinel. Using polycrystalline specimens Pettit et al. [23] found that at 1200°C, the spinel grew to 3 μ thickness in $\approx$50 h and more rapidly at higher temperatures. It was concluded that spinel formation occurred by counterdiffusion of nickel and aluminum ions with an activation energy of 115 kcal/mol, which equals that for diffusion of aluminum through polycrystalline alumina. This probably reflects a lower value for activation energy in alumina because of the contribution of the grain boundaries to apparent self-diffusion in polycrystalline alumina. Since self-diffusion in single-crystal alumina whiskers may require a higher activation energy, their stability may be improved sufficiently for use at 1200°C in nickel in oxidizing atmospheres.

To promote wetting and inhibit interfacial reactions with the matrix, whiskers have been coated with both metals and ceramics. Any improved behavior depends on the coating not being dissolved, and the maximum time for dissolution (resulting if diffusion but not chemical reaction occurs) can be estimated for the diffusion distance x:

$$x \approx \sqrt{Dt} \quad (6)$$

where D is the diffusion coefficient of the matrix into the coating and t is the time. The variation of D with composition and temperature is discussed in Reference 24. For self-diffusion in metals, $D \sim 10^{-9}$ cm^2/sec near the melting temperature T_m, and 10^{-12} cm^2/sec at about three-fourths of the melting temperature. If diffusion of the matrix into the coating is approximated by the self-diffusion values (a fair approximation if the structure and melting temperature of the matrix and coating are similar), a 0.1-μ-thick (10^{-5} cm) coating on the whisker would be dissolved in $\sim$0.1 sec at the T_m and $\sim$100 sec at $0.75T_m$. Reported dissolution of coatings by nickel (tungsten on silicon carbide [18], tungsten-titanium on sapphire [25], and boron nitride, carbon, alumina, and boron carbide on boron [26])

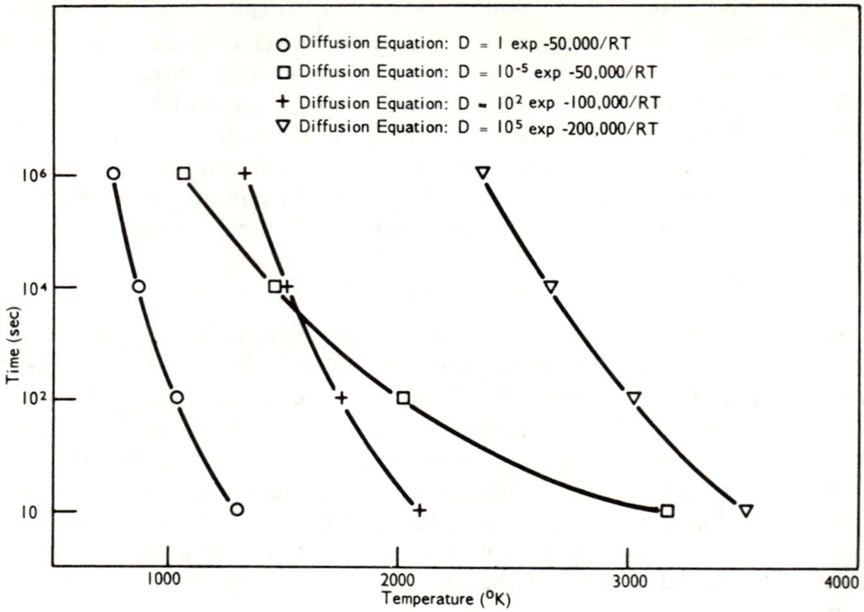

Figure 8.8 Diffusion time-temperature relations calculated from $X^2\ Dt = D_o t \exp(Q/RT)$. Curves are for $X = (2 - \sqrt{2})\ \mu$. (Approximate diffusion into the outer 50% (by area) of a cylinder with initial radius equal to 2 μ.)

verify that diffusion cannot be neglected. Diffusion times at various temperatures were calculated from Eq. 6 for typical diffusion equation in metals ($D \sim 1 \exp\ -50{,}000/RT$) and ceramics ($D \sim 10^2 \exp\ -100{,}000/RT$) and are plotted in Figure 8.8. This shows that diffusion through metallic coatings less than 1 μ thick will occur in a few seconds at temperatures above 1000°K. The coating will dissolve even faster if it reacts with the matrix, unless a fully dense and stable reaction product is formed to envelop the fiber. Instances of protective reaction products have not been reported. The rapid decrease in diffusion and reaction kinetics with decreasing temperature together with the reported stability of whisker coatings in aluminum [27] suggest that for long-term, elevated-temperature applications, stable composites may be more easily developed with lower melting metals.

MECHANICAL STABILITY

Mechanical stability of the interface is particularly important in fiber-reinforced composites since the interface must transfer the applied load from the matrix to the fibers. Efficient utilization of the tensile strength (σ) of cylindrical fibers requires a minimum interfacial shear strength (τ):

$$\tau \geqslant \frac{d}{2l} \tag{7}$$

The largest percentage of whiskers have aspect ratios (l/d) from 100 to 1000, so that the interfacial strength need only be about 0.5% that of the whisker, or about 5000 psi for whisker strengths of 10^6 psi.

Although most engineering metals have shear strengths exceeding this, residual stresses from fabrication or deformation may cause flow or fracture at the interface. For example, residual interfacial stress in a 2024 aluminum alloy-boron composite found from X-ray line broadening to be 9000 psi (tension) exceeded the tensile yield strength (7000 psi) of a control specimen of the matrix [28]. Although plastic deformation in metal matrices can accommodate large interfacial stresses and strains, such is not the case with resins. For example, shrinkage of resins around fibers greater than a critical length (for interfacial failure) caused the fiber to bend into a sine-wave shape (Fig. 8.9) [29]. When the matrix cannot deform plastically, the equality sign in Eq. 7 applies, that is, τ should only be great enough to fully load the fibers to the breaking strength, so that the strain energy released when the fiber breaks can be adsorbed by debonding at the fiber matrix interface [30]. For larger τ, the strain energy is released through crack propagation to adjacent fibers causing catastrophic failure of the composite.

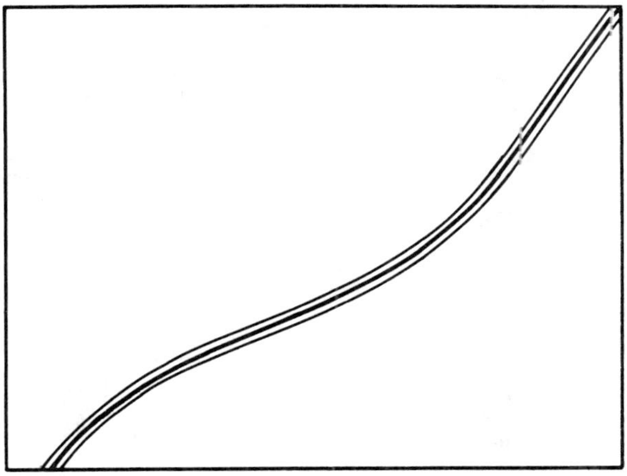

Figure 8.9 Sinusoidal character of an alumina whisker in GE LTV-615 resin (Mag. 318X) [29].

Although resins generally wet and bond well to ceramics, it has been found that the bond between untreated, continuous, high-strength carbon filaments and resin matrices can be too weak for effective load transfer. This load transfer can be greatly improved by a process known as "whiskerizing," whereby very fine graphite or silicon carbide whiskers are grown, by controlled, gaseous, high-temperature exposure, perpendicular to the carbon filament axis. When these whiskerized filaments are infiltrated by the resin matrix, the whiskers provide excellent sites for the mechanical gripping of the fiber by the matrix. By variation of the thermal treatment parameters, various whisker morphologies can be obtained. Figures 8.10*a*, *b* [31] show whiskerized Thornel (a carbon, filament produced by National Carbon Company) with "barbed wire," and with a "fox tail" type of silicon carbide whisker growth. This process has raised the interlaminar shear strength of carbon fiber reinforced epoxy composite from approximately 3000 psi to well over 12,000 psi [31].

The matrix must deform sufficiently to protect the fibers during initial fabrication and subsequent shaping treatments. Pressing loose fibers and metal powders causes substantial fiber breakage [18, 32], except when the matrix readily accommodates the fiber (such as when the matrix is partially liquid during pressing). Although mechanical deformation may strengthen composites, matrix strain will generally be so much greater than that of the fibers that fiber breakage will occur unless the interface and adjacent matrix

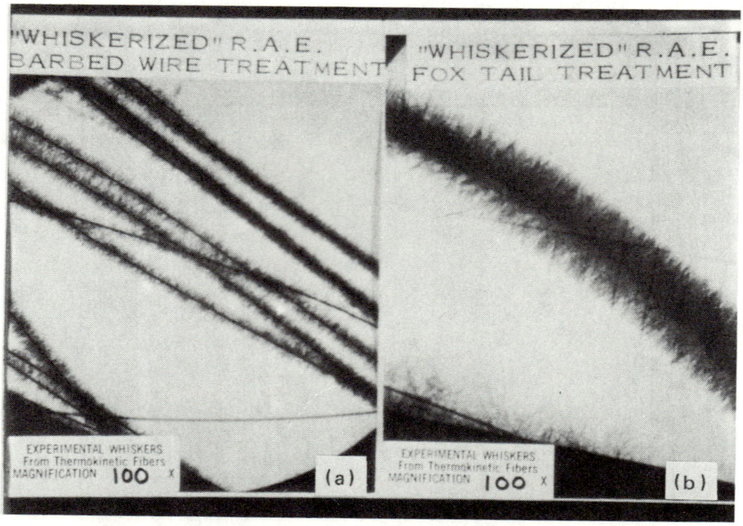

Figure 8.10 Whiskerized carbon filaments. (a) "barbed wire," (b) "fox tail." Courtesy of J. V. Milewski, Thermokinetic Fibers, Inc.

can deform sufficiently to protect the fibers. Both Al-Al_3Ni [33] and Al-SiC [11] composites have been cold-rolled and hot-rolled to large (>50%) cumulative reductions in directions perpendicular to the fiber axes without excessive fiber breakage. However, small reductions (<15%) when rolling parallel to the fiber axes resulting in both fracture and edge cracking. Cladding of composites reduces edge cracking in rolling.

The response of whisker-reinforced composites to applied loads is dependent on the strength and deformability of the interfacial zone as well as stress concentrations at the interface. Weakly bonded interfaces may delaminate under applied stress or reversal thereof and act as cracks. When the load is changed from tension to compression, the fiber stress goes through zero. Although the fiber deforms elastically, the matrix will undergo plastic deformation in these reversals. Bonding between metal matrices and ceramic whiskers is usually improved by coating the whisker with a metal prior to forming the composite. Tensile strengths of infiltrated composites using coating whiskers exceed those using uncoated whiskers (Fig. 8.11) [34]. While uncoated whiskers were poorly bonded to the matrix and were pulled out in tension tests (Fig. 8.12*a*) [9], coated whiskers were well bonded to the matrix and failed in tension under tensile loading (Fig. 8.12*b*).

Improved interfacial bonding obtained with coated whiskers also increases resistance to creep under sustained or cyclic loading (Fig. 8.13). Limited fatigue data for aluminum reinforced with copper aluminide platelets, nickel aluminide whiskers, steel wires, and silica fibers suggest that the ratio of the fiber modulus to matrix modulus must exceed unity and that good interfacial bonding is required for strengthening [21]. The importance of interfacial bonding may also be seen in microstrain behavior. For example, the magnesium boron composites using continuous boron filaments showed a decrease in various moduli after annealing, after cycling in the microstrain range, and after stressing through the microstrain range; this behavior suggests possible debonding [35].

The interface must be able to withstand plastic deformation in its vicinity. Figure 8.14 illustrates this for a high-purity aluminum sessile drop on high-purity polycrystalline silicon carbide. Because deformation of the drop provided an effective contact angle of 90°, the applied stress of about 8000 psi can be considered a shear stress (Fig. 8.7). From Eq. 7 it can be seen that the interfacial strength exceeds that required to effectively transfer the load to a whisker having a strength of 1 million psi and an $l/d > 100$.

In the aluminum-nickel aluminide system, except for specimens having fibers parallel to the load axis, the mechanism of failure is matrix shear on planes 45° to the applied loads. Slip directions are predominantly parallel to the fibers when the fiber axes are oriented at 45° to the load axis and perpendicular to the fibers (on planes 45° to the load axis) when they are oriented

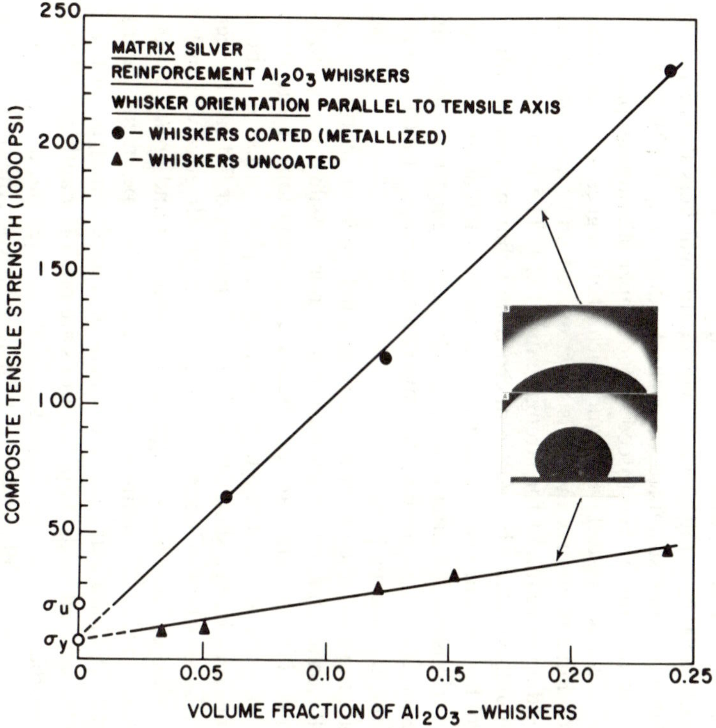

Figure 8.11 Effect of interfacial bonding on strength of silver-Al_2O_3 composites at various fiber volume fractions. Insert photos: sessile drops of molten silver on (A) Ni-coated Al_2O_3 plaque ($\theta = 47°$), and (B) uncoated plaque ($\theta = 122°$), temperature 1100°C, pressure 5×10^{-6} torr [3]. (Courtesy of General Electric Space Sciences Laboratory.)

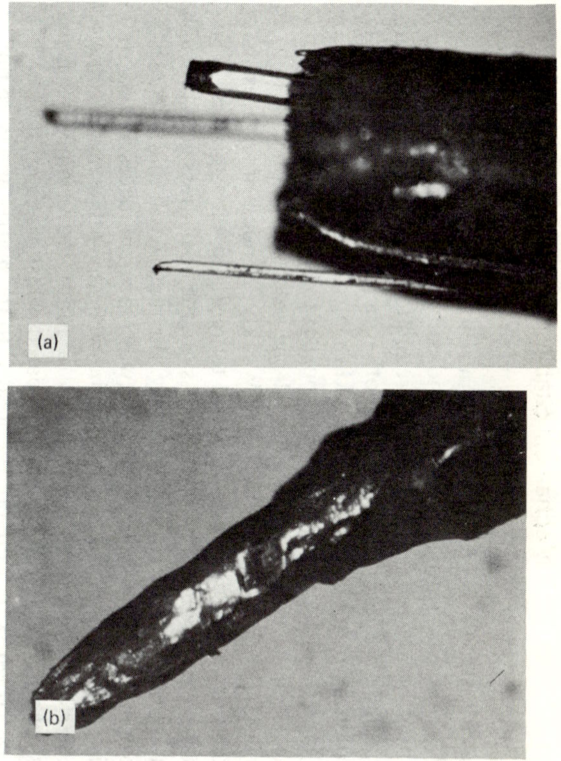

Figure 8.12 Effect of nickel coating Al_2O_3 whiskers at the fracture region of Ag-Al_2O_3 composite wires [9]. (a) Weak composite, no coating; showing whisker "pull-out" (60X); (b) Strong composite, Ni coating; no pull-out (35X).

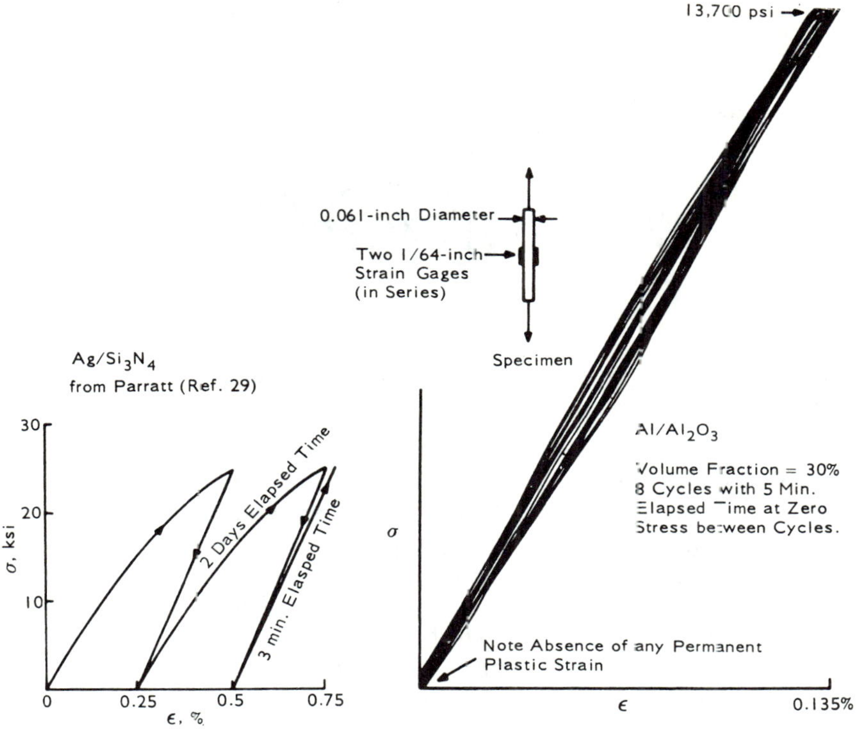

Figure 8.13 Cyclic stability at room temperature of Ag/Si_3N_4 and Al/Al_2O_3 composites [27].

at 90° to the load direction. Tensile elongations greater than 5% were observed. Thus, considerable plastic deformation in the matrix occurred prior to failure of the composite without apparent damage to the interface. For loading directions parallel to the whisker axis, after failure of the whiskers in tension a small crack formed in the matrix surrounding each whisker, and the accumulation of these cracks caused failure; still, elongations of a few percent were obtained and no whisker pullout was observed. Impact toughness of these alloys also indicated that the interfaces as well as the matrix were able to withstand local deformation without failure. Other whisker-metal composites should at least qualitatively show similar behavior. However, no others are currently available in sufficient quantity to permit comparable experiments.

At elevated temperatures, the matrix strength will decrease and thus the whisker length required for efficient load transfer (Eq. 7) will increase. For example, in the unidirectionally solidified composite $Ta\text{-}Ta_2C$, most

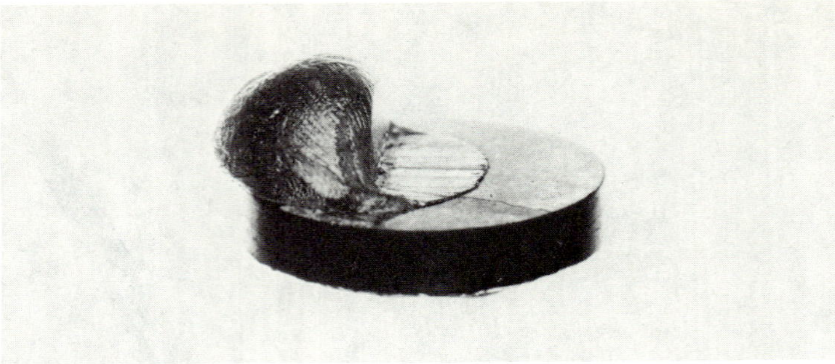

Figure 8.14 Partially sheared high-purity aluminum sessile drop on polycrystalline SiC plaque. True plaque diameter is 3.8 in. (Mag. 4X).

of the room-temperature strengthening is lost at 1370°C although the whiskers were chemically stable at that temperature. Similar results were noted in the Nb-Nb_2C [36] and Al-SiC [10] systems. Decreasing composite strength with increasing temperature may be the result of both matrix softening and reduced mechanical gripping of the fiber by the matrix due to differential thermal expansion. (The fiber strength is reduced much less than the matrix strength in these temperature ranges.)

SPINELS AS INTERFACES

The possibility of spinel formation as interfacial reaction products in metals reinforced with oxide fibers has already been noted. Because the relatively open anion (oxygen) structure of a spinel permits changes in its bivalent and trivalent cation (metal) structure, its physical, chemical, and mechanical properties may be varied. Thus, spinels may be used to promote interfacial bonding, since they potentially can form strong chemical bonds with both metals and ceramics. At the oxide-spinel boundary, certain lattice planes of the spinel anion structure may be coherent with that of the oxide fiber. A gradient in cations within this anion structure may result from reactions of the type

$$MeO + Al_2O_3 \rightarrow MeAl_2O_4$$

where *Me* represents a metal. At the spinel-metal boundary, the cations of the spinel may form a continuous bond with the metal lattice.

Although the properties of many spinels are not fully known, they generally possess high melting points, good mechanical strength, and are chemically stable. As noted earlier, these characteristics are highly desirable. Table 8.5 shows some spinels which may be considered for use as interfaces in oxide whisker-metal composites. It may be seen that the spinel structure permits a very wide range of metal substitutions. Therefore, it may be possible to tailor a strong, stable, spinel interface to a whisker-metal matrix composite. Further research is needed to delineate the value of these interesting materials for interfacial bonding.

TABLE 8.5 REPRESENTATIVE SPINELS THAT HAVE POTENTIAL AS INTERFACES (FROM J. STIGLICH, AMMRC, UNPUBLISHED COMPILATION)

Spinel	Type	Spinel	Type
$AlFe_2O_4$	Inverse	$MgAl_2O_4$	Normal
$CoCr_2O_4$	Normal	$MgFe_3O_4$	Approximate Inverse
$CoFe_2O_4$	Inverse	$MnCo_2O_4$	Inverse
$CuCo_2O_4$	Normal	$NiCo_2O_4$	Approximate Inverse
$CuCr_2O_4$	Normal	$NiAl_2O_4$	Normal
$FeCo_2O_4$	Inverse	$NiCr_2O_4$	Normal
$FeCr_2O_4$	Normal	$NiFe_2O_4$	Approximate Inverse
Fe_3O_4	Inverse	$TiCo_2O_4$	Inverse
$MgCr_2O_4$	Normal	$ZnCo_2O_4$	Normal

CONCLUSIONS

Interfacial considerations for wetting, bonding, chemical, and mechanical stability in whisker composites have been discussed, emphasizing practical rather than theoretical aspects. Available data indicate that molten metals will not generally wet ceramic whiskers even when specific wetting agents such as titanium are added to the base metal. These additions generally increase interfacial reactions which tend to consume the whiskers. Metallic coatings have promoted wettability, but dissolution of the coating in the matrix may permit dewetting during liquid-phase fabrication and debonding during solid-phase fabrication or applications. However, coatings that are effective diffusion and dissolution barriers have been found for nickel and aluminum matrices. Results of mechanical deformation of whisker-reinforced metal composites indicate that whisker breakage will readily occur unless the areal reduction per operation is very small, the matrix flows perpendicular to the whisker axis, and no interfacial pores or whisker-whisker contact occurs. However, careful processing has enabled total reductions $>50\%$ to be achieved without catastrophic edge cracking or whisker breakage. An understanding of the role of the interface and control of the factors affecting it are essential to the development of useful whisker-reinforced composites. Generally, interfacial structures cannot be explained by current theories and more fundamental research is needed to establish the nature of whisker-metal interfaces. Present data suggest that interfacial bonding results more from mechanical gripping than chemical bonding and that coherently bonded interfaces seem unlikely. Although current data can be used to determine the stability or instability of a given system under specified conditions, they do not indicate the specific reactions causing chemical instability. Such reactions and their kinetics require further study and clarification. In summary, an understanding of the interface and the factors which affect it are essential to optimizing the fabrication and application of whisker-reinforced composites.

ACKNOWLEDGMENT

The authors appreciate the helpful suggestions of and discussions with Drs. Howard Aaron, George Bishop, and Eric Kula.

REFERENCES

[1] D. L. Harrod and R. T. Begley, "Some General Aspects of Interface in Composites," *Proc. 10th National Symposium SAMPE*, **10** (1966), E1.

[2] L. J. VanVlck, "The Metal Ceramic Boundary," *Metals Eng. Quart.*, **5,** X (1965), 7–12.

[3] H. W. Rauch, Sr., W. H. Sutton, and L. R. McCreight, "Survey of Ceramic Fibers and Fibrous Composite Materials," Air Force Materials Laboratory Report AFMLTR-66-365, October 1966.

[4] S. M. Wolf, A. P. Levitt, and J. Brown, "Wetting and Bonding Between Aluminum Alloys and Sapphire," Army Materials and Mechanics Research Center, AMRA TR 66-19, July 1966.

[5] S. Glasstone, *Textbook of Physical Chemistry*, D. Van Nostrand Co., Inc., New York, 2nd Ed., 1946, p. 1324.

[6] B. V. Allen and W. D. Kingery, "Surface Tension and Contact Angles in Some Liquid Metal-Solid Ceramic Systems at Elevated Temperatures," *Trans. AIME*, **215,** 30 (1959), 30–37.

[7] W. H. Sutton and E. Feingold, *Role of Interfacially Active Metals in Apparent Adherence of Nickel to Sapphire*, General Electric Company, Materials Science Research, Vol. 3, Plenum Press, 1966.

[8] W. E. Sauer, W. H. Sutton, and E. Feingold, "Investigation of Bonding in Oxide-Fiber (Whisker) Reinforced Metals," General Electric Company, Contract DA-19-066-AMC 330(X), Interim Report, AMRA CR 66-01/4, 18 April 1966–18 April 1967.

[9] M. Nicholas and D. M. Poole, "The Influence of Oxygen on Wetting and Bonding on the Copper-Tungsten System," *J. Mater. Sci.*, **2,** (1967), 269–274.

[10] W. H. Sutton, J. Chorné, R. L. Mehan, and W. E. Sauer, "Development of Composite Structural Materials for High Temperature Applications," General Electric Company, Contract NoW-65-0176-c, 1st Progress Report 23 November 1964–22 February 1965.

[11] P. J. Lare, A. P. Divecha, and H. Hahn, "Research on Whisker Reinforced Metal Composites," Melpar, Incorporated, Contract NoW 66-0213, Final Report, April 1966–April 1967.

[12] J. Chorné, C. A. Bruch, and W. H. Sutton, "Development of Structural Materials for High Temperature Applications," General Electric Company, Contract NoW 66-0443-d, January 1967.

[13] E. F. Brush, Jr. and C. M. Adams, Jr., "Vapor Coated Surfaces for Brazing Ceramics." *Welding J.* **47** (1968), 106s.

[14] M. H. Richman and R. D. French, "Investigation of the Role of the Interface in Composite Materials," Final Report No. AMMRC CR 68-13(F), Dec. 1968, DAAG 46-68-0036 with Brown University.

[15] M. J. Warbisi, "Wetting and High and Low Energy Surfaces by Liquid Adhesives and its Correlation to Bond Strength," Picatinny Arsenal Technical Report 3456, January 1967.

[16] W. C. Jones and M. C. Porter, "Measurement of Contact Angles on Filaments to Indicate the Effects of Surface Treatments," *Proc. 10th Nat. Symposium SAMPE*, **10** (1966), 17.

[17] M. Nicholas, R. Foran, and D. M. Poole, "The Adhesion of Metal Alumina Interfaces," Metallurgy Division and Research Division, U.K.A.E.A. Research Group, Atomic Energy Research Establishment, Report AERE-R5258, September 1966.

[18] A. Divecha, P. Lare, and H. Hahn, "Research on Silicon Carbide Whisker-Metal Matrix Composites." Melpar Incorporated, Contract AF33 (615) 5431, Second Quarterly Report, 1 October 1966–1 January 1967.

[19] L. A. Warnes, "The Compatibility of Silicon Nitride Whiskers with Metallic Matrices at Elevated Temperatures," Ministry of Aviation, Explosive Research and Development Establishment, Report E.R.D.E. 17/R/65, July 1965.

[20] D. W. Petrasek and J. W. Weeton, "Alloying Effects on Tungsten-Fiber Reinforced Copper-Alloy or High Temperature Alloy Matrix Composites," NASA Technical Note NASA TN D 1578, October 1963.

[21] M. J. Salkind, et al., "Investigation of the Creep, Fatigue, and Transverse Properties of Al_3Ni Whisker and CuA_2 Platelet Reinforced Aluminum," United Aircraft Corporation, Final Report under Contract NoW 65-0384-d, 11 May 1966.

[22] F. D. Lemkey and M. J. Salkind, "The Growth and Properties of Carbide Whisker-Reinforced Refractory Metals from the Melt," United Aircraft Corporation, 11 May 1966, presented at the International Conference on Crystal Growth, 1966.

[23] F. S. Pettit, E. H. Randklev, and E. J. Felten, "Formation of NiA_2O_4 by Solid-State Reaction," *J. Am. Ceram. Soc.*, **49** (1966), 199–203.

[24] P. G. Shewmon, *Diffusion in Solids*," McGraw-Hill, New York, 1963.

[25] J. Chorné, C. A. Bruch, and W. H. Sutton, "Factors Affecting the Preparation and Strength of Nickel-Alumina Whisker Composites," paper presented at Symposium on Metal-Matrix Composites, 70th Annual ASTM Meeting, Boston, 29 June 1967.

[26] AVCO Corporation, "Research on Boron Filament-Metal Matrix Composites Materials," AVCO Corporation, Space Systems Division Monthly Letter Report No. 7, Contract AF33 (615)-3154, 7 May 1966.

[27] R. Mehan, "Fabrication and Evaluation of Sapphire Whisker-Reinforced Aluminium Composites," 70th Annual Meeting of the ASTM, June 25–30, 1967, Boston.

[28] H. P. Cheskis and R. W. Heckel, "X-Ray Line Broadening and Residual Stress Studies on Composite Materials," Drexel Institute of Technology, 70th Annual Meeting of the ASTM, June 25–June 30, 1966.

[29] G. A. Schmidt, "Whisker Matrix Bonding Studies," General Electric Company, Contract NoW 64-0330-d, Quarterly Progress Report No. 1, 27 May 1964–27 August 1967.

[30] J. Mullin, J. M. Berry, and A. Gatti, "Some Fundamental Fracture Mechanisms Applicable to Advanced Filament Reinforced Composites," *J. Comp. Mater.*, **2,** 1 (January 1968), 82–103.

[31] J. J. Shyne and J. V. Milewski, "Whiskerizing—A Physical and Chemical Change to the Interface of Graphite Fibers," Preprint of 24th Annual Technical Meeting of the Society of Plastics Industries, Reinforced Plastics and Composites Division, February 7, 1969.

[32] N. J. Parratt, "Reinforcement of Plastic and Metals by Silicon Nitride Whiskers," E.R.D.E., Waltham Abbey, England, 58th Annual Meeting.

[33] M. J. Salkind and F. D. George, "Matrix Strengthening Mechanisms in A_3Ni Whisker-Reinforced Aluminum," United Aircraft Research Laboratories, Contract NoW 66-0206-D, 2nd Quarterly Progress Report, 12 December 1966.

[34] W. H. Sutton and J. Chorné, "Factors Affecting the Tensile Strength of Metals Reinforced with Strong Fibers," Strengthening Mechanisms Metals and Ceramics, Syracuse University Press, Syracuse, 1965.

[35] M. R. Pinnel, A. Lawley, and D. R. Hay, "Microstrain Behavior of Metal Matrix Composites," Drexel Institute of Technology, 70th Annual Meeting of the ASTM, June 25–June 30, 1966.

[36] F. D. Lemkey and M. J. Salkind, "The Growth and Properties of Carbide Whiskers-Reinforced Refractory Metals from the Melt," Paper No. A24, *Proceedings of an International Conference on Crystal Growth*, Edited by H. Steffen Peiser, Pergamon Press, 1967.

Willard H. Sutton

9 Principles and Methods for Fabricating Whisker-Reinforced Composites

INTRODUCTION

The basic purpose of the fabrication process is to combine the whiskers with the matrix in such a manner that the strength, modulus, and other desirable properties* of the whiskers are utilized to maximum advantage. This means that the whiskers must be aligned, uniformly spaced, and adequately bonded to the matrix. Like other composite materials, the whisker composites should be considered as a materials system that consists of at least two constituents: the whiskers and the matrix. Each constituent must perform several specific functions if the desirable composite properties are to be achieved. In order to perform these functions, which will be discussed later, the whiskers must be combined with the matrix in a carefully prescribed manner, and they must not be weakened by chemical reactions or be fragmented during the fabrication process.

The technology involved in the preparation of whisker composites is highly complex and involves far more than the elementary concept of simply joining two phases together. For example, in powder compacts, the achievement of theoretical (100%) density through a pressure and heat-treating process may be all that is required. However, in a whisker composite the attainment of complete densification can still result in a weak product.

* In this chapter, only mechanical properties are discussed, although optical, electronic, magnetic, thermal, and other whisker properties are promising areas for research and development.

Thus, in order to produce high-strength whisker composites, one must be aware of considerably more factors than just the fabrication parameters themselves. These will be discussed in greater depth. The current state of the composite-fabrication art is in an early stage of development, and there is much to be learned. However, many of the problem areas have been uncovered, so that more scientific and systematic approaches can be used now toward their solution.

This chapter is intended to present some of the broader aspects and the principles of composite fabrication and to review the progress made thus far on the preparation of whisker composite materials.

FACTORS TO BE CONSIDERED IN COMPOSITE PREPARATION

A unique feature of fiber composites is their ability to be made into structural materials having a variety of desirable, tailor-made properties. This feature, therefore, gives the designer far greater flexibility to create new structures, particularly where directional properties are important, along with superior specific properties, namely, high strength-to-weight and high stiffness-to-weight ratios. However, the achievement of superior properties is critically dependent upon the manner in which the constituents are combined.

Since these materials are tailor-made, and since the desired (directional) properties can be achieved only when the constituents are combined in a prescribed manner, the functions of each constituent, the mechanical and geometrical requirements, and the physical and chemical requirements will be reviewed prior to the discussions on fabrication methodology.

Function of the Constituents

The strengthening of fiber composite materials is based on the principle of reinforcement. Thus, the prime function of the fibers or whiskers is primarily that of supporting externally applied loads. On the other hand, the main role of the matrix is to bind the whiskers together and to transmit and distribute the loads to the individual whiskers. Many times, however, it is difficult to combine the matrix with the fiber, unless the fibers are first coated with another material (i.e., a wetting or bonding promoter). Thus, the coating becomes a third constituent, whose function is to aid in the fabrication process and to serve as the stress-transfer region between the fibers and the matrix.

Figure 9.1 summarizes several mechanical functions required of each of the constituents. The relative importance of each function will vary with the

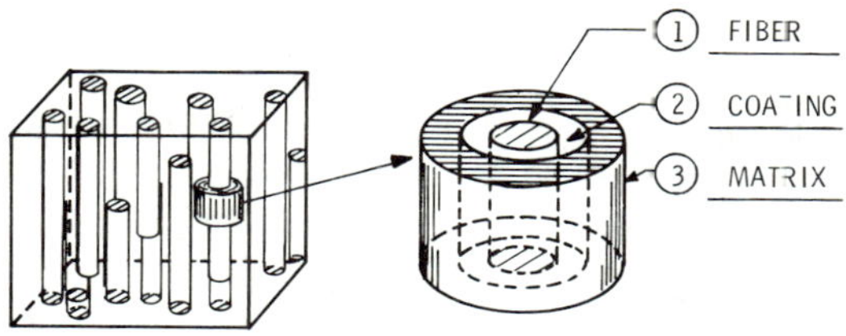

CONSTITUENT	FUNCTION
(1) FIBER (WHISKER)	a) PRINCIPAL LOAD BEARING ELEMENT
	b) RESTRAINS CRACK PROPAGATION IN MATRIX
	c) IMPARTS STIFFNESS TO MATRIX
	d) LOWERS COMPOSITE DENSITY (METAL MATRIX)
(2) FIBER COATING	a) PROMOTES WETTING AND BONDING
	b) PROTECTS FIBER FROM MECHANICAL DAMAGE
	c) SERVES AS DIFFUSION BARRIER
	d) CONTROLS FIBER SPACING AND VOLUME FRACTION
	e) AIDS IN FIBER ALIGNMENT (ie, MAGNETIC)
	f) TRANSFERS STRESSES FROM MATRIX TO FIBERS
(3) MATRIX	a) TRANSFERS EXTERNAL LOADS TO FIBERS
	b) BINDS FIBERS TOGETHER
	c) SPACES FIBERS
	d) PROTECTS FIBERS FROM EXTERNAL ENVIRONMENT
	e) CRACK ARRESTOR AT FIBER BREAKS

N502-965

Figure 9.1 Function of the constituents in fiber composite materials.

particular composite application and with the composition of the constituents. This will be further elaborated upon when the preparation and properties of specific composite systems are discussed.

Mechanical and Geometrical Considerations

Under ideal conditions, the elastic modulus and the tensile (and yield) strength of a unidirectionally reinforced composite* are often viewed as being linearly dependent on the fiber concentration, and indeed, this has been observed in several composite systems [1–3]. Thus, the rule-of-mixtures is frequently used to predict both composite strength and modulus,

* It is assumed that all the fibers are aligned along the tensile direction.

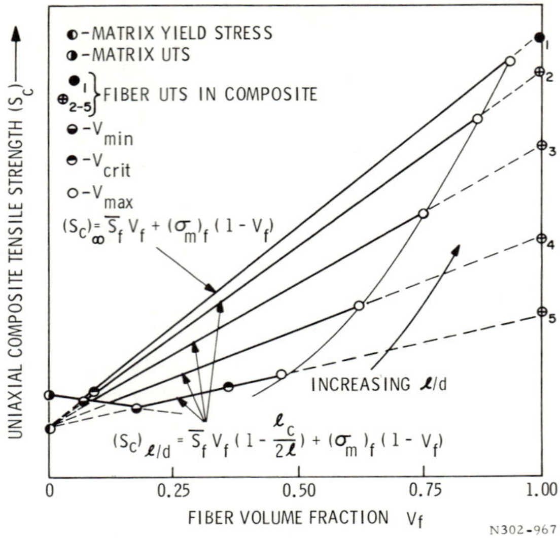

Figure 9.2 Effect of fiber volume fraction, fiber strength, and fiber aspect ratio on composite strength,

and thereby provides a convenient means for comparing predicted with observed values.

Figure 9.2 illustrates the use of the rule-of-mixtures (which was discussed in Chapter 7) in predicting composite strength when the fibers are uniaxially aligned. This oversimplified approach assumes equal strains in both the matrix and fibers, so that the composite stress, σ_c (and strength, S_c), at any given strain can be predicted from the stress-strain curves of the individual constituents. Thus, when the fibers are continuous in length throughout the specimen, the composite strength can be expressed as follows:

$$(S_c)_\infty = \bar{S}_f V_f + (\sigma_m)_f (1 - V_f); \qquad (V_f > V_{\min}) \tag{1}$$

where $(S_c)_\infty$ = strength (UTS) of composite having continuous fibers

$\bar{S}_f$ = average strength of fibers

V_f = fiber volume fraction

$(\sigma_m)_f$ = stress in matrix at fiber breaking strain

$V_{\min}$ = Minimum fiber volume fraction to cause fiber reinforcement

However, when the fibers are broken, or when they are discontinuous (which is the case for most whisker composites), the stress along the fiber is no longer uniform, as shown in Figure 9.3. Assuming that the fiber is shorter than the specimen length, and assuming that both the fibers and matrix elongate elastically to fracture, the shear stress at the interface will rise rapidly from the fiber ends (where $\tau = 0$) to a maximum value at a relatively short distance from the fiber ends. Then continuing further toward the center of the fiber, the shear stress will decay rapidly, and if the fiber is sufficiently long, its value will be zero, as shown in Figure 9.3*b*. The tensile stress in the fiber (Fig. 9.3*c*) will also build up along the fiber length, from zero at the ends to a maximum value toward the fiber center. On the other hand, if the fiber is continuous in length (viz., $l \cong \infty$), then the interfacial shear stress will be zero, and the tensile stress will be constant along the fiber axis. Thus, it can be seen that for short reinforcing fibers, there is a portion along the fiber's ends where the stresses are less than that of a fiber which is continuous in length. This region is often called the fiber ineffective length [4]. In order to take into account the effect of this length

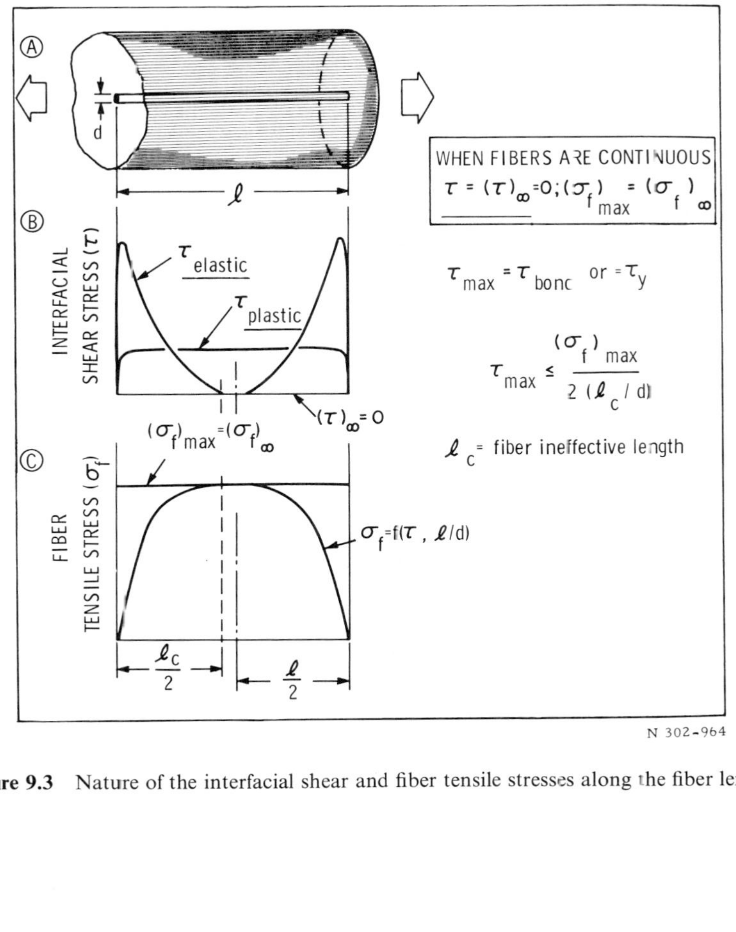

Figure 9.3 Nature of the interfacial shear and fiber tensile stresses along the fiber length.

on the composite strength, Kelly and Davies [5] have modified Eq. 1 as follows:

$$(S_c)_{l/d} = \bar{S}_f\left(1 - \frac{lc}{2l}\right)V_f + (\sigma_m)_f(1 - V_f); \qquad V_f > V_{\min}\,(\text{chap. 11}) \quad (2)$$

where $(S_c)_{l/d}$ = strength in composite containing short fibers
l_c = critical fiber length
l = total length of fiber

Thus, as the fiber length approaches an infinite value (viz., a continuous length), $(S_c)_{l/d} \to (S_c)_\infty$. In Figure 9.2, a family of strength curves are shown for composites having fibers of different aspect ratios (length-to-diameter ratios) l/d. Assuming a linear tensile stress buildup from ends of the fiber, Kelly and Davies [5] showed that when the fibers were all of critical length, lc, then $(S_c)_{l/d} = 0.5(S_c)_\infty$. When the fiber length was ten times l_c, $(S_c)_{l/d} = 0.95\,(S_c)_\infty$.

Referring also to Figure 9.2, the family of curves shown could also represent the strength of composites where the average strength level of the reinforcing fibers is different. Thus $(S_c)_1 \gg (S_c)_5$, where the average strength of the fibers in composite number 1 is much greater than in composite number 5. This also means that if the fiber strength is degraded, for example, from a value of $(S_f)_1$ to $(S_f)_5$, the composite strength would be decreased proportionately. Also, both the values of $V_{\min}$ and V_{crit} would be increased proportionately. In other words, when the fibers in a composite are either broken or weakened, the composite strength will be decreased accordingly, and the critical volume fraction of fibers (V_{crit}) required to exceed the strength of unreinforced matrix will also be increased. When high-strength whiskers having high-aspect ratios (i.e., $l/d \approx 1000$) are used to reinforce soft metals, the value of $V_{\min}$ may be less than 0.01. However, if the whiskers are greatly weakened (or broken) during the fabrication process, $V_{\min}$ and V_{crit} may approach much larger values as shown in Figure 9.2. For example, in composite number 5, the value for $V_{\min}$ is about 0.2, and for V_{crit}, about 0.35.

Another problem also arises when shorter and shorter (or smaller and smaller diameter) fibers are used as the reinforcements. This arises from practical considerations, where it becomes increasingly difficult to fully align the fibers. Thus, as the fibers become shorter or finer, there is a tendency to have increasing fiber overlap and bunching, which in turn substantially increases the overall spacing between fibers, and reduces the value of the maximum packing fraction.

A sharp reduction in the value of $V_{\max}$ with decreasing fiber aspect ratio (l/d ratio) is shown in Figure 9.2. The exact shape of the curve passing through the V_{max} points will depend on the specific details of each composite system, on the fabrication procedure, and on the actual values of l and d.

For circular fibers, the maximum (theoretical) packing density, V_{max}, is about 0.91. In the case of glass-reinforced plastics, where the continuous fibers can be filament-wound under rather precise conditions, $V_{max} \cong 0.86$ [6]. However, for whisker composites, the measured value of V_{max} generally is much lower, being about $V_{max} \sim 0.4$ to 0.5. In a few exceptional cases, $V_{max} \sim 0.6$. However, as the fiber packing becomes large ($V_{max} > 0.3$), there is an increasing tendency to cause fiber breakage during composite fabrication or during mechanical forming.

The problem of alignment becomes far more acute when short, thin fibers are used instead of continuous ones. Misalignment reduces fiber packing density, and it also reduces the strength of the composites. For example, if all the fibers are misaligned at a given angle, the composite strength will be reduced by a predictable amount. Kelly and Davies [5] along with others [7–9] have shown this, and the relationships are shown in Figure 9.4. The value of Φ_{crit} is generally small for many metal-matrix composites (i.e., $\Phi_{crit} = 5°$ or less when $V_f = 0.5$ or less).

Below the value of Φ_{crit}, the composite strength is governed by the fracture strength of the fibers. However, when $\Phi > \Phi_{crit}$, failure occurs in

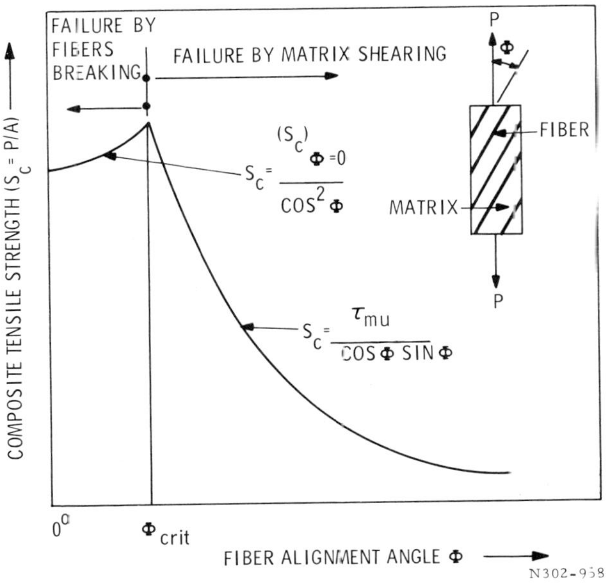

Figure 9.4 Effect of fiber alignment angle on the tensile strength of composites containing unidirectionally aligned fibers.

the composite by shear deformation of the matrix between the fibers. Since the strength of the composite is primarily dependent on the matrix ultimate shear strength, τ_μ (Fig. 9.4), once Φ exceeds Φ_{crit}, the composite strength can be increased by increasing τ_μ (by metallurgical methods in the case of a metal matrix), or by altering the fiber configuration. For example, composite strengths and the values of Φ_{crit} have been increased considerably by utilizing cross-ply fiber configurations [7] which greatly restrain the shear deformation mode of the matrix.

In summary, there are many mechanical and geometrical factors affecting composite strength and, from a fabrication viewpoint, many trade-offs between such things as processing conditions, fiber strength, fiber alignment, and fiber packing, must be considered. Also, there are many other complicating factors, such as the scatter in fiber strength, fiber taper, poor fiber distribution, matrix porosity, and formation of brittle phases in the matrix. These are not discussed, although the effects of some of these factors will be shown later.

While the nature of the actual stresses in fiber composites is far more complicated than that just discussed, the relationships shown do give the composite fabricator some useful, simple guidelines regarding the mechanical and geometrical factors that control composite strength. For a detailed analysis of these stresses and their effects, see Chapter 7.

Physical and Chemical Considerations

Physical and chemical reactions, such as changes in fiber shape, recrystallization, grain growth, formation of new phases, diffusion of atomic species, and formation of voids, are likely to occur during the fabrication process. The results of these and other changes can have profound effects on the microstructure and on the properties of the composite material. For example, any physical or chemical change that might adversely affect the transfer of stresses between the whiskers and the matrix, or any changes that might reduce the strength of the whiskers, must be taken into account in the fabrication procedure.

Thus, any of these factors that exert a critical influence in the mechanical performances of the composite materials should be of prime concern to the fabricator. In the preparation of composites, at least four factors should be considered:

1. Contact surface between whiskers and matrix (wetting).
2. Interfacial bond strength.
3. Chemical reactions at the interface.
4. Surface perfection of the whiskers.

These factors are strongly dependent on one another, but for the purposes of discussion, it is convenient to treat them (or their effects) separately.

Contact Surface Between Fibers and Matrix (Wetting). Whether fluid or solid-state methods are used to fabricate whisker composites, the whiskers must be fully embedded in the matrix and their entire surface should be in contact (on a molecular scale) with the matrix. Otherwise, the transfer of stresses from the matrix to the whiskers will be adversely affected, and the composite strength will be reduced.

From a fabrication viewpoint, wetting (when the contact angle θ is less than 90°) between the matrix and whiskers is desired since this favors maximum contact surface area under equilibrium conditions (Chapter 8). Thus, wetting conditions tend to promote maximum contact between the whiskers and matrix. Conversely, if the whiskers are not wetted by the matrix, the probability of voids and regions of poor contact at the interface becomes very great.

The relations between wetting (and nonwetting), microstructure, and properties of polycrystalline, multi-phase materials has been reviewed rather extensively in the literature [10–16]. However, much is yet to be learned about wetting (and sintering), so that new knowledge and principles can be applied effectively to the fabrication of all composite materials. This is particularly true for the case of fiber composites, where chemical, mechanical, and geometric interactions become increasingly complex.

The present discussion will briefly highlight some of the basic principles that can be used in composite fabrication procedures. Figure 9.5 shows the effects of wetting and nonwetting conditions between a fiber surface and a matrix which is in the molten state. The matrix may be added initially to the fibers as a powder, as a coating, or by other techniques (to be discussed later). However, once the matrix becomes molten, irrespective of its initial state, it will either wet or not wet the fiber as shown in Figure 9.5*a* or *b*. If chemical reactions, diffusion, or preferential absorption do not occur at the liquid-solid boundary, the liquid matrix should immediately take on an equilibrium shape shown in Figure 9.5*b*. However, if reactions do occur with time, where the surface tension of the liquid may change or where the solid-liquid interfacial tension may change (due to dissolution of the fiber, due to the formation of new phases, or due to impurity or atmosphere effects), the equilibrium shapes shown in Figure 9.5*b* may change to those shown in Figure 9.5*c*. Systems that were initially wetting may remain wetting; or those that were initially nonwetting may also remain the same; or the reverse effect may occur: nonwetting systems may become wetting and vice versa. The consequences of these changes may be desirable in some cases, but usually they add to the problems in composite preparation and adversely

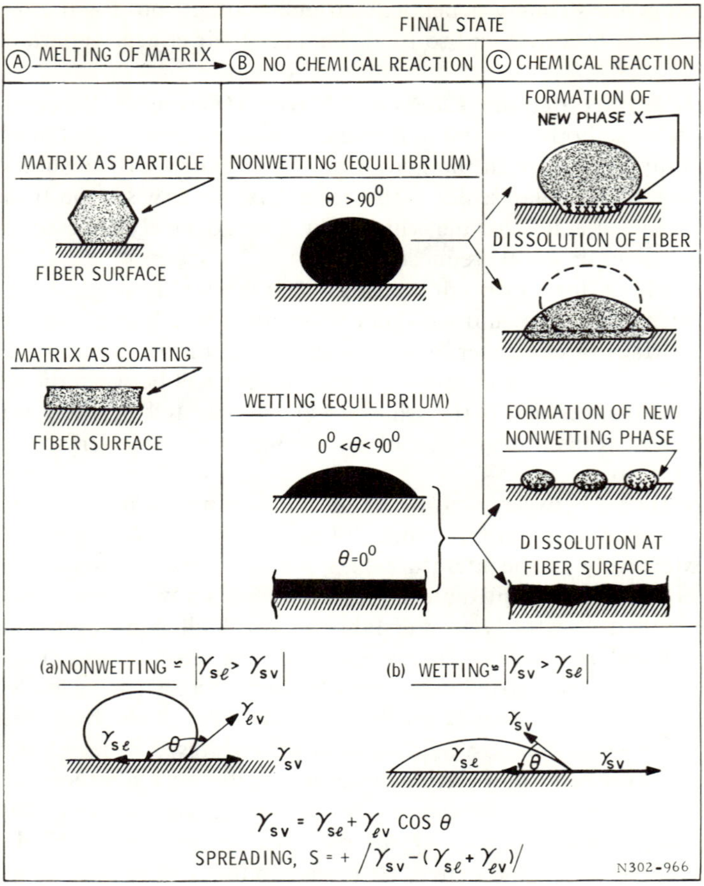

Figure 9.5 Effect of wetting conditions on the contact region between the molten matrix and fiber surface.

affect composite properties. Thus, it is desirable to know in advance the wetting characteristics of the composite constituents as a function of temperature and atmosphere, how to control or change the conditions, and what the consequences will be on both short- and long-term composite properties.

As described in Chapter 8, the "degree" of wetting can be determined conveniently by using the sessile drop method, where a drop of liquid rests on a solid substrate. As shown in the lower portion of Figure 9.5, wetting occurs when the contact angle, θ, is less than 90°. Perfect, or spontaneous, wetting occurs when $\theta = 0°$. Whether or not wetting occurs is determined by the magnitude of the horizontal components of the solid-liquid, liquid-vapor, and solid-vapor surface tensions, as shown also in Figure 9.5. Thus, when the solid-liquid interfacial tension, γ_{SL}, is large with respect to that of the solid-vapor, γ_{SV} (and γ_{LV}), nonwetting conditions are favored. Conversely, when γ_{SL} is small relative to γ_{SV} (γ_{LV} remaining the same), wetting conditions predominate. In order to have spreading of the drop occur, the value of the spreading coefficient, $S = |\gamma_{SV} - \gamma_{SL} - \gamma_{LV}|$, must be positive. These conditions are discussed in detail elsewhere (see Chapter 8) and will not be discussed further.

Some effects of wetting conditions on the fiber-matrix contact surfaces and on the resulting composite microstructure are illustrated schematically in Figure 9.6. In Case I, the matrix becomes fully molten during fabrication, and two situations—wetting and nonwetting—are shown. When nonwetting conditions prevail, the final composite material is likely to contain voids between the fibers (Fig. 9.6). On the other hand, wetting conditions will tend to minimize the occurrences of voids. In Case II, the fiber and matrix are shown along with a third, liquid phase. Assuming the matrix to be initially in the particulate form (as indicated schematically by a hexagonal cross-section in Fig. 9.6), this phase might be the coating on the matrix particle (or in the particle), or it might be the result of alloying effects, where a liquid phase exists in equilibrium with the solid phase (i.e., a liquidus region in a phase diagram). However, several possibilities exist as shown in Figure 9.6. If neither the matrix nor fiber are wetted by the liquid phase, this phase will tend to form droplets, and a large void density is likely to be observed in the composite microstructure. If this composite were initially fabricated by powder metallurgical methods, namely, by densely compacting the fiber-powder mix together, then once the liquid phase forms during the heating process, it would be ejected from the compact as droplets. The same phenomenon is observed in the "sweating" of powder compacts when a nonwetting liquid phase is formed [10]. On the other hand, the liquid phase may wet only the particle,* in which case, it would tend to form a continuous film (or low contact angle) over the matrix particles, and still form droplets on the fiber. The resulting composite microstructure (shown highly idealized in Fig. 9.6) would be an array of particles fully coated by the liquid phase. The surface tension of the liquid matrix would also tend to draw the particles together. The regions surrounding the exposed fiber

* In all the drawings in Case II, Figure 9.6, the matrix particle is shown to be hexagonal in cross-section, and no changes were assumed to occur which would alter its shape or size, that is, dissolution, grain growth, etc.

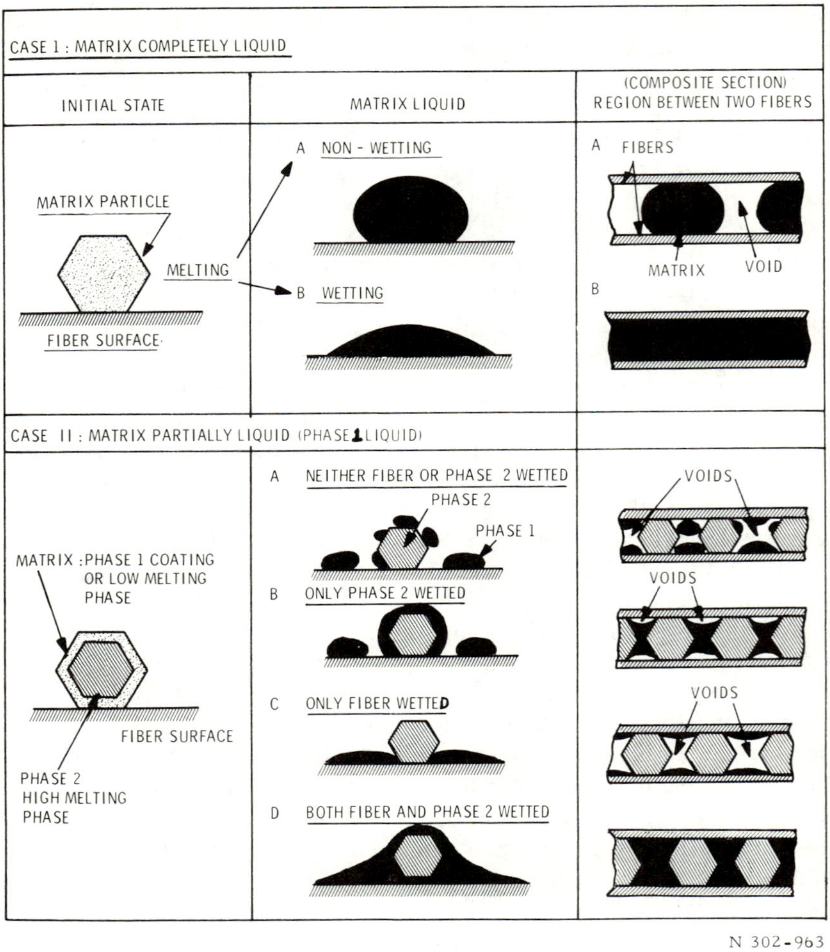

Figure 9.6 Schematic diagram showing wetting conditions in matrix-fibers system where the matrix becomes fully liquid (Case I) or partially liquid (Case II) during the composite fabrication process.

surfaces would be voids, since the fibers were not wetted by the liquid phase. If the matrix particles were in contact with the fibers, as shown in Figure 9.6, and if they were not free to turn, the liquid phase would not penetrate between the two solid phases. This would also hold true if only the fibers were wetted by the liquid phase, except that the voids would now occur between the particles, which is illustrated also in Figure 9.6. The ideal situation occurs when both the fiber and the matrix particles are wetted by the liquid phase. Then the liquid penetrates between both the fiber-particle and particle-particle points of contract, as well as between the interstital regions.

The cases shown in Figure 9.6 are for relatively simple and idealized conditions. In most composite systems, these situations are complicated by additional factors, such as the dissolution of solid phases in the liquid; the presence of impurities in the matrix or adsorbed species on the particle and fiber surfaces; and chemical reactions, diffusion, and vaporization of species. Thus, a detailed treatment of any of the conditions discussed here would have to be based on the specifics of each system under consideration. It is these details which are important to the successful fabrication of whisker composites.

Thus far, only equilibrium (static) conditions have been discussed. However, the dynamic conditions are also equally important, since they are active during most of the fabrication process. In considering the infiltration process, for example, when the matrix is in a liquid state and penetrates a bundle or array of packed fibers, additional factors become important. The degree of penetration of the liquid matrix into the spacings or channels between the packed fibers is proportional to the surface tension of the matrix (γ_{LV}), and to the degree of wetting (measured by the contact angle, θ), and is inversely proportional to the effective diameter of the channel, d_c. For purposes of discussion, it is assumed that the fibers are all aligned and are all uniformly spaced, so that the channels are all of the same size and run parallel to the fiber length.

If the liquid matrix wets the fibers, and if the fibers are bundled together and positioned vertically over the matrix, then the liquid will rise by capillary action upon contact, up through the channels to a height that is determined by the weight of the column. Thus, a pressure difference, P_c, due to the surface tension of the liquid will be balanced by the hydrostatic pressure of the liquid column. If the channel diameter is d_c, the capillary pressure, P_c, is given by the expression:

$$P_c = \frac{4\gamma_{LV}\cos\theta}{d_c} \tag{3}$$

The channel spacing (d_c) is dependent on the diameter of the fibers (d) and on the packing density of the fibers (or volume fraction, V_f). If the fibers are arranged so that their cross-sections in a given plane show cubic packing, then the channel diameter can be approximated by the following:

$$d_c \approx \left(\sqrt{\frac{\pi}{2V_f}} - 1\right)d \tag{4}$$

By substituting the value of d_c in Eq. 4 into Eq. 3, the capillary pressure of

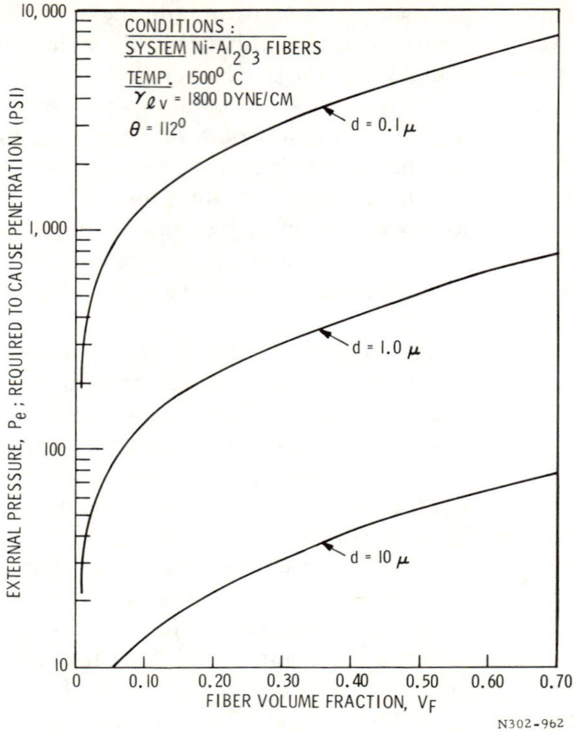

Figure 9.7 Calculated external pressures (P_e) required to force molten nickel into the capillaries between whiskers having different diameters and different packing fractions (V_f).

any liquid penetrating the fiber bundle will be given approximately as

$$P_c \approx \frac{4\gamma_{LV} \cos \theta}{(\sqrt{(\pi/2V_f)} - 1)d} \tag{5}$$

If wetting occurs ($\theta < 90°$), P_c will be positive and the liquid matrix will penetrate into the fiber bundle. If nonwetting conditions prevail ($\theta > 90°$), P_c will be negative, and an additional, external pressure, P_e, will be required to force the liquid into the channels. Figure 9.7 shows values of P_e ($P_e \geq P_c$) necessary to force liquid nickel into channels between Al_2O_3 whiskers. Liquid nickel does not wet Al_2O_3 at 1500°C [16–20], so that positive external pressures are necessary for infiltration to occur. In the example shown in Figure 9.7, the whisker diameters range from 0.1 to 10 μ,* and

* 1 $\mu = 4 \times 10^{-5}$ in.

the whisker volume fractions vary from less than 0.1 to 0.7. The results indicate that for very fine whiskers of Al_2O_3 ($d \sim 0.1\ \mu$), high pressures exceeding 1000 psi will be required to cause liquid nickel to penetrate into whiskers bundles where V_f exceeds 0.1. For any given volume fraction, each tenfold increase in whisker diameter will bring about a corresponding decade decrease in the value of P_e. Thus, for a bundle of aligned, 10 μ-diameter whiskers with a packing fraction of 0.5, about 50 psi pressure (vs. about 500 psi when $d = 1\ \mu$) would be required to cause penetration by the liquid nickel.*

Unfortunately, the application of external pressure alone does not solve the penetration problem for all whiskers of small size. Other factors are involved: for example, once the metal begins to solidify, the solidification shrinkage will be sufficient to cause voids or apparent dewetting in some of the smaller channels. Thus, it is necessary to have wetting conditions prevail (viz., $\theta < 90°$) during the entire composite fabrication process. Since molten nickel does not wet Al_2O_3, other approaches that will promote wetting are necessary. From thermodynamic considerations, the condition for spontaneous wetting ($\theta = 0°$) is

$$[\gamma_{SV} - \gamma_{SL}] > 0 \tag{6}$$

This relationship provides a useful guide for developing approaches and for selecting compositions that will promote wetting in systems that are inherently nonwetting. Wetting can be enhanced by any change which increases the (positive) difference of $\gamma_{SV} - \gamma_{SL}$. Two common methods which can be implemented toward this end are (1) by increasing γ_{SV} by changing the surface at the solid-gas (S-V) boundary; for example, by depositing a metallic coating ($\gamma_{SV} \sim 1600$ ergs/cm^2) on the Al_2O_3 surfaces ($\gamma_{SV} \sim 900$ ergs/cm^2), or (2) by decreasing γ_{SL} by doping the liquid with interfacially active impurities which tend to concentrate at the liquid-solid boundary and lower the L-S interfacial energy. Thus far, the best wetting conditions have been achieved when the whiskers were precoated with a thin, adherent film of a metal [21]. However, the stability of the coating during fabrication becomes a limiting problem, especially when high-temperature metals such as nickel are used. For example, nearly all metals are soluble in liquid nickel to the point where various coatings investigated [22] were dissolved in a matter of seconds. Under these conditions, the infiltration periods must be limited to short times. Therefore, it is necessary to examine the rate of infiltration, or the dynamics of the infiltration process. In the ideal situation, Poiselles' equation can be used, and Washburn [23] was able to show that the rate of

* For fibers having hexagonal instead of cubic packing, the values of P_e would run about 40% greater since the channels are smaller for comparable values of d.

TABLE 9.1 FACTORS AFFECTING THE INFILTRATION OF MOLTEN METAL INTO THE CHANNELS BETWEEN PACKED FIBERS

Conditions	Factors	
	Chemical	Physical
Static (Thermodynamic)	Composition of each phase (L, V, S). Chemical potential and phase relationships. Temperature of molten metal and fibers. Influence of impurities.	Relative surface tensions (L-S, S-V, L-V).[a] Wetting and capillary forces. Relative thermal expansion of matrix and fibers.
Dynamic (Fluid Dynamic)	Solubility rate of fibers (or fiber coating) in metal matrix. Oxidation or degradation of coating on fibers prior to infiltration. Dissolved gases in matrix. Influence of impurities on the three above factors. Influence of time, temperature, pressure, or chemical potential of phase and diffusion of species.	Geometry, packing density, and orientation of fibers which determine the shape, size, and uniformity of channel spacings. Infiltration rate, volume, and direction of metal flow. External driving pressure. Erratic channeling of metal column due to inhomogenieties in wetting, channel spacing, etc. Formation of pores due to break in metal column. Viscosity and temperature of metal. Vapor pressure of the metal. Solidification shrinkage of the metal.

[a] L-S = Liquid-solid interface.
L-V = Liquid-vapor interface.
S-V = Solid-vapor interface.

infiltration is

$$\frac{dl}{dt} = \frac{(P_e + P_c)d_c^2}{32\eta l} \qquad (7)$$

where l = capillary (channel) length

t = time

P_e = external hydrostatic pressure applied to infiltrating metal

P_c = capillary pressure

d_c = effective capillary diameter

η = viscosity of the infiltrant

Additional factors affecting the relationships shown in Eq. 7 are summarized in Table 9.1. It would seem that the rate of infiltration, dl/dt, could be significantly increased by utilizing a high driving external pressure, P_e. However, experience [22] has shown that this is not the prime consideration, since the flow of the infiltrant tends to follow the path of least resistance, namely, along the capillaries having the largest diameter. Since dl/dt is proportional to the square of d_c, the effect of variations in capillary diameter becomes highly significant in actual cases because the channels are seldom of uniform size. As a result, the larger channels are penetrated rapidly, and the smaller ones tend to be passed by. In addition to nonuniform penetration under actual infiltrating conditions, the dissolution of the coating, and hence, dewetting (negative P_c), can also play a dominant role in preventing complete infiltration. Thus, when $P_e - P_c = 0$, $dl/dt = 0$, and penetration will cease.

However, in systems where the coating is not dissolved by the matrix, experiments have demonstrated the relative ease in which the coated fibers can be penetrated. For example, notable differences were observed in the degree of penetration of silver into bundles of coated and uncoated Al_2O_3 whiskers [24]. In this case, the coating was nickel, which was wetted by the silver. The coating made it possible to fully penetrate the whiskers with silver, which resulted in much stronger composites, as shown in Figure 8.10.

Interfacial Bond Strength. As mentioned previously, whiskers generally have very short lengths (0.01 to 0.50 in. long), and in order for them to carry the applied loads, the matrix must transfer these loads to them via the interface. This can be done effectively only when they are adequately bonded along their lengths to the matrix. If the fibers are in partial contact with the matrix, the degree of bonding will be limited, and premature failure is likely to occur by separation of the fibers and matrix—a condition called fiber "pullout." This situation occurred when molten silver was infiltrated into a bundle of uncoated Al_2O_3 whiskers packed into a fused silica tube.

Although the silver did not wet the uncoated whiskers ($\theta = 118°$) [24], penetration into the channels occurred. However, the final contact between the whiskers and silver was poor.* The composites were weak, as shown in Figure 8.11. Fracture occurred by the whiskers pulling out of the matrix with little or no silver adhering, as shown in Figure 8.12. On the other hand, the use of a thin nickel coating† (see inset photo, Fig. 8.11) caused the molten silver to spread over the coated whisker surfaces and maintain an intimate contact with them during cooling and solidification. The net result was the formation of strong composites in which the shear strength between the Ag-Ni-Al_2O_3 interfaces was sufficiently strong for the effective transfer of the applied stresses to the whiskers. Failure occurred by plastic deformation in the silver matrix, and the fracture surface of the composite showed no whiskers exposed (Fig. 8.12).

A precise value of the interfacial shear strength cannot be easily determined analytically because of many complicating factors. Furthermore, the value of the shear stress varies nonuniformly along the fiber surfaces, as shown schematically in Figure 9.3*b*. The peak shear stress (τ_{max}) near the fiber ends depends on several factors, including the shape of the fiber ends, the elastic constants of the constituents, and the strength of the interfacial bonds [25–29]. The values of τ are also affected by the proximity of other fibers and fiber-breaks, by differences in Poissons ratio, by thermal expansion coefficients of the constituents, and by temperature [5, 30–32]. If the matrix deforms plastically, the maximum interfacial shear stress will be limited by the yield stress of the matrix where $\tau_{max} = \frac{1}{2}\sigma_y$‡ as shown in Figure 9.3*b*. Thus, the maximum value of τ shown in Figure 9.3*b* will be determined either (1) by the strength of the interfacial bond, τ_b, or (2) by the yield stress of the matrix, σ_y, whichever one is the weaker (assuming that the reinforcing fibers are sufficiently long). Under these conditions, where the matrix undergoes plastic deformation and where τ remains at a constant value, the critical fiber aspect ratio can be determined from the following relationship:

$$\frac{l_c}{d} \cong \frac{S_c}{2\tau_y} \qquad (8)$$

Using Eq. 8, it is possible to estimate values of l_c/d for given values of S_f and τ. These values are upper, rather than lower, limits for l_c/d since it is assumed that no strain hardening occurs.§ However, in real composite

* The whiskers used in these experiments were relatively coarse ($d = 25\ \mu$) so that silver penetration into the channels was essentially complete, and did not require high infiltration pressure (see Fig. 8.11).

† In this case, the coating was applied by sputtering.

‡ In this case, it is assumed that the matrix undergoes no strain hardening.

§ Actually, strain hardening effects in the matrix will occur to varying degrees, depending on the structure and composition of the matrix, and on the proximity (spacing) of adjacent fibers. Thus, τ_y must be associated with the ultimate strength of the matrix (i.e., $\tau_y = \frac{1}{2}\sigma$ ultimate).

materials, a metal matrix is very likely to strain harden in the vicinity of the fibers, so that the flow stress of the matrix will increase. As a result, the values of l_c/d and l/d will be correspondingly lower than those for the same fiber stress in a nonstrain-hardening matrix.

Table 9.2 shows values of the critical fiber aspect ratio (l_c/d) required for the peak fiber stress (σ_{max}) to build up. Also shown are values of l/d required to produce an average fiber stress ($\bar{\sigma}_f$) equal to 95% of S_f for fibers embedded in various matrices.

In matrices such as copper and nickel (assuming no work hardening), $l_c/d \cong 100$ when the average fiber stress is 1 million psi. For weaker matrices such as aluminum or polymers, the l_c/d values should be 2 to 3 times greater than those for copper and nickel (see Table 9.2). For whiskers reinforcing a resin matrix, the values l/d should be of the order of 2000 to 3500 to achieve an average fiber tensile stress in the 500,000 to 1 million psi range. Experimental composites of Al_2O_3 whiskers in an epoxy matrix have shown values of $S_f \cong 750{,}000$ psi when $l/d \approx 800$–2000 [33].

In the reinforcement of silver, Table 9.2 indicates that the value of l/d should fall between 630 to 1250 in order to achieve an average whisker stress of 950,000 psi. Experimental studies have shown that Al_2O_3 whiskers, where $l/d = 1300$, have achieved stresses in a silver matrix averaging $S_f \cong 970{,}000$ psi [34].

Since the values of τ_{max} decrease with increasing temperature, even greater aspect ratios than those indicated in Table 9.2 will be required in elevated temperature applications. For example, silver reinforced at room temperature would require that $l/d \cong 130$ when $\bar{\sigma}_f \approx 100{,}000$ psi. Results of elevated temperature tests (870°C) on silver reinforced with Al_2O_3 whiskers have shown that a value of $l/d = 300$ is sufficient to have the whiskers support average stresses of 100,000 psi [34].

Mehan and co-workers [35] have studied the elevated temperature properties of aluminum reinforced with nichrome-coated Al_2O_3 whiskers. They found that for composites containing 23*v/o* whiskers, where $l/d \sim 2000$ and $\tau_{max} = 400$ psi, the creep resistance was an order of magnitude better than the unreinforced matrix at 500°C. Furthermore, failure appeared to be abrupt, showing no tertiary creep, and was caused largely by whisker fracture and not by fiber pullout. This was also observed by Kelly and Tyson in silver composites reinforced with short tungsten wires [36]. The results of the creep tests on the Al_2O_3 whisker/aluminum composites indicate that even though the test temperature was about 83% of the aluminum melting point, there was a sufficient degree of interfacial bonding to allow for sufficient stress buildup in the whiskers to cause them to fracture. In Figure 9.8, this is shown in the electron scanning photomicrographs of the fracture surface of a composite specimen. The white regions represent bare Al_2O_3 surfaces, which are mostly fractured ends of the whiskers.

TABLE 9.2 ESTIMATED VALUES OF FIBER ASPECT RATIOS BASED ON THREE LEVELS OF FIBER STRENGTH, AND FOUR MATRIX MATERIALS HAVING DIFFERENT VALUES OF τ_{max}.

Matrix	τ_{max} (psi)*	$S_{f(max)}$ (psi)	(ℓ_c/d)	(ℓ/d)***
Resin	1,400**	100,000	36	360
		500,000	179	1790
		1,000,000	357	3570
Aluminum	2,000	100,000	25	250
		500,000	125	1250
		1,000,000	250	2500
Silver	4,000	100,000	13	130
		500,000	63	630
		1,000,000	125	1250
Copper Nickel	5,000	100,000	10	100
		500,000	50	500
		1,000,000	100	1000

* Based on $\tau_{max} = \tau_y \cong \frac{\sigma_y}{2}$ = const. along fiber length.

** $\tau_{max} \cong \tau_{frict.}$ (see p. 35, ref. 5).

*** Values of avg. fiber stress when $\ell/d = 10\,\ell_c/d$.

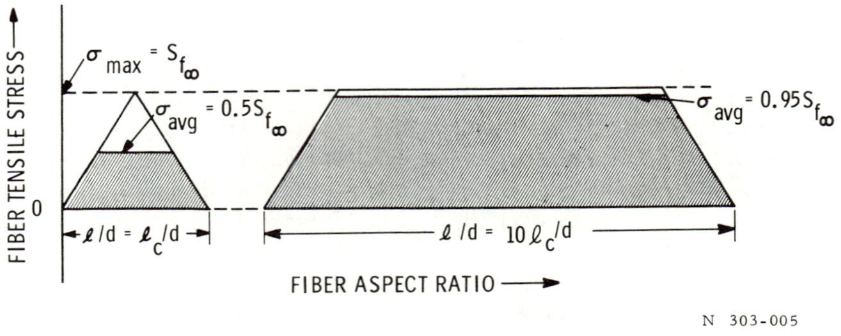

N 303-005

A small amount of debonding is also evident along the surface of some of the exposed whiskers.

Chemical Reactions at the Interface. Because of the high temperatures that are likely to occur in whisker-reinforced metals and alloys during fabrication or during service, chemical reactions and interdiffusion (and

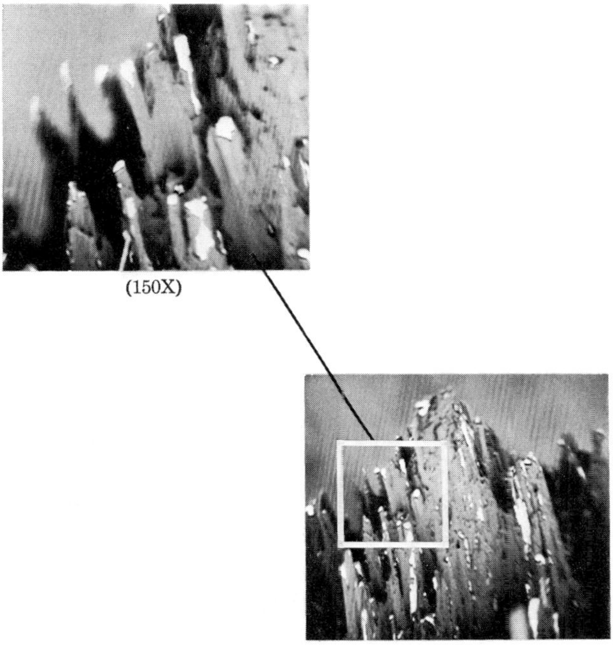

(150X)

(50X)

Figure 9.8 Fracture appearance of a composite of aluminum reinforced with coated Al_2O_3 whiskers and creep-tested at 500°C for a period of 23 h. Light areas indicate exposed (uncoated or fractured) whiskers (from Mehan [35]).

solution, if a liquid phase is present) are likely to occur. At the present time, few details are available on the interactions between metals and whiskers, and on their effects on mechanical properties. The very small size of the whiskers makes experimental studies and observation most difficult and time consuming. However, some studies have been conducted on the interactions between: Al and Al_2O_3 whiskers [35, 37, 38]; Al, Ni and Si_3N_4 whiskers [39]; Ni and Ni alloys and Al_2O_3 whiskers [40–43]; and Ti, Ni, Ni-alloys, and SiC whiskers [44].

The results of these investigations, in general, have shown that the chemical reactions at the interface tend to degrade and weaken the whiskers. Although quantitative data are not available, the results can be divided into three categories: (1) reactions that strengthen the whiskers; (2) reactions that have little or no effect; and (3) reactions that weaken the whiskers. Since the first category is rare (with the exception of whiskers that are chemically etched) [45, 46], the latter two will be considered as being

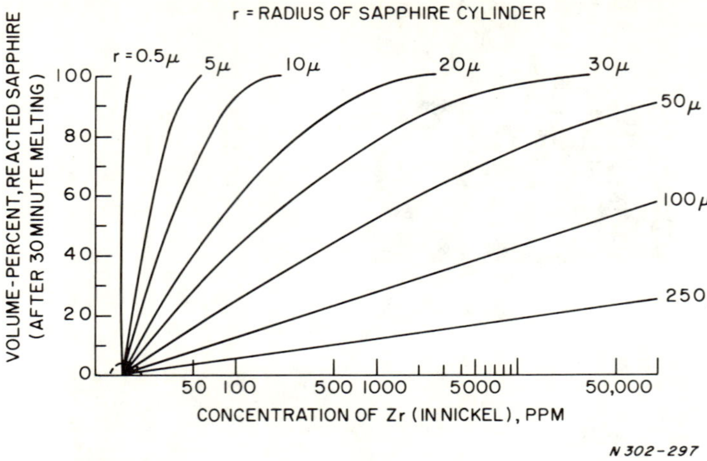

Figure 9.9 Reacted volume of sapphire cylinders of various radii as a function of the concentration of zirconium in molten nickel after 30 min (from [47]).

critical parameters and must be taken into account during the fabrication process.

In considering the nature of the chemical reactions, two situations may arise: (1) where the whiskers (because of their very small dimensions) are entirely consumed by a reaction with the matrix, and (2) where the reactions penetrate a finite distance into the whisker.

Studies of the chemical reaction between single-crystal α-Al_2O_3 and molten Ni, Ni-alloys have shown that small additions of reactive* elements to the nickel are sufficient to seriously degrade the crystal's strength [42]. Indeed, the results of studies by Feingold, Sutton, and Sauer show that if the nickel contains only 50 parts per million of zirconium, a 10-μ-diameter Al_2O_3 whisker will be entirely consumed in the Ni-alloy matrix within a half-hour exposure at 1500°C (in vacuo) [47]. Smaller diameter whiskers would require even lower Zr-concentrations to be totally reacted, as shown in Figure 9.9. Recent studies, based on the use of Hf (which behaves similarly to Zr), have also shown comparable deleterious effects on Al_2O_3 whiskers [42]. Sapphire whiskers, coated first with HfC and then with W (to promote wetting), were infiltrated with a NiCrFe alloy at 1500°C for 30 sec in a H_2 atmosphere [42]. The results showed that the surface of the whiskers (about 20 μ in diameter) were severely degraded (as shown in Fig. 9.10), and that HfO_2 had formed. Similarly coated whiskers, shown in Figure 9.11,

* Elements which have a lower free energy of formation of the oxide than Al.

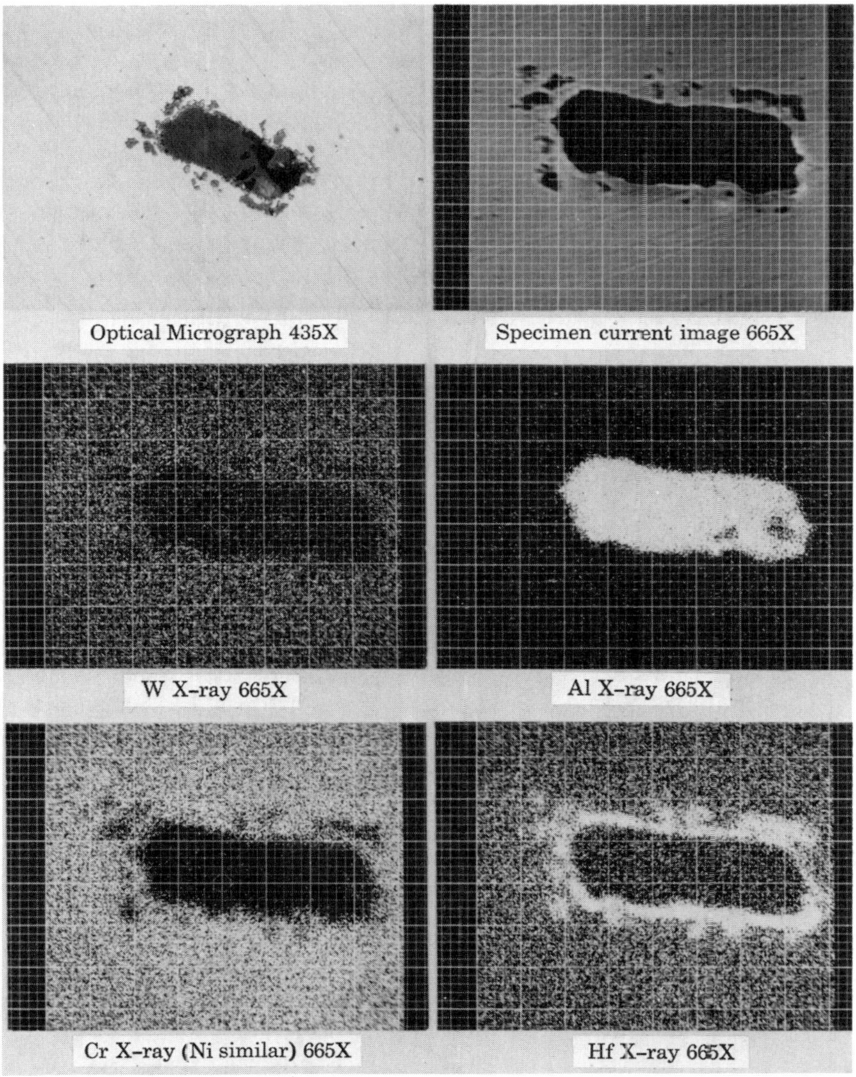

Figure 9.10 Optical micrographs of W/HfC-coated Al_2O_3 whiskers infiltrated with molten NiCrFe (at 1500°C for $\frac{1}{2}$ min), and associated scanning electron-probe photomicrographs showing distribution of the elements surrounding it. Note absence of W at the interface (from [42]).

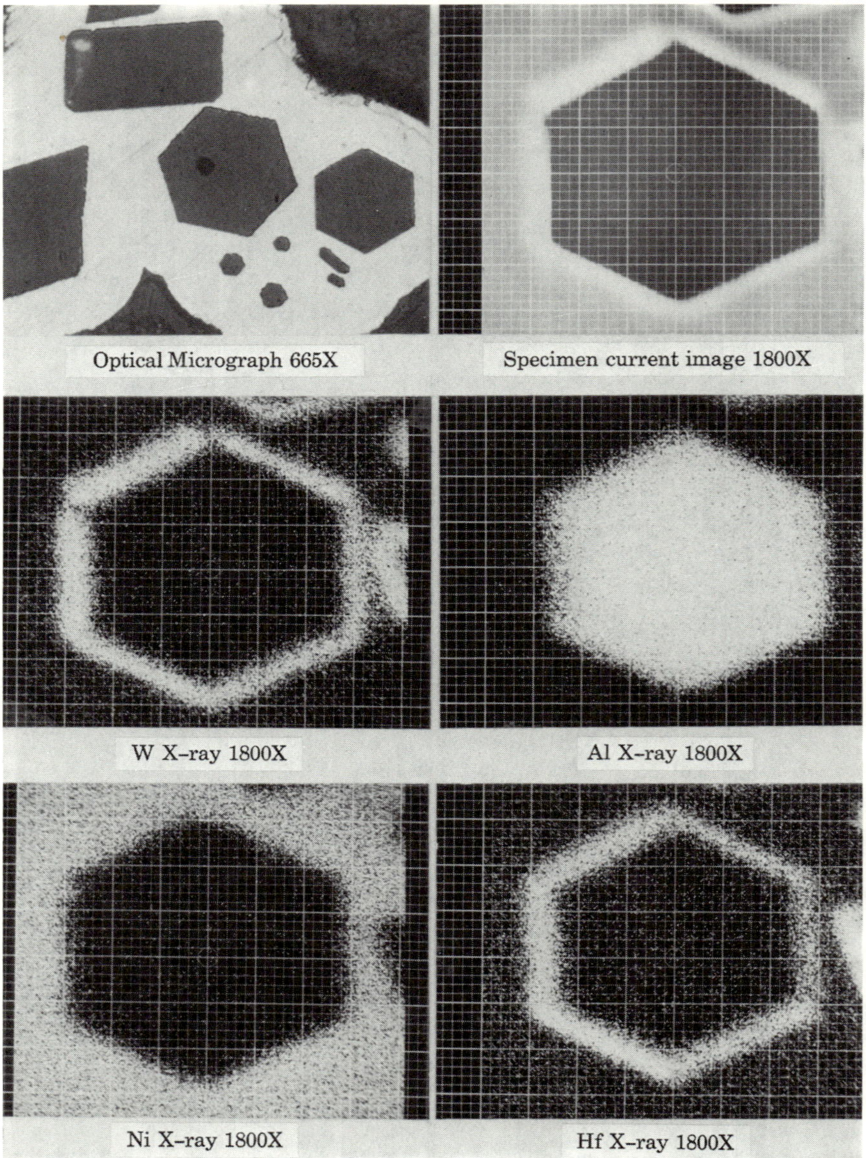

Figure 9.11 Optical micrograph and associated scanning electron-probe photographs of Al_2O_3 whiskers coated with HfC and W, and then heated to 1500°C for $\frac{1}{2}$ min. (Specimen was then electroplated with Ni to enable a polished section to be made.)

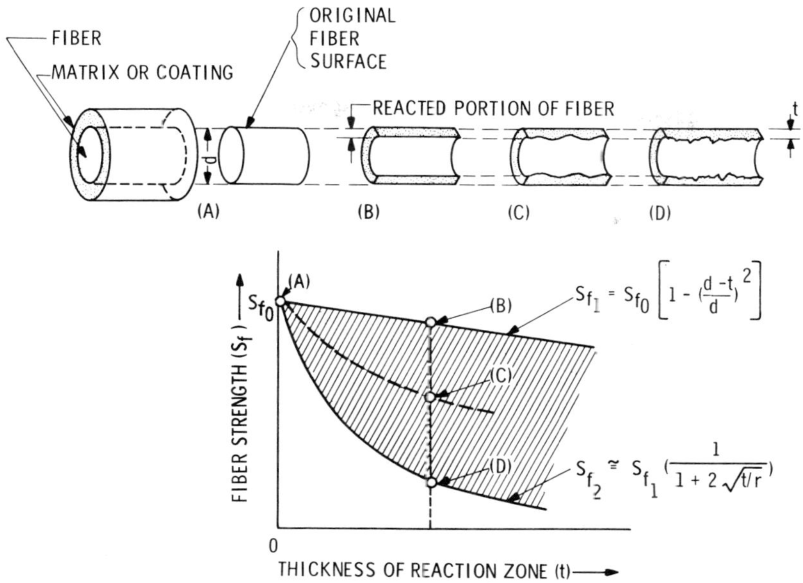

Figure 9.12 Effect of interfacial chemical reactions on the strength of a reinforcing fiber.

heated to 1500°C, but without the infiltration of the alloy,* did not show severe attack, although some HfO_2 was detected. It thus appears that the presence of a liquid phase enhances the reactions between Al_2O_3 and chemically active species (Zr, Hf) at high temperatures.

Not all reactions are severe, and their occurrence may be difficult to ascertain under the microscope. However, their presence may be observed readily if they serve as stress risers. Figure 9.12 shows schematically two types of reactions occurring between a fiber (whisker) and the matrix or coating: (1) a reaction that proceeds uniformly into the whisker (Fig. 9.12*b*), and (2) a reaction that proceeds nonuniformly (Fig. 9.12*c* and *d*). If the unreacted portion of the whisker forms a smooth interface with the reacted portion, the strength of the fiber may decrease according to the upper curve, indicated by points A and B. This assumes that the reacted portion is weak and carries no load. However, if the interface is uneven (jagged) due to regions of relatively deep penetration, such as sharp slits or cracks (the ratio of t/r being large)†, then the stress concentrations arising from this

* An outer nickel layer was electroplated on the heat-treated whiskers to aid in metallographic mounting and polishing.

† In this case, t would be the crack length or penetration depth, and r would be the crack tip radius.

type of reaction are likely to be significant, and are likely to greatly weaken the whisker, as indicated in Figure 9.12 by the curve labeled A–D. An intermediate situation is shown, whereby the interface is wavy (r is large with respect to t), and the effects on whisker strength are also intermediate, as shown by curve A–C. This type of uneven interface has been observed in studies on the wetting of Al_2O_3 by Ni-Ni alloys [16, 42, 47]. Push-off tests on solidified sessile drops indicate that the strength of the brittle phase (the Al_2O_3) follows qualitatively the relationships shown in Figure 9.12 [16, 42, 47]. The main point to be emphasized, however, is that while the reaction layer may be thin relative to the whisker diameter ($t/d \ll 1$), the uneven penetration of the layer may prove to be far more important to whisker strength, especially when cracklike protrusions ($t/r \gg 1$) are formed.

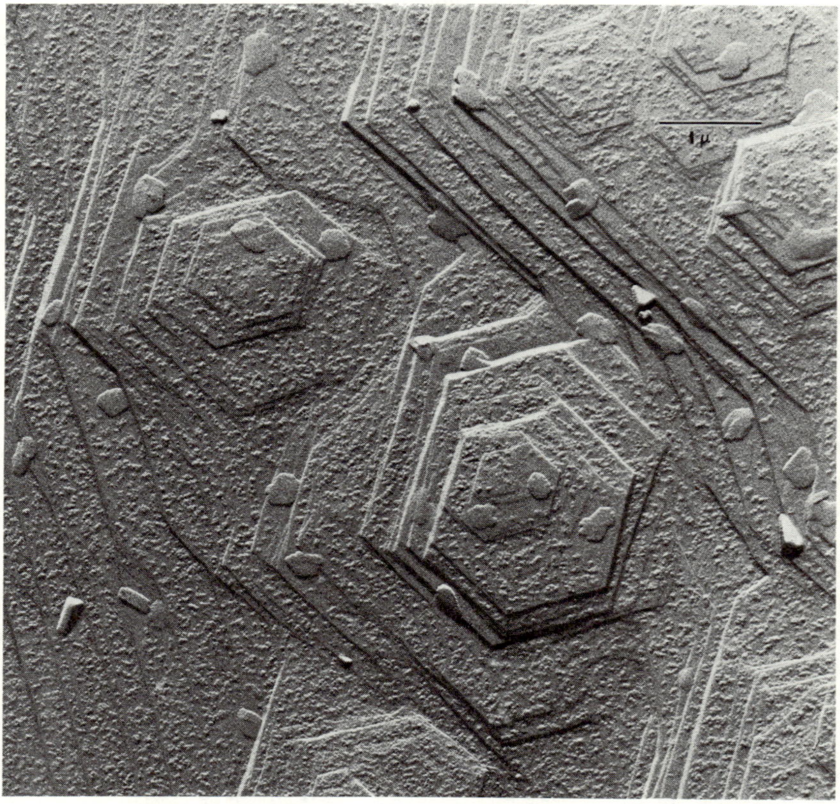

Figure 9.13 Electron photomicrograph of a surface replica from an A-Type α-Al_2O_3 whisker (mag. 13,334X). (Courtesy of E. Feingold.)

The main problem, therefore, is to control or minimize the interfacial reactions by (1) choosing a stable whisker-matrix combination; (2) choosing favorable fabrication conditions (i.e., where temperatures do not get too high); (3) eliminating any elements in the matrix (or coating) that will attack the whiskers; or (4) coating the whiskers with a thin, protective barrier layer or film.

Surface Perfection of the Whiskers. Many of the whiskers produced today contain a wide variety of surface imperfections which either weaken them, or provide sites which tend to react more readily with the matrix or coating materials. For example, the A-type, $\langle hk\bar{o}\rangle$, Al_2O_3 whiskers frequently contain hexagonal growth steps, which can act as stress risers [48, 49]. Such steps are shown in Figure 9.13, and the effect of their presence (or absence) on strength is discussed in Chapter 6. However, smaller whiskers are less likely to contain these flaws, and therefore they are able to support greater stresses than larger ones. An electron photomicrograph of a nearly perfect A-type Al_2O_3 whisker is shown in Figure 9.14. While smaller whiskers are stronger than larger ones, they are also more difficult to handle and align, and have a larger surface-to-volume ratio, which means that they have a proportionately larger surface area over which chemical reactions and diffusion can occur. Thus, the best size of whiskers (as measured by diameter or aspect ratio) for handling is not necessarily the smallest, but more likely, a size which optimizes the trade-offs between strength, handling and packing characteristics, and coating thickness, which will be discussed later.

SELECTION OF CONSTITUENTS AND FABRICATION PROCESS

Several decisions need to be made in the preparation of whisker-composite materials and these will be discussed briefly. These decisions arise from the fact that there is a wide range of constituent materials available; that there are numerous functional contributions that each constituent can make; and that there are several fabrication procedures that can be followed.

Essentially five decisions need to be made:

1. The choice of matrix material.
2. The choice of whiskers (composition and size).
3. The choice of whisker orientation(s).
4. The choice of coating material.
5. The choice of fabrication method.

The factors influencing these choices can be divided into two types, as shown in Table 9.3. The constituent factors include the composition and

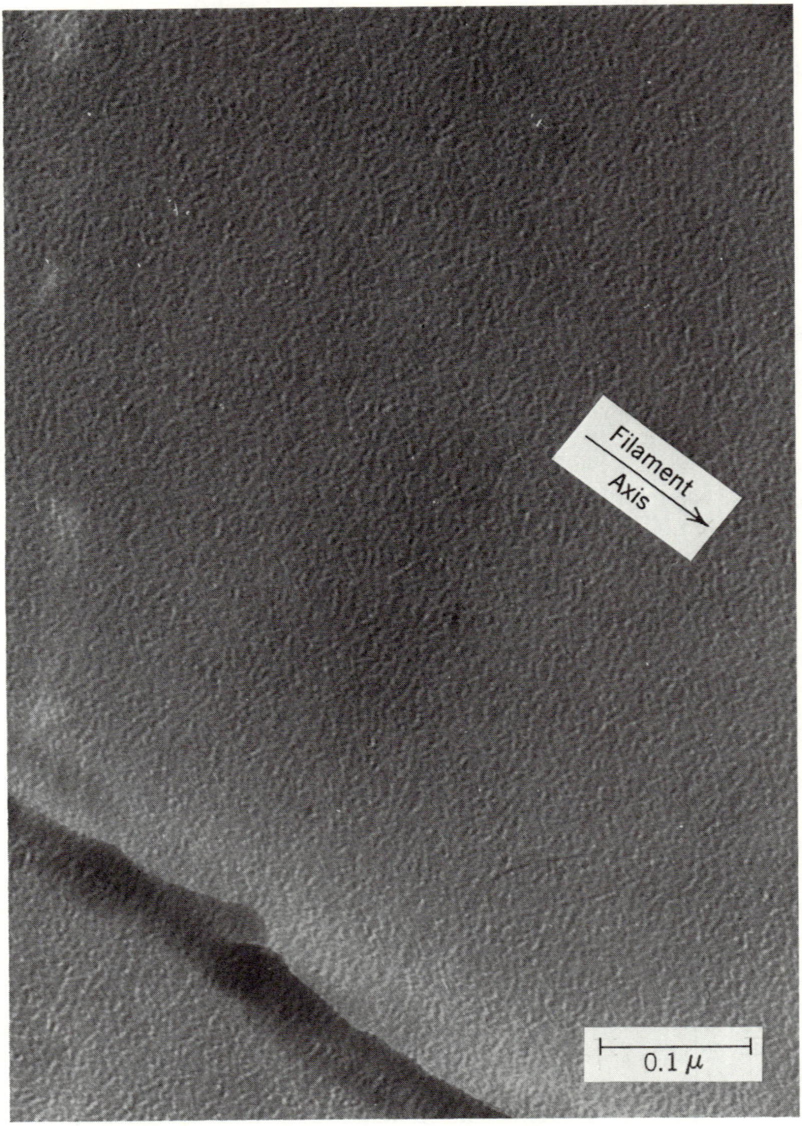

Figure 9.14 Electron photomicrograph (mag. 225,000 ×) of a surface replica prepared from a typical sapphire whisker surface. The surface is very smooth and any surface protuberances are less than 20*a*.

TABLE 9.3 FACTORS INFLUENCING THE CHOICE OF CONSTITUENTS AND FABRICATION PROCESS

Selection Type	Factors Affecting Decision: Constituent Factors	Factors Affecting Decision: Composite Factors
Matrix	Density Melting point Ductility (toughness) Shear strength Corrosion resistance	Design requirements Application Environment Constituent functions
Whiskers	Strength Modulus Density Melting point Size and aspect ratio Availability (and cost)	Chemical and mechanical compatibility of whisker/coating/matrix Economics
Whisker coating	Wetting Bonding Stability	Performance
Whisker configuration (orientation)		Volume fraction Spacing Overlap Alignment direction(s)
Fabrication process	Damage to whiskers Ease of process Automation	Methods which optimize constituent properties with composite performance

properties of the whiskers, matrix, and coating, and their chemistry (wetting, bonding, and stability). The composite factors are concerned more with the total material, and they include: the design and performance requirements; the mechanical and chemical compatibility of the constituents; and the production costs.

The choice of the fabrication process depends on the following: the specific constituent materials to be used; the procedure(s) which optimizes composite properties; and the economics of production (costs, degree of automation, efficiency of handling whiskers, etc.) and of utilization, or cost effectiveness.

Although several decisions must be made from the time the constituents are selected to the time the final composite is produced, the number of choices will be reduced greatly as the performance requirements of the composites are specified: that is, as the desired mechanical properties, the service temperatures and environments, and so forth, are specified, since these conditions largely determine the composition of the constituent

TABLE 9.4 TYPES OF WHISKERS CONSIDERED FOR REINFORCING MATERIALS (Data taken from Ref. 50)

Whisker	Melting Temperature (°F)	Density (lb/in.^{3})[a]	Modulus (10^6 psi)	U.T.S. (10^3 psi) (max. meas.)	Strength/Density (10^6 in.)	Modulus/Density (10^7 in.)	Cost ($/lb)
Ceramic							
Graphite	6600$^{(1)}$	0.06$^{(1)}$	102$^{(1)}$	2850$^{(3)}$	47$^{(1)}$	170$^{(1)}$	n.a.[b]
SiC	4870$^{(2)}$	0.12$^{(4)}$	70$^{(2)}$	3000$^{(2)}$	26$^{(2)}$	61$^{(3)}$	$200–9000
BeO	4660$^{(3)}$	0.10$^{(2)}$	50	1900	18	49$^{(4)}$	n.a.
B_4C	4440$^{(4)}$	0.09$^{(3)}$	70$^{(2)}$	2000$^{(4)}$	22$^{(3)}$	80$^{(2)}$	n.a.
Al_2O_3	3700	0.14	62$^{(3)}$	4000$^{(1)}$	21$^{(4)}$	43	$7500
Si_3Ni_4	3450	0.12$^{(4)}$	45$^{(4)}$	2000$^{(4)}$	17	48	$900
Metallic							
Cr	3430	0.27	35	1290	5	13	n.a.
Fe	2800	0.23	29	1900	8	13	n.a.
Cu	1980	0.32	18	430	1	6	n.a.

Rank:
$^{(1)}$ First
$^{(2)}$ Second
$^{(3)}$ Third
$^{(4)}$ Fourth

[a] Ranked according to lowest density.
[b] n.a. = not available commercially.

materials. When whisker-reinforced resins are used, the choice of the resin is likely to be based on the specific type of whisker selected. However, when metals are used as the matrix, the choice of the metal (or alloy) is more likely to be based on the application temperature and on other factors such as the degree of corrosion resistance and ductility required. After the metal is selected, the whisker type may then be chosen. This choice will be influenced largely by chemical compatibility requirements, particularly if the temperatures to be encountered are high, that is, 1000°C or greater.

It is beyond the scope of this chapter to discuss further the choice of matrix materials, whisker orientations, and coating compositions, other than to illustrate some specific examples. However, the types of whiskers and the fabrication procedures that are currently available will be discussed in the next two sections.

WHISKER TYPES

While many types of organic, metallic, and ceramic whiskers have been reported in the literature [50], only relatively few are available currently for reinforcing purposes. Table 9.4 shows some of the whiskers that have been studied the most. For comparison, a ranking of one to four has been given to the four whisker types shown in Table 9.4 having the highest values in each of the property columns. It is readily evident that only the ceramic whiskers rank on this basis. The metallic whiskers are not as strong or stiff (lower modulus) and they are also more dense, which makes their specific strengths (strength/density ratio) and specific moduli (modulus/density ratio) very low when compared to the ceramic whiskers. Furthermore, metallic whiskers, such as those of iron or copper, are weakened on exposure to moisture or by handling, and by dislocations that can be introduced readily into their surface. Thus, they are not well suited as reinforcements.

On the other hand, the ceramic whiskers are more stable on exposure to moisture and to abrasion, and they also possess better high-temperature mechanical properties, which make them prime candidates for reinforcing metallic matrices. However, the simple ranking of the whiskers in Table 9.4 can also be misleading, since other factors of importance were not included. For example, graphite whiskers are ranked first in all categories except in ultimate tensile strength. However, in metal matrices at elevated temperatures, they are more likely to be degraded chemically than the other whiskers such as Al_2O_3 and SiC. The Al_2O_3 and SiC whiskers are produced in the greatest commercial quantities today.

Once the whisker type has been selected, the choice of whisker size (length, diameter) would seem to be straightforward, since the finest-diameter whiskers generally have the greatest strengths (see Chapter 6). However, other factors must also be considered in the selection of an optimum size for best composite properties. Such factors include:

1. Cost.
2. Aspect ratio and ineffective length.
3. Alignment and packing density.
4. Snarling and twisting.
5. Matrix wetting.
6. Chemical reactivity.
7. Matrix embrittlement.
8. Coated packing density.

At present, the cost for short, small-diameter whiskers is less than for the larger and longer ones. For example, SiC whiskers (1 μ-dia. × 200 μ) are priced at about \$200/lb, while the larger ones (5 μ-dia. × 2000 μ) are priced at about \$9000/lb [50], although it is expected that the price will be greatly reduced in the future (Chapter 1). However, the comparison is not strictly valid, since different growth processes were used, and the morphology and crystal structures are also different. Although both types may have comparable tensile strengths, the larger type with the greater length-to-diameter ratio (2000 vs. 200) is likely to have a higher reinforcing efficiency (see Table 9.2). For a given whisker volume fraction, the shorter ones would have a greater number of ends (which serve as stress-risers in the matrix) and these whiskers would have proportionately a larger fraction of their lengths ineffectively stressed.

The smallest whiskers are usually more difficult to handle, to align, and to pack densely in a composite material. They are more likely to become twisted or entangled during handling. Also, it is more difficult to bring the matrix into total contact with all whiskers (wetting) since the smaller ones have a much greater surface area per unit volume. In the reinforcement of metals, small diameters become critically important since chemical reactions are likely to occur. For example, if a 1-μ diameter whisker reacts with the matrix to form an 0.5-μ thick penetration zone, then approximately 75% of the whisker will have been affected, and presumably damaged. However, if the same reaction occurs for a whisker 10-μ in diameter, then only about 10% of the whisker will have been reacted.

As smaller and smaller diameter whiskers are used as reinforcements, the spacing between the whiskers becomes proportionately smaller for a constant volume fraction. This will affect the mechanical properties of the matrix. If the whiskers are sub-micron in diameter, their spacing may be so close that a dispersion strengthening effect in a metallic matrix may

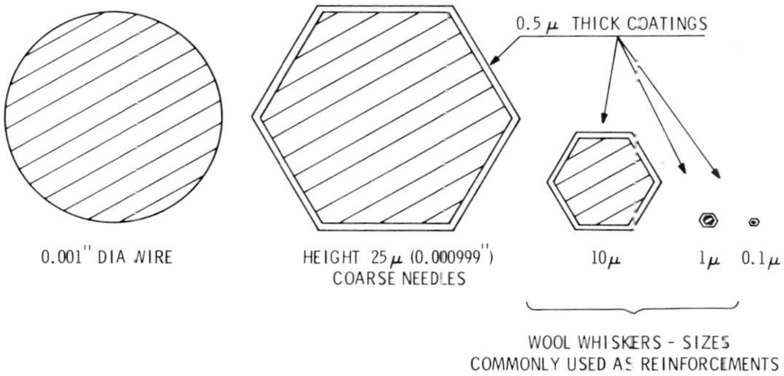

Figure 9.15 Relative thickness of 0.5 μ coatings on whiskers of different sizes.

result. However, smaller spacings can adversely affect composite toughness or work of fracture [51].

Most whiskers used to reinforce metals and polymers have diameters ranging from 1 to 10 μ, and lengths ranging from 100 to 25,000 μ. A comparison of the diameter (or cross-sectional height) of whiskers is shown in Figure 9.15. A 0.5 μ-thick coating is shown on each whisker. Such a coating might serve as a diffusion barrier in order to protect the whiskers from chemically reacting with the matrix; or may be used to promote wetting between the whiskers and the matrix. As the whiskers become finer, the coating shown in Figure 9.15 becomes an increasingly significant portion of the total cross-sectional area. Figure 9.16 shows the effect of

Figure 9.16 Maximum whisker volume fraction achievable in composite containing whiskers of different widths (W) and different coating thicknesses (t). The curves are based on the assumption that the coated whiskers are packed together in a perfect rectangular manner.

coating thickness on the maximum packing fraction of square whiskers having different widths. For whiskers 1 μ in width, the maximum packing fraction would be about 70% by volume for a 0.1 μ-thick coating. If the coating was 1 μ thick, then only about 10 vol% could be achieved. On the other hand, a 10-μ wide whisker could be packed to volume fractions of about 95% and 70% for coating thicknesses of 0.1 μ and 1.0 μ, respectively (assuming perfect rectangular packing). In some cases, coatings 1 μ thick, or thicker, are necessary when high-temperature (greater than 1100°C) casting processes are used, because thinner coatings will dissolve too rapidly in the molten matrix.

In summary, the choice of whisker and whisker size will depend on mechanical, or chemical, and on processing requirements.

WHISKER PROCESSING AND COMPOSITE FABRICATION METHODS

The fabrication procedure for preparing whisker composites is not simple and requires several steps, which can be divided into two categories: (1) whisker processing and (2) composite fabrication. These categories are shown schematically in Figure 9.17 and will be discussed further. As many as eight separate steps may be required to produce the final composite material.

Whisker Processing

Whiskers, as grown, usually exhibit a large variation in their size, in their degree of crystalline and surface perfection, and in their mechanical properties (especially their strength). Often, they are entangled among themselves and are interdispersed with other growth debris. For this reason, the growth product in many cases should be regarded as an ore, in which high-quality,

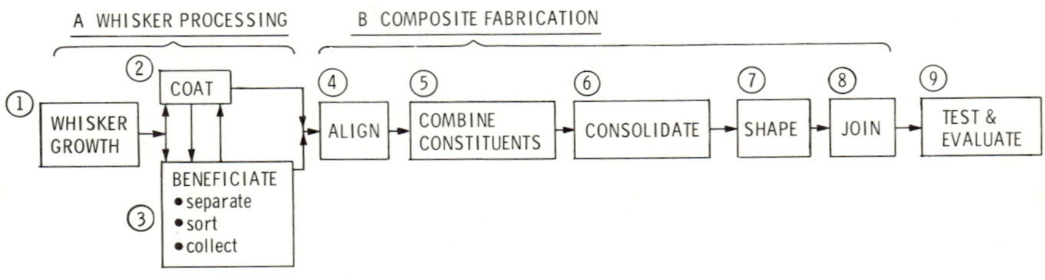

Figure 9.17 Flow diagram showing steps in the preparation and evaluation of whisker composites.

high-strength whiskers of varying dimensions are intermixed with weaker ones, with overgrowths, and with other growth debris. Thus, in order to utilize the whiskers, the growth product must first be beneficiated (cleaned, sorted, and collected according to size). A coating may be required, and this may be accomplished prior to or after beneficiation. If no coating is needed, then the whiskers may be aligned after beneficiation, as shown in Figure 9.17.

Whisker Growth. The growth of discontinuous whiskers has been discussed in Chapters 2 and 3. In most cases, growth is accomplished by a batch process or by a modified batch process whereby growth capsules are moved continuously through a controlled atmosphere furnace. Since the whiskers are so small, there are a very large number of whiskers in one pound (i.e., over 1 billion if they are less than 10 μ in diameter). Thus, the problem of handling is further complicated by the large number of whiskers to be handled.

In the case of Al_2O_3 whiskers grown in a hydrogen atmosphere at about 1300°C, where the gaseous reactants originated from molten aluminum and from silica-bearing refractories (which provide the source of oxygen) [52], the range of cross-sectional areas and lengths for three types of growth products [53] are shown in Figure 9.18. Also shown is the mean tensile strength observed for the whiskers as a function of their cross-sectional area. Each type of whisker product—wool, long-wool, or needles—can be grown reproducibly by controlling the growth parameters (temperature, atmosphere, time, etc.). However, the range of sizes is still too large for any one product type. This is shown schematically in Figure 9.18, where the range of sizes for any product type covers at least two orders of magnitude. Thus a more rigid control over the growth parameters, such as the temperature gradient within a growth chamber, is necessary. Another solution to the large variation in whisker size is to sort and classify them according to their dimensions. While the most desirable situation is to grow all the whiskers having the same dimensions, the most practical approach will probably be some combination of growth control and sorting.

The overgrowths on whiskers, such as branching, steps, crystalline material, and other defects, such as dislocations, fissures, and jagged edges, not only weaken the whiskers and cause a wide variation in strength dispersion (for a given size), but they greatly complicate further handling and effective classification. Thus, a great premium results from optimized growth processes which control not only whisker diameter but also whisker perfection. Some typical defects found in Al_2O_3 whiskers are shown in Figure 9.19. Under idealized growth conditions, the whiskers will be relatively straight and free from overgrowths as shown in Figures 9.20

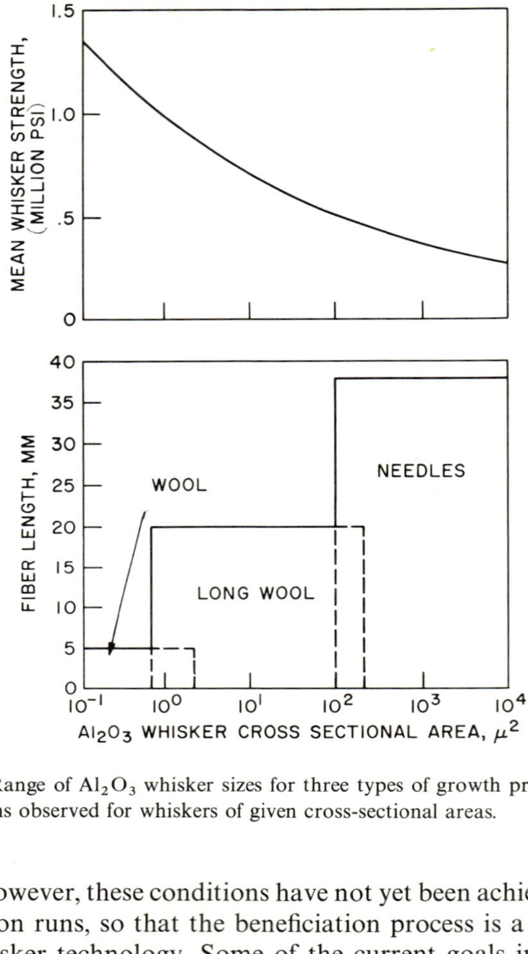

Figure 9.18 Range of Al_2O_3 whisker sizes for three types of growth products, and the mean tensile strengths observed for whiskers of given cross-sectional areas.

and 9.21. However, these conditions have not yet been achieved reproducibly in production runs, so that the beneficiation process is a necessary step in current whisker technology. Some of the current goals in whisker growth are summarized in Table 9.5.

Whisker Coating. The role of the coating on whiskers has been discussed previously and the functions were outlined in Table 9.1. Also, the optimum coating thickness for a given diameter whisker was discussed. The most desirable features of a coating process are listed in Table 9.6. Two of the chief requirements* of the coating are stability and compatibility. Thus, the

* Many whiskers, such as Al_2O_3, require coatings in order to promote wetting (contact) with the matrix. However, other whiskers, such as SiC are wetted directly with metals such as nickel [62]. They still require a coating under these conditions, since molten nickel greatly degrades the whisker surface of the SiC [55].

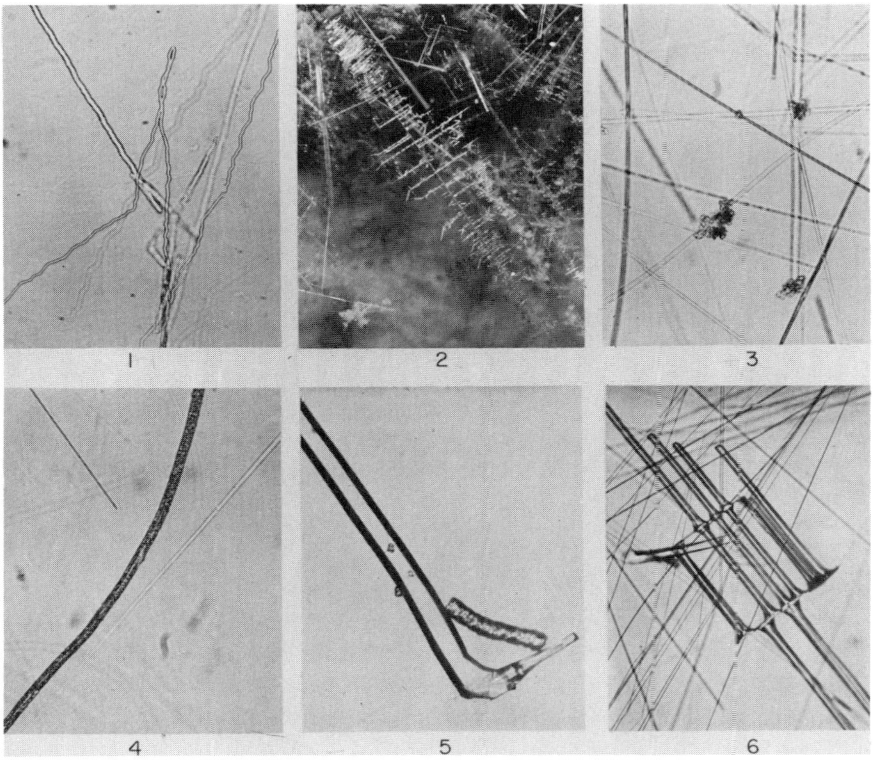

Figure 9.19 Some types of defects observed in Al_2O_3 whiskers: kink growths (1); crystalline overgrowths (3 and 4); branching and parallel growths (2 and 6); and bends (5).

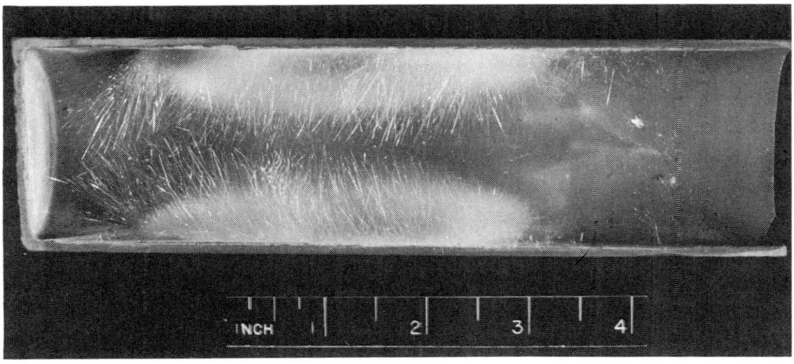

Figure 9.20 Example of Al_2O_3 whiskers (long-wool) grown under ideal conditions in the laboratory.

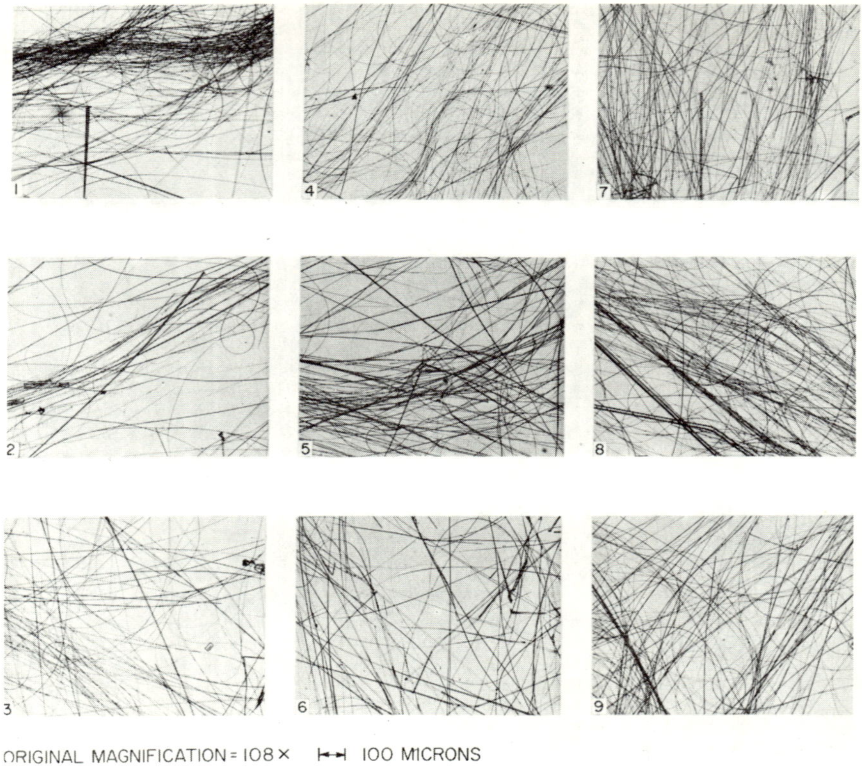

Figure 9.21 Photomicrograph showing high aspect ratio whiskers that are relatively free from defects. These whiskers were taken directly from the growth chambers and were not classified according to size.

coating must not be totally dissolved by or diffused into the matrix. Furthermore, it must not react chemically with the whiskers and thereby weaken them. These and other requirements have been reviewed in a recent paper [54].

Coatings can be applied by several techniques: chemical vapor-phase deposition [55, 56], sputtering [57], vacuum evaporation [60, 61], electroless plating [58, 61], and dipping in metal-resinoid liquids followed by burning-off of the organic base [59]. The first two methods have received the greatest effort.

TABLE 9.5 CURRENT GOALS IN THE WHISKER GROWTH PROCESS

High average strength
High aspect ratio
Minimize debris
Platelets
Defective whiskers
Maximize
Quantity
Quality
Uniformity
Reduce cost
Continuous process

Sputtering by ion bombarding a metal target in an argon atmosphere under reduced pressures has proven to be a versatile technique, since the whisker surfaces can first be cleaned by ion-bombarding prior to the deposition of the coating material. Furthermore, when the whiskers are placed in the "dark zone," the deposition of the metal is not by line-of-sight, so that the as-grown whiskers (which are loosely packed) can be coated uniformly throughout. A wide variety of metals and compounds have been deposited by this technique, including Ni, Ti, Cu, W, Re, Mo, Ta, W-Ti, Ni-Ti, Ni-Cr, steel, Cr, NiFeCr, NiAl, HfC, WC, WB, and VC [54].

An electron photomicrograph of a replicated Al_2O_3 single-crystal plaque coated with 0.1 μ thickness of titanium and then with about 0.5 μ thickness nickel is shown in Figure 9.22. Note that the surface is not smooth, which is typical of the sputtered deposits. The smoothness of a similar uncoated plaque is shown in Figure 9.23. The uniformity of the coatings deposited on the whiskers by the sputtering process is indicated in Figure 9.24. In this

TABLE 9.6 DESIRABLE FEATURES OF THE COATING PROCESS

Uniformity of coating
High whisker/coating bond strength
Stable throughout service and fabrication temperature range
Whiskers not degraded by
Coating
Coating process
Coating not dissolved by matrix

Figure 9.22 Electron photomicrograph of a sputtered titanium-nickel film which had been deposited on an ion-cleaned sapphire plaque (mag. 54,000X).

case, a Pt-coating, about 1 μ thick, was applied to the whiskers. The whiskers were next aligned, and the bundle containing 45 vol% whiskers was then infiltrated with molten silver at about 1050°C.

Coatings of Ni, Ti, W have also been successfully applied by chemical vapor deposition from carbonyl or halide gases [55, 56]. However, if the whisker mats are too densely packed, the metal will not penetrate into the central portions of the mat, and thus uniformity of coating thickness becomes a major problem.

The choice of coating material(s) depends on the composition of the whiskers and of the matrix, and on the temperatures (and time) encountered

Figure 9.23 Electron photomicrographs of a sapphire plaque after ion (Ar^+) cleaning (mag. 42,000 X).

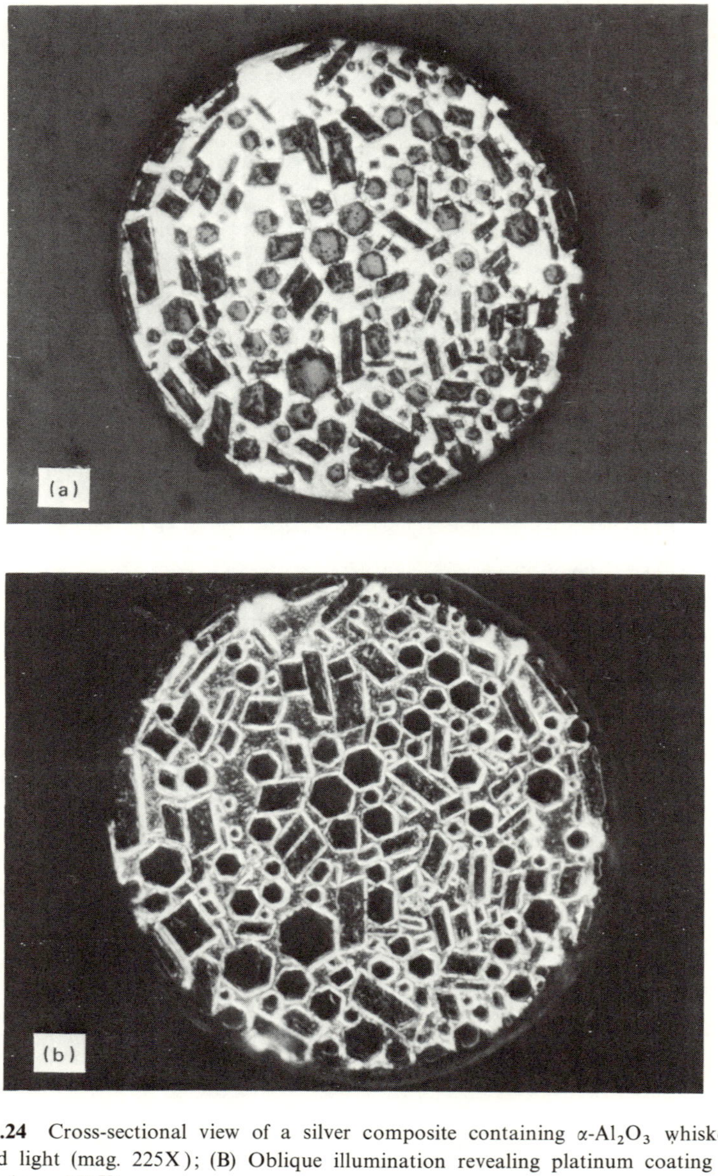

Figure 9.24 Cross-sectional view of a silver composite containing α-Al_2O_3 whiskers. (A) Reflected light (mag. 225X); (B) Oblique illumination revealing platinum coating on the whiskers (mag. 250X).

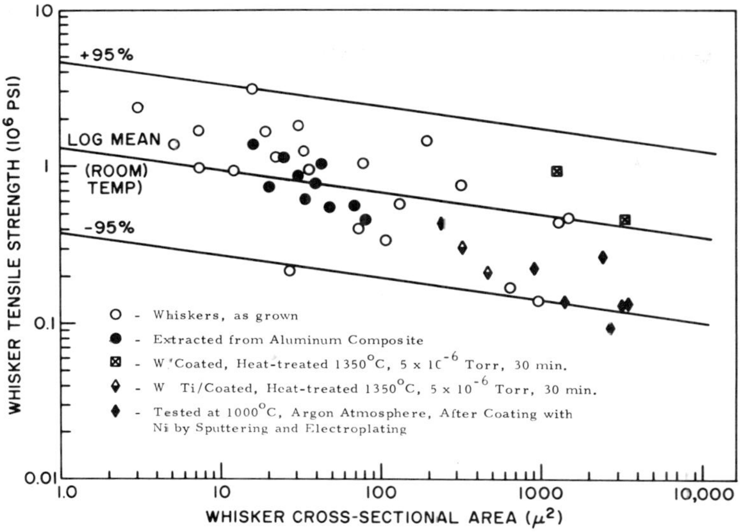

Figure 9.25 Effect of coating and thermal treatment on the room temperature strength of Al_2O_3 whiskers.

during fabrication and testing. Sapphire whiskers appear to be stable when coated with tungsten at heat-treating temperatures as high as 1350°C (30 min at 5×10^{-6} torr), as shown in Figure 9.25. They are also stable when coated with nickel or with a duo-layer of titanium and nickel, and then infiltrated with molten aluminum at 750°C (Fig. 9.25). However, a thin titanium layer next to the whiskers may cause degradation of the whiskers at temperatures above 1000°C, as shown by the data in Figure 9.25 on the W-Ti coated whiskers. The deleterious effect of Ti, Cr, or Zr is more pronounced at higher temperatures, and this has been shown by coating sapphire plaques and rods with these metals and heat treating them at elevated temperatures [42, 54].

The nature of the deposited coating, its stability in various matrices, and its effect on whisker strength still requires considerable research, especially where application temperatures exceed 1000°C. Stable coatings, which also serve as effective diffusion barriers between the whiskers and matrix, will undoubtedly be one of the most important factors in the development of high-strength, high-temperature, whisker-reinforced alloys. Considerable progress has been made to date [35, 40, 41] but there are still several problems to be solved.

Whisker Beneficiation. The whisker beneficiation step starts with as-grown or with coated whiskers and then (a) cleans and separates the whiskers individually, (b) sorts and classifies them according to size, and (c) collects the various size fractions for further processing. The goals for the whisker beneficiation process are shown in Table 9.7. If there is a significant amount of weak whiskers (containing overgrowths, branches, etc.), they are purposely broken down and are removed in the classification process. The weak whiskers may be broken by tumbling, by stirring in a liquid slurry, by agitation in an air elutriator, or by textile processes (carding, etc.). During the process of breaking down the weaker whiskers, the stronger ones can be removed and separated individually. Once in the separated state, the whiskers are next collected according to their cross-sectional area

TABLE 9.7 CURRENT GOALS IN THE WHISKER BENEFICATION AND CLASSIFICATION PROCESS

Separate whiskers individually
Mechanically break down defective whiskers
Collect various size fractions
Debris = Fine and coarse fractions
Broken whiskers and overgrowths
Develop continuous, automatic process

or to their lengths, or both. Whether liquid or air classification is used depends on the dimensions of the whiskers. The liquid process works well for the finest whiskers (1 to 5 μ in diameter $\times$ 100 to 1000 μ long). This process has been described by Parratt [63] and has led to the development of a commercially available whisker (and short fiber) classifier [64]. A schematic diagram of this classifier, which operates continuously, is shown in Figure 9.26. Basically, a water slurry, containing 1 to 5% whiskers and a dispersing agent, is passed through two rotating, concentric screens. The longer fibers are collected and then washed off the screens into a product receiver, while the shorter whiskers are passed through both screens into another collector. The collected size fractions can be reprocessed, if necessary, to narrow the spread or overlap in lengths. Multi-sets of screens can be used if desired. This classifier has effectively separated SiC and Si_3N_4 whiskers by their lengths, since the as-grown whiskers had a remarkably low variation in their diameters [63].

In the case of the Al_2O_3 whiskers, which generally have a larger spread in diameters, air elutriation techniques have worked well [37, 40]. In this case, jets of air are used to separate the growth product into individual whiskers, and then a steady flow of air is used to float the whiskers up a

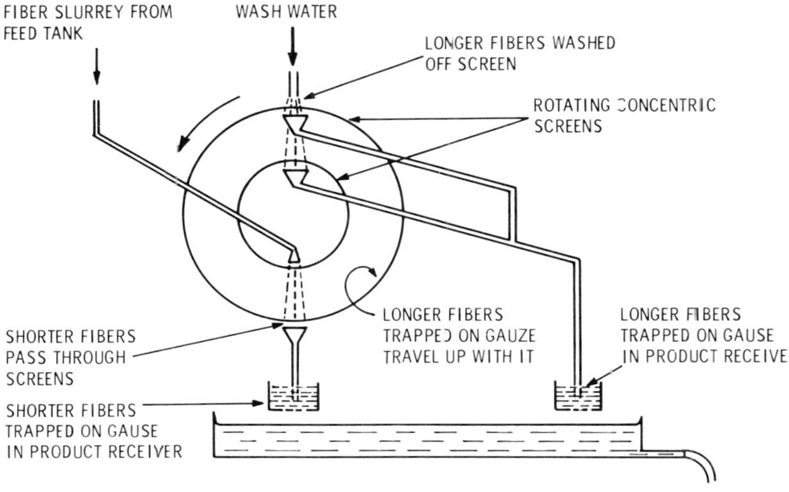

Figure 9.26 A simple fiber classifier. (Courtesy of R. W. Gooding and N. J. Parratt, Crown Copyright, 1967.)

narrow column and onto an inverted collecting screen. The collected size of the whiskers is determined by the velocity of the air moving up through the chamber. The greater the velocity, the larger are the whiskers collected. The finest fractions are therefore collected first, with progressively larger fractions being collected after each incremental increase in air flow through the chamber. Typical sizes collected are less than 1 μ, 1 to 10 μ, 10 to 25 μ, and greater than 25 μ. The Al_2O_3 whiskers processed by air elutriation are usually coated with a metal or alloy, and such a coating prevents static charges from building up on the whisker surfaces.

Figure 9.27 summarizes schematically the processing events for Al_2O_3 whiskers from growth (by a continuous process) to classification. Several classification techniques other than those just discussed are suggested. Currently, the growth of whiskers is not yet at the stage of full automation.

Composite Fabrication

The fabrication of whisker composites usually involves at least three steps: (1) alignment, (2) addition of the matrix to the whiskers, and (3) consolidation of the combined constituents into a fully dense body. Subsequent forming or shaping may be utilized, depending on the specific requirements. In the case of preparing test specimens, the composites may be machined or ground to specified shapes for various mechanical property tests. The specimens may involve more complex shapes, or may

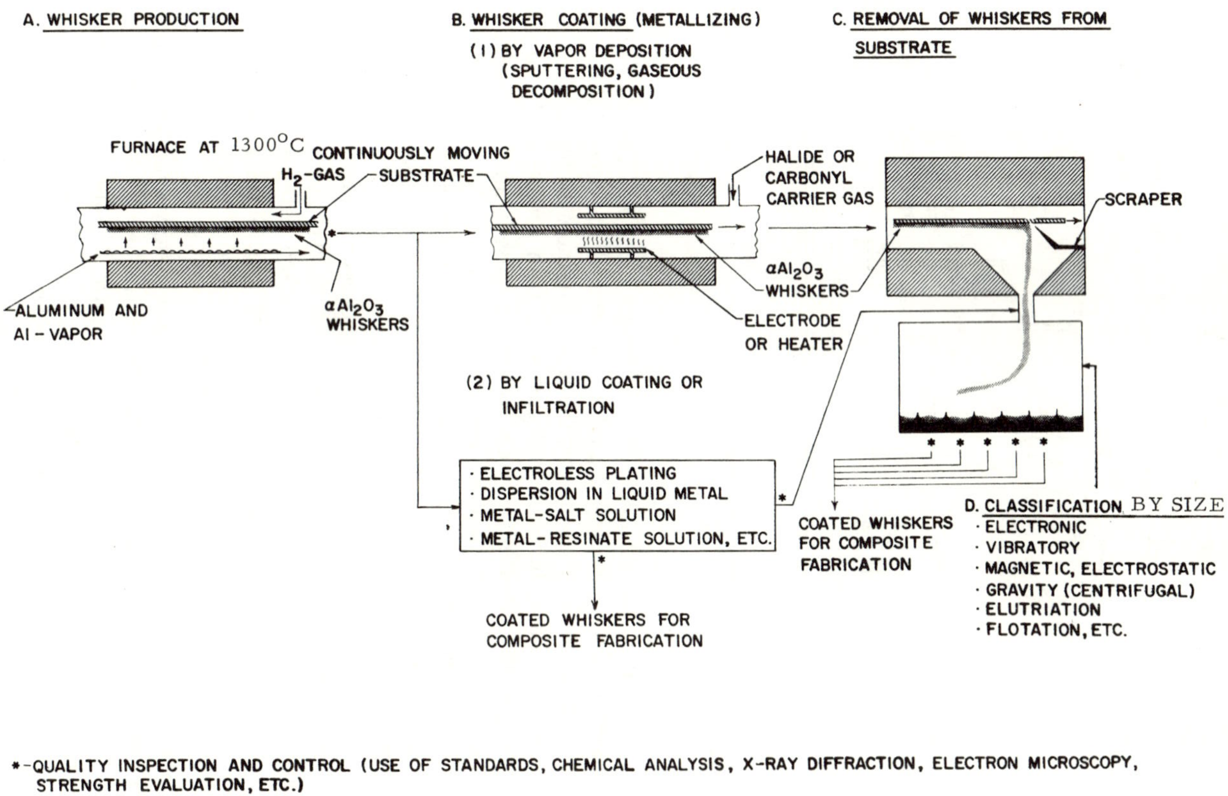

Figure 9.27 Schematic chart showing whisker processing from growth to classification (Ref. [92]).

even be joined to special grips or tabs by brazing or welding for elevated temperature tests.

Figure 9.17 indicated several steps that may be used in the fabrication of composite specimens. Although each step is numbered (from 4 to 8), the sequence is not necessarily in chronological order. For example, the whiskers may be aligned* either prior to or after the incorporation of the matrix, or they may be aligned when the composite material is consolidated and further shaped by extrusion or by other processes. Thus, the steps shown in Figure 9.17 can be arranged in several sequences: that is

④—⑤—⑥—⑦—⑧; ⑤—④—⑥—⑦—⑧;

⑤—⑥—(7, 4)—⑧; ⑤—(6, 7, 4)—⑧

and so on. The circles containing more than one number indicate that all of the enclosed steps occur during one process.

Whisker Alignment. This step is of major importance, since it directly affects the efficiency of whisker loading and the ultimate strength achieved in the composite. As Figure 9.4 indicated, small angles of misalignment in unidirectionally oriented whisker composites can result in drastic decreases in composite strength.

The ease with which whiskers can be aligned depends on numerous related factors, which include: whisker size, whisker aspect ratio, scatter in size and aspect ratio, branching and overgrowth, degree of entanglement prior to alignment, viscosity and flow characteristics of carrier medium (or matrix), whisker volume fraction, and method of alignment. While a high degree of alignment can be achieved by any one of several methods (such as extrusion), the effects of the alignment process on whisker damage and breakage must be taken into consideration. Many of the problems encountered during the alignment process result from the whiskers' small size and the large number per unit volume (i.e., up to 10^9 whiskers per pound). Furthermore, whiskers smaller than about 10 μ in diameter become very flexible and are easily twisted or bent. Hence, even though the whiskers may be perfectly aligned, they still may be twisted or may be wavy rather than straight, so that their effective loading in the composite is not possible. The deleterious effects of slightly twisted graphite fibers 7 μ in diameter in the reinforcement of nickel, for example, has been shown by Niesz [65]. Therefore, in order to achieve maximum composite strength, the orientation process should not only align the whiskers, but should also stretch them to

* Alignment is not always desired, and hence may be omitted from the fabrication procedure. Random alignment is used in the preparation of some whisker papers, felts, and mats.

the extent that they are straight. In the final composite, the whiskers should also be distributed uniformly and should not directly touch one another.

As mentioned previously, the alignment step can be performed prior to or after composite consolidation, or even during the stage where the matrix is combined with the whiskers. Numerous processes for whisker alignment have been investigated which include: extrusion and/or drawing [66, 71–74, 77]; use of carrier fluids (elutriation, channeled or directed flow, etc.) [43, 75, 78, 89]; magnetic or electrostatic (or dielectric) [44, 69, 70, 75, 78, 79] fields; paper-making methods (use of screens) [66–69]; textile methods (spinning, combing, carding, etc.) [75, 76]; vibration [78]; rolling [75]; and even by hand [25, 75, 78]. A detailed discussion of these methods could easily fill another book. However, some of these processes will be discussed briefly.

Extrusion and/or drawing methods have worked well and are being scaled up for continuous processing. One very successful technique involves the dispersion of a few percent whiskers in a viscous medium (ammonium alginate and water), and then extruding the slurry through a specially designed nozzle into a dilute acid bath [72]. The bath causes the alginate slurry to gel. This action holds the aligned fibers in place, and then the gelled tape (about $\frac{1}{4}$ in. in diameter) is further stretched by drawing and winding onto a rotating drum at a rate of approximately 5 f/sec. After several layers are built up on the drum, a cut is made across the drum length, and the layers are flattened out to form a large sheet (sizes as large as 24 in. × 54 in.). This sheet is then dried, and the organic binder is burned off. The sheet is next impregnated with a resin, placed in a mold and compacted to the desired whisker volume fraction by squeezing out the excess resin from the mold. Epoxy composites containing as much as 55 vol% of well-aligned whiskers have been prepared by this procedure [72]. Bend tests have shown strengths as high as 125,000 psi and elastic moduli of 14.7×10^6 psi. The whiskers used were Si_3N_4 and have an elastic modulus of about 45×10^6 psi. This process is now in the pilot plant stage of development at the Explosive Research and Development Establishment (British Ministry of Technology Establishment) [72].

Wohrer and Economy [73, 74] have also developed an extrusion process as well as a spinning process which results in highly aligned SiC whiskers in continuous tape form. They previously investigated several alignment techniques, which are listed in Table 9.8 [74]. However, it was concluded that improved methods would be needed to provide the necessary degree of alignment. The two methods were finally used to prepare continuous organic tapes containing up to 15 vol% whiskers. The organic carrier in the tape was burned off (without loss of alignment), and a resin was then sprayed onto the yarn. A dilute resin solution was used to promote complete

TABLE 9.8 CONVENTIONAL PROCESSING TECHNIQUES INVESTIGATED FOR PREPARING ALIGNED, WHISKER-REINFORCED RESINS [74]

Technique	Whisker Loading Attainable (vol %)	Degree of Orientation	Reinforcing Effectiveness
Casting	to 3%	Poor to fair	None
Centrifugal casting	to 6%	Fair	Slight
Compression molding	to 15%	Poor	None
Transfer molding	to 12%	Fair	Good
Extrusion	to 7%	Good	Slight
Extrusion and cold drawing	to 7%	Excellent	Slight
Injection molding	to 5%	Fair	Slight

penetration, and composites containing as much as 50 vol % whiskers were prepared by compression molding layers of the impregnated tape.

Metallic composites have also been prepared by incorporating a fine metal powder into the organic carrier prior to tape formation. The two methods, spinning and extrusion, are illustrated in Figure 9.28.

Calow and Wakelin [66] have further extended the alginate extrusion process by adding finely divided silver, nickel, or iron powders to a whisker-alginate bath (plasticized material) and then by extruding a gelled tape

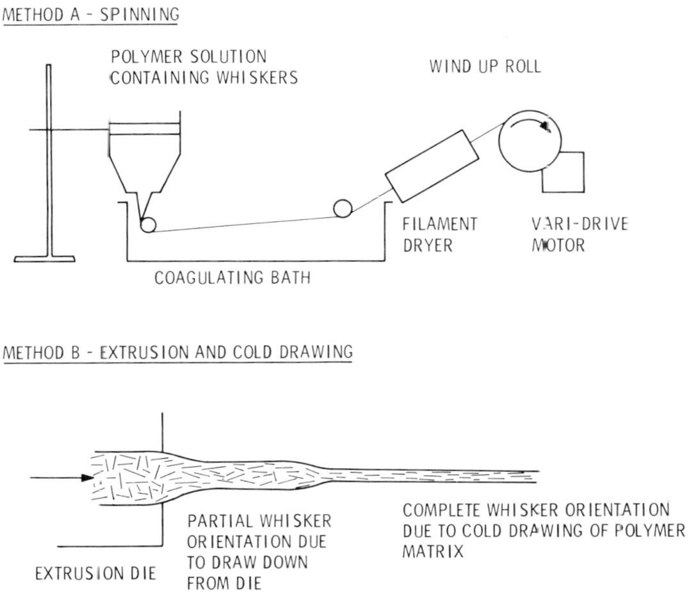

Figure 9.28 Two methods used to produce highly oriented, continuous whiskers tapes. (From Wohrer and Economy, Ref. [74], Courtesy of *SAMPE Journal*.)

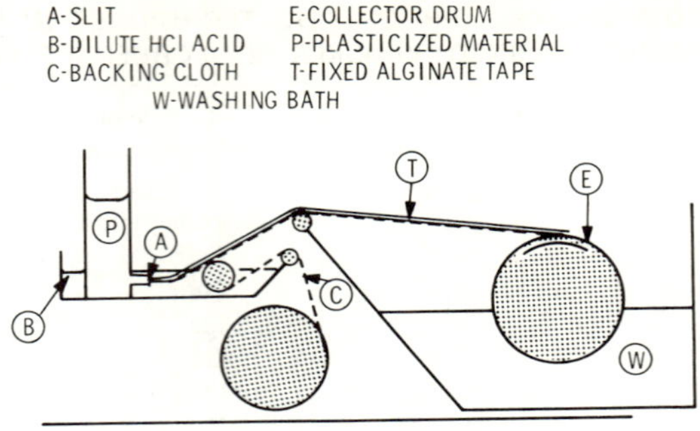

Figure 9.29 Schematic drawing of the alginate-extrusion process used to prepare whisker tapes. (From Calow and Wakelin, Ref. [66], Courtesy of the Institute of Metals.)

containing both whiskers and metals. The process is shown schematically in Figure 9.29. The tape is then washed and dried, and the alginate is burned off. Densification is achieved by hot pressing layers of the whisker/metal tape. Composites of Si_3N_4 whiskers in silver were prepared by this method and also by the hot extrusion of silver powder—whisker mixtures that had first been pressed into compacts [66]. The results are shown in

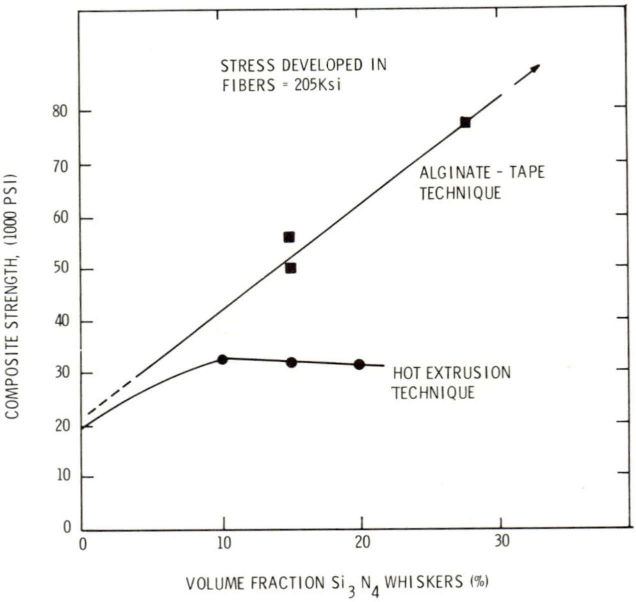

Figure 9.30 Comparative strength values of silver composites prepared by hot extrusion and alginate extrusion techniques. (From Calow and Wakelin, Ref. [66], Courtesy of the Institute of Metals.)

Figure 9.30. From the curves shown, it seems likely that the leveling off in composite strength with increasing whisker content for hot-extruded specimens is due to mechanical damage and breakage of the whiskers. On the other hand, the alignment achieved during the alginate process is performed when the whisker concentration is low, so that the whiskers do not interfere. The final whisker content is increased up to 30 % by subsequent consolidation (hot pressing).

Another promising approach is that of aligning the fibers in a dielectric medium [68–70]. By applying an electric potential, the whiskers become polarized and align themselves between the electrodes. At TRW Systems [70], a process has been developed whereby one electrode is a moving, continuous aluminum foil, and as the foil is raised from the liquid, capillary action draws the aligned whiskers onto the foil. After drying, the foils containing the whiskers are stacked and laminated by hot pressing. The formation of stringers containing aligned SiC whiskers is shown in Figure 9.31 and cross sections of the composite material are shown in Figure 9.32.

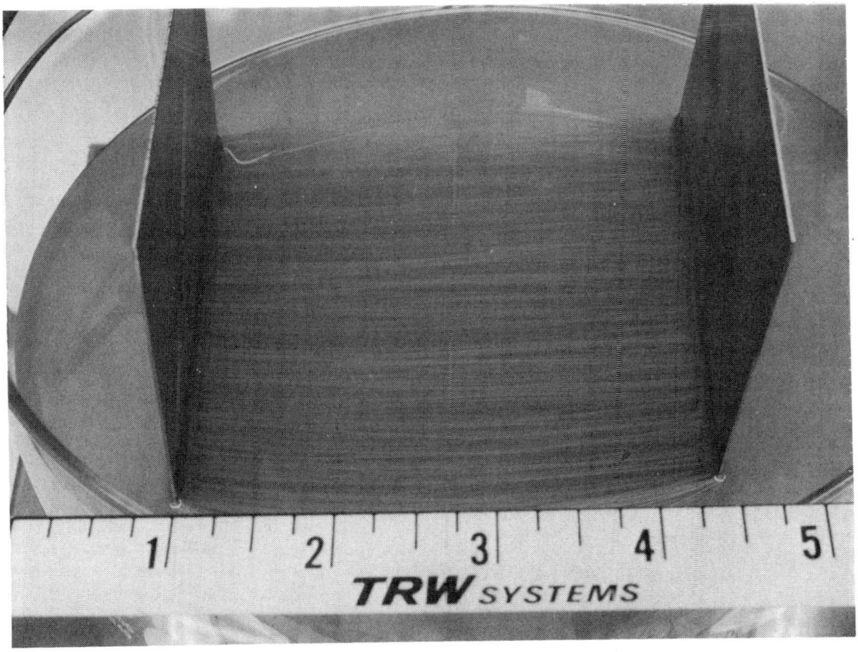

Figure 9.31 Whisker alignment in a dielectric solution between two electrodes. (From A. Toy, Ref. [70].)

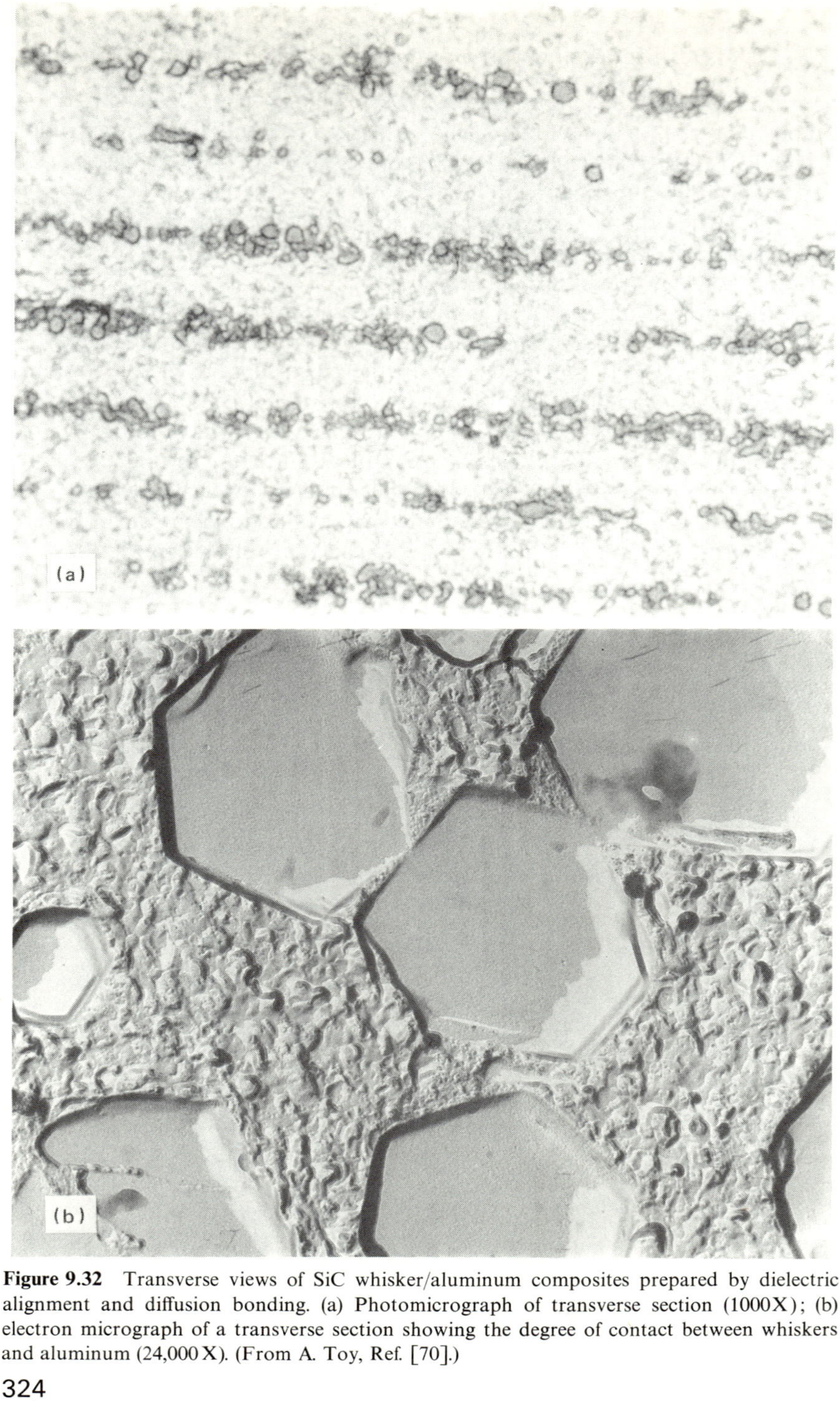

Figure 9.32 Transverse views of SiC whisker/aluminum composites prepared by dielectric alignment and diffusion bonding. (a) Photomicrograph of transverse section (1000X); (b) electron micrograph of a transverse section showing the degree of contact between whiskers and aluminum (24,000X). (From A. Toy, Ref. [70].)

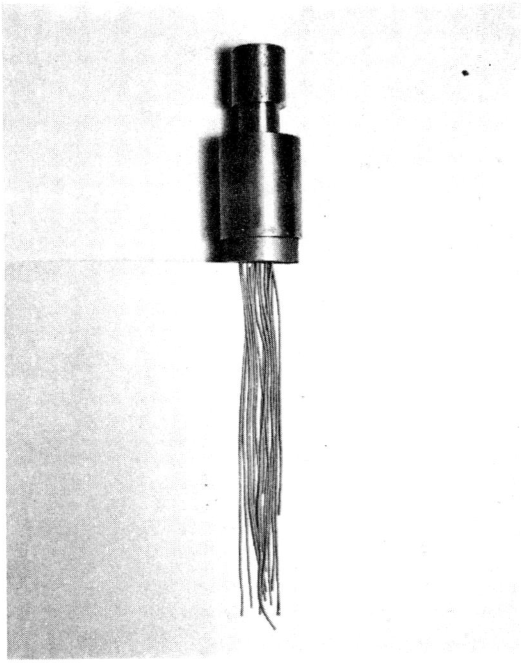

Figure 9.33 Strands of aligned SiC whiskers in a slurry containing an aluminum alloy powder and a binder. (From H. N. Barr, Hittman Associates, Ref. [77]).

Hittman Associates [77] have developed a technique for producing tapes of aligned SiC whiskers in an aluminum alloy. The whiskers and alloy powder are blended in a slurry containing a small amount of binder, and then the mixture is extruded through a die to form continuous strands, such as those shown in Figure 9.33. These strands are next formed in a variety of shapes (sheets, rods, tubes, cones, etc.), are dried and fired to remove the binder, and are pressed at temperatures where a liquid phase in the matrix is formed. The presence of the liquid aids in matrix penetration and composite densification.

The alignment processes described thus far have been for small diameter whiskers (i.e., 1 to 5 μ in diameter). However, for larger whiskers (diameters of 10–15 μ and greater) where lengths are as great as 20 mm (20,000 μ) air elutriation techniques have worked well. At the Space Sciences Laboratory of the General Electric Company, for example, classified whiskers of Al_2O_3 (long wool, see Figures 9.18 and 9.21) were separated individually by air blasts in a chamber, and then were gently floated up through a vertical

Figure 9.34 Strands of metal-coated Al_2O_3 tapes prepared by air elutriation and collimation. (a) Whisker tape-making machine; (b) individual whisker tapes (mag. 12X). (From Refs. [35, 41]).

column in a controlled, continuous flow of air [82]. Near the top, the column is divided into compartments that taper down and terminate at small rectangular slits, through which the whiskers exit. As the whiskers pass through the compartments, their direction is changed from horizontal to vertical, and thus they merge in collimated tape form through the slits, all aligned vertically. The emerging whiskers are then drawn onto a moving screen (with a slight vacuum behind it). The motion of the moving screen causes the emerging whisker tapes to lay down on the screen in parallel rows.

The continuous tapes are then rolled off the end of the moving screen, as shown in Figure 9.34*a*, and are wound onto a rotating drum. Strands of individual Ni-coated Al_2O_3 whisker tapes are shown in Figure 9.34*b*, where the alignment of the coated whisker can be seen.

Combination and Consolidation of Constituents. Numerous methods have been used to combine the matrix with the whiskers. In some cases, alignment of the whiskers has been achieved during the combining and consolidating processes, which were discussed previously. Basically, the combining processes can be divided into three categories:

1. Deposition (molecular) processes.
2. Liquid processes.
3. Solid processes.

Several examples of these processes are shown in Table 9.9 along with the consolidating methods for a variety of whisker composites.

Deposition (Molecular) Processes. In these processes, the matrix is usually deposited chemically from the vapor or electrolytically from a solution onto the whiskers surfaces. The coatings are usually metallic, and may be thin (1 μ or less in the thickness) for promoting wetting or for providing an electrically conductive surface on a nonconducting whisker, or they may be relatively thick (tens or even hundreds of microns in thickness) and thereby serve as the matrix upon consolidation. Chemical vapor deposition (CVD) methods have been used successfully to deposit relatively thick coatings of Ni, Ni-W, and NiCr onto Al_2O_3 whisker mats by using fluidized bed techniques [78, 79]. However, consolidation of the thickly coated whisker mats has not been successful because of gross whisker breakage [79].

Electrolytic processes have been used frequently: (1) to coat the whiskers as they settle onto an electrode (electro co-deposit); (2) to coat the whiskers in a loose mat or thread (electroplate), which requires further consolidation; and (3) to completely coat and fill in all interstices in a whisker bundle (electroform). In some cases, the whiskers can be aligned rather well as they settle in the electro co-deposition process by causing the electrolyte

TABLE 9.9 METHODS USED TO

Method Used to Combine Constituents	Matrix	Whisker[a] Composition	Alignment	Coating	Consolidation Process[b]
Deposition (molecular)					
Chemical Vapor Deposition	Ni or NiCr	Al_2O_3, SiC	R	C	H.P. or L.P.H.P.
Electro Co-Deposit	Ni	Al_2O_3	R	C	H.P.
	Ni	Al_2O_3, SiC	R	C, U	—
	Ni	Al_2O_3	A	C, U	As deposited or hot forged
Electroplate	Ni	Al_2O_3	A	C	H.P.
	Ni	Al_2O_3, SiC	A, R	C	Cold and Hot Roll, H.P.
Electroform	Ni	Al_2O_3	A	C	—
Liquid (matrix)					
Alloy (Eutectic)	Al, Cb	$NiAl_3$, Cb_2C	A	U	Unidirectional solidification
Infiltration	Resin	Al_2O_3	A	C	Polymerize matrix
	Ag	Al_2O_3	A	C	Solidification of matrix
	Cu, Ni-alloys	Al_2O_3	A	C	Solidification of matrix
	Al	B_4C	A	C	Solidification of matrix
	Al-alloys	Al_2O_3	A	C	Solidification of matrix
Melting of powdered matrix	Fe, NiCr	Al_2O_3	R	C, U	L.P.H.P.
Casting of melt and whiskers	Co-alloys	Al_2O_3, SiC	R	C, U	Solidification of melt
Spin, extrude, draw slurry of whisker, matrix powder, and carrier solution	Ag, Fe, Ni	Si_3N_4	A	U	Dry, burn off organic carrier, H.P.
	Al	SiC	A	U	Dry, burn off organic carrier, H.P.
	Al-alloy	SiC	A	U	Dry, burn off organic carrier, L.P.H.P.
	Cu, Al, Mg	SiC	A	U	Dry, burn off, sinter, H.P.
	Al-alloy	SiC	A	U	Dry, burn off, H.P.
Filter slurry or settle out whiskers and matrix	Ag-alloy	Si_3N_4	R	U	L.P.H.P.
	Cu, Al-alloy	Al_2O_3, SiC	R	U	H.P.
	NiCr, Al-alloy	SiC	A	C, U	L.P.H.P.
	NiCr, Al-alloy	Al_2O_3, SiC	A, R	C, U	L.P.H.P.
	Mg-alloy	Al_2O_3, SiC	A, R	C, U	L.P.H.P.
Impregnation into whisker strands and tapes	Resin	SiC	A	U	H.P. and squeeze out excess resin
	Resin	Al_2O_3	A	U	H.P. and squeeze out excess resin
	Resin	Si_3N_4	A	U	H.P. and squeeze out excess resin
Solid state					
Powder matrix and whiskers	Ni	Al_2O_3	R	C	H.P.
	Al	Si_3N_4	A	U	Extrude
	Ni, Ti	Al_2O_3, SiC	R	U	High-energy rate forming
Deposit whiskers on Al-foil	Al	SiC	A	U	Diffusion bond

[a] R = random whisker alignment
A = whiskers unidirectionally aligned
C = coated whiskers
U = uncoated whiskers

Forming or Shaping Process	Remarks	Reference
Hot roll	SiC whiskers reacted chemically with matrix	[78, 79]
—	Problem of voids and low whisker content	[78]
—	Problem of voids and low whisker content	[88]
—	Align whiskers via flow in electroplating bath	[89, 78, 43]
—	Problem of consolidation and whisker breakage	[40, 41, 90]
—	SiC whiskers reacted with matrix, also whisker breakage	[75]
—	Excellent properties, limited to very small specimens	[90]
Extrude, roll	Whisker content fixed by alloy composition	See Chapter 10
—	Achieved high-strength composites	[33]
—	Achieved high-strength composites	[59, 92]
—	Achieved varying degrees of success; coating stability a problem	[22, 41, 90]
—	Achieved varying degrees of success; coating stability a problem	[91]
—	Achieved varying degrees of success; coating stability a problem	[35, 37, 93]
—	Problems of interfacial reactions and whisker segregation	[83, 84, 94]
—	Whiskers concentrated at grain boundaries, low whisker content	[95]
—	Excellent whisker alignment and packing	[66]
—	Excellent whisker alignment and packing	[96]
—	Excellent whisker alignment, fabrication of complex shapes	[77]
—	Excellent whisker alignment	[74]
—	Excellent whisker alignment, problem of carbon burn-off	[85]
—	Problem of matrix porosity and wetting	[63, 67]
Hot roll	Little whisker alignment, much whisker breakage	[85]
Clad & roll	Ni-coated whiskers were aligned magnetically during settling in slurry	[44, 87]
Clad & roll	Ni-coated whiskers were aligned magnetically during settling in slurry	[78–80, 87]
Hot extrusion	Successfully rolled composites with aligned whiskers	[81, 87]
—	Achieved excellent alignment, little whisker breakage	[73, 74]
—	Achieved excellent alignment, little whisker breakage	[97]
—	Achieved excellent alignment, little whisker breakage	[63, 67]
—	Many whiskers broken	[90]
—	Extended long rods, but many whiskers broken	[99]
—	Chemical reactions minimized, but whiskers still broken	[86]
—	Achieved excellent alignment of whiskers	[70, 98]

[b] H.P. = Hot pressing
L.P.H.P. = Liquid phase hot pressing

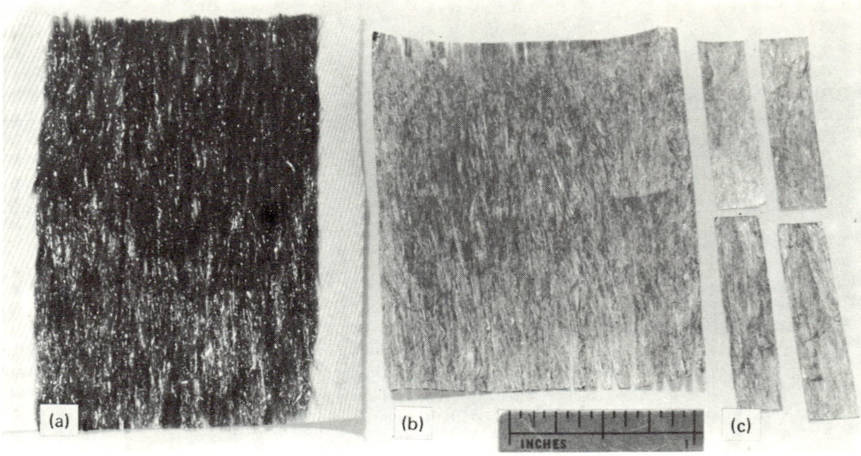

Figure 9.35 Preparation of nickel-reinforced composites by electroplating. (a) Manually oriented metallized whiskers; sheet of loose W-coated, aligned Al_2O_3 whiskers. (b) Electroformed prepreg sheet (after electroplating in nickel bath. (c) Electroformed prepreg loads; $1\frac{1}{4}$ in. × $\frac{1}{4}$ in. strips ready for stacking and hot pressing. (From Ref. [41]).

containing the whisker to flow directionally across the cathode. Examples of these methods are listed in Table 9.9.

The deposition processes nearly always leave voids in the matrix so that a consolidation process is required. Figure 9.35 shows some steps used in preparing nickel composites for electroplating. The whiskers are first given a thin conductive (tungsten) coating by a sputtering (vapor ion-plating) process and then are aligned and placed on a glass woven fabric (Fig. 9.35*a*) [41]. Next, the whiskers are placed in a special holder and are attached to an electrode, and nickel is allowed to electroplate and fill in the interstices between the coated whiskers (Fig. 9.35*b*). The electroplated prepreg sheet is next cut into strips (Fig. 9.35*c*), which are stacked and hot pressed to form a composite specimen. The cross-section of a typical composite specimen containing 23 *v/o* whiskers is shown in Figure 9.36.

Liquid Processes. Processes involving the matrix in a liquid form are often preferred, since they usually involve low pressures and are least likely to break the whiskers. During composite fabrication the matrix may be entirely molten, may be partially molten (as in liquid-phase hot-pressing, LPHP), or it may be solid as a finely divided powder dispersed with the whiskers in an appropriate carrier medium. Table 9.9 shows several methods used to fabricate composites which use the matrix in liquid form.

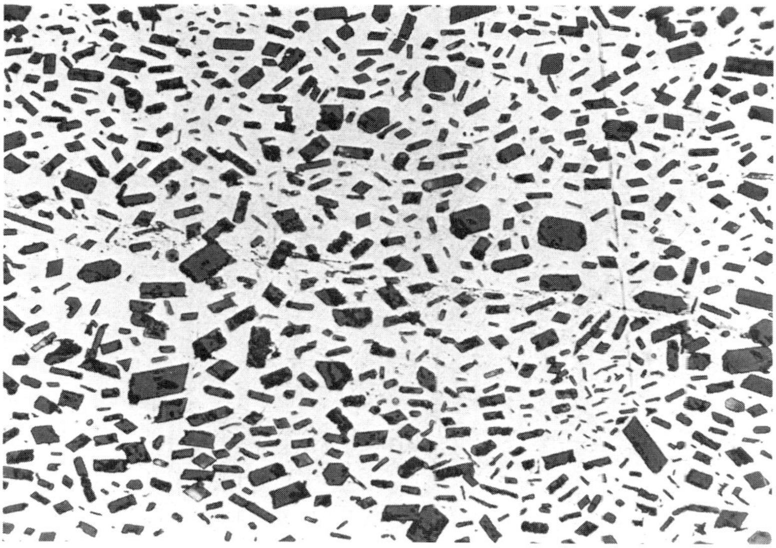

Figure 9.36 Microstructure of a transverse section of an electroplated, hot-pressed composite of nickel containing 23 vol% Al_2O_3 whiskers. The specimen was hot pressed at 1175°C under 3000 psi for 20 min (mag. 290X). (From Ref. [41]).

The case for the solidification of certain eutectic alloys to form both the matrix and reinforcing fibers is discussed in Chapter 10.

Infiltration [22, 33, 35, 54, 90, 92] has been a method used extensively, since little whisker breakage occurs during the penetration of the molten matrix into random mats or bundles of aligned whiskers. However, the higher processing temperatures, required by the use of molten metals has thus far limited the use to lower melting-point metals and alloys, such as Al, Cu, and Ag. An aluminum specimen containing about 20 *v/o* of nichrome-coated Al_2O_3 whiskers is shown in Figure 9.37. This specimen was prepared by vacuum infiltration of molten aluminum into a packed bundle of aligned whiskers. Although molten aluminum will not wet uncoated whiskers, it did wet and penetrate the coated whiskers by capillary action. Figure 9.37*a* shows the end of the specimen where the aluminum and the coating were dissolved away to expose the Al_2O_3 whiskers; and Figure 9.37*b* shows a view of the mid-section of the same composite.

Metals or alloys having melting points greater than 1100°C, such as iron or nickel-base alloys, present much greater problems, since in molten form they will either chemically degrade (weaken) the whiskers, or they will dissolve away the protective coatings on the whiskers during the infiltration

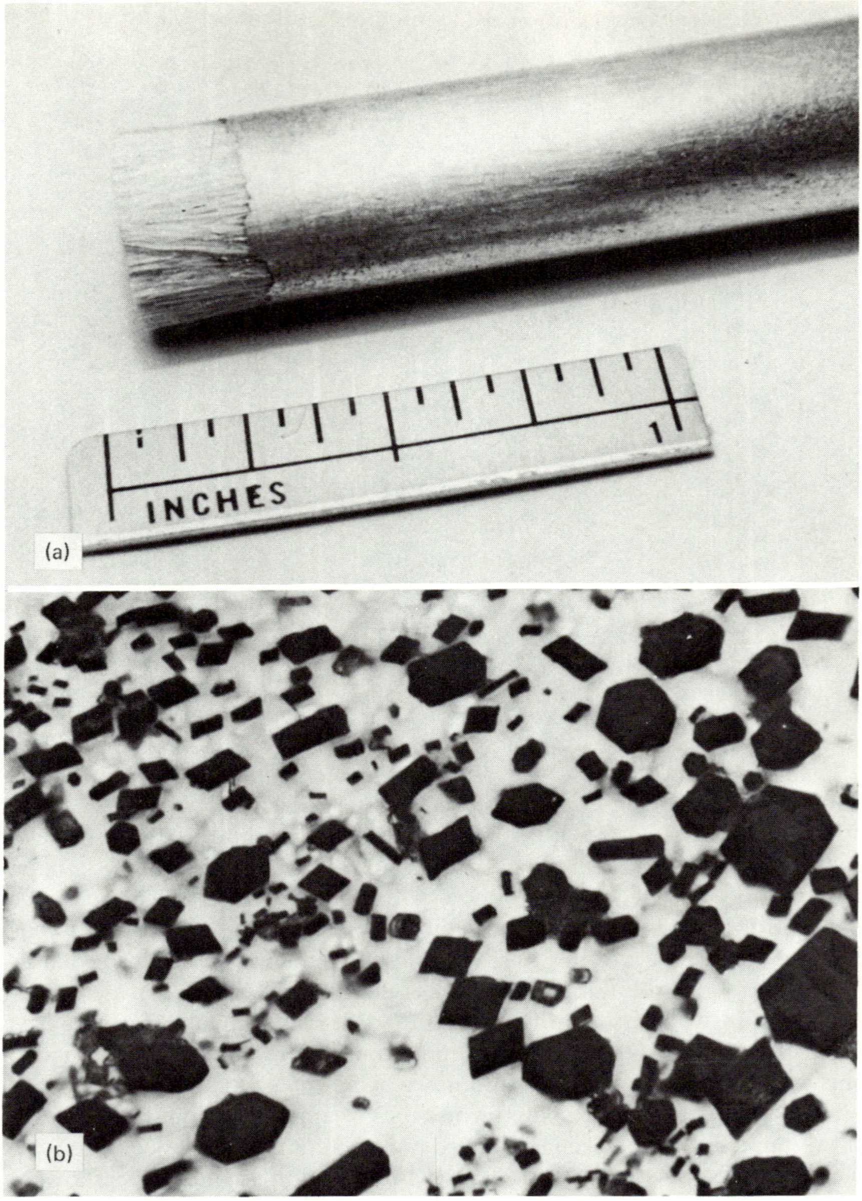

Figure 9.37 A composite of aluminum reinforced with 27 vol% Al_2O_3 whiskers which was prepared by vacuum infiltration. (a) The aluminum matrix and coating were etched away from one end exposing the whiskers. (b) Cross-section view of the whiskers in the aluminum matrix (mag. 675 X). (From Ref. [100]).

process. Some limited success has been achieved with these latter alloys when small (capillary) specimens were prepared [41, 90].

In the case of resins, the low processing temperatures make the infiltration or impregnation of oriented whisker threads or tapes a relatively easy task; a task which is well suited for automated, continuous processing. In addition, a relatively open whisker thread or tape, which has been impregnated with a resin, can be warm pressed or molded so that the excess resin can be squeezed out, thereby leaving a thread or tape containing a high whisker concentration (up to 50 *v/o* or more). However, in the case of mats or tapes of whiskers containing liquid metals, the whiskers tend to segregate due to large density differences between the whisker and the metals. Thus, the squeezing out of excess molten metal is not an easy or desirable procedure [78, 79, 83, 84].

Greater success has been achieved by dispersing the whiskers and matrix (in finely divided, powdered form) in a suitable liquid, or carrier, medium. The whisker/matrix mixture or blend can then be filtered or settled out to form a mat [44, 63, 67, 85, 87], or the blend can be spun, drawn, or extruded to form continuous, aligned strands or tapes [66, 74, 77, 85, 96]. The carrier medium is usually removed by drying and burning (oxidation). Consolidation can be accomplished by hot pressing (HP) or liquid-phase hot-pressing (LPHP). The combination of these procedures generally results in highly oriented, void-free whisker composites, as indicated in Table 9.9.

When the whisker content is very low, specifically, less than 1%, the whiskers can be dispersed in a molten alloy and pressure or die cast into specified shapes [95]. During solidification, the whiskers may concentrate in the grain boundaries of the alloys, which improves both composite strength and elongation [95].

Solid Processes. Solid-state procedures have also been used to prepare whisker composites, but generally with somewhat less success. The most common procedure is to consolidate a powdered metal whisker blend by hot pressing [90]. Sintering usually does not work well, since the whiskers tend to spring apart after they are densely compacted [63]. An unusual method for depositing aligned whiskers onto a moving thin foil of aluminum was developed by TRW Systems [98]. Stacks of these "coated" foils are then layed up and are consolidated into dense composites by diffusion bonding [70, 98] (see Figs. 9.31 and 9.32).

Extrusion of powdered-metal/whisker blends have met with only limited success, since the whiskers are usually fractured severely during the extrusion process [99]. Similar effects have been observed in blends of whiskers and metals powders which have been consolidated by hot pressing, or hot autoclaving, and then by exposing the compacts (placed in a hermetically sealed container) to high-energy rate-forming conditions, such as pneumatic compaction [86]. The advantage of the latter process is that the time period required for compaction is very short so that undesired chemical reactions between the whiskers and matrix will be minimized.

Once whisker-metal mixtures are densified, they may be subjected to further forming or shaping operations by hot rolling, extrusion, or forging. However, great care is required to minimize whisker breakage. For example, it is most difficult to further roll or extrude a composite containing randomly aligned whiskers without severely fracturing the whiskers [85, 87]. On the other hand, if the whiskers are highly aligned and do not cross or contact each other, extrusion and rolling in a direction perpendicular to the whisker lengths have been successfully achieved [80, 87].

DISCUSSION

The preparation and fabrication of whisker composites is complex and involves several individual steps. Each step is of critical importance to the following one. Thus, the cleaning and the sorting of the whiskers play a dominant role in the subsequent alignment of the whiskers which, in turn, has a profound influence on the combining and consolidating processes. Thus each step affects the final structure and the properties of the composite material.

While there are many approaches available to fabricate composites, as indicated in Table 9.9, only a few have produced composites which effectively utilize the whisker properties. Better processing procedures will undoubtedly be developed in the future. However, the greatest success most likely will result from an appreciation of the total composite technology, which includes the design and mission of the final product; the mechanics, physics, and chemistry of the constituents; and the behavior of the combined constituents.

Just as the constituents are arranged in a prescribed geometrical manner to provide tailor-made composite properties, the individual processing steps must be selected and arranged so that the total fabrication procedure can be tailor-made to produce composites having the desired whisker-matrix combinations which best utilize the whisker properties.

Each whisker type, with its own characteristic set of mechanical, physical, and chemical properties, is best suited to certain processing procedures which may not be suitable for other whisker types. For example, very small whiskers (1 to 3 μ in diameter and 300 to 900 μ long; $l/d \sim 300$) will handle differently from much longer varieties (viz., 10 to 50 μ in diameter and 10,000 to 50,000 μ long, $l/d \approx 1000$). The smaller whiskers can be mixed with a viscous carrier medium and then spun, drawn, or extruded. They are also amenable to paper-making processes. The longer whiskers, on the other hand, are better handled by elutriation and textile processing methods.

Once the whiskers are aligned, the degree of alignment (or percentage of whiskers aligned in a given direction) plays a major role in subsequent processing operations. If the whiskers are perfectly aligned, the composites can be hot-pressed, forged, hot-rolled (perpendicular to the whiskers' length) without significant breakage [87]. However, if the whiskers are misaligned or are randomly oriented, the same amount of pressing or rolling will severely fracture the whiskers [41, 87]. The degree of whisker alignment will also control the maximum packing fraction; in randomly oriented composites, the whisker concentration will rarely exceed 15 vol% whereas in unidirectionally aligned composites, the whiskers can be packed in excess of 50 to 60 vol% [67, 74, 96].

One of the most promising methods for aligning the whiskers without significant breakage and for achieving high whisker loadings in the final composite material, is the extrusion or spinning of whisker-matrix slurrys and viscous suspensions. In these suspensions, the whisker concentration must be low, so that the spacing between individual whiskers is relatively large. Thus, during the extrusion or spinning process, the whiskers are relatively free to move independently and align themselves in the extrusion or drawing direction. Once aligned, the whiskers can be drawn together (concentrated) through the removal of the carrier liquid by evaporation. The matrix can be added, if it was not done so previously, by impregnation or infiltration. Thus, impregnated strands of aligned, whisker tape or yarn can be woven, can be made into sheets, or can be made into a variety of shapes that can be subsequently consolidated to form the final end product.

In summary, no simple fabrication process (involving several unit steps) is universally applicable to all whisker composites. The choice of specific procedures involving various combinations of individual steps, depends on the nature of the whiskers, on the desired degree of whisker packing and alignment, on the matrix, and on the allowable processing conditions (time, temperature, pressure, etc.). For metals having relatively low melting points, such as aluminum, copper, or silver, casting or infiltration techniques can be emphasized. However, for metals having higher melting temperatures, such as iron, nickel, or cobalt, prime emphasis must be devoted to controlling the degrading chemical reactions between the whiskers and the matrix. Unfortunately, the selection of a given set of processing steps involving relatively low temperatures does not necessarily solve all of the problems, since these composites will likely be used at elevated temperatures. For example, Figure 9.38 illustrates some of the difficulties encountered in the fabrication of nickel composites containing Al_2O_3 whiskers. The whiskers are coated to promote wetting and bonding to the matrix and also to minimize

PROCESS	WHISKER-COATING-MATRIX STABILITY	COMPOSITE DENSITY	MATRIX STRENGTH, DUCTILITY	WHISKER ASPECT RATIO: ℓ/D	WHISKER VOLUME FRACTION	WHISKER ALIGNMENT
	ASSESSMENT OF FACTORS IMPORTANT TO COMPOSITE STRENGTH — DEGREE OF DIFFICULTY: + MINOR, O ORDINARY, – MAJOR					
LOW TEMPERATURE / SOLID STATE						
CO-ELECTRO FORM	+	+	+	+	–	–
CO-VAPOR FORM	+	+	+	+	–	–
ELECTRO PLATE	+	–	+	+	O	+
VAPOR PLATE	+	–	+	+	O	+
HIGH TEMPERATURE / SOLID STATE						
CO-PRESS + SINTER	O	–	–	–	+	–
CO-EXTRUDE + SINTER	O	–	–	O	O	O
PLATE + HOT PRESS	O	+	+	–	+	+
CO-HOT PRESS	O	O	+	–	+	–
HIGH TEMPERATURE / LIQUID PHASE(s)						
CO-PRESS + SINTER	–	O	+	–	+	–
CO-EXTRUDE + SINTER	–	O	+	O	O	O
PLATE + HOT PRESS	–	+	+	–	+	+
CO-HOT PRESS	–	+	+	–	+	–
INFILTRATE	–	O	+	+	+	+

Figure 9.38 Assessment of factors important to the composite strength of nickel specimens reinforced with Al_2O_3 whiskers.

chemical reactions. Figure 9.38 shows that the low temperature forming processes (such as electrodeposition or CVD) are advantageous with respect to minimum damage to the coating and to the whiskers, but they present other problems in whisker packing density and alignment. On the other hand, the high-temperature processes, involving the matrix in either solid or liquid state, will generally favor desirable matrix properties, such as ductility and elimination of voids, and will favor higher whisker concentrations, but the likelihood of degrading chemical reactions presents a serious problem.

CONCLUSIONS

The preparation and manufacture of whisker composites consists of several individual steps which include: whisker processing (coating and benefication), whisker alignment, incorporation of the matrix, and consolidation of the whisker-matrix combination. Subsequent steps involve the shaping of the composite to a final form and then joining to a structural member or component. However, the success achieved in these latter steps depends largely on the success achieved in all of the previous ones.

An ultimate goal of the composite fabricator is to be able to select the unit steps such that the entire process can be tailor-made to produce specific composites which best utilize the whisker properties. At the same time, the composite properties must be predictable (reproducible). Thus, in order to effectively use the reinforcing capabilities of the whiskers in a given matrix, (1) the whiskers must exceed a critical volume fraction, (2) they must be fully aligned (the misalignment angle must be less than a critical value), and (3) they must exceed a critical aspect ratio. Furthermore, the whiskers must not be significantly weakened by chemical reactions during processing at the expense of adequate wetting and bonding to the matrix which is necessary.

Whisker technology is a relatively new field, and there is much to be learned. Factors that will greatly advance this technology include:

1. Development of suitable analytical techniques for designing and accurately predicting properties of composites.
2. Development of processes which will grow larger quantities of whiskers (free from gross defects and branching) at reasonable costs.
3. Development of automated processes which will clean and grade the whiskers to specific sizes (according to length and diameter).
4. Development of automated processes for continuously aligning the whiskers and for incorporating them into the matrix.
5. Development of consolidating processes which will not fracture the whiskers, nor will chemically degrade the whiskers.
6. Establishment of procedures for further shaping the composites to final form, if necessary, and for joining the composites to other structures.
7. Establishment of acceptance criteria for specifying whiskers, tapes, prepregs, and composites.
8. Development of suitable quality-control procedures for each processing step and the final product.

One of the major potential applications of the ceramic whiskers is the reinforcement of high-temperature alloys. Before this can be achieved satisfactorily, suitable coatings and coating procedures must be developed.

These coatings need to provide the required degree of wetting and bonding between the whiskers and matrix, they must be stable and not be dissolved or be diffused completely into the matrix, and they must serve as an effective diffusion barrier between the whiskers and matrix.

ACKNOWLEDGMENTS

The author is grateful for the help of many colleagues who provided material and suggestions during the preparation of this chapter. In particular, the author wishes to acknowledge the assistance of J. Chorné, C. A. Bruch, E. Feingold, R. L. Mehan, M. J. Noone, A. Gatti, and R. Jakas. Thanks are also extended to H. W. Rauch, Sr. and L. R. McCreight. Several of the photomicrographs were provided by M. Birenbaum. Also the continued support of R. A. Schmidt, Naval Air Systems Command; J. A. Herzog, formerly of the Air Force Materials Laboratory; A. P. Levitt, Army Materials and Mechanics Research Center; and J. Gangler, NASA-Headquarters, in various whisker programs is gratefully acknowledged.

The opinions expressed in this chapter are those of the author and do not necessarily represent the views of the General Electric Company.

REFERENCES

[1] D. L. McDanels, R. W. Jech, and J. W. Weeton, "Analysis of Stress-Strain Behavior of Tungsten-Fiber-Reinforced Copper Composites," *Trans. AIME*, **233** (1965), 636–642.

[2] D. Cratchley, "Factors Affecting the UTS of a Metal/Metal-Fiber Reinforced System," *Powder Met.*, **11** (1963), 59–72.

[3] A Kelly and W. R. Tyson, "Fiber-Strengthened Materials," in *High Strength Materials*, V. F. Zackay, Ed., John Wiley and Sons, New York, 1965, pp. 578–602.

[4] B. W. Rosen, "Mechanics of Composite Strengthening," in *Fiber Composite Materials*, ASM, Metals Park, Ohio, 1965, pp. 37–75.

[5] A. Kelly and G. J. Davies, "The Principles of the Fiber Reinforcement of Metals," *Met. Rev.*, **10** (1965), 1–77.

[6] D. V. Rosato and C. S. Grove, Jr., "Filament Windings: Its Development, Manufacture, Application and Design," Interscience, New York, 1964.

[7] P. W. Jackson and D. Cratchley, "The Effect of Fiber Orientation on the Tensile Strength of Fiber-Reinforced Metals," *J. Mech. Phys. Solids*, **14** (1966), 49–64.

[8] E. Z. Stowell and T. S. Liu, "On the Mechanical Behavior of Fiber-Reinforced Crystalline Materials," *J. Mech. Phys. Solids*, **9** (1961), 242–260.

[9] S. W. Tsai, "Strength Characteristics of Composite Materials," NASA Report CR224, April 1965.

[10] M. Humenik, Jr. and T. J. Whalen, "Physicochemical Aspects of Cermets," in *Cermets*, J. R. Tinklepaugh and W. B. Crandall, Eds., Reinhold, New York, 1960, pp. 6–49.

[11] W. D. Kingery, "Role of Surface Energies and Wetting in Metal-Ceramic Sealing," *Bull. Am. Ceram. Soc.*, **35** (1956), 108–111.

[12] B. C. Allen and W. D. Kingery, "Surface Tension and Contact Angles in Some Liquid Metal-Solid Ceramic Systems at Elevated Temperatures," *Trans. AIME*, **215** (1959), 30–37.

[13] C. S. Smith, "Grains, Phases, and Interfaces: An Interpretation of Microstructures," *Trans. AIME*, **175** (1948), 15–51.

[14] O. K. Riegger and L. Van Vlack, "Microstructures of Intermetallic Compound-Liquid Systems," *Trans. ASM*, **56** (1963), 513–527.

[15] A. Bondi, "The Spreading of Liquid Metals on Solid Surfaces," *Chem. Rev.*, **52** (1953), 417–458.

[16] W. H. Sutton and E. Feingold, "Role of Interfacially Active Metals in The Apparent Adherence of Nickel to Sapphire," in *Materials Science Research*, Vol. 3, Plenum Press, New York, 1966, pp. 577–611.

[17] C. R. Kurkjian and W. D. Kingery, "Surface Tension at Elevated Temperatures, III, Effect of Cr, In, Sn and Ti on Liquid Nickel Surface Tension and Interfacial Energy with Al_2O_3," *J. Phys. Chem.*, **60** (1956), 961–963.

[18] W. M. Armstrong, A. C. D. Chaklader, and J. F. Clarke, "Interface Reactions Between Metals and Ceramics, I. Sapphire-Nickel Alloys," *J. Am. Ceram. Soc.*, **45** (1962), 115–118.

[19] J. E. Ritter, Jr. and M. S. Burton, "Adherence and Wettability of Nickel, Nickel-Titanium Alloys, and Nickel-Chromium Alloys to Sapphire," *Trans. AIME*, **239** (1967), 21–26.

[20] M. Nicholas, R. D. Forgan, and D. M. Poole, "The Adherence of Metal-Alumina Interfaces," *J. Matls. Sci.*, **3** (1968), 9–14.

[21] W. H. Sutton, "Role of the Interface in Metal-Ceramic Ceramic (Whiskers) Composites," Soc. Plastics Inc., presented at 21st Annual Technical Meeting, Chicago, Illinois, February 8–10, 1966.

[22] W. H. Sutton, J. Chorné, A. Gatti, and W. E. Sauer, "Development of Composite Structural Materials for High Temperature Applications," Final Report, Contract NOw-60-0465d, February 1965.

[23] W. E. Washburn, *Phys. Rev.*, **17** (1921), 273.

[24] W. H. Sutton, A. Talento, and J. Chorné, "Development of Composite Structural Materials for High Temperature Applications," Sixth Progress Report, Contract NOw-60-0465d, November 1961.

[25] N. F. Dow, "Study of Stresses Near a Discontinuity in a Filament-Reinforced Composite Metal," General Electric Co., Report R63SD61, August 1963.

[26] H. L. Cox, "The Elasticity and Strength of Paper and Other Fibrous Materials," *Brit. J. Appl. Phys.*, **3** (1952), 72.

[27] D. M. Schuster and E. Scala, "The Mechanical Interaction of Sapphire Whiskers With a Birefrigent Matrix," *Trans. AIME*, **230** (1964), 1635–1640.

[28] W. R. Tyson and G. J. Davies, "A Photoelastic Study of the Shear Stresses Associated with the Transfer of Stress During Fiber Reinforcement," *Brit. J. Appl. Phys.*, **16** (1965), 199–205.

[29] R. W. Jech and R. A. Signorelli, "The Effect of Interfiber Distance and Temperature on the Critical Aspect Ratio in Composites," NASA TND-4548, May 1968.

[30] E. Scala, "Design and Performance of Fibers and Composites," in *Fiber Composite Materials*, ASM, Metals Park, Ohio, 1954, pp. 131–156.

[31] D. E. W. Stone, "An Experimental Determination of the Stresses Around a Reinforcing Fibre," Royal Aircraft Establishment, Report TR67015, January 1967.

[32] A. R. T. DeSilva, "A Theoretical Analysis of Creep in Fiber Reinforced Composites," *J. Mech. Phys. Solids*, **16** (1968), 169–186.

[33] W. H. Sutton, B. W. Rosen, and D. G. Flom, "Whisker Reinforced Plastics for Space Applications," *Soc. Plastics Engrs. J.* **20** (1964), 1203–1209.

[34] W. H. Sutton and J. Chorné, "Potential of Oxide-Fiber Reinforced Metals," in *Fiber Composite Materials*, ASM, Metals Park, Ohio, 1965, pp. 173–222.

[35] R. L. Mehan, R. Jakas, and C. A. Bruch, "Behavior Study of Sapphire Wool Aluminum and Aluminum Alloy Composites," Tech. Rpt. AFML-TR-68-100, May 1968.

[36] A. Kelly and W. R. Tyson, "Tensile Properties of Fiber Reinforced Metals—II: Creep of Silver-Tungsten," *J. Mech. Phys. Solids*, **14** (1966), 177.

[37] R. L. Mehan, C. A. Bruch, E. Feingold, "Evaluation of Sapphire Wool and Its Incorporation into Composites of High Strength," Tech. Rpt. AFML-TR-66-357, December 1966.

[38] R. L. Mehan and E. Feingold, "Room and Elevated Temperature Strength of α-Al_2O_3 Whiskers and Their Structural Characteristics," *J Mater.*, **2** (1967), 239–270.

[39] E. H. Andrews, "Destruction of Silicon Nitride Whiskers by Reaction with Metals at High Temperatures," *J. Mater. Sci.*, **1,** 4 (1966), 377–382.

[40] J. Chorné, C. A. Bruch, and W. H. Sutton, "Development of Composite Structural Materials for High Temperature Applications," Final Report, Contract NOw-66-0443-d, January 1967.

[41] J. Chorné, C. A. Bruch, R. Jakas, and W. H. Sutton, "Development of Composite Structural Materials for High Temperature Applications," Final Report, Contract NOOO19-67-C-0243, January 1968.

[42] M. J. Noone and W. H. Sutton, "Investigation of Bonding in Oxide-Fiber (Whisker) Reinforced Metals," Final Report, AMRA CR66-01/7, Contract DA-19-066-AMC-330(X), May 1968.

[43] P. Grenier and M. Marchal, "Les Materiaux Composites Renforcés Par Des Fibres. Application Au Cas D'Une Matrice De Nickel Renforcée Par Des Trichites D'Alumine," *Memoires Scientifiques Rev. Metallurg.*, **65,** 4 (1968), 345–354.

[44] A. Divecha, P. Lare, and H. Hahn, "Silicon Carbide Whisker-Metal Matrix Composites," Tech. Rpt. AFML-TR-67-321, October 1967.

[45] R. F. Register, P. D. Gorsuch, and L. Girifalco, "Structure and Properties of Aluminum Oxide Whiskers, ' *Mater. Res. Std.* (May 1967), 203–206.

[46] P. D. Bayer and R. E. Cooper, "The Effects of Chemical Polishing on the Strength of Sapphire Whiskers," *J. Mater. Sci.*, **2,** 4 (1967), 347–353.

[47] E. Feingold, W. H. Sutton, and W. E. Sauer, "Investigation of Bonding in Oxide-Fiber (Whisker) Reinforced Metals," Report AMMRA-CR-66-01/2, Contract DA-19-066-AMC-330(X), October 1966.

[48] D. M. Marsh, "Stress Concentration At Steps On Crystal Surfaces and Their Role in Fractures," *Fracture of Solids*, AIME, New York, 1963, p. 119.

[49] R. L. Mehan, W. H. Sutton, and J. A. Herzog, "A Review of the Method of Measuring the Strength of Whiskers and Their Role in Reinforcing Ductile Metal Matrices," *AIAA J.* **4,** 11 (1966), 1889–1898.

[50] H. W. Rauch, Sr., W. H. Sutton, and L. R. McCreight, *Survey of Ceramic Fibers and Fibrous Composite Materials*, Academic Press, New York, 1968, 436 pp.

[51] G. A. Cooper and A. Kelly, "Tensile Properties of Fibre-Reinforced Metals: Fracture Mechanics," *J. Mech. Phys. Solids*, **15** (1967), 279–297.

[52] E. Feingold and A. Gatti, "The Role of Mullite Furnace Refractories in the Growth of Sapphire Whiskers," General Electric Co., Report R65SD11, April 1965.

[53] C. A. Bruch, data taken from References 35 and 40.

[54] M. J. Noone, E. Feingold, and W. H. Sutton, "Interfaces in Composites," ASTM STP No. 452, San Francisco, 1969, 59–89.

[55] H. Hahn, "Progress in Preparation and Working of Whisker Reinforced Metal Composites," paper 67-DE-35, presented at Design Engineering Conference, ADME, May 15–18, 1967, New York, N.Y.

[56] J. Schmitt, "Situation Actuelle Des Trichites Et Leurs Applications Potentielles: Deuxième Partie: Materiaux Composites Armés De Trichites," *Verres et Refractaires*, **21,** 6 (1967), 530–538.

[57] J. Chorné, C. A. Bruch, R. Jakas, and W. H. Sutton, "Development of Composite Structural Materials for High Temperature Applications," First Quarterly Report, Contract NOOO19-67-C-0243, April 1967.

[58] W. H. Sutton, J. Chorné, R. L. Mehan, E. Sauer, "Development of Composite Structural Materials for High Temperature Applications," First Progress Report, Contract NOw-65-0176-C, February 1965.

[59] W. H. Sutton, A. Talento, J. Chorné, and R. Regester, "Development of Composite Structural Materials for High Temperature Applications," Seventh Progress Report, Contract NOw-60-0465-d, February 1962.

[60] See Reference 50, Contact Report No. 23, p. 148.

[61] R. H. Duff, F. Mansur, and E. G. Wolff, "Metal-Coated Whiskers," in *Advanced Fibrous Composites*, SAMPE, **10,** Paper F-27, 1966.

[62] W. H. Sutton, unpublished data.
[63] N. J. Parratt, "Whisker Reinforced Plastics and Metals," *Chem. Eng. Progr.*, **62,** 3 (1966), 61–67.
[64] Brochure on Fiber Classifying Unit, Glass Developments Ltd., 148/180 Stewards Road, Battersea, London, England.
[65] D. E. Niesz, "Development of Carbon-Filament Reinforced Metals," Final Report Contract NOOO19-67-C-0342, June 1968.
[66] C. A. Calow and R. J. Wakelin, "The Alignment of Whiskers In Metallic Materials," *J. Inst. Metals*, **96** (1968), 147–154.
[67] N. J. Parratt, "Reinforcing Effects of Silicon-Nitride Whiskers in Silver and Resin Matrices," *Powder Met.*, **7,** 14 (1964), 152–167.
[68] R. A. Keppel and R. D. Walker, Jr., "Paper from Inorganic Fibers," University of Florida, Tech. Paper No. 237, *Eng. Progr.*, **16,** 8 (August 1962).
[69] R. A. Keppel and S. N. Subbanna, "Development of Oriented Silicon Carbide Whisker Bodies," Appendix to Final Report, AFML-TR-67-321 (see Reference 44).
[70] A. Toy, "Mechanical Properties of Fiber Reinforced Aluminum Base Composites," presented at the Fall Meeting of the Metallurgical Society of AIME, Cleveland, Ohio, October 15–19, 1967.
[71] Anon, "A Big Squeeze Aligns Fibres in Metal," *New Scientist*, Nov. 23, 1967.
[72] R. P. Ayerst, C. Bailey, M. T. Dupress, and M. E. N. Evans, "A Pilot Plant for the Alignment of Asbestos Fibres and Whiskers by the Alginate Extrusion Process," British Ministry Tech., Explosives Res. and Dev. Establishment, Report ERDE 22/R/67, November 1967.
[73] L. C. Wohrer, F. J. Frechette, and J. Economy, "Whisker-Reinforced Plastics," *Machine Design*, **38,** 30 (1966), 138, 140, 141.
[74] L. C. Wohrer and J. Economy, "Single Crystal Fiber Reinforcements," in *Advanced Fiber Composites*, SAMPE, **10** (1966), B-69.
[75] J. A. Alexander, J. C. Whithers, and B. A. Macklin, "Investigation of Three Classes of Composite Materials for Space Vehicle Application," NASA Contractor Report NASA CR-785, May 1967.
[76] J. J. Shyne and R. G. Shaver, "Whisker Composite Technology," in *Advanced Fibrous Materials*, SAMPE, **10** (1966), paper B-6.
[77] H. N. Barr, "Investigation of Silicon Carbide Whisker Reinforced Metallic Matrix Composites," Final Technical Report AFML-TR-67-296, September 1967.
[78] H. Hahn, A. P. Divecha, P. Lare, and B. Dennison, "Research on Whisker Reinforced Metal Composites," Final Report Contract NOw-65-0387-8, April 1966
[79] P. J. Lare, A. P. Divecha, and H. Hahn, "Research on Whisker-Reinforced Metal Composites," Final Report, Contract No. NOw-66-0213-d, May 1967.
[80] P. J. Lare, A. P. Divecha, H. Hahn, "Research on Whisker-Reinforced Metal Composites," Final Report, Contract No. NOOO19-67-C-0404, July 1968.
[81] R. A. Hermann, A. P. Divecha, P. J. Lare, and H. Hahn, "Whisker-Reinforced Metals," Final Tech. Report Contract No. NOOO19-67-C-2404, June 1968.
[82] R. P. Jakas and J. Chorné, "Method and Means for Separating Individual Fibers from a Fibrous Mass," U.S. Pat. 3, 421, 619; Jan. 14, 1969.
[83] B. C. Raynes, "Studies of the Reinforcement of Metals with Ultra High Strength Fibers (Whiskers)," Final Report, Contract No. NOw-61-0207-c, November 1961.
[84] B. C. Raynes and R. H. Kelsey, "Studies of the Reinforcement of Metals with Ultra High Strength Fibers (Whiskers)," Final Report, Contract NOw-62-0235-c, Oct. 1962.
[85] W. Wolkowitz and T. S. Stanislow, "Extrusion and Hot Rolling-Two Advanced Fabrication Techniques for the Preparation of Whisker-Metal Composites," in *Advanced*

Techniques for Material Investigation and Fabrication, SAMPE, **14** (November 1968), paper 11-4A-3.

[86] E. A. Snajdr and J. F. Williford, Jr., "Investigation of Fiber-Reinforced Metal Matrix Composites Using a High-Energy-Rate Forming Method," Proc. 14th Ref. Composite Wkg. Group, pp. 133–161, AFML TR-68-129, May 1968.

[87] A. P. Divecha, P. J. Lare, R. A. Hermann, and H. Hahn, "Development of Whisker-Reinforced Aluminum and Magnesium Composites," Proc. 14th Ref. Composite Wkg. Group, AFML-TR-68-129, pp. 163-177, May 1968.

[88] I. Ahmad, V. P. Greco, and J. M. Barranco, "Reinforcement of Nickel with Some High Strength Filaments," *J. Composite Materials*, **1,** 1 (1967), 18–29.

[89] E. G. Wolff, "Electroformed Discontinuous Fiber Composites," Proc. 14th Ref. Composite Wkg. Group, AFML TR-68-129, pp. 391–402, May 1968.

[90] J. Chorné, W. H. Sutton, C. Bruch, and E. Feingold, "Development of Composite Structural Materials," Final Report, Contract NOw-65-0176c, November 1965.

[91] A. Gatti, C. Mancuso, E. Feingold, and R. Jakas, "Study of the Growth Parameters Involved in Synthesizing Boron Carbide Filaments," Final Report, Contract NASw-1205, March 1966.

[92] W. H. Sutton, A. Gatti, and E. Feingold, "Development of Composite Structural Materials for High Temperature Applications," 13th Prog. Rpt., Contract NOw 60-0965d, August 1963.

[93] R. L. Mehan, "Fabrication and Evaluation of Sapphire Whisker Reinforced Aluminum Composites," in *Metal Matrix Composites*, ASTM, Spec. Tech. Pub. No. 438, pp. 29–58, 1968.

[94] R. W. Sicka, T. Harkulich, M. S. Vukasovich, and R. H. Kelsey, "Reinforcement of Nickel Chromium Alloys with Sapphire Whiskers," Final Report, Contract NOw 64-0125c, Oct. 1964.

[95] R. H. Rosenberg and J. J. Ryan, "Investigation of Fiber Strengthened Cast Cobalt-Base Alloys," Final Report, Contract No. NASw-1531, June 1957.

[96] O. D. Deex and R. G. Schierding, "Silicon Carbide Whisker Reinforced Aluminum Composites," Monsanto Res. Corp., Report HPC 68-65, Contract N-00014-67-c-0218, ARPA Order No. 873, July 1968.

[97] D. G. Flom, V. F. Mazzio, and E. Friedman, "High Strength, High Modulus, Whisker Reinforced Plastic Composites," Final Report, AFML-TR-66-362, Feb. 1967.

[98] M. E. Kirkpatrick, M. E. Staudhammer, and J. L. Reger, "A Continuous Process for the Production of Whisker-Reinforced Composites," presented at the Third Symposium on High Performance Composites, St. Louis, Mo. 1967.

[99] J. E. Gordon, "Whisker Reinforced Materials," presented at the Int. Cont. on the Mechanics of Composite Materials, Philadelphia, Pa., May 1967.

[100] W. H. Sutton, "Preparation, Microstructure, and Properties of Inorganic Fibrous Composite Materials," in *Ceramic Microstructures*, John Wiley and Sons, New York, 1968, Ch. 39.

M. J. Salkind, F. D. Lemkey, and F. D. George

10 Whisker Composites by Eutectic Solidification

INTRODUCTION

The virtues of whisker reinforcing have been adequately described in this book and elsewhere, but the significant problems associated with economically fabricating substantial quantities of composite material with consistent properties remain as major obstacles to the development of this technology. Separately grown whiskers must be separated, sorted, aligned, and incorporated into a matrix material in a uniform distribution with well bonded interfaces in sufficient quantity to provide significant reinforcement. The problems associated with handling microscopic crystals during these operations have solutions, but they are generally costly and slow. An alternative method for producing whisker composites is to grow them directly from the melt. This can be accomplished by unidirectionally solidifying certain eutectic alloys so that the two phases grow simultaneously from the melt in a parallel array as seen in Figure 10.1. The process produces, in one step, a finished whisker composite with a uniform distribution of aligned, well bonded whiskers in the matrix.

The word eutectic is derived from the Greek "eutektos" which means "easily melted." The name results from the fact that the eutectic composition has the lowest melting temperature of any combination of the two reacting phases. Eutectics were well known to the ancient Greeks and Romans because of their unique low melting behavior. Today, eutectic brazes and solders still find use in industry because of this characteristic. For the same reason, eutectic alloys are often used for automatic fire detection and sprinkler systems. Many eutectic reactions exist in nature, each having a precise

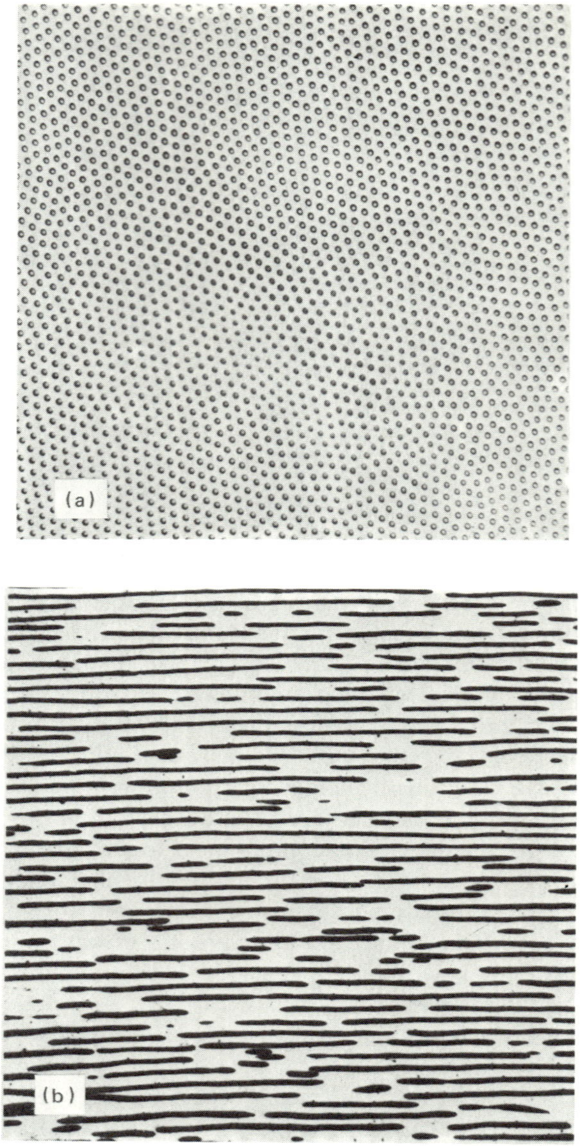

Figure 10.1 Microstructure of unidirectionally solidified Al-Al_3Ni eutectic exhibiting rodlike whiskers (after Lemkey, Hertzberg, and Ford [53, 54]). (a) Transverse sections (mag. 1000X) [5]; (b) longitudinal sections (mag. 500X) [6].

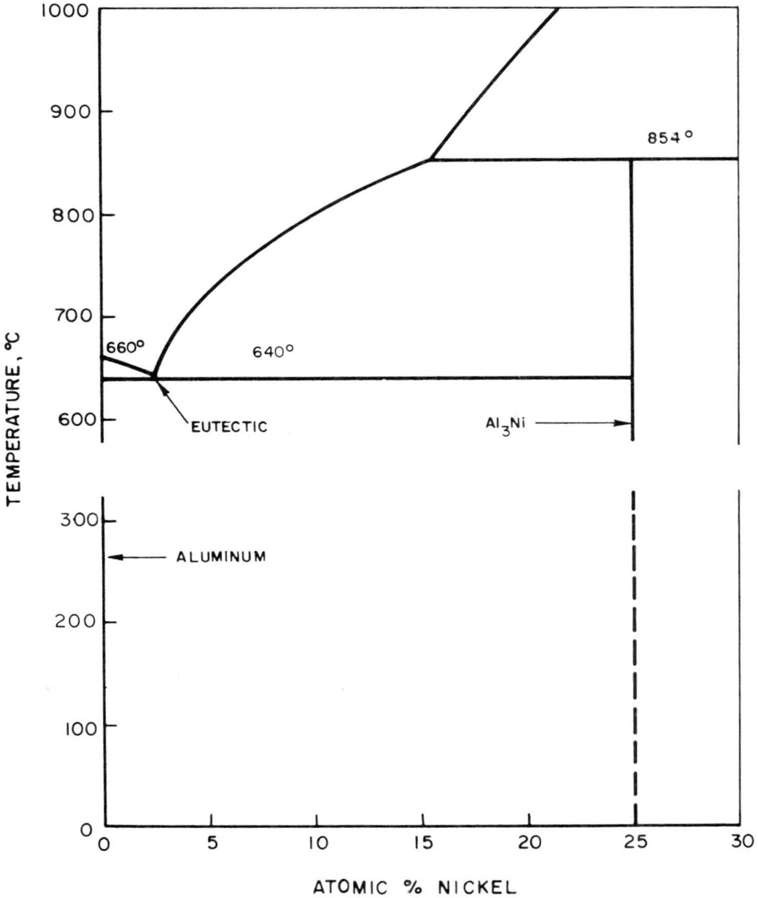

Figure 10.2 Al-Al_3Ni phase diagram (after Wiley [86]).

chemical composition and temperature. Figure 10.2 gives an example of an eutectic that exists in the aluminum-nickel alloy system. Aluminum containing 2.8 at.% nickel exists as a homogeneous liquid solution above the eutectic or melting temperature (640°C) and as two distinct solid phases below this temperature. The reversible eutectic reaction can be written:

$$\underset{(2.8\text{ at.\% Ni})}{100\text{ vol\% liquid (Al-Ni solution)}} \underset{\text{Heating}}{\overset{\text{Cooling}}{\underset{\longleftarrow}{\overset{\longrightarrow}{640^\circ\text{C}}}}} \underset{(0.02\text{ at.\% Ni})}{89\text{ vol\% }\alpha\text{ (Al-Ni solid solution)}} + \underset{(25\text{ at.\% Ni})}{11\text{ vol\% }Al_3Ni}$$

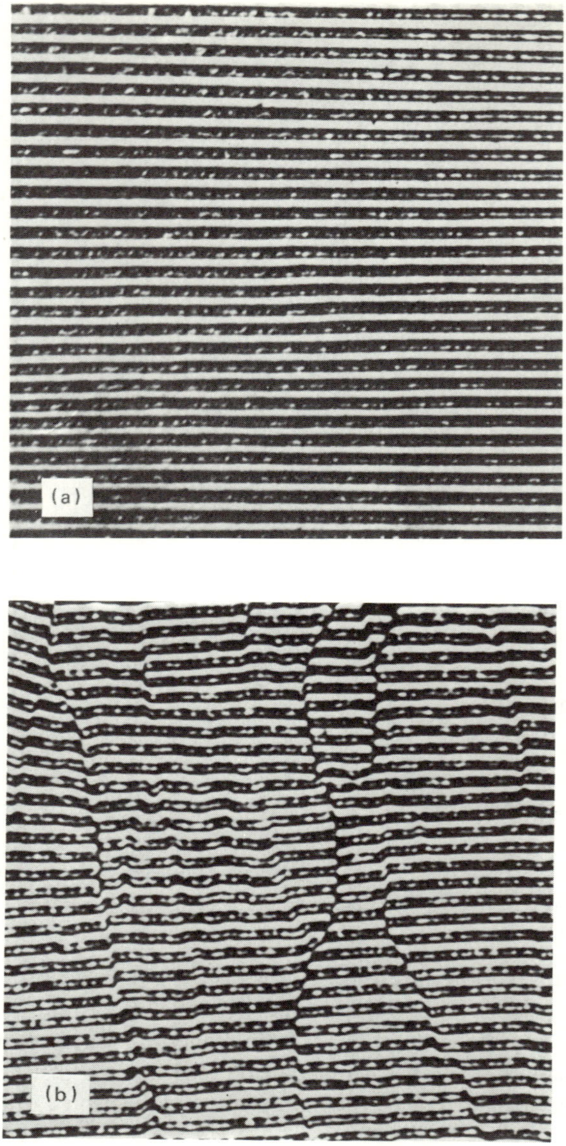

Figure 10.3 Microstructure of unidirectionally solidified Al-$CuAl_2$ eutectic exhibiting lamellar structure. (a) Longitudinal section (mag. 500X); (b) transverse section (mag. 500X).

Because the composition of the liquid is fixed and lies between the composition of the two solid phases, which are also fixed for any temperature, the relative amount of each of the two solid phases is fixed.

Upon solidification, the two solid phases of an eutectic frequently separate into a regular geometric structure, the nature of which is dependent upon the relative amounts of each phase, the crystal structure and growth behavior of each phase, the relative interfacial energy between phases, and the thermodynamics of the system. Two of the most common eutectic microstructures are the rodlike structure as seen in Figure 10.1, and the lamellar structure as exhibited by the Al-$CuAl_2$ eutectic and seen in Figure 10.3. In addition, globular, script, and several other structures have been observed.

The process of unidirectional solidification can be accomplished by any one or a combination of methods which cause the solid to grow into the liquid unidirectionally with a plane interface. Normal solidification, depicted in Figure 10.4, has been used most commonly. A crucible containing the

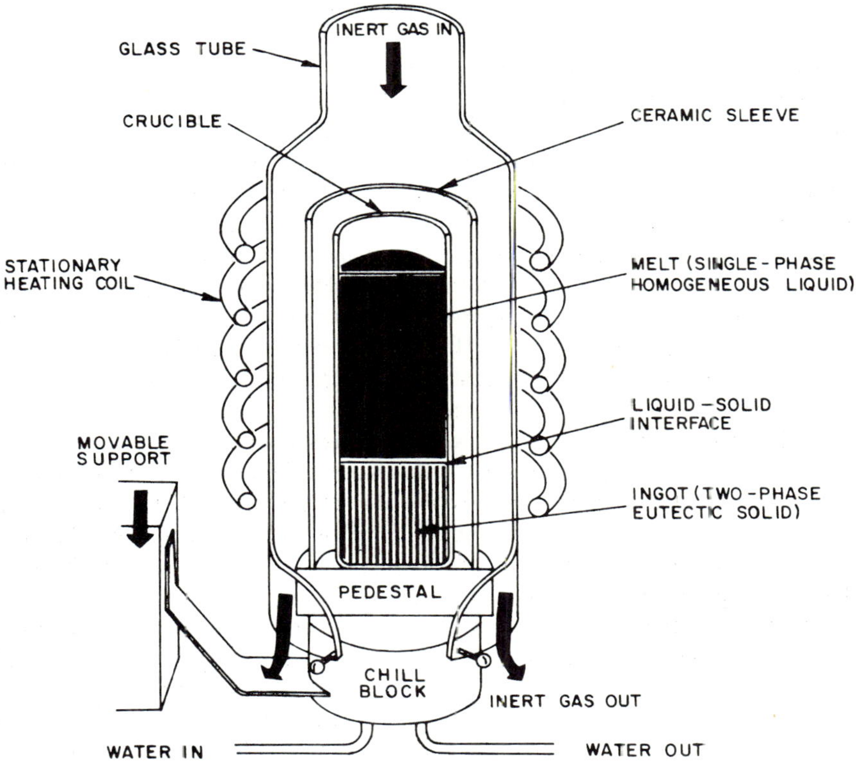

Figure 10.4 Unidirectional solidification of eutectic composites (after Salkind and Lemkey [87]).

eutectic alloy is heated until the alloy is completely molten, then withdrawn downward from the heat source at a constant rate. The material in the bottom of the crucible freezes first and solidification proceeds upward at a rate dependent upon the crucible withdrawal rate and heat flow conditions. Heat flow can be controlled by properly choosing the apparatus materials and geometry and by an end quench at the bottom of the crucible. Vertical processing is preferable to horizontal solidification as the heat flow is more symmetrical and the problems of convection currents in the liquid metal are minimized [1]. Unidirectional solidification can be accomplished by zone melting or by casting the material in an open-bottomed crucible on a chill block and reducing the heat input to the upper part of the crucible at a controlled rate. If the solidification rate, purity of the melt, and thermal gradient in the liquid are properly controlled, a plane interface between the liquid and solid can be maintained and a uniform structure can be produced.

HISTORICAL DEVELOPMENT

The history of eutectics is one of discovery, obscurity, and rediscovery. Recently, Kerr and Winegard [2] have thoroughly reviewed the history of eutectic alloys and have shed considerable light on the origins of our present understanding. Although eutectic alloys were known to the ancient Romans, the concept was rediscovered separately in both Germany and England during the latter part of the nineteenth century. Excavations near Silchester in England uncovered old plumbing which had been made by the Romans after their conquest of Britain. The solder used in the plumbing was found to be a low melting eutectic of lead and tin. The name "eutectic" was coined by Guthrie in England in 1884 [3]. The presence of two phases was noted in 1895 by Ponsot [4], and the lamellar nature of eutectic microstructures was discovered by Rosenhain and Tucker in 1909 [5]. It was at that time that one of the first of many controversies arose. The German metallurgist, Tammann, proposed incorrectly that the lamellar structure was the result of alternate nucleation and growth of each phase parallel to the moving liquid-solid interface. Although Vogel demonstrated in 1912 [6] that the phases grow perpendicular to the liquid-solid interface, the influential Tammann persisted in his error until 1925 [7].

The present classifications of eutectic microstructures according to their appearance have their origins in the work of Brady [4] and Portevin [8] in the early 1920s. Microstructures were classified as globular, lamellar, angular, or crystallographic and it was suggested that the determining factor was the surface tension of the pure phases. These arguments were subject to modification when the profound influence of dilute impurities on the surface tension of pure metals was later discovered. One factor that retarded the understanding of eutectic microstructures was the relative impurity of the elements used in early studies of eutectics. The development of the zone refining process by Pfann [9] in the 1950s allowed the study of eutectic microstructures without the modifying effects of impurities.

Recently, Hunt and Jackson [10] have developed a scheme for classifying eutectic microstructures based upon thermodynamic considerations. They have shown that three groups of microstructures exist and are dependent upon the entropy of fusion of each phase. If both phases have low entropies of fusion, they will freeze simultaneously perpendicular to a plane liquid-solid interface, and normal or controlled microstructures such as those seen in Figures 10.1 and 10.3 will result. If one phase has a high entropy of fusion, it will grow into the liquid ahead of the interface and likely branch or grow in varied directions. If both phases have high entropies of fusion, they will both grow rapidly into the melt ahead of the nominal liquid-solid interface, resulting in a system of intertwined, branched, or angular crystals. Neither of the last two categories results in the regular aligned microstructures desired for useful eutectic composites. Fortunately, many eutectics fall into the first category.

The first unidirectionally solidified aligned eutectic microstructure was produced by Winegard, Majka, Thall, and Chalmers [11] in 1951. Their purpose however, was the study of eutectic solidification phenomena and it was Kraft [12, 13], then at the United Aircraft Research Laboratories and now at Lehigh University, who first envisioned the production of aligned anisotropic materials based on unidirectional solidification of eutectics. Kraft's expectations concerning the strength of eutectic whiskers were based on Webb and Forgeng's observations [14] on *in-situ* grown whiskers.

In 1927, Hargreaves [15] published a paper which described colonies, an important aspect of eutectic microstructure. As seen in Figure 10.5, a colony eutectic microstructure consists of separated areas of common microstructure. Hargreaves noted that colonies are different from metallic grains or crystals, but could not identify their origin. Despite the fact that Hargreaves' paper was published in the *Journal of the Institute of Metals*, one of the most prominent metallurgical journals, the work was apparently unrecognized in the literature, and the concept of colonies was subsequently rediscovered in the late 1950s [16]. It is now recognized that colonies are the result of solidification with a nonplane liquid-solid interface and results from an inability to diffuse impurities ahead of the interface rapidly enough [16]. As mentioned earlier, the solidification rate, purity, and thermal gradient must be controlled in order to maintain a plane interface and produce an aligned microstructure; but this understanding, and its relationship to the colony microstructure, has only been with us for less than a decade.

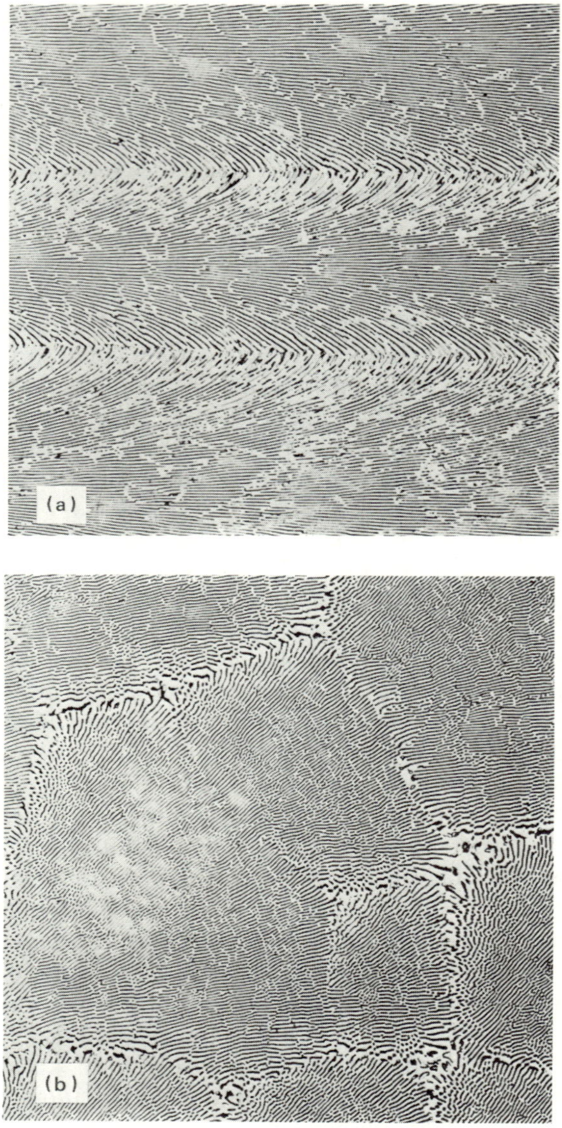

Figure 10.5 Colony microstructure in unidirectionally solidified $Mg_{17}Al_{12}$-Mg eutectic (after Kraft [43]). (a) Longitudinal section (mag. 200X); (b) transverse section (mag. 200X).

An important feature of aligned eutectics is that many of them grow with a preferred crystallographic relationship and preferred crystallographic interfaces between the two phases. Straumanis and Brakss [17] first demonstrated the preferred crystallography experimentally in the 1930s. An important contribution to the understanding of this phenomenon was made by Kraft and his associates [18–21] in the 1960s. Kraft developed a technique for simultaneously determining the orientation of both phases by X-ray diffraction and demonstrated that unidirectionally solidified eutectics grow with a preferred crystallography and preferred crystallographic interfaces. Recently, techniques have been developed [22, 23] for electrothinning foils of eutectic composites which allow simultaneous study of the crystallography and microstructure by transmission electron microscopy and diffraction. This technique has brought a precision to these studies previously unobtainable by X-ray diffraction and light microscopy. Many new studies, including a determination of the crystallographic facets on the whiskers in the Al-Al_3Ni system, have been facilitated by this technique. Figure 10.6 is a transmission electron micrograph of the Al-Al_3Ni eutectic

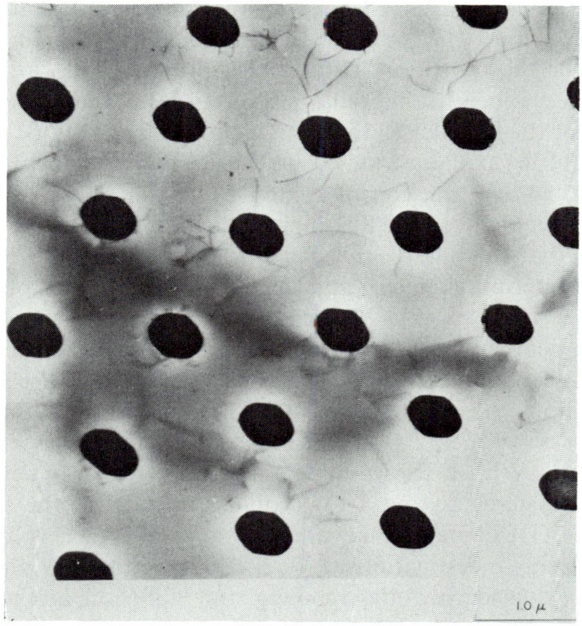

Figure 10.6 Transmission electron micrograph of Al-Al_3Ni; transverse section (after Tice, Lasko, and Lemkey [23]).

exhibiting the preferred faceted interfaces. A recent review by Kraft and Lemkey [24] summarizes the results of many of these studies.

A significant contributor to the understanding of eutectic solidification has been Tiller [25]. He laid the foundations for a quantitative description of eutectic growth and derived a simple linear relationship between the square of the interlamellar or inter-fiber spacing and the solidification rate which is predicted by the diffusion-controlled coupled growth of the two phases. This relationship was based upon a previous analysis by Zener of diffusion-controlled eutectoid growth [26] and the concepts of constitutional supercooling developed by Chalmers and his associates [27, 28]. The relationship between solidification rate and structure is extremely important since the properties of eutectic composites are often determined by the size and spacing of the aligned phase.

Since unidirectionally solidified eutectics most often exhibit regular lamellar or rodlike microstructures and sometimes both, there has been considerable speculation as to whether or not these are alternate forms of the same structure, and about the factors which control the lamellar-rod transformation. Tiller [25] predicted that the lamellar structure would transform to rods at high solidification rates and Yue [29–32] observed this behavior and suggested that it was due to constitutional supercooling caused by impurities. Chadwick [33] also suggested that the impurities were responsible for the transition. Hunt and Chilton [34, 35] argue, however, that orientation effects are more important and that the curved interface is responsible for the lamellar to rod transition. Cooksey, Munson, Wilkinson, and Hellawell [36] explained the occurrence of rodlike structures in systems having a low volume fraction of second phase and lamellar structures in systems having a high volume fraction on the basis of a geometric argument. They argued that the relative interfacial area (and therefore energy) of the system is the determining factor and since it varies with the volume fraction, that structure which exhibits the least interfacial area would occur. This argument is demonstrated schematically in Figure 10.7 along with the structures of several eutectic systems. Since the interfacial energy is often quite anisotropic, the argument must be modified to account for the fact that the interfacial energy is not necessarily directly proportional to the interfacial area. With a few notable exceptions, the argument accurately predicts that systems having a low volume fraction of the second phase (less than 30%) will be rodlike and those with higher volume fraction, lamellar. Recent investigations [37, 22] have modified this argument somewhat by considering undercooling and diffusion at the liquid-solid interface in addition to the interfacial area.

Although we have some understanding of the causes of lamellar or rod microstructures, there remain several unexplained phenomena. The Ta-Ta_2C

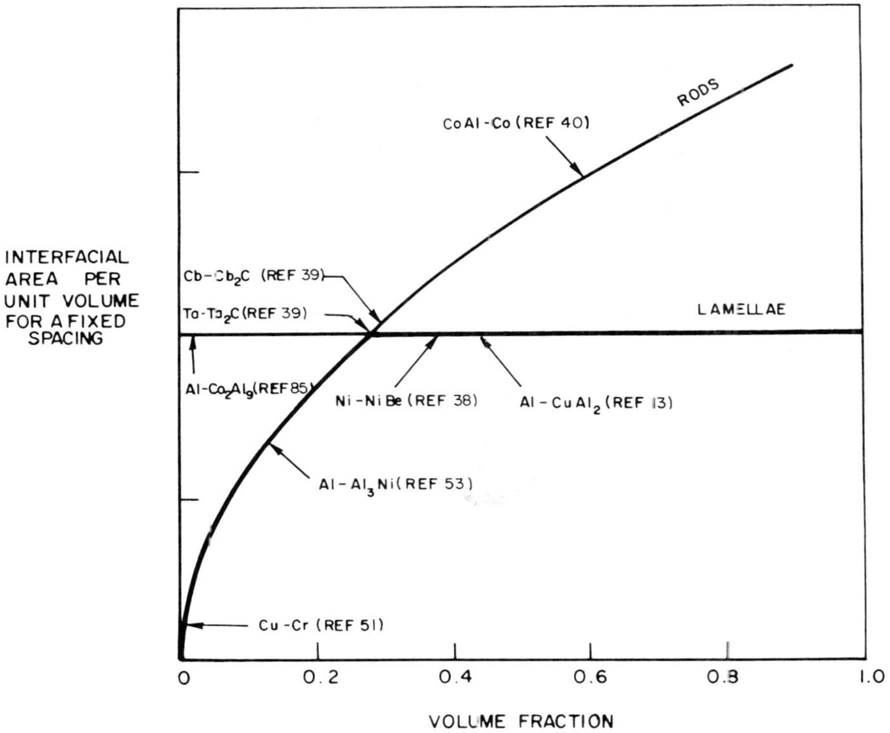

Figure 10.7 Relative interfacial areas of normal rod and lamellar eutectics as a function of volume fraction.

system [38, 39] exhibits bladelike (or "perforated lamellar") Ta_2C whiskers when solidified once, and rodlike whiskers when solidified more than once under the same conditions (Fig. 10.8). The CoAl-Co system [40], as seen in Figure 10.9, exhibits both the lamellar and rod structure in an ingot presumably produced under conditions of plane front growth. This system is also unique in that it exhibits a rodlike structure with such a high content (65 vol%) of second phase. This can be attributed to the fact that at the eutectic temperature where the structure is formed, the content of second phase is lower; but due to changes in solid solubility upon cooling to room temperature, the volume fraction of second phase increases. As will be discussed later in this chapter, the difference between rod and lamellar structures is important because the two structures exhibit profoundly different properties. For a more thorough treatment of eutectic solidification phenomena, the reader is referred to several review articles [2, 25, 41–43].

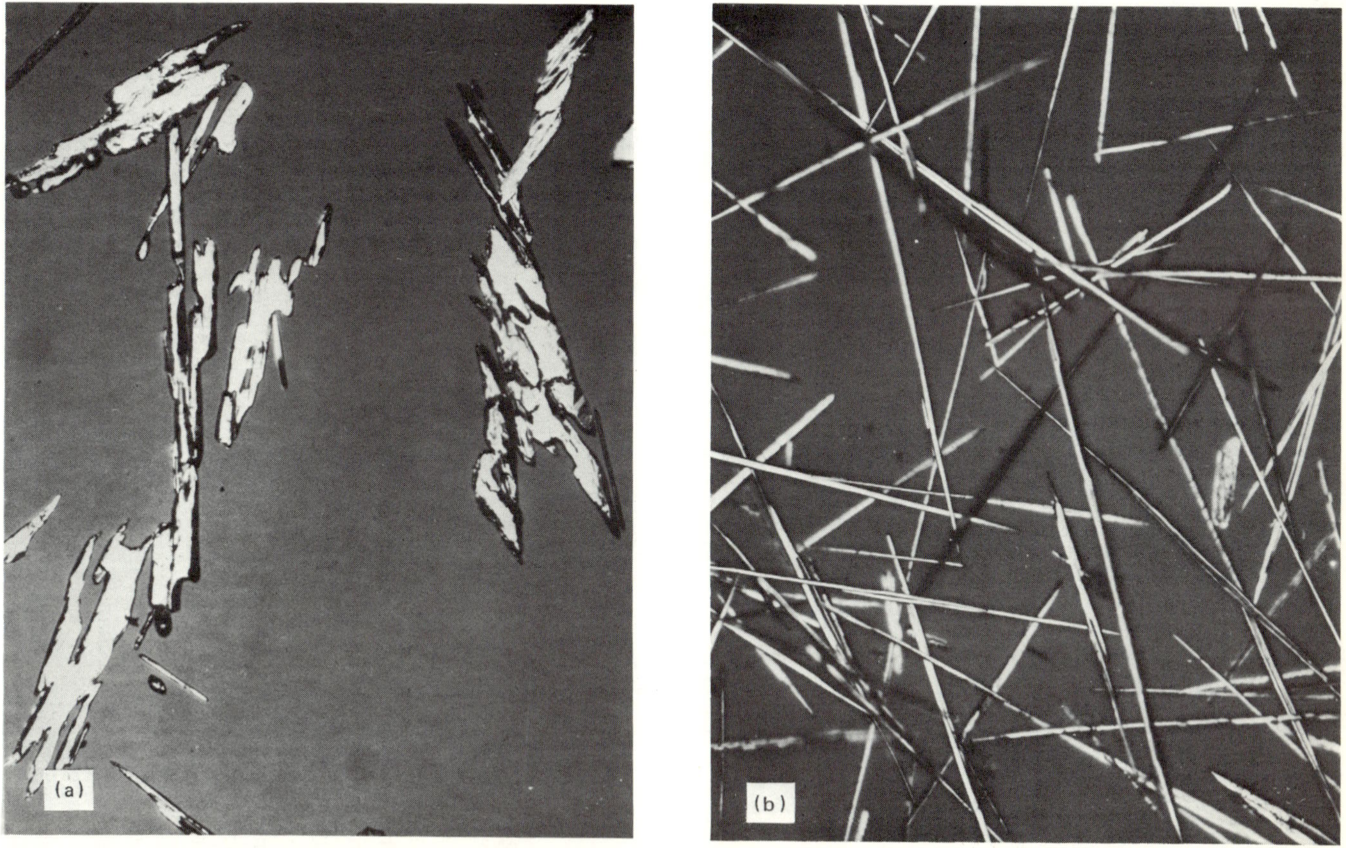

Figure 10.8 Ta_2C whiskers extracted from two different unidirectionally solidified Ta-Ta_2C eutectic specimens (after Lemkey and Salkind [39]). (a) Bladelike whiskers from low-strength specimen (Mag. 200 X); (b) needlelike whiskers from high-strength specimen (Mag. 200 X).

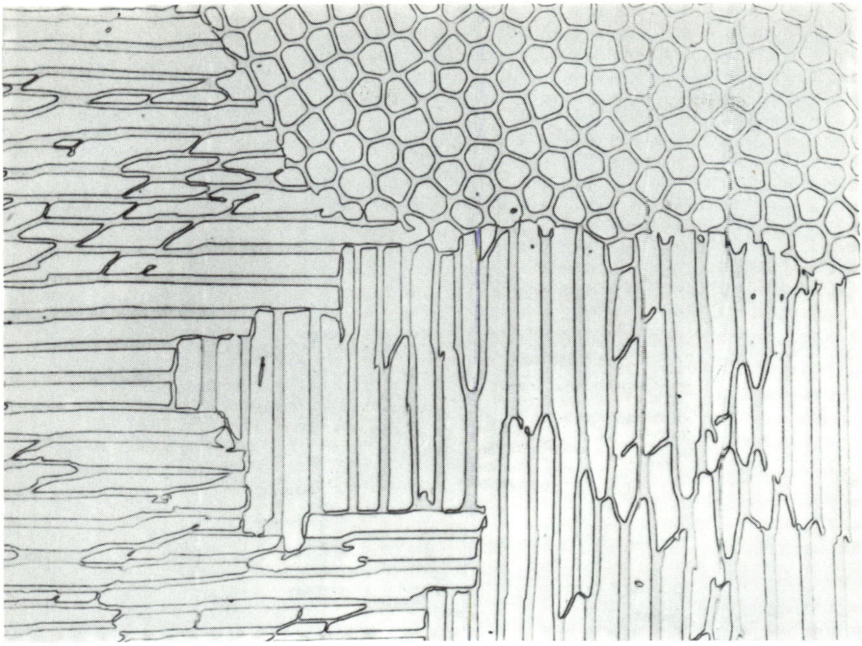

Figure 10.9 Transverse section of unidirectionally solidified CoAl-Co eutectic exhibiting both rod and lamellar microstructure (mag. 150 X) (after Cline [40]).

MECHANICAL BEHAVIOR OF ALIGNED EUTECTICS

The origin of the concept of using *in-situ* grown whiskers to provide strengthening can be found in an article by Webb and Forgeng published in 1958 [14]. They measured the properties of several species of whiskers produced by precipitation in the solid state as well as chromium whiskers extracted from the Cu-Cr eutectic (not unidirectionally solidified) and found them to be quite strong. They also discussed the strengthening of alloys containing strong dispersed particles in terms of a volume average of the properties of each phase. This principle of combined action has been stated many times [44–48] and has come to be known as the rule of mixtures:

$$\sigma_c = \sigma_f V_f + \sigma_M V_M \tag{1}$$

where σ is the strength, V the volume fraction, and the subscripts $c, f,$ and M refer to, respectively, the composite, fiber, and matrix. The matrix strength, σ_M, is the stress in the matrix at the failure strain of the composite, rather than

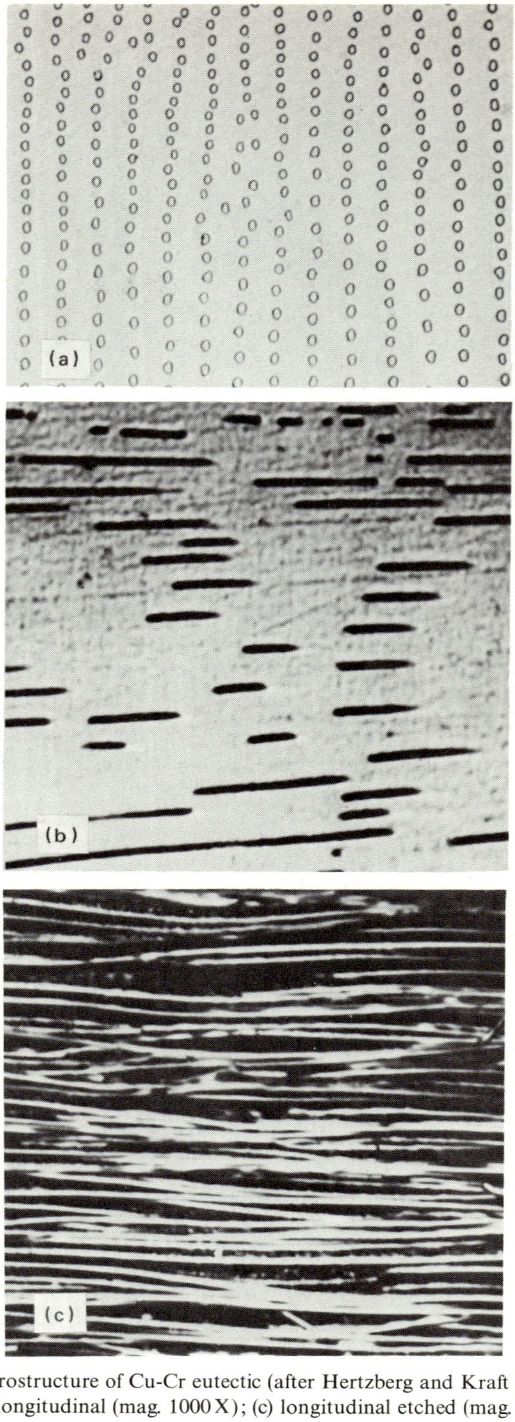

Figure 10.10 Microstructure of Cu-Cr eutectic (after Hertzberg and Kraft [5]). (a) Transverse (mag. 1000 X); (b) longitudinal (mag. 1000 X); (c) longitudinal etched (mag. 1000 X).

its ultimate tensile strength. With some modification [48] the rule of mixtures has been found to adequately describe the tensile stress-strain behavior of *aligned* fiber composites in the direction of fiber alignment. Interestingly, it does not describe the behavior of a composite containing randomly oriented fibers or particles which is the case that Webb and Forgeng considered.

In the late 1950s, Kraft and his colleagues at the United Aircraft Research Laboratories initiated a research program aimed at the production of *aligned* composite materials based on unidirectionally solidified eutectic alloys [12, 13]. This effort led to the development of the lamellar Al-$CuAl_2$ [13, 19, 49] and the whiskerlike Cu-Cr [49–52] and Al-Al_3Ni [49, 53–63] eutectic composites which have been studied extensively.

Whisker Composites

The first whisker composite produced by unidirectional solidification of an eutectic alloy was the copper-chromium system [49, 51] which contains approximately 1.6% by volume of aligned chromium whiskers in a copper matrix. The microstructure of this system is seen in Figure 10.10. Because the longitudinal sectioning plane is not exactly parallel to the whiskers, the true length of the whiskers cannot be ascertained from a longitudinal section (Fig. 10.10*b*). Dissolving the matrix with acid exposes the chromium fibers, as seen in Figure 10.10*c*, and gives a better idea of the fiber length. Although this system contains too low a whisker content to provide significant reinforcement, it has served as a vehicle for the study of the mechanics and fracture behavior of composites. Individual chromium whiskers extracted by dissolving the matrix in acid were found to exhibit linear stress-strain behavior, failure strains of several percent, and tensile strengths in excess of 1 million psi [49, 50]. These experiments demonstrated that whiskers grown *in-situ* by unidirectional eutectic solidification were true “whiskers,” exhibiting the characteristics previously demonstrated for vapor-grown whiskers [64, 65].

Whisker-matrix bonding and load transfer in eutectic whisker composites were studied by observing the behavior of partially exposed whiskers on the tensile surface of a bend specimen of Cu-Cr [51]. As seen in Figure 10.11, the tensile surface was parallel to the fibers and was polished and etched for examination in the microscope during bending. As the specimen was loaded, the whiskers broke in tension (see Fig. 10.11*b*). The broken segments, however, remained bonded to the matrix so that they could be subsequently loaded by shear at the interface and broken into even smaller segments (Fig. 10.11*c*). This demonstrated the fact that although only the lower part of each fiber was bonded to the matrix, the bond strength was sufficient to transfer enough load to fracture the whiskers in tension.

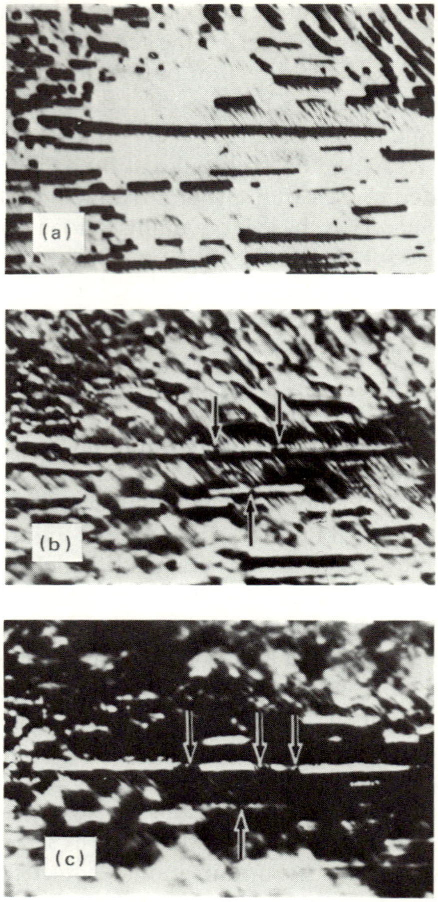

Figure 10.11 Whisker fracture in Cu-Cr eutectic. Arrows denote fracture of Cr whisker (after Hertzberg and Kraft [51]). (a) Unstressed (mag. 500 ×); (b) partially stressed (mag. 500 ×); (c) fully stressed (mag. 500 ×).

The fracture of the Cu-Cr eutectic is a classic example of the behavior of a whisker-reinforced ductile metal [49, 51]. As seen schematically in Figure 10.12, when the fibers are loaded to failure in tension, they fracture and form voids in the ductile matrix (Fig. 10.12*b*). An alternative failure mechanism is based upon the fibers breaking in tension as the result of the stress concentration ahead of the tip of a propagating crack. Upon further application of load, the voids grow (Fig. 10.12*c*) until they coalesce, resulting in failure (Fig. 10.12*d*). The resulting fracture surface (Fig. 10.12*e* and Fig. 10.13) consists of skewed dimples each containing a broken whisker. The

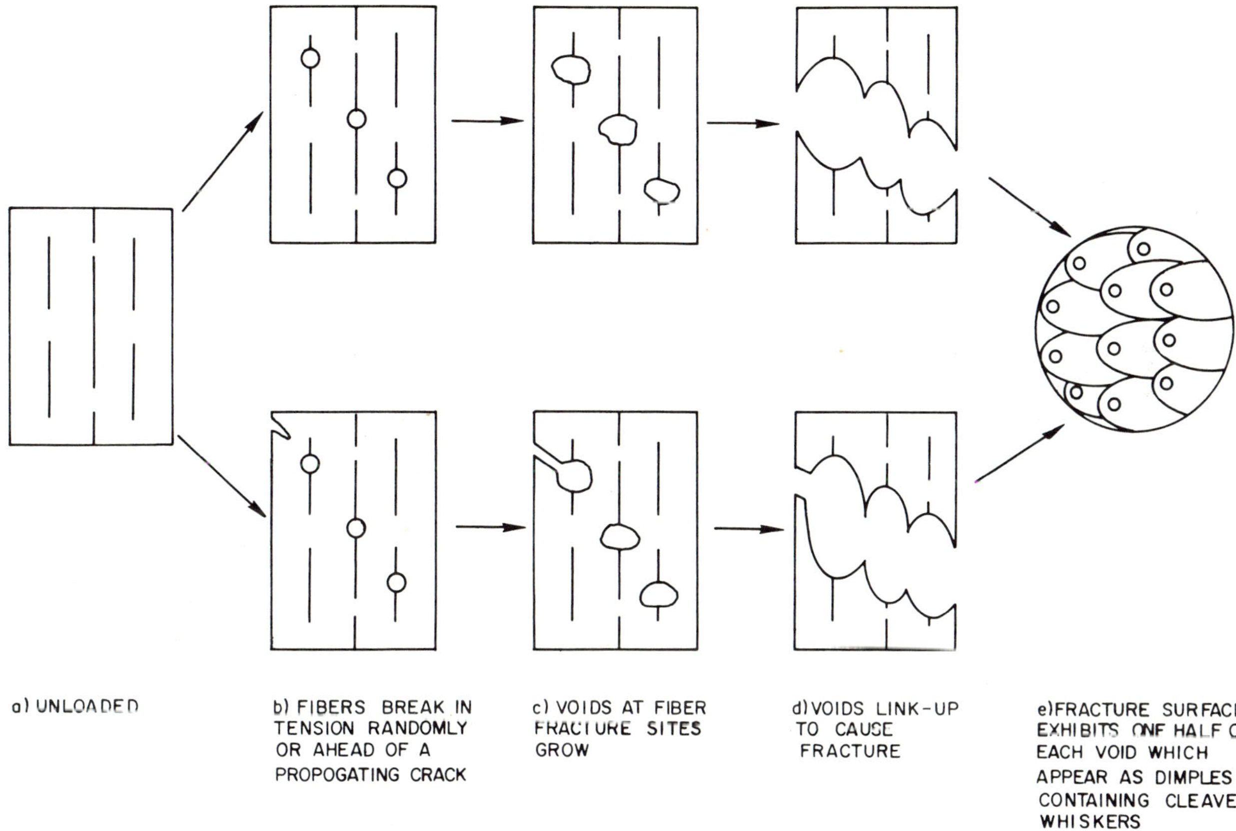

Figure 10.12 Fracture of a whisker-reinforced ductile metal.

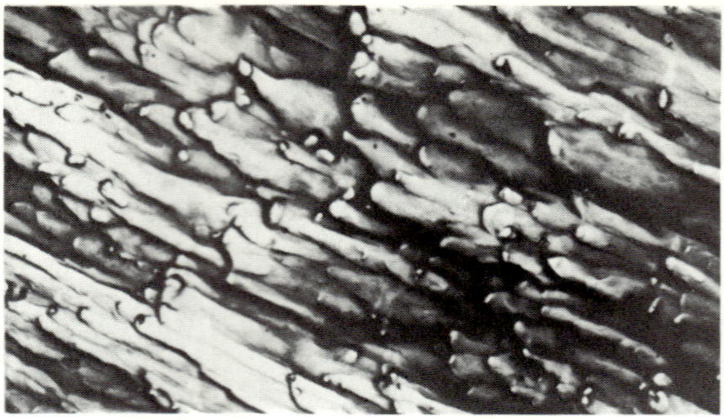

Figure 10.13 Fracture surface of Cu-Cr eutectic (mag. 2000×) (after Hertzberg and Kraft [51]).

dimples are one half of the voids and are skewed if the fracture surface is at an angle to the whiskers (as in Figs. 10.12 and 10.13) or equiaxed if the fracture is perpendicular to the whiskers. (See Fig. 7 in [54].)

The first eutectic whisker composite to exhibit significant whisker reinforcement was the Al-Al_3Ni system [49, 53–63], the microstructure of which is seen in Figure 10.1. The stress-strain behavior of this material which contains 11% by volume of aligned Al_3Ni whiskers in an aluminum-nickel solid solution matrix is seen in Figure 10.14 along with that of the matrix material and the conventionally cast eutectic for comparison. The theoretical composite tensile strength can be calculated using the rule of mixtures (Eq. 1). Using a value of 400,000 psi for the whisker tensile strength [54] and 4200 psi for the matrix tensile strength [56, 63], the composite strength is (400,000)(.11) + (4200)(.89) = 48,000 psi. This value is in good agreement with experimentally determined values [49, 54–57]. The fracture behavior of the Al-Al_3Ni eutectic is similar to that of the Cu-Cr system as seen in Figure 10.15. The broken fibers ahead of the crack tip, as seen in Figure 10.15*a*, indicate a mechanism similar to that described in Figure 10.12. The fracture surface (Fig. 10.15*b*) contains skewed dimples, each of which contains a cleaved whisker. The skewed dimples, indicating fracture at a steep angle to the whiskers, is typical of the fracture behavior of Al-Al_3Ni.

The tensile behavior of Al-Al_3Ni as a function of whisker orientation [56, 63] is seen in Figure 10.16. As expected, the maximum strength and minimum elongation was exhibited for the 0° orientation where the whiskers control the tensile behavior. The specimens oriented at 45° to the load axis exhibited the largest values of elongation and low strengths. This behavior has also been noted for lamellar Al-$CuAl_2$ [49, 54] and has been attributed to the fact that an active slip plane oriented approximately parallel to the

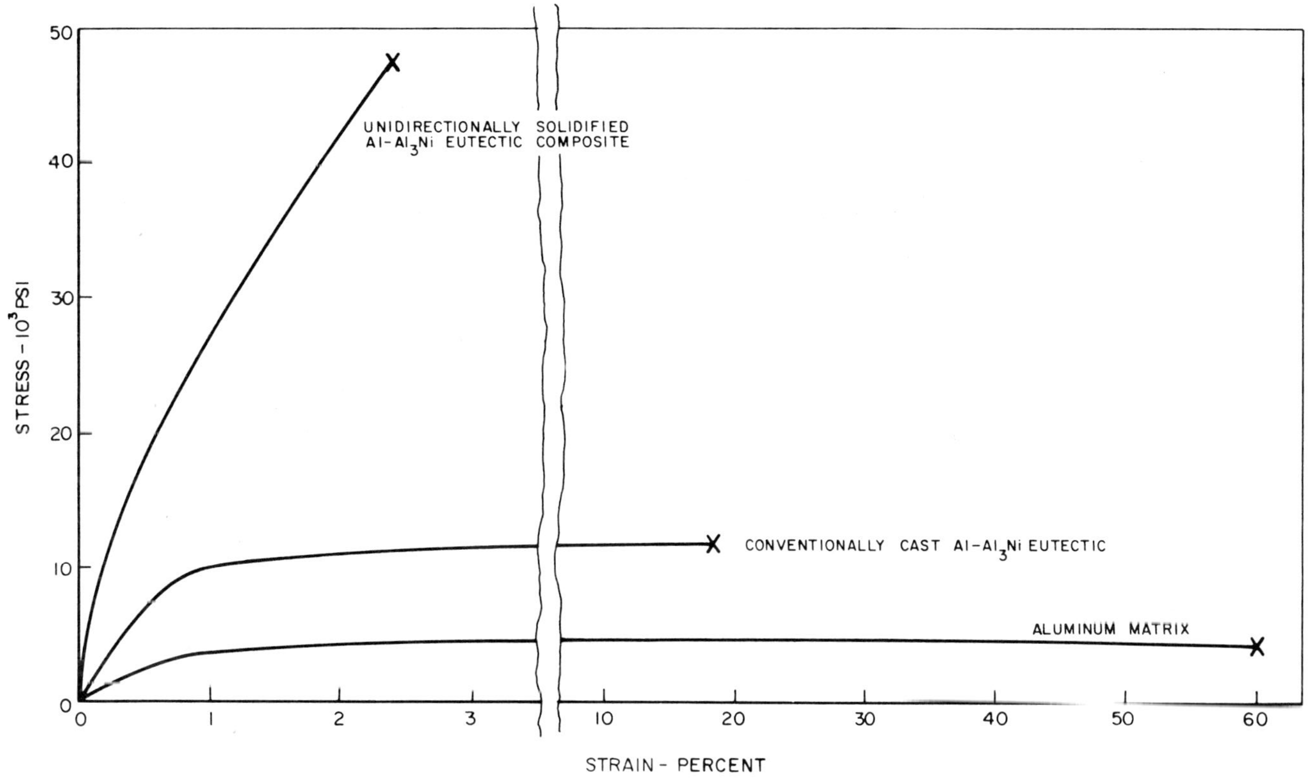

Figure 10.14 Tensile behavior of Al-Al_3Ni (after Lemkey, Hertzberg, and Ford [53]).

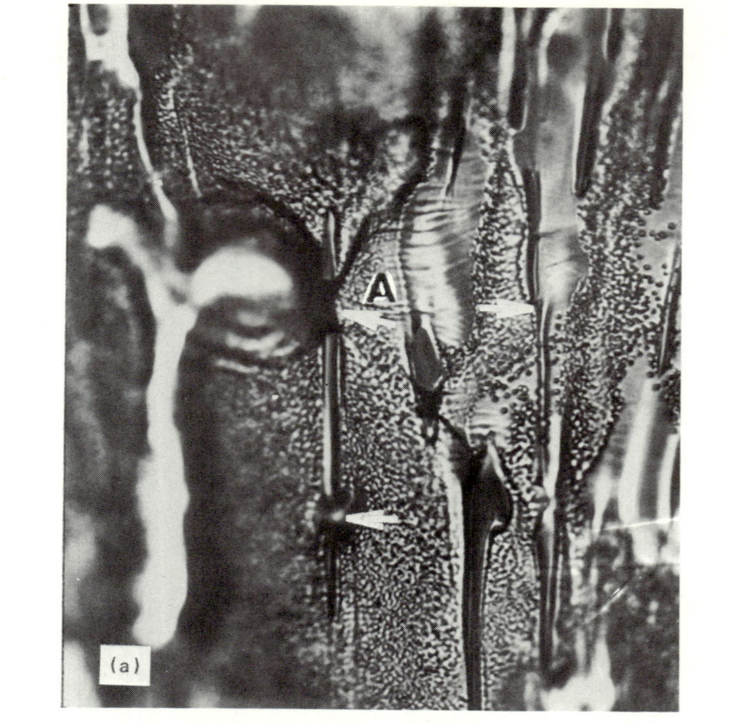

Figure 10.15 Fracture of Al-Al_3Ni. (a) Whisker fracture and void formation. Arrows denote broken whiskers (mag. 1000X) (after Hertzberg, Lemkey, and Ford [54]). (b) Replica of fracture surface (after Salkind et al. [60]).

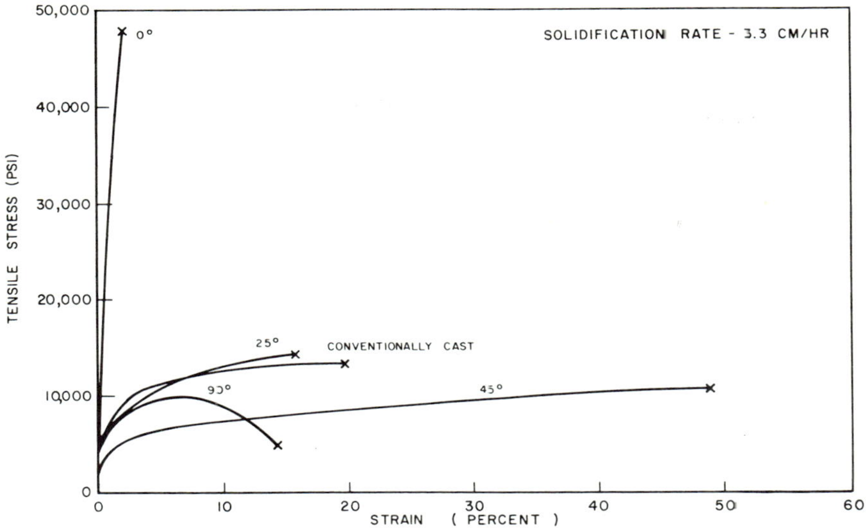

Figure 10.16 Effect of whisker orientation on stress-strain behavior of Al-Al_3Ni eutectic alloy (after George, Ford, and Salkind [63]).

whiskers can generate a large amount of shear without loading the whiskers significantly. The transverse tensile behavior (90°) was found to be dependent upon the whisker morphology [63] which is determined by the solidification rate [53, 56, 63]. As seen in Figure 10.17, this system exhibits bladelike whiskers when solidified slowly and rodlike whiskers at higher solidification rates. The bladelike whiskers were found to block transverse flow when tested at 90°, and little ductility (approximately 15% reduction in area) was noted. The rodlike whiskers, on the other hand, allowed significant transverse slip and reductions in area of more than 85% were noted for 90° tensile tests. In the latter case, the specimens necked down to an elliptical shaped fracture surface with the whiskers parallel to the major axis of the ellipse. This behavior was also noted for Cu-Cr [52] and is due to the fact that slip and therefore transverse flow, can readily occur perpendicular to but not parallel to the whiskers.

The impact behavior of Al-Al_3Ni has been investigated using both Charpy [62] and ballistic impact tests. As seen in Figure 10.18, six orientations were studied in an effort to determine the anisotropy of the material. For material exhibiting needlelike whiskers (solidified at 2 cm/hr) only two (Type A and C) of the six orientations fractured completely. The other four exhibited incomplete fracture and extensive bending. The Type A specimens

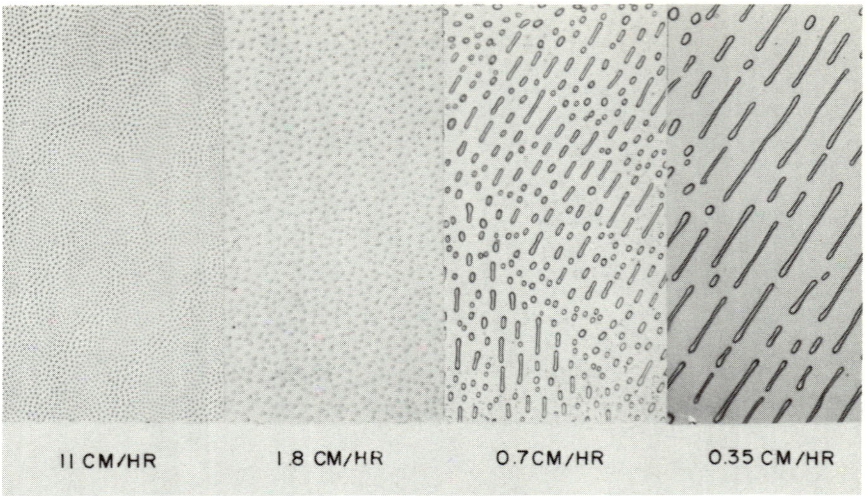

Figure 10.17 Transverse microstructure of Al-Al_3Ni as a function of solidification rate (mag. 375X) (after George, Ford, and Salkind [63]).

exhibited impact energies of 8 to 10 ft-lbs and the Type C, nearly 20 ft-lbs. The material containing bladelike whiskers (solidified at 0.6 cm/hr) exhibited fracture for all orientations. Values varied from 28 ft-lbs for the toughest orientation (Type F) to 10 ft-lbs for the least tough orientation (Type E). These values compare quite favorably with the best cast aluminum alloys. Ballistic impact testing using a 0.7 gm steel pellet impacted at 1000 ft/sec produced no cracking or penetration (as seen in Fig. 10.19) in 0.1 in. thick specimens.

The elevated temperature properties of Al-Al_3Ni have been studied both in short time tensile [55, 60] and creep tests [56, 61]. The tensile strength retention at elevated temperature is superior to that of conventional aluminum alloys and comparable to sintered aluminum powder as seen in Figure 10.20. The creep behavior of Al-Al_3Ni exhibits the three stages of creep with a low steady-state creep rate and a short period of tertiary creep with little rupture ductility. This is similar to the behavior reported for tungsten fiber-reinforced nickel alloys [66]. It was found that the creep behavior of Al-Al_3Ni was strongly dependent upon the solidification rate. As seen in Figure 10.21, material solidified at 11 cm/hr and containing finer, more closely spaced whiskers exhibited creep-rupture lives two orders of magnitude greater than that for material solidified at 2 cm/hr and containing coarser, more widely spaced whiskers. This behavior was surprising in the light of the fact that both materials exhibited approximately the same room temperature tensile strength [55]. Several tentative explanations for this behavior have been proposed [56]. Among these are (1) dislocation substructure-microstructure interaction analogous to dispersion strengthening; (2) the effect of whisker aspect ratio; and (3) the effect of whisker creep and fracture. Recently, Al_3Ni whiskers have been shown to creep at 350°C [67].

An important consideration for the use of composite materials at elevated temperature is the long-time thermal stability of the structure. Fiber

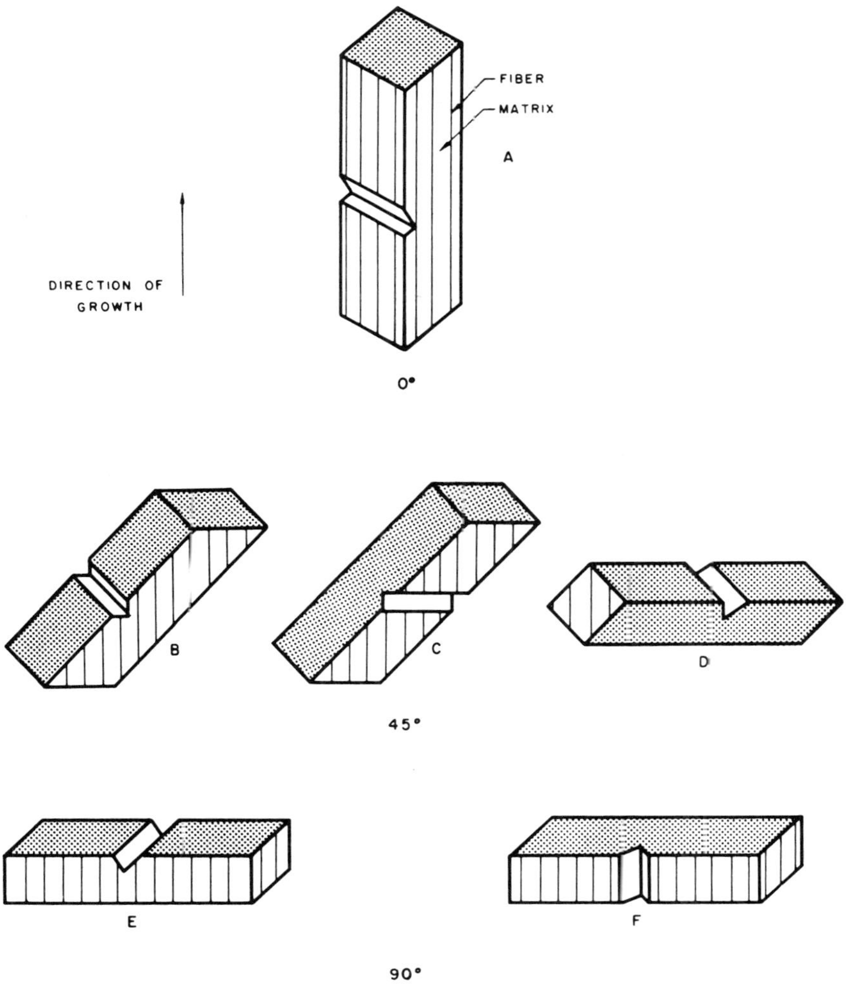

Figure 10.18 Orientation of Charpy impact specimens (after Salkind and George [62]).

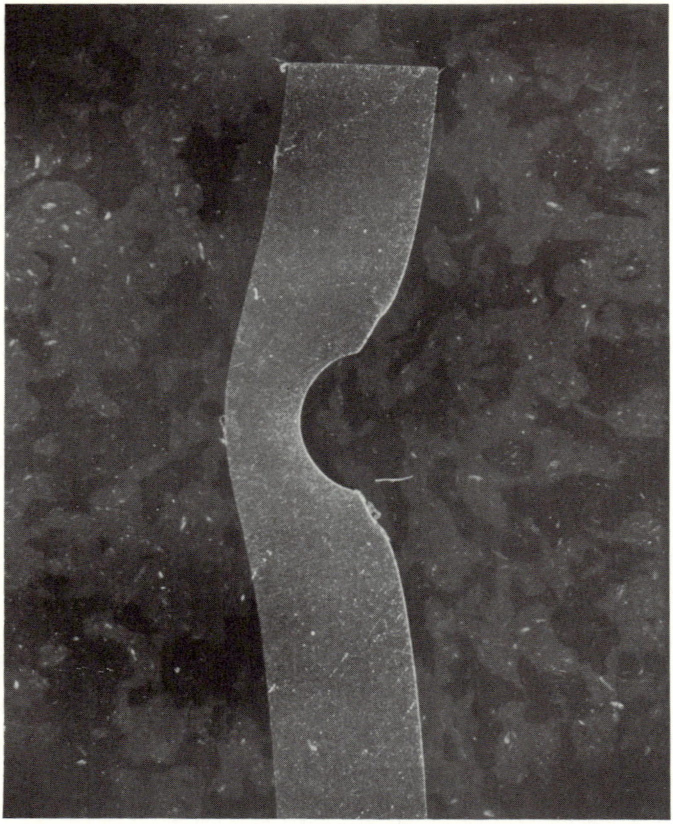

Figure 10.19 Ballistic impact behavior of $Al\text{-}Al_3Ni$ (mag. 6 X) (after Salkind and Lemkey [87]).

composites are subject to two types of instability at elevated temperatures [58]. The first results from chemical interaction between the fiber and the matrix or diffusing gases such that dissolution of one phase or formation of a new phase or phases occurs. The properties of the reinforcement are significantly reduced, although the structure may be retained. The second type of instability occurs in systems in which the two phases are chemically compatible with each other, but because of the large amount of interfacial area between phases, the fibers or whiskers spheroidize and agglomerate. The $Al\text{-}Al_3Ni$ whisker composite has been shown to exhibit remarkable microstructural stability at elevated temperature [55, 58] as seen in Figure 10.22. With the exception of small changes in solid solubility, the two phases are in

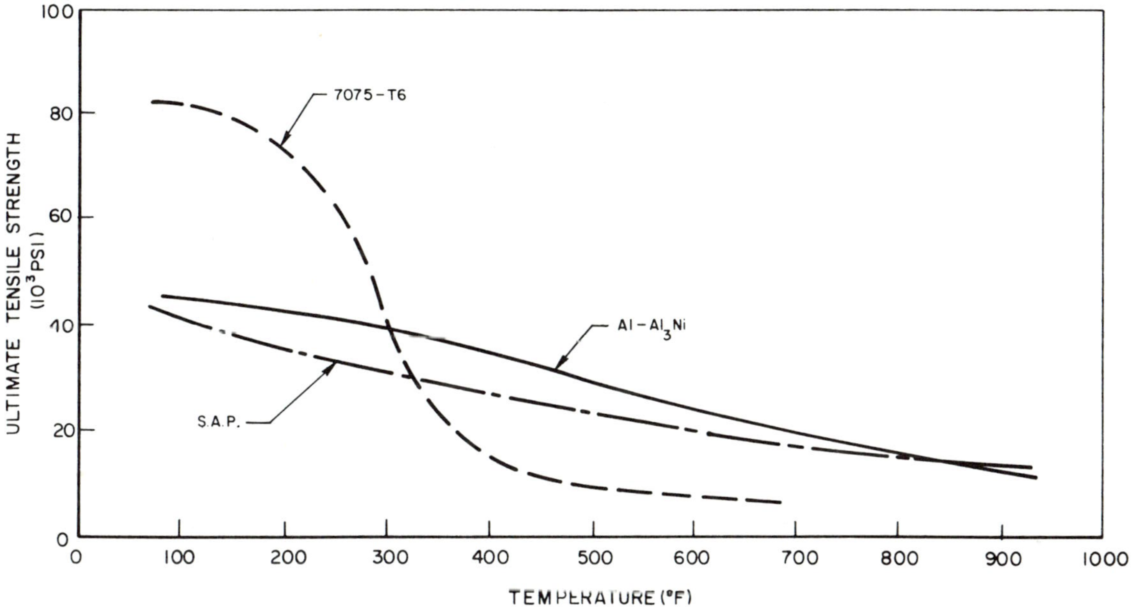

Figure 10.20 Elevated temperature tensile strength of Al-Al_3Ni and several Al alloys (after Salkind, Bayles, George, and Tice [55]).

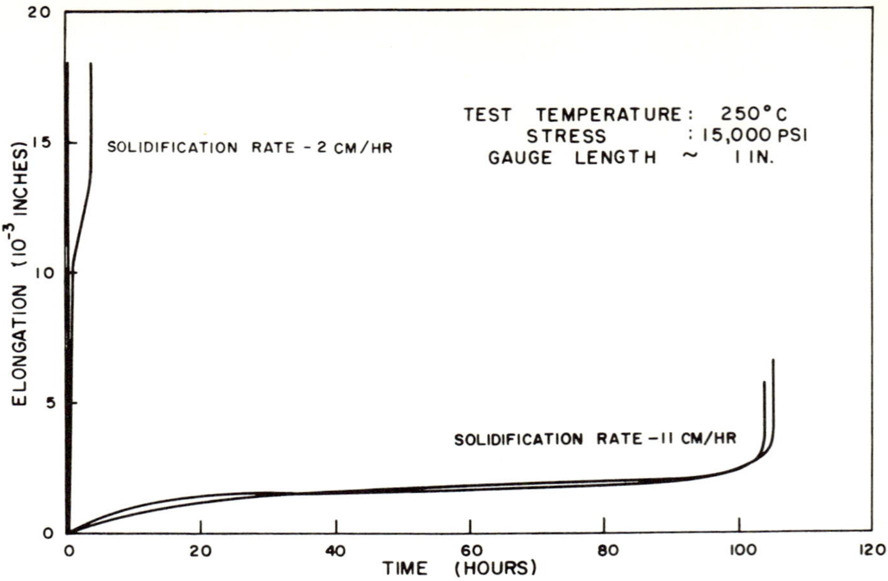

Figure 10.21 Creep behavior of Al-Al_3Ni (after Salkind, Bayles, George, and Tice [55]).

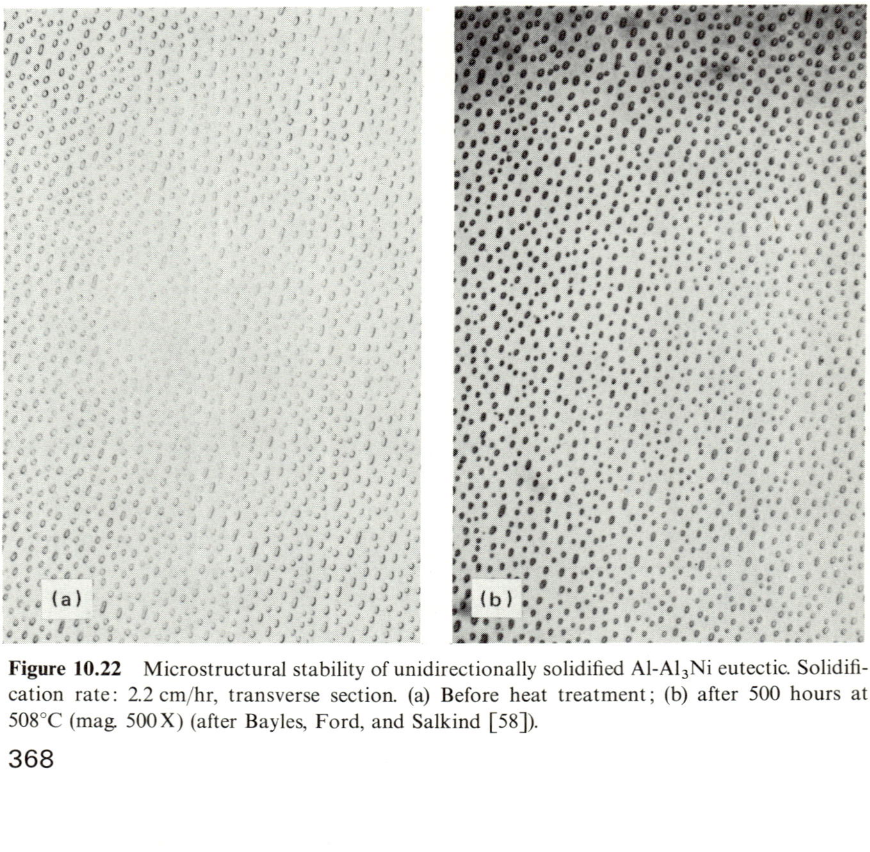

Figure 10.22 Microstructural stability of unidirectionally solidified Al-Al_3Ni eutectic. Solidification rate: 2.2 cm/hr, transverse section. (a) Before heat treatment; (b) after 500 hours at 508°C (mag. 500X) (after Bayles, Ford, and Salkind [58]).

chemical equilibrium at elevated temperature. The excellent physioco-chemical stability is due to the fact that, although there is a large amount of interfacial area between the phases, the specific interfacial energy is low because the phases grow with a preferred crystallographic relationship between phases [53] and preferred crystallographic interfaces [23]. As seen in Figure 10.23, the whiskers do coarsen at very high temperatures but they do not spheroidize or foreshorten. They maintain the preferred crystallography and interfaces and the strength of the composite is not impaired [58]. As seen in Figure 10.24, the room-temperature tensile strength of Al-Al_3Ni after a 100-hr heat treatment at 96% of its absolute melting temperature is the same as that for untreated material. For comparison, the behavior of several conventional alloys [68] and a sintered aluminum product [69] are included.

The fatigue behavior of Al-Al_3Ni has been studied in both tension and flexure [56, 70]. The low-cycle tension fatigue behavior has been found to follow the so-called "Coffin-Manson" relationship [71, 72] modified by Sachs, et al. [73];

$$N^{1/2}\epsilon_T = \epsilon_f - \epsilon_o \tag{2}$$

where N = the number of cycles to failure

ϵ_T = the total strain range per cycle

ϵ_o = the mean strain

and ϵ_f is a derived quantity obtained by extrapolation to $\frac{1}{4}$ cycle (or $\frac{1}{2}$ cycle, whichever is equivalent to a monotonic tensile test) and is related to the tensile ductility. The value of ϵ_f obtained for Al-Al_3Ni was found to be greater than 10%, which is considerably larger than the tensile ductility (about 2%). This was taken to mean that the local ductility in the matrix ahead of the crack tip is the controlling factor, rather than the tensile ductility of the composite.

The flexural fatigue behavior of Al-Al_3Ni [70] is seen in Figure 10.25 with that of the matrix (Al-0.005 wt% Ni) for comparison. The endurance limit of the composite at 10^8 cycles is 4 to 5 times that of the matrix alone. It is suggested that the matrix fatigue failure is the controlling factor and since the fibers carry most of the load, the actual stress in the matrix is a small fraction of the total stress. In addition, it has been found that the fibers divert fatigue cracks [70].

The effect of cold-rolling on the structure and properties of Al-Al_3Ni has been studied recently [57]. It was found that the material could not be rolled parallel to the fibers, but that by rolling perpendicular to the fibers, cold reductions of 99% were achieved, without cracking. The fibers were found to be undamaged and both the longitudinal and transverse

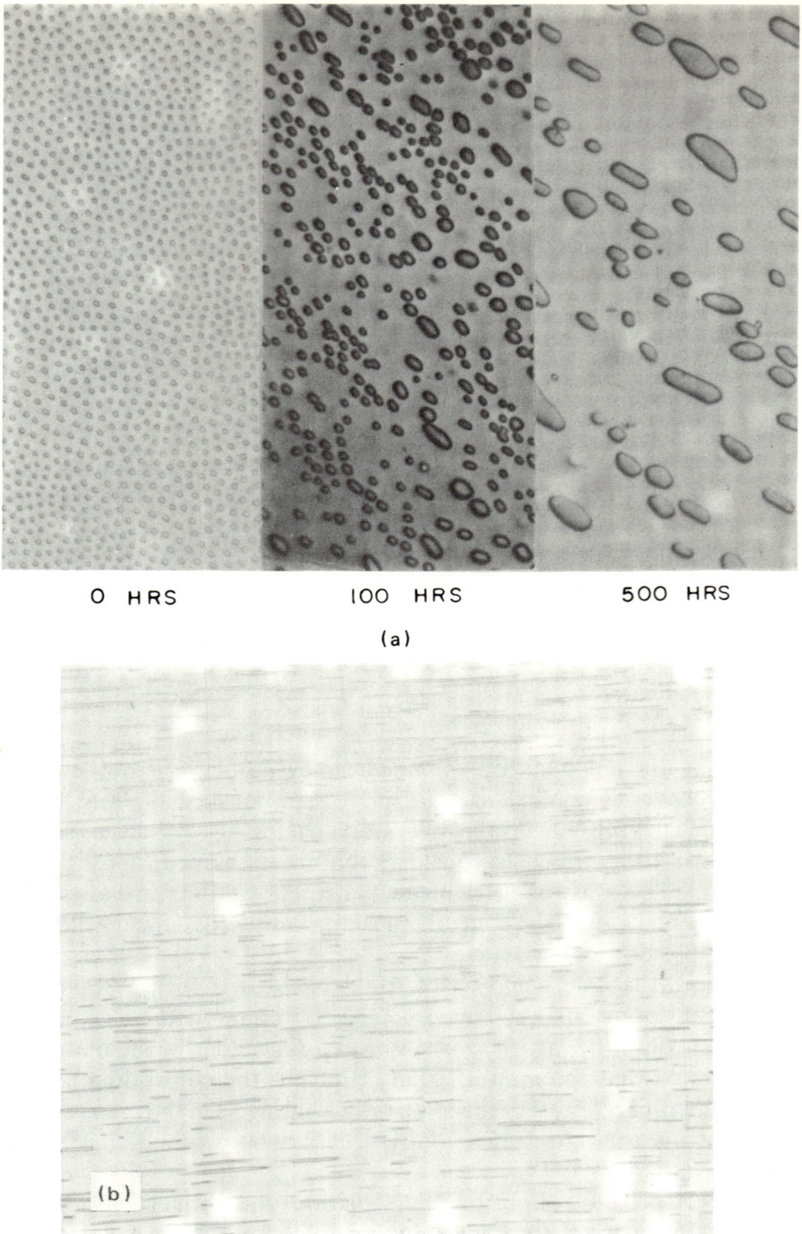

Figure 10.23 Microstructure of heat-treated Al-Al_3Ni; solidification rate: 10.6 cm/hr. (a) Transverse sections after heat treatment at 608°C (mag. 1000X); (b) longitudinal section after 500 hrs at 608°C (mag. 200X) (after Bayles, Ford, and Salkind [58]).

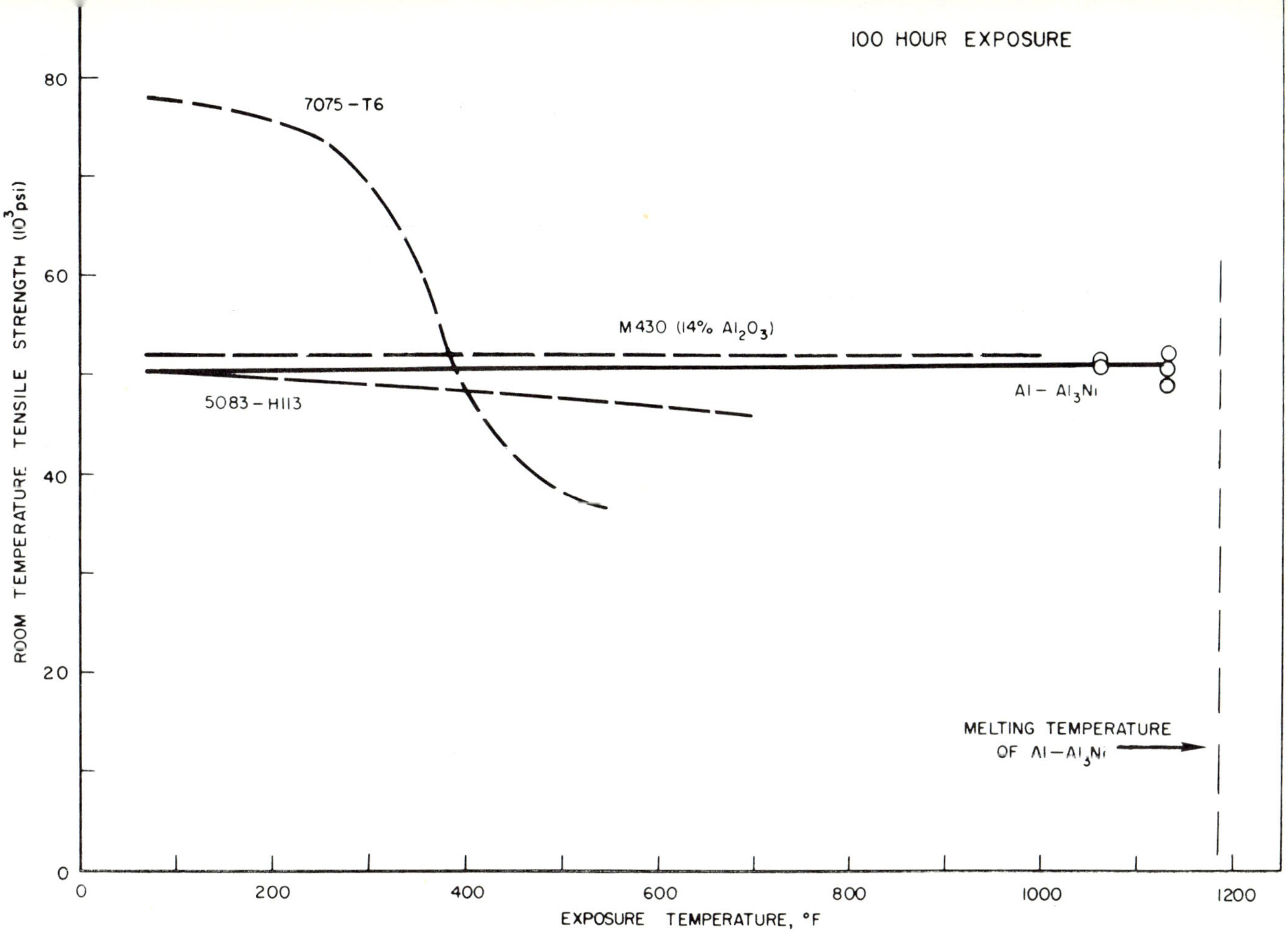

Figure 10.24 Effect of elevated temperature exposure on room temperature tensile strength (after Bayles, Ford, and Salkind [58]).

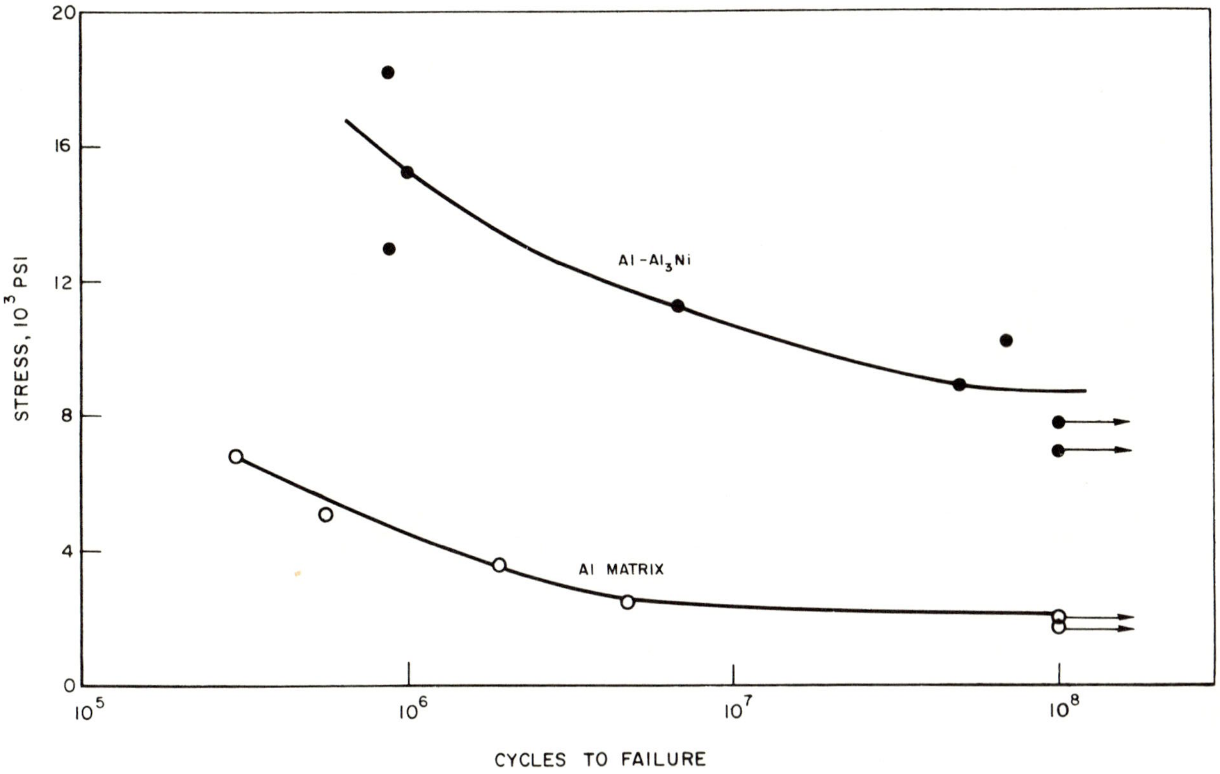

Figure 10.25 Flexural fatigue behavior of Al_3Ni whisker-reinforced Al unnotched (after Salkind and George [70]).

tensile strengths increased with reductions up to 70%. An increasing waviness of the fibers was noted as the cold work was increased. At very high reductions (greater than 85%) the whiskers began to fracture and rotate and the longitudinal strength declined to below that for unrolled material. The material cold-rolled more than 95% was isotropic because the fractured whisker segments were rotated until they were nearly randomly oriented.

Preliminary experiments have been made in an effort to join Al-Al_3Ni [61, 113]. Since the microstructure would be destroyed by fusion welding, solid-state diffusion bonding techniques were used. Because of the excellent microstructural stability at elevated temperature [58], bonding was readily accomplished at 525°C. Since the load in a fiber matrix must be transferred from fiber to fiber by shear in the matrix, joints perpendicular to the fibers are unsatisfactory. Successful scarf joints at 5 and 10° to the fibers have been made [113]. When tensile tested, these specimens failed away from the joint. In addition, cross-bonded Al-Al_3Ni, which provides reinforcement in two directions, has been made [61].

Lamellar Composites

In addition to the rodlike eutectic microstructure which results in a whisker composite, the lamellar eutectic is also important as it can result in a composite consisting of sheets of a strong, brittle reinforcing phase aligned in a matrix of another phase. It has been shown that thin films of many materials are quite strong, although there is some question as to whether the strength is analogous to that of whiskers. Little data on the mechanical properties of thin films are available because of gripping problems and edge tearing associated with testing. It is, however, known that films are often strong, and that lamellar composites exhibit reinforcing. In fact, since the lamellar reinforcement is two-dimensional compared with whiskers which are one-dimensional, lamellar composites offer the possibility of having a less anisotropic material; that is, reinforced in two directions. In addition, since lamellar composites generally have a higher volume fraction of reinforcement than eutectic whisker composites, there is a better chance of a lamellar composite being strong.

Lamellar eutectics have probably been studied more extensively than rodlike eutectics. The earliest efforts involving unidirectional solidification of eutectics were with lamellar microstructures [11–13] and the first system which Kraft studied was the Al-$CuAl_2$ eutectic which consists of nearly 50% by volume of $CuAl_2$ lamellae in an aluminum-copper solid solution matrix [13]. The microstructure of this system is seen in Figure 10.3. The mechanical behavior of this material has been studied extensively [40, 54–56, 60, 61] and it serves as an excellent model system for studying the behavior of lamellar composites.

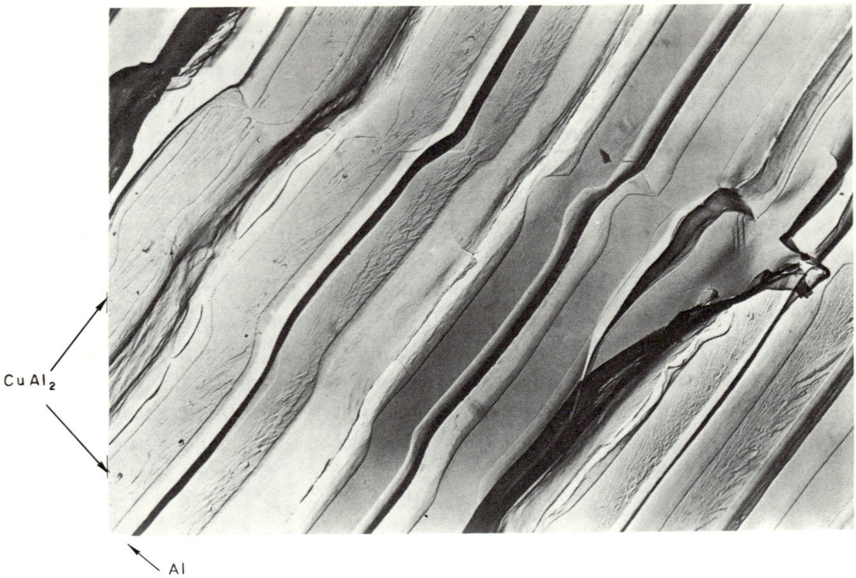

Figure 10.26 Surface of Al-$CuAl_2$ tensile specimen fractured at room temperature (mag. 6000X) (after Salkind, Bayles, George, and Tice [35]).

The ultimate tensile strength of Al-$CuAl_2$ at 20°C was found to be approximately 38,000 psi with a tensile elongation of less than 0.7% [55]. This was surprising in view of the fact that Al-Al_3Ni, which contains less than one-fourth as much reinforcement, is measurably stronger. It was speculated that because the $CuAl_2$ lamellae are larger and more irregular in shape than the Al_3Ni whiskers, they are less perfect and therefore weaker. Although the understanding of the behavior of these composites is far from complete, there appears to be sufficient evidence to indicate that even though lamellar eutectics contain more reinforcement, the lamellae do not reinforce as effectively as do whiskers in eutectic whisker composites. A replica of the fracture surface of an Al-$CuAl_2$ tensile specimen is seen in Figure 10.26. The surface consists of cleaved $CuAl_2$ lamellae with either necked or sheared aluminum layers between. This appearance was explained [49, 54, 55] as being the result of cleavage failure of the $CuAl_2$ followed by ductile failure of the aluminum. The aluminum can fail by necking if the $CuAl_2$ failures are at the same level and by shear if they are not.

The flexural behavior of Al-$CuAl_2$ has been studied [49, 54] using microbend samples and the results are summarized in Figures 10.27 and 10.28. As seen in Figure 10.27, the maximum strength was obtained parallel to the lamellae and low values between 45° and 90°. This is somewhat similar to the

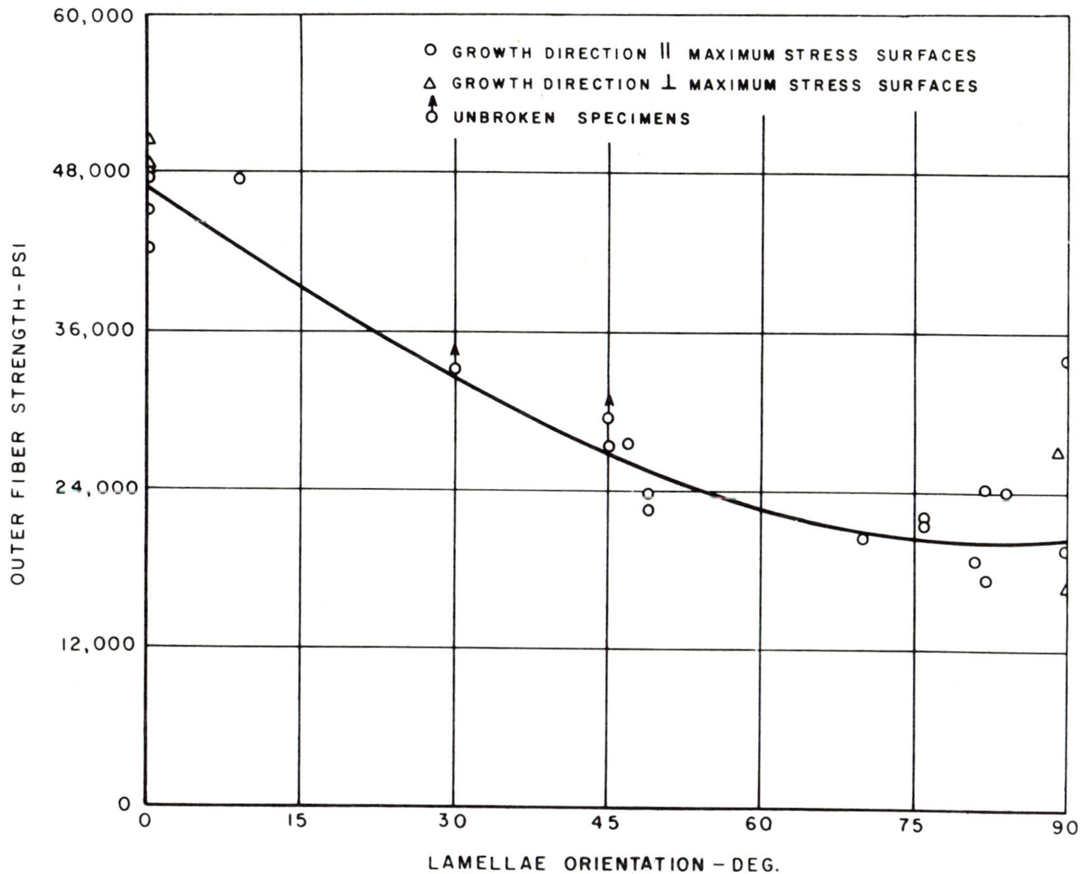

Figure 10.27 Outer fiber strength versus lamellae orientation of Al-$CuAl_2$ alloy (after Hertzberg, Lemkey, and Ford [54]).

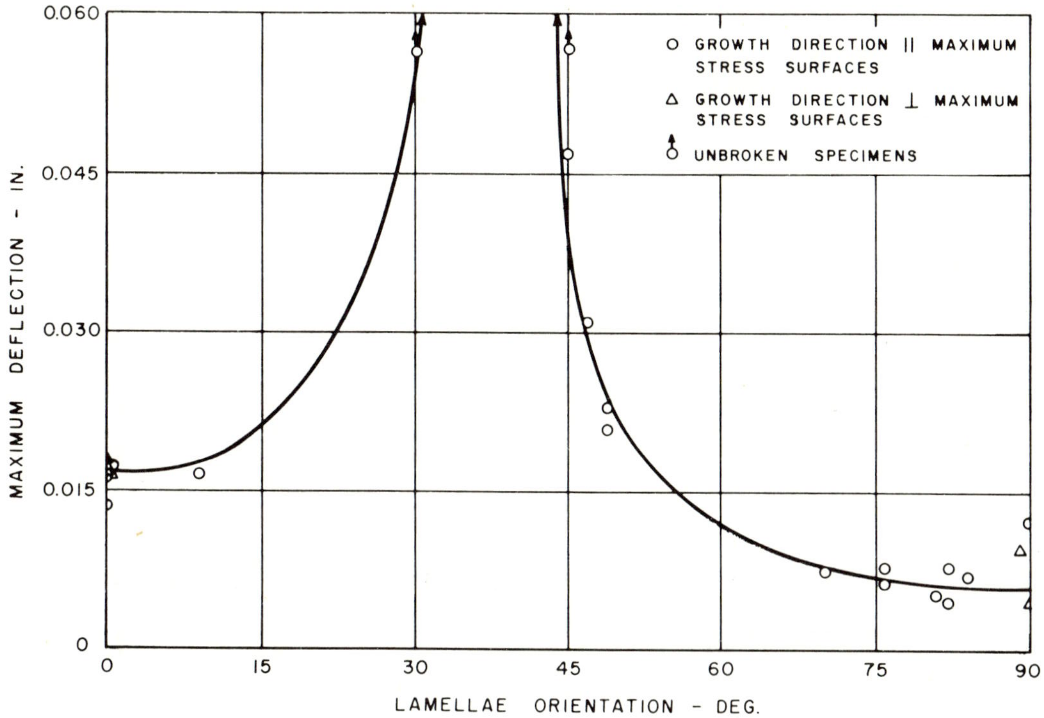

Figure 10.28 Maximum deflection versus lamellae orientation of Al-$CuAl_2$ alloy (after Hertzberg, Lemkey, and Ford [54]).

behavior of a fiber composite and indicates that the lamellae are reinforcing by a load transfer mechanism parallel to the lamellae as in whisker reinforcement. It should be noted that the deflection (Fig. 10.28) of Al-$CuAl_2$ reaches a maximum near 45° for the same reason as the Al-Al_3Ni whisker composite (see Fig. 10.16); the maximum resolved shear stress is nearly parallel to the lamellae and large amounts of shear can occur without intersecting the $CuAl_2$. The more brittle behavior at greater orientations is due to the fact that cracks readily run through the $CuAl_2$ phase or the interface between phases.

The elevated temperature tensile behavior (Fig. 10.29) of Al-$CuAl_2$ [55, 60] was similar to that of Al-Al_3Ni in that the strength was maintained well at elevated temperatures. There was a marked increase, however, in the tensile elongation at temperatures above 450°F as seen in Figure 10.29. Considerable flow in the aluminum occurred parallel to the $CuAl_2$, and since the platelets prevented transverse flow, large elongations with little necking were noted. The creep behavior of Al-$CuAl_2$ was found to be inferior to that of Al-Al_3Ni [56] although considerable rupture ductility (approx. 30%) was noted. The Al-$CuAl_2$ microstructure was found to be quite stable at elevated temperatures [55, 74, 75]. Recession of the $CuAl_2$ platelets from growth "faults" [74] were found to be the major instability.

The low cycle tension-tension fatigue behavior of Al-$CuAl_2$ was found to be superior to the Al-Al_3Ni whisker composite [56]. The stress-strain loops of the Al-$CuAl_2$ were found to be considerably narrower than those for the Al-Al_3Ni, and it was proposed that the $CuAl_2$ platelets blocked transverse flow, thus reducing the plastic strain per cycle and enhancing the fatigue behavior.

Al-$CuAl_2$ was found to be extremely brittle when tested in both Charpy and ballistic impact [76]. This behavior has been observed for other lamellar eutectics and is thought to be a major drawback for their use in structural applications. It can be argued that a crack initiating on the surface or interior of a whisker composite must pass through the ductile matrix many times in order to sever a part. The ductile matrix can plastically deform and absorb energy from the crack, thus slowing or stopping it. The brittle sheets in a lamellar composite, on the other hand, offer easy low-energy fracture paths for a crack. Although one can achieve high impact values for lamellar composites by judiciously orienting the specimen [77], in actual practice one cannot count on the crack taking the path of most resistance. Ballistic impact tests of Al-$CuAl_2$ [76] indicates brittle behavior for any orientation.

Practical Eutectic Composites

Experiments with low-temperature eutectic systems have been sufficiently encouraging so that efforts have been made to produce so-called practical

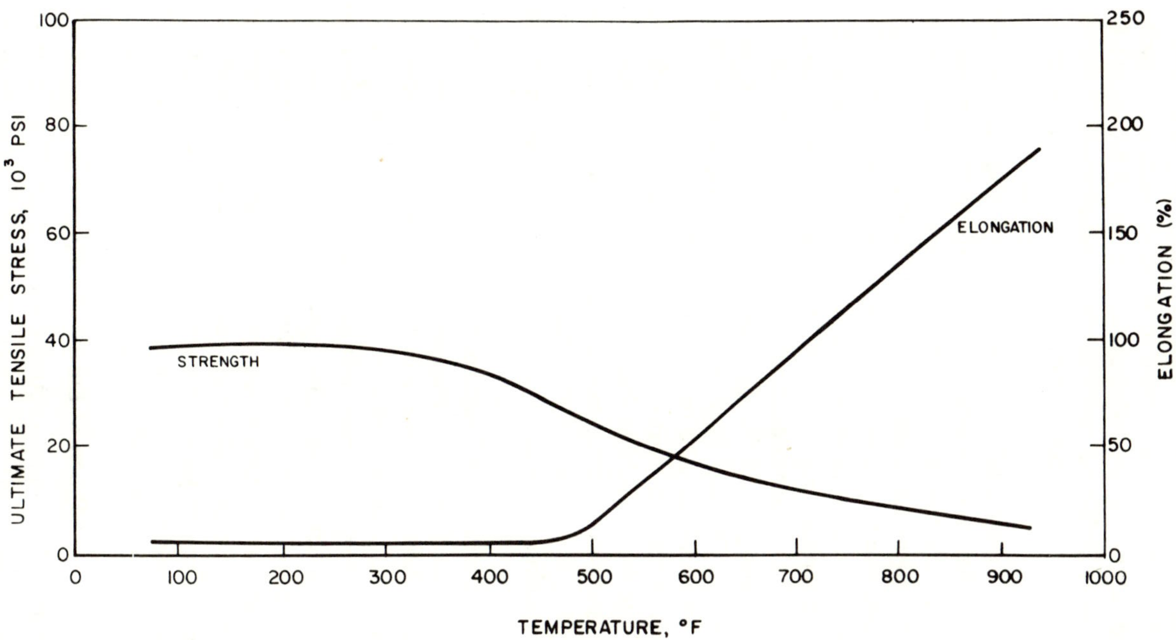

Figure 10.29 Effect of temperature on the tensile strength and ductility of unidirectionally solidified Al-$CuAl_2$ (after Salkind, Bayles, George, and Tice [55]).

eutectics, that is, materials to be considered as items of commerce. One of the first such studies involved the investigation of the Ni-NiBe, Ni-Ni_3B, Ti-TiB, and Ta-Ta_2C eutectics [38]. The Ni-NiBe system which contains approximately 40% by volume NiBe lamellae in a nickel-beryllium solid solution matrix was found to exhibit little more strength than the matrix and less that of the conventionally cast eutectic. The Ni-Ni_3B eutectic [38, 78] contains 35% by volume nickel lamellae in a matrix of Ni_3B. This system was studied in an effort to determine whether the presence of ductile lamellae would improve the ductility of the intermetallic compound. As expected, it did not for reasons which are similar to the arguments in the foregoing section. The Ti-TiB eutectic was found to be anomalous, that is, it would not grow in an aligned structure when unidirectionally solidified.

The Ta-Ta_2C eutectic and an analogous system, Nb-Nb_2C, were the first practical eutectic whisker composites produced by unidirectional solidification [38, 39]. Both systems contain approximately 30% by volume of carbide whiskers in a refractory metal matrix. The tensile strength of Ta-Ta_2C is 150,000 psi at room temperature and, as seen in Figure 10.30, is superior to one of the best commercial tantalum alloys, Ta-10W [79], at temperatures above 2000°F. The Nb-Nb_2C composite has more than five times the strength of pure Nb and, as seen in Figure 10.31, is superior to any commercial Nb-base alloy [80] on the basis of specific strength. The thermal stability of Nb-Nb_2C was found to be excellent [39]. After more than 250 hours at 2500°F there was found to be no change in either the microstructure or tensile behavior.

The Ni-NiMo system [56, 81] consists of more than 40% by volume of NiMo lamellae in a matrix of nickel saturated with molybdenum. At 20°C, this material exhibits a tensile strength of more than 160,000 psi, a Young's modulus of 40×10^6 psi, and a tensile elongation of 0.5%. The elevated-temperature tensile strength of this material is compared to that of several superalloys [82] in Figure 10.32. This material is stronger than any nickel base superalloy at 2000°F. Although this composite is quite strong, it is relatively brittle and is inferior to superalloys in oxidation resistance at elevated temperature.

The NiAl-Mo eutectic, consisting of 11% by volume of reinforcing molybdenum fibers in an intermetallic NiAl matrix, has been recently investigated [83]. This system has been studied in the hot-worked condition and, although it is considerably stronger than NiAl, the oxidation resistance is poor due to the presence of the Mo fibers. An interesting system is the CoAl-Co eutectic [40] which, when unidirectionally solidified, contains 65% by volume of Co rods in a CoAl matrix (see Fig. 10.9) but when hot-rolled changes to a structure consisting of elongated CoAl particles in a cobalt matrix. This hot-rolled material exhibits strengths approaching 200,000 psi

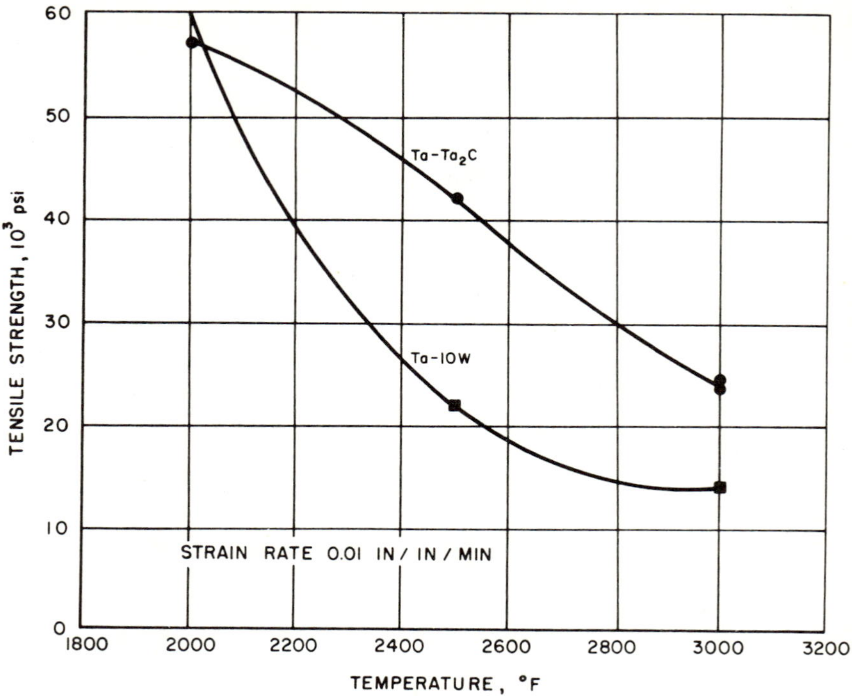

Figure 10.30 Strength of Ta-Ta_2C and Ta-10W (after Lemkey and Salkind [39]).

at room temperature with 14% elongation. As seen in Figure 10.33, the tensile strength of this material falls rapidly as the temperature is increased and the elongation becomes quite large. In fact, elongations greater than 850% were noted at 1200°C.

A recent study of the Ni-Cr-Si-C system [84] has been directed toward the production of aligned structures which are not simple binary eutectics. Although this study has not yet been successful in producing a practical material, there appears to be considerable merit in considering such systems for further development.

PROPERTIES OF EXTRACTED WHISKERS

In order to understand the mechanical behavior of a eutectic whisker composite, it is necessary to determine the properties of the whisker phase isolated from the matrix. It is possible, however, that the strength of individual

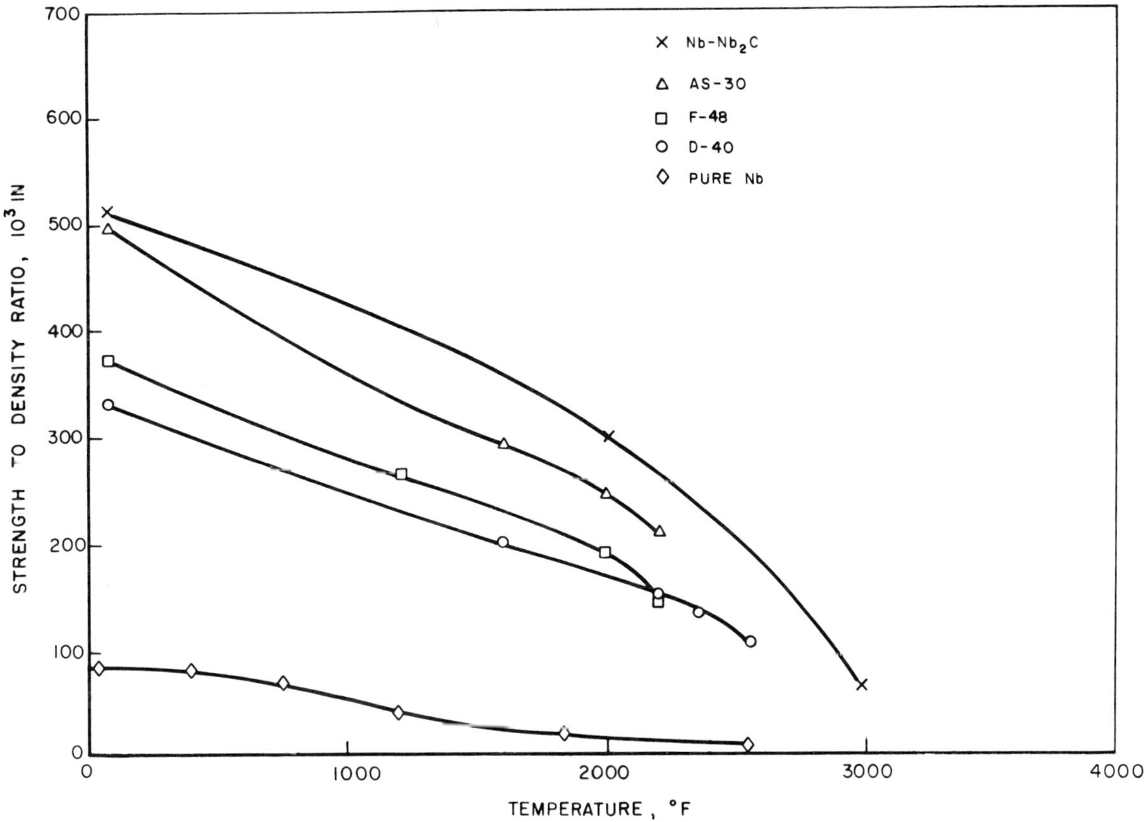

Figure 10.31 Specific tensile strength of Nb-base alloys (after Lemkey and Salkind [39]).

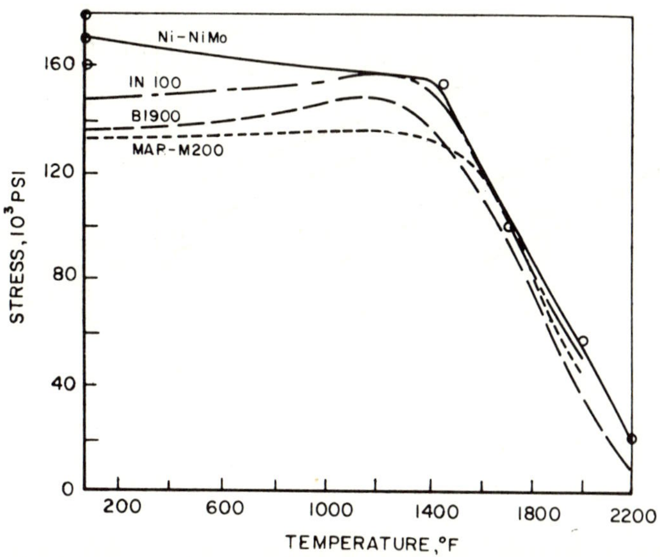

Figure 10.32 Tensile strength of Ni-NiMo and several nickel-base superalloys (after Thompson [81]).

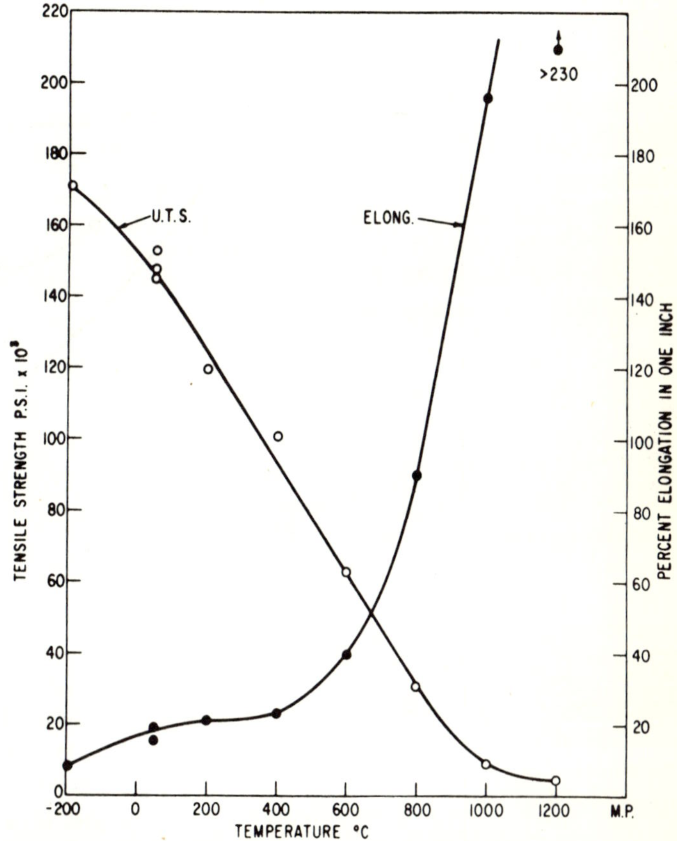

Figure 10.33 Tensile strength and elongation of CoAl-Co samples rolled at 1000°C and annealed 2 hours at 1000°C (after Cline [40]).

whiskers extracted from the eutectic may be quite different from their behavior *in-situ*. For example, the whisker may break up within the matrix but continue to reinforce, provided the segment lengths exceed the critical aspect ratio for the system. On the other hand, it is possible that a crystal, which displays nearly theoretical strength when isolated from the matrix, may be deformed at a relatively low stress in the matrix by dislocation pile-up at the interface. Determination of the intrinsic whisker strength is made difficult by chemical damage during dissolution from the matrix or by handling during testing. On the other hand, during chemical separation, surface steps may be removed, resulting in higher apparent strength, especially in hard phases.

To a great degree the effectiveness of the whisker phase in reinforcing the matrix should depend upon its intrinsic strength. The theoretical ultimate strength of a whisker is determined by the interatomic binding forces. The elastic modulus is related directly to these cohesive forces and, as a first approximation, the theoretical strength can be computed as being between 3 and 17% of the Young's modulus [65]. As mentioned previously, Webb and Forgeng [14] first concluded from the results of bend tests that microcrystals isolated from solid alloys could be considered as whiskers in that they exhibited strengths approaching theoretical values.

Tensile tests of extracted eutectic whiskers have been performed using a specially devised soft tensile apparatus designed by Lemkey and Kraft [50]. Bend tests have been performed on ground glass slides using a small microprobe to move the whisker ends so that the whisker was pulled into successively tighter loops. A suitable fluid such as water or mineral oil was used to retain the whisker under stress prior to optical strain measurements. The maximum strain was computed by measuring the mean diameter of the whisker by optical or transmission electron microscopy and dividing by twice the minimum radius of curvature. Figure 10.34 shows a typical test performed on an isolated niobium carbide whisker.

Chromium Whiskers

The chromium phase of the Cu-Cr eutectic was isolated from ingots unidirectionally solidified at 2.1–3.5 cm/hr without observable chemical attack by dissolution of the copper phase in dilute nitric acid. A few of the needlelike chromium whiskers were found to be twisted around their growth axis. The mean transverse dimension of the whiskers varied from 0.25 to 1.25 μ with aspect ratios of nearly 10,000:1. Replicas of composite transverse sections indicated a faceted circular cross-section as seen in Figure 10.10. The chemical composition of extracted Cr whiskers determined by spectrographic analysis is given in Table 10.1.

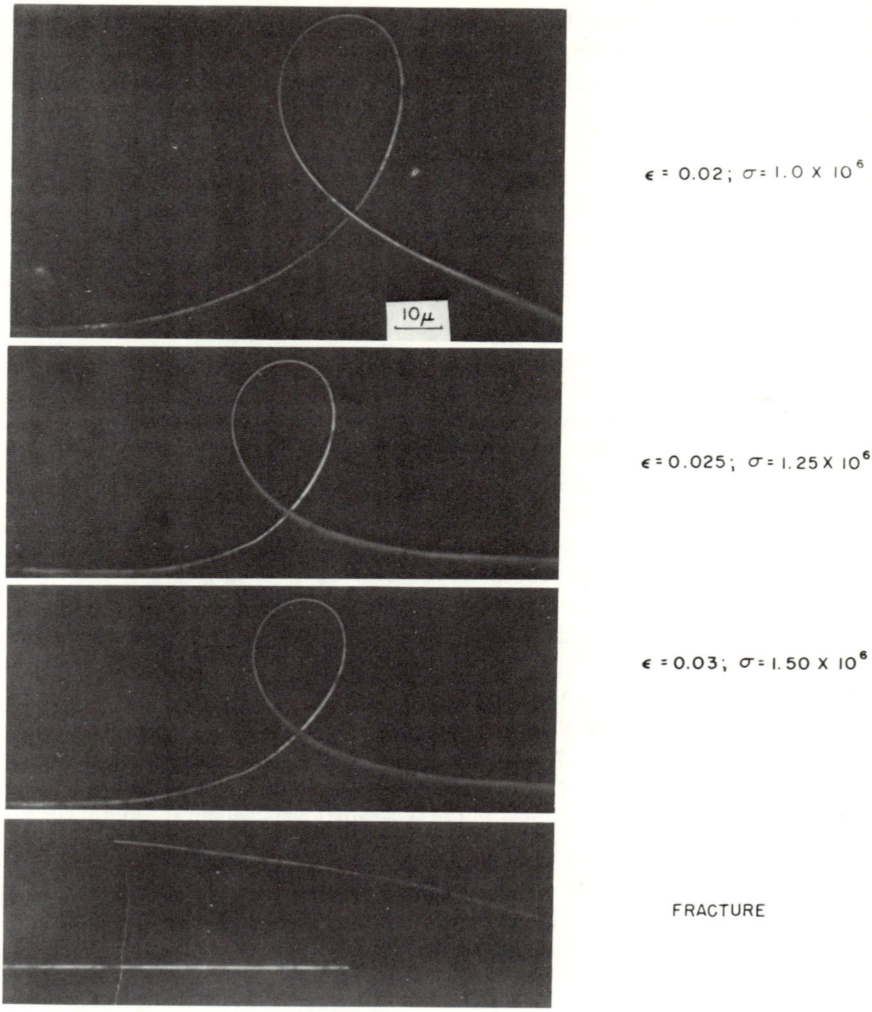

Figure 10.34 Bend test of extracted niobium carbide whisker.

The solubility of copper in chromium has not yet been thoroughly investigated but seems to be very small [88]. Selected area electron diffraction experiments revealed that most of the chromium whiskers grew with their long axis parallel to the ⟨112⟩ direction. Indexing of the orientations was complicated by the presence of Kikuchi lines superimposed on the spot diffraction patterns. Two whiskers displayed a ⟨111⟩ growth direction a few

TABLE 10.1 COMPOSITION OF EXTRACTED Cr WHISKERS

	Cr	Fe	Si	Mg	Al	Cu	Ti
Percent by weight	Bal	.08	.70	.01	.06	.30	.01

microns away from one end changing to a ⟨112⟩ orientation along the rest of the whisker length.

The thickness of each of the fractured whiskers was measured in the electron microscope and the stress computed on the basis of a circular cross-sectional area. The results of 24 tensile tests are summarized in Figure 10.35. From the data it is difficult to determine the exact relationship between size and strength because of the large scatter; however, a curve is drawn to suggest the size effect. The principle source of error is believed to be associated with the measurement of the whisker cross-sectional area. Size effects have been previously reported for tests performed on much thicker whiskers grown from the vapor [64, 65, 89, 90] and extracted from a unidirectionally solidified eutectic alloy [91]. A typical stress-strain curve is seen in Figure 10.36. Strain was measured in six of the tests and experimental values of the elastic modulus are reported in Table 10.2.

Young's modulus for chromium in the ⟨112⟩ direction was calculated from the reported compliance constants to be 38.4×10^6 psi [92]. This modulus value is in good agreement with the values in Table 10.2 in the light of the precision of the cross-sectional measurement. At elastic strains greater than 2%, deviation from Hooke's Law was observed. This effect has been observed previously [65, 93] and is in agreement with theoretical predictions.

Many of the whiskers fractured with observable necking as seen in Figure 10.37. After fracture, some of these whiskers were observed to be kinked in several regions. The kinks were removed spontaneously when heated to 285°C. Several other whiskers fractured with such force that they disintegrated.

The whiskers exhibited bending strains considerably greater than those measured in tensile tests. The largest strain measured in bending was approximately 5.5%. Previously, Wlodek et al. [94] reported that chromium whiskers extracted from conventionally cast copper-chromium alloys were capable of withstanding in excess of 3% elastic deformation at room temperature and about 4% elastic deformation at −196°C in bending. Furthermore, the chromium whiskers often deformed plastically by kinking at liquid nitrogen temperatures. Because only a small fraction of the total whisker volume is subjected to the maximum strain, strength values obtained by bend tests are not considered as reliable as those obtained in axial tension.

An additional point of interest was the fact that after extraction from the matrix the chromium filaments were pyrophoric at approximately 285°C.

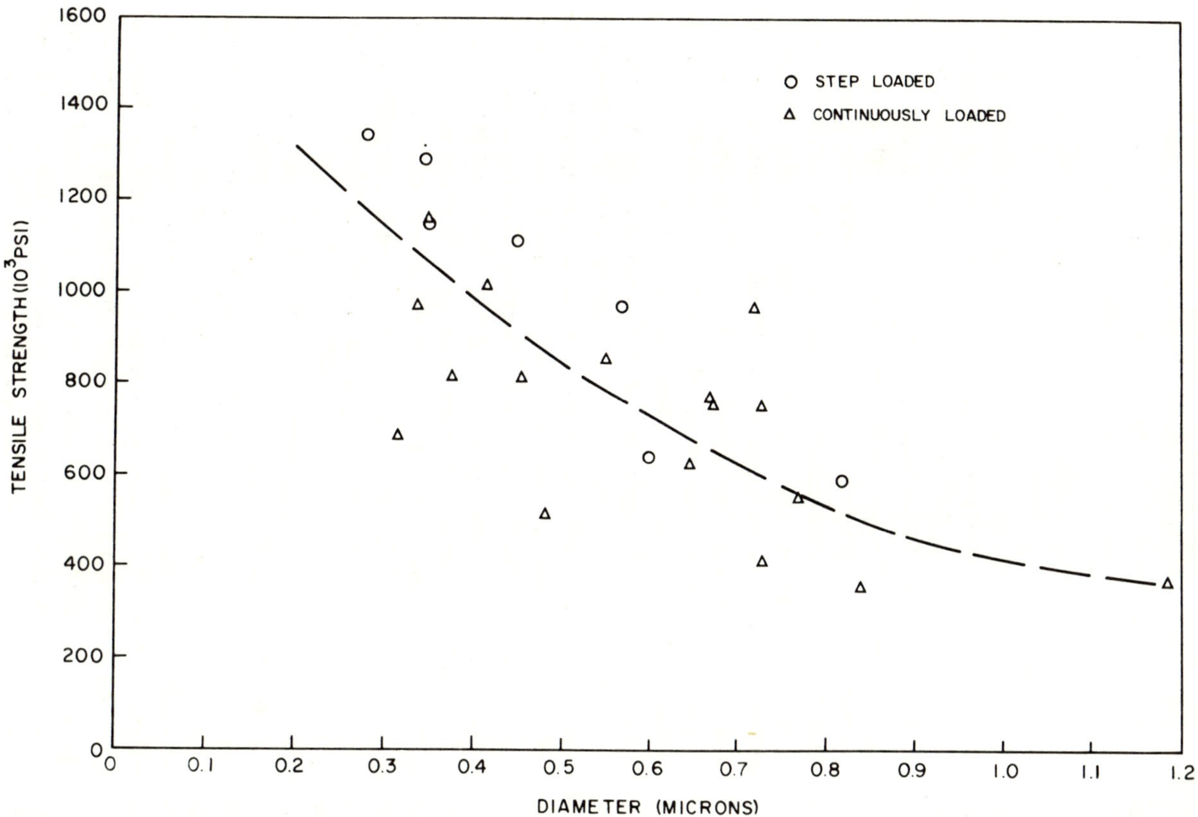

Figure 10.35 Variation of tensile strength with diameter of Cr whiskers.

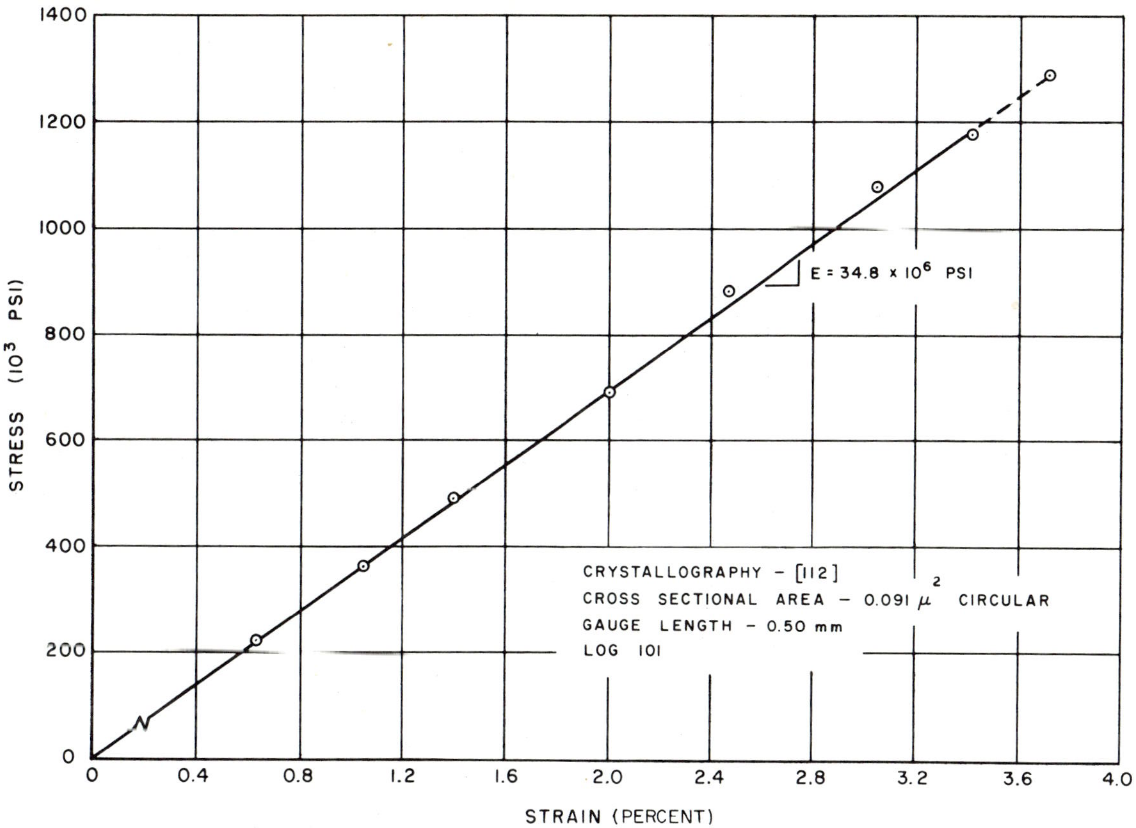

Figure 10.36 Stress-strain curve for Cr whisker.

TABLE 10.2 SUMMARY OF RESULTS OF TENSILE TESTS OF EXTRACTED CHROMIUM WHISKERS

Specimen Number	Diameter (μ)	Gauge Length (mm)	Maximum Elastic Strain (%)	Tensile Strength (10^3 psi)	Modulus of Elasticity (10^6 psi)	Fiber Axis	Fracture Mode
101	0.34	0.50	3.42	1290	34.8	⟨112⟩	ductile, kinked
105	0.35	0.56	2.27	1150	45.7	—	—
107	0.60	0.49	1.68	620	34.0	—	—
151	0.45	0.53	1.93	1110	40.0	⟨112⟩	kinked
154	0.57	0.63	2.10	969	37.5	—	shear
155	0.82	0.71	1.68	585	34.0	—	shear

When stored for several days in still air the chromium whisker would no longer spontaneously burn at this temperature; however, no evidence of the oxide was observed by X-ray or transmission electron diffraction on whiskers stored for more than one year.

Al_3Ni Whiskers

The Al_3Ni intermetallic phase of the Al-Al_3Ni eutectic was isolated by dissolving the matrix in a 3% aqueous solution of hydrochloric acid. The whiskers, when separated, were usually of high optical reflectivity and uniform thickness. When extracted from ingots grown at solidification rates between 2 and 18 cm/hr, several whiskers were found to be more than $\frac{1}{2}$ in. long. This was equivalent to the original length of the specimen dissolved. Wavy whiskers of variable thickness were commonly found among the larger Al_3Ni whiskers extracted from ingots grown at rates below 2 cm/hr. These whiskers also exhibited multiple branching during growth [53]. A typical spectrographic chemical analysis of extracted Al_3Ni whiskers is given in Table 10.3.

Al_3Ni has an orthorombic DO_{20} structure [95]. Both X-ray and electron diffraction indicated that the fiber axis was always [010]. The whisker cross-section is a distorted hexagonal shape as seen in Figure 10.6 which reflects the internal symmetry of the Al_3Ni crystal structure [23].

TABLE 10.3 IMPURITY CONTENT OF EXTRACTED Al_3Ni WHISKERS

	Pb	Mg	Si
Percent by weight	0.15	0.01	0.15

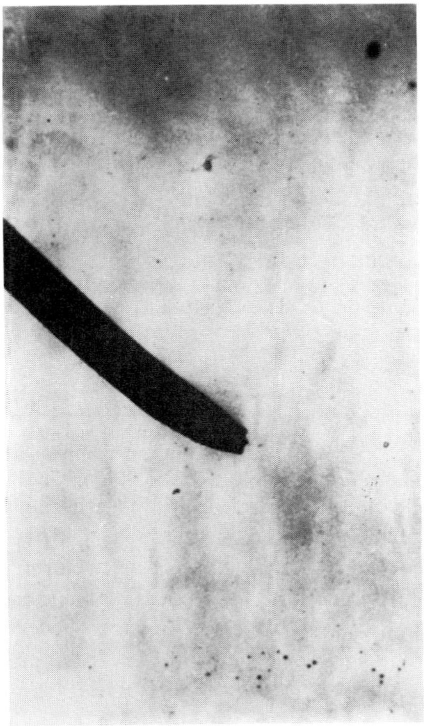

Figure 10.37 Electron micrograph of fractured Cr whisker exhibiting necking (mag. 30,000 X).

The microhardness of Al_3Ni in the [010] direction was 510 kg/mm^2 as measured in the eutectic. The room temperature uniaxial tensile properties of extracted Al_3Ni whiskers were determined as previously described for chromium whiskers [50]. The ultimate tensile strength varied from 241 to 401 $\times$ 10^3 psi and Young's modulus ranged from 19 to 22 $\times$ 10^6 psi [54]. Using the latter value for the whisker modulus and 10.3 $\times$ 10^6 psi for the modulus of Al in the $\langle 110 \rangle$ lattice direction, good agreement was obtained between the calculated modulus of Al-Al_3Ni and the experimentally determined values [54]. The extracted whiskers did not exhibit any observable plasticity and fracture occurred by cleavage. A number of whiskers initially failed at stresses less than 100,000 psi. These premature failures were attributed to either chemical attack during extraction or surface growth defects which acted as stress concentrators.

Bend tests were also conducted and elastic behavior to fracture was observed. It was noted that the whiskers did not always fail at the point of maximum stress, indicating the presence of a defect in the loop. Measured strains were approximately twice those observed in tensile tests [54]. These results support the contention that defects in the whiskers significantly affect their mechanical properties.

Nb_2C Whiskers

Nb_2C whiskers were extracted from unidirectionally solidified specimens of Nb-Nb_2C by dissolution of the matrix in a 10% alcoholic bromine solution. A typical spectrographic analysis of the impurity content of

TABLE 10.4 IMPURITY CONTENT OF EXTRACTED Nb_2C WHISKERS

Element	Fe	Si	Cr	Mg	Al	Cu	Ti	Ta
Percent by weight	.07	.03	.1	.005	.02	.02	.05	.1

extracted diniobium carbide is given in Table 10.4. The whiskers were uniform in thickness along the length except at the ends which were tapered. Aspect ratios of approximately 430 were measured for whiskers extracted from specimens solidified at 15 cm/hr [96]. Electron diffraction measurements revealed that the whiskers grew preferentially in the $[10\bar{1}0]$ direction and were bounded by faces parallel to the prismatic Type II and basal planes [39]. The diffraction patterns were complex due to the polymorphism of Nb_2C [97, 98]. The low temperature modification, α-Nb_2C has a structure similar to that of ζ-Fe_2N and transforms at 1200°C to the hexagonal β-Nb_2C which is similar to the ϵ-Fe_2N structure. The presence of two phases within the extracted whisker is inferred from a transmission electron micrograph (Fig. 10.38).

Bend tests of Nb_2C whiskers indicated them to be quite strong. Several were stressed numerous times above 1.5×10^6 psi without failure. The modulus was calculated from the measured modulus of the eutectic using the rule of mixtures [96]. Fracture occurred by cleavage and no observable plastic deformation was noted.

It is interesting that high strength in this carbide was obtainable without evidence of internal perfection or near perfection. Further investigations of the strength of whiskers in both single-phase modifications of Nb_2C and the two-phase state, and its effect on the mechanical properties would be of interest.

NONSTRUCTURAL APPLICATIONS FOR ALIGNED EUTECTICS

In addition to unidirectionally solidified eutectics exhibiting high strength, there is also the possibility of using these aligned anisotropic structures for a variety of other applications such as electronic, magnetic, optical, and

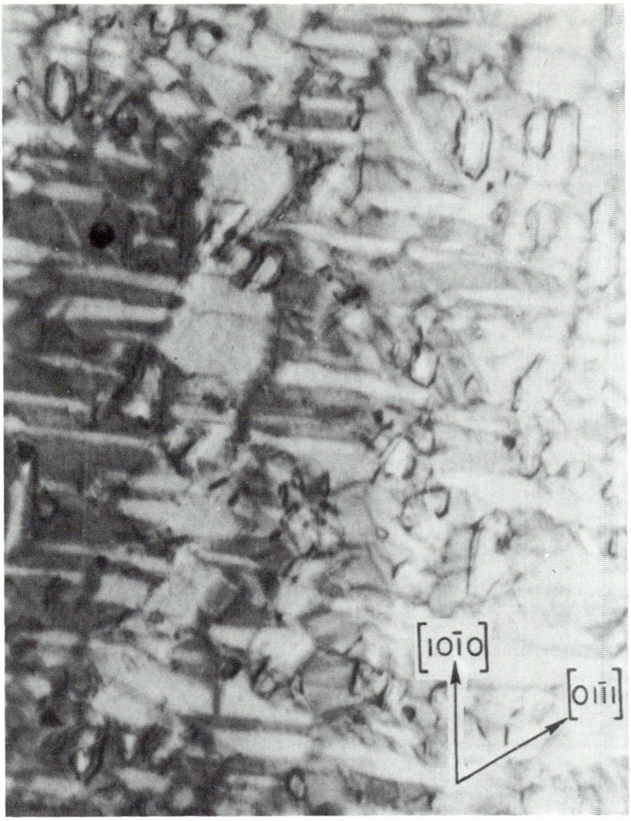

Figure 10.38 Transmission electron micrograph of extracted Nb_2C whisker (mag. 46,000 X).

porous devices. Extensive studies of the use of eutectics for such applications are being made at the United Aircraft Research Laboratories, Siemens Central Research Laboratory, Lehigh University, and the Lockheed Palo Alto Research Laboratory. This work has been the subject of a recent review [99].

The first eutectic to be studied for its electical properties was the InSb-Sb system [100]. This system contains triangular antimony rods in a matrix of indium antimonide. The electrical conductivity parallel to the fibers was more than ten times that in the transverse direction. In addition, both the thermoelectric power and the thermal conductivity were found to be anisotropic, although not to the same extent as the electrical conductivity as seen in Table 10.5.

A significant amount of work has been done with the InSb-NiSb eutectic [101–105] and several similar systems [104, 106, 107]. The InSb-NiSb

TABLE 10.5 THERMOELECTRIC PROPERTIES OF InSb-Sb EUTECTIC (AFTER LIEBMANN AND MILLER [100])

Sb Rod Size (cm)	Direction of Measurement	Electrical Resistivity (ohm-cm)	Thermoelectric Power (microvolts per degree)	Thermal Conductivity (watts per degree)
4.3×10^{-4}	parallel to Sb rods	3.05×10^{-4}	−18	0.1113
	perpendicular	33	−71	0.0822
8.5×10^{-4}	parallel	3.00	−12	0.1235
	perpendicular	33	−53	0.0940
28×10^{-4}	parallel	3.15	−8.2	0.1430
	perpendicular	34	−28	0.1059

structure consists of aligned, conducting NiSb rods in a semiconducting InSb matrix. Because the conducting fibers short circuit the Hall effect, the electrical resistance in a magnetic field is quite anisotropic [101] as seen in Figure 10.39. When the electric current, magnetic field, and fibers are mutually orthogonal, the resistance of the eutectic is an order of magnitude higher than that for a configuration in which the fibers are parallel to either the magnetic field or the current.

One of the many new devices made possible by such materials is a contactless variable resistance [105]. By rotating the eutectic composite in a fixed magnetic field, resistance changes of more than an order of magnitude can be made. The device is free from the problems of contact wear and sparking common in existing variable resistors. A related device is a contactless switch. By placing a current-limiting transistor in series with the contactless variable resistor, the current may be made to turn off completely when the eutectic composite is oriented such that the current is below a certain value. A DC to AC converter which is an analogue of the mechanical chopper uses the same principle as the contactless switch to "chop" an incoming DC signal. A brushless DC motor [105] can be made by orienting three eutectic composites at 120° to each other in a rotating permanent magnetic field such that only one of the three exhibits a low resistance at any one time. If each eutectic is connected to one of three field pole coils, then only one will be magnetized at a time, the effect produced by using brushes in a conventional DC motor. Finally, the magneto resistively anisotropic eutectic composite makes an excellent probe for determining and mapping magnetic fields [105]. The resistance is readily measured in a bridge circuit and, because the material is much more sensitive than the semiconducting matrix alone, very accurate measurements can be made.

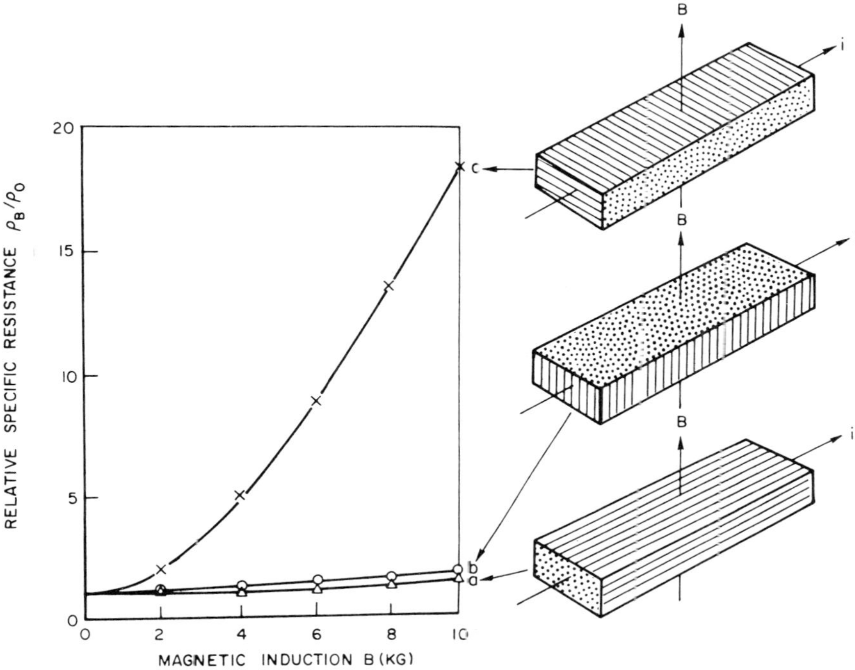

Figure 10.39 Magnetoresistive anisotropy of InSb-NiSb eutectic (after Weiss and Wilhelm [101]).

Similar eutectic systems have been used to make infrared polarizing devices [102]. The eutectics are unidirectionally solidified so that the conducting fibers act as aligned dipoles in a relatively transparent matrix. As seen in Figure 10.40, when the structure is oriented such that the dipole fibers are parallel to the electric vector, they draw power from the alternating electromagnetic field. When the dipoles are perpendicular to the electric vector, no power is drawn from the field. A polarizing power of more than 0.99 for infrared radiation has been achieved with the InSb-NiSb eutectic [102].

Several eutectic systems have been prepared with the idea of producing superior ferromagnetic materials [99, 107, 108]. Because the eutectic fiber sizes are generally of the order of 1 μ, it is hoped to produce an aligned structure with single domain ferromagnetic fibers. Two such systems, the Fe-FeS eutectic [107] and the Fe-Fe_xSb eutectic [99, 108], containing respectively 9 and 15 vol% iron fibers, exhibit superior coercivity to iron. The data are summarized in Table 10.6.

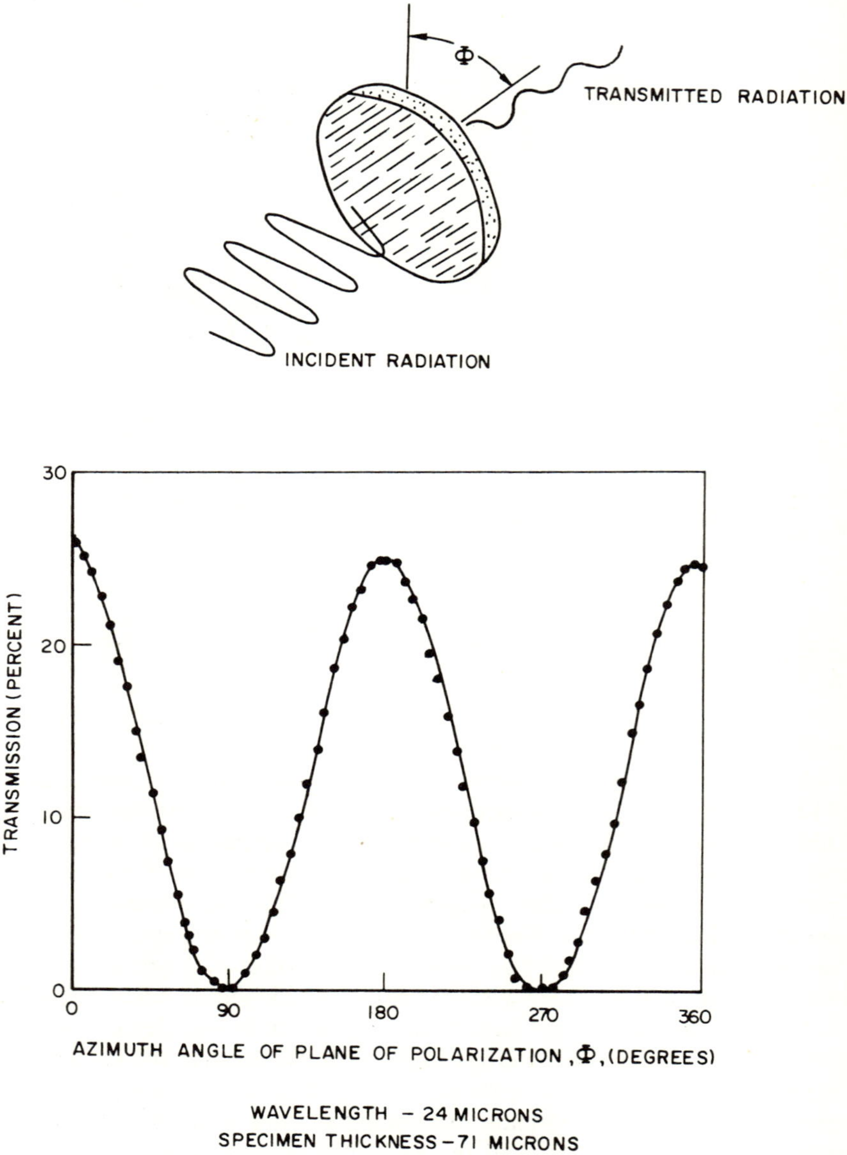

Figure 10.40 Polarization of infrared radiation by InSb-NiSb eutectic (after Paul, Weiss, and Wilhelm [102]).

TABLE 10.6 MAGNETIC PROPERTIES OF EUTECTICS

Material	Fiber Content (%)	Average Fiber Diameter (cm)	Coercivity (oersteads)	Saturation Magnetization (gauss)	Reference
Fe-FeS	9	2×10^{-3}	10.5	1800	107
Fe-FeS	9	1.5	10.3	2011	108
Fe-Fe_xSb	15	1.5	12.5	3381	108
Fe-Fe_xSb	15	1.0	17.9	3243	108

It should be noted that the coercivity was considerably increased when the iron fiber size in the Fe-Fe_xSb systems was reduced by more rapid unidirectional solidification. This was thought to be due to the fact that the fiber size was approaching the size expected for single domains. Moreover, the saturation magnetization was found to be proportional to the fiber content.

In addition to these applications, aligned eutectics are also being considered for optical devices and an electronic image dissector [99]. Studies have also been made of superconductivity [109, 110] with the intention of preparing hard superconductors by pinning vortices at the interfaces in a very fine microstructure. The results have been disappointing thus far, probably because the structures produced are not fine enough.

A possible application that combines both the unique structural and nonstructural aspects of composites is for strong electrical conductors. Alloying or working aluminum, copper, and silver to increase their strength results in significant reduction in conductivity so that a compromise between strength and conductivity must be made. Because it is possible to reinforce an unalloyed, unworked metal matrix with strong fibers, both high strength and high conductivity can be combined in one composite material. The resistivity and strength of Al-Al_3Ni and several Al alloys [111] is seen in Figure 10.41. If one uses a figure of merit based upon the product of tensile strength and conductivity, the Al-Al_3Ni eutectic composite is superior to all aluminum alloys. Composites of this type may find use in transmission wires and switch contacts where an optimum combination of strength and electrical conductivity is required.

A unique porous material with precisely sized and spaced pores has been produced by controlled solidification of monotectic alloys [112]. The monotectic reaction is analogous to the eutectic reaction except that upon cooling, the liquid transforms to a solid plus another liquid, instead of two solids. By unidirectionally cooling such a system, a structure consisting of aligned liquid "fibers" in a solid matrix can be produced, and the liquid blown out to form channels. Since the size and spacing of the liquid "fibers"

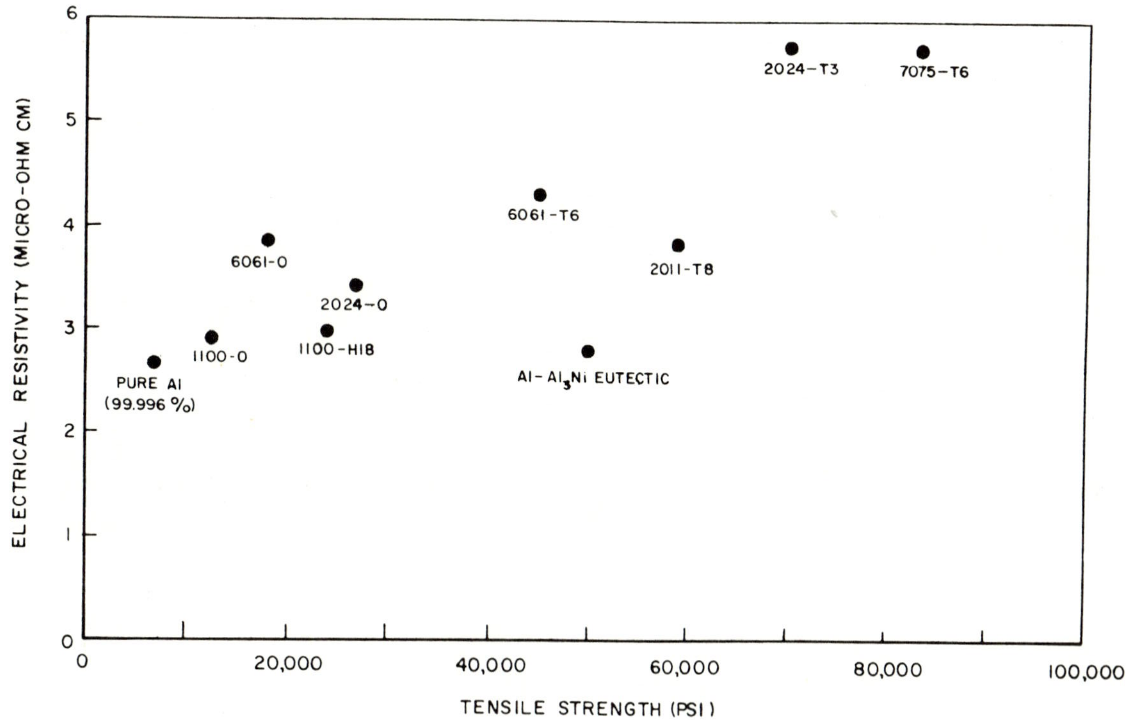

Figure 10.41 Strength and resistivity of aluminum alloys.

can be controlled by adjusting the solidification rate [25, 26], a device with uniformly spaced pores of controlled size can be produced.

OUTLOOK FOR EUTECTIC COMPOSITES

The fact that a finished composite material can be produced in a simple one-step operation is the greatest advantage that unidirectional solidification of eutectic alloys has over other methods of composite fabrication. The greatest drawback is the fact that one cannot arbitrarily choose the matrix and reinforcement separately or the amount of reinforcement; but instead must live with the eutectics which occur in nature. Fortunately, the eutectic reaction is one of the most common and a majority of alloy systems contain at least one eutectic reaction. There are thousands of known eutectics and thousands more as yet undiscovered. This restriction affects the method which one uses to choose a eutectic composite for a given application. One must choose the matrix material suitable for the application (based upon melting

Figure 10.42 Unidirectionally solidified Ni-NiMo turbine blade configuration (after Salkind and Lemkey [87]).

point, density, toughness, corrosion resistance, etc.) and then seek eutectics of this base material which have sufficient reinforcement of any kind compatible with the application.

Because eutectic composites exhibit good high-temperature strength coupled with good microstructural stability, there is a significant effort presently being made to develop high-temperature eutectics suitable for such applications as jet engine turbine parts. Since the high-strength properties are obtained in the cast material, it seems entirely feasible to unidirectionally solidify parts as close as possible to their final shape. The results of early experimentation in this area is seen in Figure 10.42. The turbine blade was made from the Ni-NiMo eutectic in a mold of that shape by unidirectional solidification parallel to the long axis of the airfoil. In addition, it also appears feasible to cold or hot work and diffusion bond eutectic composites [38, 57, 61, 113].

The economics of the unidirectional eutectic solidification process is yet to be thoroughly explored; however, present experience indicates that it may be quite reasonable. An important economic tradeoff that must be made when this process is put into production is that between melt purity and solidification rate. Purer starting materials allow faster solidification without liquid-solid interface instability; however, the purer starting materials are more expensive. This tradeoff will probably be different for each alloy system and production facility.

ACKNOWLEDGMENTS

The authors are indebted to the following staff members of the United Aircraft Research Laboratories: W. Tice for electron microscopy and diffraction, W. Hermann for metallography, M. Atwell for chemical analysis, and R. Fanti, Dr. E. Breinan, Dr. F. Galasso, Dr. K. Kreider, and Dr. E. Thompson for reviewing the manuscript. The authors would also like to thank R. North of the Pratt and Whitney Aircraft Division of the United Aircraft Corporation for his assistance in fatigue testing and ballistic impact testing. Some of the work described herein has been sponsored by the Naval Air Systems Command, Office of Naval Research, The Army Materials and Mechanics Research Center, and the Air force Materials Laboratory. In addition, permission of the United Aircraft Research Laboratories for the preparation and publication of this chapter is gratefully acknowledged.

REFERENCES

[1] G. Cole and W. Winegard, *J. Inst. Metals*, **93** (1964–65), 153–164.
[2] H. Kerr and W. Winegard, *J. Metals* (May 1966), 563–570.
[3] F. Guthrie, *Phil. Mag., Ser.* 5, **17** (1884), 462.

[4] F. Brady, *J. Inst. Metals*, **28** (1922), 369.
[5] W. Rosenhain and P. Tucker, *Trans. Roy. Soc.* (*London*), **209A** (1909), 89.
[6] R. Vogel, *Z. Anorg. Allgem. Chem.*, **76** (1912), 425.
[7] G. Tammann, *A Textbook of Metallography*, Chem. Cat. Co., New York, 1925.
[8] A. M. Portevin, *J. Inst. Metals*, **29** (1923), 239.
[9] W. Pfann, *Trans. AIME*, **194** (1952), 747.
[10] J. Hunt and K. Jackson, *Trans. AIME*, **236** (1966), 843–852.
[11] W. Winegard, S. Majka, B. Thall, and B. Chalmers, *Can. J. Chem.*, **29** (1951), 320.
[12] R. W. Kraft, U. S. Patent 3,124,452, Mar. 10, 1964.
[13] R. W. Kraft and D. Albright, *Trans. AIME*, **221** (1961), 95.
[14] W. Webb and W. Forgeng, *Acta Met.*, **6** (1958), 462.
[15] F. Hargreaves, *J. Inst. Metals*, **37** (1927), 106.
[16] H. Weart and D. Mack, *Trans. AIME*, **212** (1958), 664–670.
[17] W. Straumanis and N. Brakss, *Z. Physik. Chem.*, **30B** (1935), 117.
[18] R. W. Kraft, *Trans. AIME*, **221** (1961), 704.
[19] R. W. Kraft, *Trans. AIME*, **224** (1962), p. 65.
[20] R. W. Kraft and D. Albright, *Trans. AIME*, **224** (1962), 1176.
[21] R. W. Kraft, *Trans. AIME*, **227** (1963), 393.
[22] D. Cooksey, M. Day, and A. Hellawell, *Crystal Growth*, Supplement to *J. Phys. Chem. Solids*, Pergamon Press, 1967, pp. 151–159.
[23] W. Tice, W. Lasko, and F. Lemkey, *50 years of Progress in Metallographic Techniques*, ASTM STP No. 430, Philadelphia, 1968, 239.
[24] L. Hogan, R. W. Kraft, and F. D. Lemkey, "Crystallography of Controlled Eutectics," to be published in *Advances in Materials Research*, **5,** H. Herman, Ed. John Wiley and Sons.
[25] W. A. Tiller in *Liquid Metals and Solidification*, American Soc. for Metals, Cleveland, 1958, p. 276.
[26] C. Zener, *Trans. AIME*, **167** (1946), 550.
[27] J. Rutter and B. Chalmers, *Can. J. Phys.*, **31** (1953), 15.
[28] W. Tiller, K. Jackson, J. Rutter, and B. Chalmers, *Acta Met.*, **1** (1953), 428.
[29] A. Yue, *Trans. AIME*, **224** (1962), 1010.
[30] A. Yue, *Trans. AIME*, **227** (1963), 64.
[31] A. Yue, *J. Inst. Metals*, **92** (1964), 248.
[32] A. Yue, *Trans. AIME*, **230** (1964), 39.
[33] G. Chadwick, *J. Inst. Metals*, **91** (1962–63), 298.
[34] J. Hunt and J. Chilton, *ibid.*, p. 338.
[35] J. Hunt, *ibid.*, **94** (1966), 125–129.
[36] D. Cooksey, D. Munson, M. Wilkinson, and A. Hellawell, *Phil. Mag.*, **10** (1964), 745.
[37] K. Jackson and J. Hunt, *Trans. AIME*, **236** (1966), 1129–1142.
[38] F. Lemkey, B. Bayles and M. Salkind, Final Report Contract DA-19-020-AMC-00434(X), July 31, 1965.
[39] F. Lemkey and M. Salkind, *Crystal Growth*, Suppl. to *J. Phys. Chem. Solids*, Pergamon Press, 1967, pp. 171–177.
[40] H. E. Cline, *Trans. AIME*, **239** (1967), 1906.
[41] W. Winegard, *Metals Rev.*, **34,** 6 (1961), 57.
[42] G. Chadwick, *Prog. Mat. Sci.*, **12** (1963), 97.
[43] R. W. Kraft, *J. Metals*, **8** (1966), 192.
[44] A. G. H. Dietz, "Design Theory of Reinforced Plastics," Ch. 9 in *Fiberglass Reinforced Plastics*, by R. H. Sonneborn, Reinhold, New York, 1954.
[45] R. Jech, D. McDanels, and J. Weeton, *Proc. Sixth Sagamore Ordnance Conf.*, August 1959, p. 116.

[46] D. McDanels, R. Jech, and J. Weeton, *Met. Progr.*, **8** (December 1960), 118.
[47] D. McDanels, R. Jech, and J. Weeton, *Trans. AIME*, **233** (1965), 636–642.
[48] A. Kelly and G. Davis, *Met. Rev.*, **10,** 37 (1965), 1–77.
[49] J. Ford, R. Hertzberg, and F. Lemkey, "Analytical and Experimental Investigations of the Fracture Mechanisms of Controlled Polyphase Alloys," United Aircraft Research Laboratories Report B910068-4, Final Report, Contract N600(19)59361, Nov. 1963.
[50] F. D. Lemkey and R. W. Kraft, *Rev. Sci. Instr.*, **33** (1962), 846.
[51] R. W. Hertzberg and R. W. Kraft, *Trans. AIME*, **227** (1963), 580.
[52] R. W. Hertzberg, *Trans. ASM*, **57** (1964), 434.
[53] F. D. Lemkey, R. W. Hertzberg, and J. A. Ford, *Trans. AIME*, **233** (1965), 334–341.
[54] R. W. Hertzberg, F. D. Lemkey, and J. A. Ford, *Trans. AIME*, **233** (1965), 342–354.
[55] M. J. Salkind, B. J. Bayles, F. D. George, and W. K. Tice, "Investigation of Fracture Mechanisms, Thermal Stability, and Hot-Strength Properties of Controlled Polyphase Alloys," United Aircraft Research Laboratories Report D910239-4, Final Report Contract NOw 64-0433d, April 1965.
[56] M. J. Salkind, F. D. George, F. D. Lemkey, and B. J. Bayles, "Investigation of the Creep, Fatigue, and Transverse Properties of Al_3Ni Whisker and $CuAl_2$ Platelet Reinforced Aluminum," United Aircraft Research Laboratories Report E910344-4, Final Report Contract NOw 65-0384d, May 1966.
[57] M. J. Salkind, F. D. George, and W. K. Tice, *Trans. AIME*, **245** (1969), 2339.
[58] B. J. Bayles, J. A. Ford, and M. J. Salkind, *Trans. AIME*, **239** (1967), 844–849.
[59] M. J. Salkind, G. Leverant, and F. George, *J. Inst. Metals*, **95** (1967).
[60] M. Salkind, F. D. George, F. Lemkey, B. J. Bayles, and J. Ford, *Chem. Eng. Progr.*, **62** (March 1966), 52–60.
[61] M. Salkind, F. Lemkey, F. D. George, and B. J. Bayles, Proc. 10th SAMPE Symposium, Symposium, Western Periodicals, No. Hollywood, Calif., 1966.
[62] M. Salkind and F. D. George, *Trans. AIME*, **242** (1968), 1237.
[63] F. D. George, J. Ford, and M. J. Salkind, Metal Matrix Composites, STP 438, ASTM, Phila. (1968), 59.
[64] S. Brenner, in *Growth and Perfection of Crystals* edited by Doremus, Roberts, and Turnbull, John Wiley and Sons, New York, 1958.
[65] S. Brenner in *Fiber Composite Materials*, ASM, Metals Park, 1965.
[66] E. Ellison and B. Harris, *Appl. Materials Res.* (January 1966), 33–40.
[67] E. Breinan and M. Salkind, *Trans. AIME*, **242** (1968), 2281.
[68] H. Voorhees and J. Freedman, ASTM STP 219, Philadelphia, 1960.
[69] D. Guy, Jr., Alcoa's Aluminum Powder Metallurgy (APM) Alloys, datasheet, March 1959.
[70] M. Salkind and F. George, "Fatigue and Bonding of Al_3Ni Whisker Reinforced Aluminum," United Aircraft Research Laboratories Report F910583-1, First Progress Report Contract N00019-68-C-0016, October 1967.
[71] J. Tavernelli and L. Coffin, *Trans. ASM*, **51** (1959), 438–453.
[72] S. Manson, NASA Tech-Note 2955, 1963.
[73] G. Sachs, W. Gerberich, V. Weiss, and J. Latorre, *Proc. ASTM*, **60** (1960), 512.
[74] R. Kraft, D. Albright, and J. Ford, *Trans. AIME*, **227** (1963), 540–2.
[75] L. Graham and R. Kraft, *Trans. AIME*, **236** (1966), 94–102.
[76] F. George and M. Salkind, private communication.
[77] J. Embury, N. Petch, A. Wraith, and E. Wright, *Trans. AIME*, **239** (1967), 114–118.
[78] S. Shapiro and J. Ford, *Trans. AIME*, **236** (1966), 536–542.
[79] F. Schmidt and H. Ogden, DMIC Report No. 189, Sept. 13, 1963.
[80] F. Schmidt and H. Ogden, DMIC Report No. 188, 1963.
[81] E. Thompson, "The Structure and Properties of the Ni-NiMo Eutectic Prepared by

Unidirectional Solidification," Proceedings of the Twelfth Meeting of the Refractory Composites Working Group, AFML TR 67–228 (1967), 214.

[82] Anon., "Nickel Base Alloys," International Nickel Co., revised 1964.

[83] A. Seybolt, E. Stover, and J. Westbrook, "Investigation of the Basic Parameters Affecting the Properties of Intermetallic Compounds," Air Force Report WADC-TR-60-184, Part VI, June 1965.

[84] H. Bates, F. Wald, and M. Weinstein, *J. Materials Science*, **4** (1969), 25.

[85] F. Lemkey, Master's Thesis, University of Connecticut, 1967.

[86] L. Willey in *Aluminum*, **1**, edited by K. R. Van Horn, ASM Metal Park, p. 377.

[87] M. Salkind and F. Lemkey, *Int. Sci. and Tech.* (March 1967), 52.

[88] M. Hansen, *Constitution of Binary Alloys*, McGraw-Hill, New York, 1958.

[89] D. M. Marsh, in *Fracture of Solids*, John Wiley and Sons, New York, 1964, p. 119.

[90] H. W. Rauch, W. H. Sutton, and L. R. McCreight, *Survey of Ceramic Fibers and Fibrous Composite Materials*. AFML TR 66–365 (1966).

[91] G. J. Davies, *Phil. Mag.*, **9** (1964), 953.

[92] R. Bechmann and R. F. S. Hearmon (Eds.), *Landolt-Bornstein Numerical Data and Functional Relationships in Science and Technology*, Vol. 1, "Elastic, Piezoelectric, Piezooptic and Electrooptic Constants of Crystals," Springer-Verlag, Berlin, 1966.

[98] S. S. Brenner, *J. Appl. Phys.*, **27** (1956), 1484.

[94] S. T. Wlodek, W. W. Webb, and W. D. Forgeng (abstract) *J. Metals*, **11** (1959), 590.

[95] A. J. Bradley and A. Taylor, *Phil. Mag.*, **23** (1937), 1049–67.

[96] F. D. Lemkey and M. Salkind, *Summary of the Eleventh Refractory Composites Working Group Meeting*, AFML TR 66–179, 1966, p. 1027–1086.

[97] A. Bowman, Los Alamos Scientific Laboratory, Los Alamos, New Mexico, private communication.

[98] E. Rudy and C. E. Barukl, *J. Am. Ceram. Soc.*, **50** (1967), 265–268.

[99] F. Galasso, *J. Metals*, **19** (June 1967), 17–21.

[100] W. Liebmann and E. Miller, *J. Appl. Phys.*, **34** (1963), 2643.

[101] H. Weiss and W. Wilhelm, *Z. Physik*, **176** (1963), 399–408.

[102] B. Paul, H. Weiss and W. Wilhelm, *Solid-State Electronics*, **7** (1964), 835–841.

[103] H. Wagini and H. Weiss, *Solid-State Electronics*, **8** (1965), 241–254.

[104] A. Muller and W. Wilhelm, *J. Phys. Chem. Solids*, **26** (1965), 2021–2028.

[105] H. Weiss, *Solid State Electronics*, **9** (1966), 443–451.

[106] A. Muller and W. Wilhelm, *J. Phys. Chem. Solids*, **26** (1965), 2029–2035.

[107] D. Albright, PhD Thesis, Lehigh University, 1965.

[108] F. Galasso, F. Douglas, W. Darby, and J. Batt, *J. Appl. Phys.*, **38** (1967), 3241.

[109] H. Cline, R. Rose, and J. Wulff, *J. Appl. Phys.*, **34** (1963), 1771–1774.

[110] S. Levy, Y. Kim and R. Kraft, *J. Appl. Phys*, **37** (1966), 3659–3665.

[111] *Metals Handbook*, 8th Ed., Vol. 1, ASM, Ohio, 1961.

[112] F. Lemkey and J. Ford (abstract), *J. Metals*, **17**, 1 (January 1955), 91.

[113] M. Salkind, *Trans. AIME*, **242** (1968), 2518.

D. M. Schuster and E. Scala

11 Properties of Whisker Composites

INTRODUCTION AND BACKGROUND

The requirements of modern technology, with its ever-increasing demands on materials, have given great impetus to the study of reinforced fibrous composites. Indeed, it would appear that some aerospace and high-temperature nuclear reactor requirements may only be satisfied by utilizing the exceptional properties of materials as they occur in filamentary or fibrous form. The extraordinary properties of whiskers, with yield strengths in excess of 1 and 2 million psi [1–3] are particularly enticing.

Whiskers, by virtue of their methods of fabrication, are generally available in relatively short lengths, of the order of $\frac{1}{2}$ in. or less. This introduces the problem of translating the properties of whiskers into useful engineering structures. Furthermore, the goal is to combine the two or more components of a composite material in such a way that the strong reinforcing phase provides mechanical strength while the matrix serves to bind the fibers together and resists oxidation and corrosive attack.

Load transfer to the strong fibers and the inhibition of crack propagation are the primary means by which composites achieve their enhanced properties. The transfer of load from the matrix to the fibers is accomplished by a shear mechanism at the fiber/matrix interface.

There are several factors that control the effectiveness with which load is transferred to the fibers. These parameters ultimately govern the properties of the composite and may be outlined as:

1. Length to diameter or aspect ratio, l/d [4–6],
2. Fiber volume fraction, V_f [4, 5, 7–9],

3. Ratio of the elastic moduli, E_f/E_m [10–13],
4. Average fiber strength and the statistics of the strength distribution [14–16],
5. Fiber orientation with respect to applied load [17–19],
6. Fiber dimensions and shape [19–21],
7. Chemistry and/or metallurgy of the fiber/matrix interface [22–26].

The references given for each of the parameters listed above is by no means exhaustive of the research that has been carried out in each category. Separation of the programs being conducted on composites into the realms of individual variables is as difficult as the separation of the effects of a single parameter in any one given experiment. In fact, the mutual interdependence of all the factors listed makes the interpretation of experimental results extremely difficult. Micks [27] has emphasized that the analysis of experimental data is further confounded by the interaction of four effects:

1. The possibility that the fabrication process has changed the properties of the reinforcing agent.
2. The fact that the fiber may exhibit a different strength in the matrix from that measured for the fiber alone.
3. The addition of fibers to a matrix material may change the inherent behavior of the matrix.
4. The fabrication process may not yield the expected conditions, that is, pores or discontinuous bonding may result.

Although there has been great technological interest in the area of fiber reinforcement in recent years, there are still a great many fabrication problems to be solved [28]. The bonding of whiskers into a matrix without damaging them and the achievement of suitable alignment is certainly no simple matter. It is for this reason, and because whisker crystals are still relatively expensive, that data on the engineering properties of whisker composites are still quite limited. However, there has been much research performed utilizing short fibers cut from continuous filamentary material. The results of investigations using short, discontinuous fibers are directly related to the whisker-reinforced materials and allow us to proceed with a discussion of the properties of discontinuously reinforced composites. In fact, if the whiskers have a high enough length to diameter ratio, their behavior approximates that of continuous filaments.

Before proceeding with a discussion of the properties of whisker composites, it is desirable to briefly review the importance of certain fiber-matrix interactions. The mechanics of whisker strengthening has already been treated in some detail in earlier chapters. The purpose here, therefore, is simply to emphasize the importance of certain ideas and parameters and provide a basis upon which the strength, toughness, fatigue, and elevated-temperature properties may be evaluated.

FIBER-MATRIX INTERACTION

As load is applied to a composite, both the whiskers and matrix initially deform elastically. However, assuming that the fiber elastic modulus, E_f, is greater than the elastic modulus of the matrix, E_m, the whiskers restrain elongation in a local region of the matrix. The effect is shown schematically in Figure 11.1 for a high modulus whisker aligned in the load direction. The elastic strain in the matrix is nonuniform with stress concentrations developing near the ends of the whisker and a reinforced region occurring along the central portion. The nonuniformity of the stress distribution decreases with increasing distance from the fiber. These effects have been observed experimentally [19, 20, 29], and have also been the subject of analytical investigations [10, 30, 31].

The normal or tensile stress in the fiber is a result of the shear stresses at the fiber/matrix interface. The implication here is that to exploit the high-strength properties of whiskers, a matrix capable of chemically or mechanically bonding to the fibers and transmitting load to them is a primary requirement. This statement is correct when discussing strength properties; however, we will see that there are some limitations when fracture toughness and fatigue are the properties under consideration.

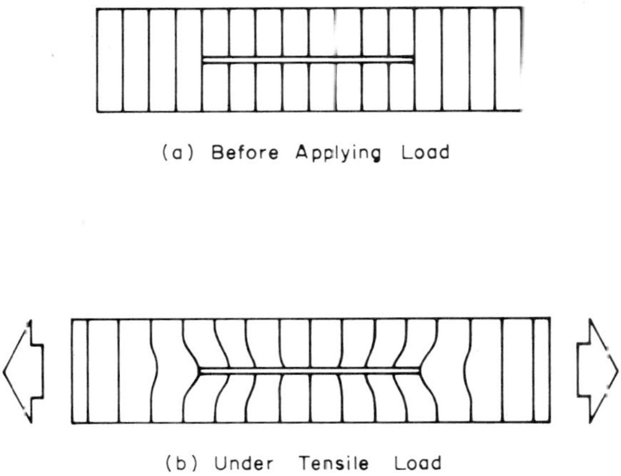

Figure 11.1 Schematic representation of the deformation occurring in the matrix of a fiber-reinforced composite under elastic loading conditions.

LENGTH TO DIAMETER OR ASPECT RATIO

The interfacial shear stresses, as discussed above, arise because of the different moduli of the fiber and matrix, and are the mechanism by which load is distributed between the two components of the composite. The normal or tensile stress in the fiber, σ_f, is applied to the fiber by the shear stress, τ. For static equilibrium in the axial direction

$$\frac{d\sigma_f}{dZ} = \frac{4\tau}{d} \tag{1}$$

where Z is distance along the whisker, and d is the whisker diameter. Since τ is not constant, σ_f increases in a nonlinear fashion. Several authors [10, 31, 33], with the aid of simplifying assumptions, have obtained solutions for the variation of σ_f with τ.

As the composite is loaded elastically, the shear stress, τ, at the fiber/matrix interface attains a maximum at either end of the whisker. The normal stress in the fiber, σ_f, is initially zero at the ends, builds up to a maximum value, and then remains constant along the length of the fiber. The normal and shear stresses increase with increasing length to diameter ratio, for a given applied load, until some critical elastic aspect ratio, $(l/d)_{ce}$, is achieved. These stress variations are represented schematically in Figure 11.2, with $\sigma_{f\infty}$ indicating the normal fiber stress that an infinite or continuous fiber would sustain under similar elastic loading conditions.

Upon further application of load, yielding in shear takes place at the fiber/matrix interface, limiting further increase in τ. With the shear stress at the

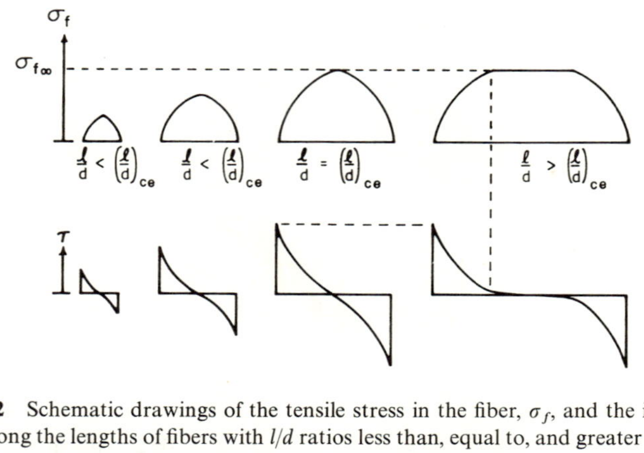

Figure 11.2 Schematic drawings of the tensile stress in the fiber, σ_f, and the interfacial shear stress, τ, along the lengths of fibers with l/d ratios less than, equal to, and greater than the critical elastic aspect ratio, $(l/d)_{ce}$. Drawn for a given applied load σ.

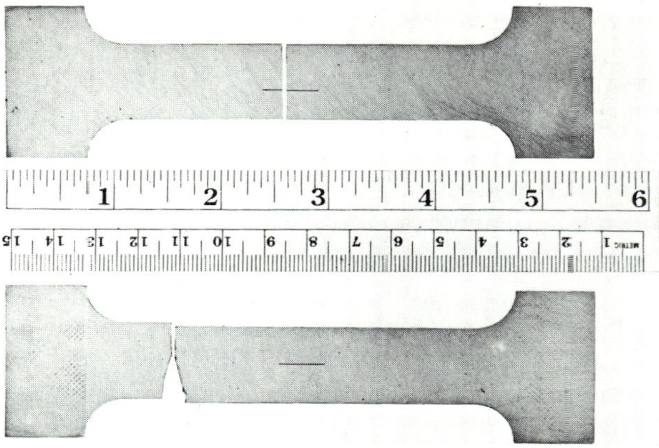

Figure 11.3 Specimens tested to failure. Top sample contains a fiber of supercritical plastic aspect ratio and bottom sample contains a subcritical aspect ratio fiber.

interface now limited to the shear strength of the matrix, τ_y, Eq. 1 is easily integrated yielding [4]

$$\sigma_f = \frac{4\tau_y}{d} Z \qquad (2)$$

Substitution of the fiber-breaking stress, σ_{fb}, for the normal stress, σ_f, in Equation 2, yields a critical aspect ratio, $(l/d)_c$, necessary to fracture the fiber at its mid point, $z = l/2$, within the matrix:

$$\left(\frac{l}{d}\right)_c = \frac{\sigma_{fb}}{2\tau_y} \qquad (3)$$

τ_y of Eqs. 2 and 3 may be interpreted as either the yield strength of the matrix in shear or the fiber/matrix interfacial shear strength. Whichever quantity is less governs the critical length to diameter ratio.

Experiments utilizing boron fibers, which had been cut into short discontinuous lengths and incorporated into an epoxy resin matrix, indicate that the limiting parameter in resin matrix systems is interfacial shear strength [32]. The critical aspect ratio necessary to load the fibers to their ultimate strength was found to be about 100. Figure 11.3 shows two fractured resin matrix specimens. The top specimen contains a supercritical aspect ratio ($l/d > 100$) boron fiber and the bottom specimen a subcritical ($l/d < 100$) fiber. As can be seen, the supercritical fiber strained with the matrix and was loaded to its ultimate strength, breaking the fiber and fracturing catastrophically through the resin. The shorter than critical length fiber was observed to debond from the matrix and became ineffective.

The specimen then failed at random along the test length. By varying the length of fiber from specimen to specimen, it was possible to determine the critical aspect ratio as 100. Inserting 100 for $(l/d)_c$ in Eq. 3 and using 3.5×10^5 psi for σ_{fb}, a value of 1750 psi was obtained for the interfacial shear strength in the boron-epoxy resin. This value is in the acceptable range for interfacial shear strengths in resin systems [34, 35]. The interfacial shear strengths in resin systems are approximately an order of magnitude less than the shear strengths of metals. Therefore, since it is the shear strength which determines $(l/d)_c$ in Eq. 3 for metals, one would expect the critical aspect ratios in metal systems to be about an order of magnitude less than in resins. This does indeed turn out to be the case [6, 39], with critical length to diameter ratios of the order of ten required in metal matrices at room temperature. At elevated temperatures, the matrix shear strength decreases and higher aspect ratios are required.

FIBER VOLUME FRACTION

Rule of Mixtures

As noted in Chapters 7 and 9, the tensile strength, σ_c, of a uniaxially aligned continuous fiber composite is approximately given in terms of the ultimate tensile strength of the fibers, σ_{fb}, and the relative volume fractions of matrix, V_m, and fiber, V_f, as [7]

$$\sigma_c = \sigma_{fb} V_f + \sigma'_m V_m \tag{4}$$

where σ'_m is the matrix stress at the breaking strain of the fibers. Since $V_m + V_f = 1$,

$$\sigma_c = \sigma_{fb} V_f + \sigma'_m (1 - V_f) \tag{5}$$

The limitations of this expression have been discussed in Chapter 7. Its validity has been demonstrated experimentally [4, 5, 7–9], provided V_f exceeds some minimum value, $V_{\min}$. A minimum volume fraction arises because the ultimate tensile strength of a ductile matrix, σ_{mu}, cannot be less than σ'_m, the stress in the matrix at the time of fiber failure. Therefore, σ_c must always be equal to or greater than $\sigma_{mu}(1 - V_f)$ and by substituting $\sigma_{mu}(1 - V_f)$ for σ_c in Eq. 5, we obtain an expression for $V_{\min}$, namely,

$$V_{\min} = \frac{\sigma_{mu} - \sigma'_m}{\sigma_{fb} + \sigma_{mu} - \sigma'_m} \tag{6}$$

$V_{\min}$ is the fiber volume fraction which must be exceeded for Eq. 5 to hold, but it is not the critical volume fraction necessary for matrix strengthening to be observed. In order for fiber reinforcement to be worthwhile, it is necessary that the strength of the composite, σ_c, be in excess of the strength of the pure matrix, σ_{mu}, and not $\sigma_{mu}(1 - V_f)$ which defines the range of validity of Eq. 5. Therefore, by substituting σ_{mu} for σ_c in Eq. 5, we obtain the expression

$$V_{\text{crit}} = \frac{\sigma_{mu} - \sigma'_m}{\sigma_{fb} - \sigma'_m} \tag{7}$$

for the critical fiber volume fraction necessary to achieve reinforcement by the fibers.

From a comparison of Eqs. 6 and 7, it becomes apparent that V_{crit} is always greater than V_{min}. It should also be noted that if a composite is to have enhanced strength over the unreinforced matrix that V_{crit} must be exceeded and not just V_{min}. It has been shown [36] that V_{crit} is quite large in some systems and can be of the order of 20% or greater.

A corresponding rule of mixtures relation which takes into account the variations of τ and the effects of length for discontinuous fibers or whiskers has been presented in Chapters 7 and 9 [36]. It indicates that when the actual whisker length, l, is greatly in excess of the critical length, l_c, composite strength approaches that of the composite reinforced with continuous filaments (Eq. 5). It should be mentioned that at high fiber volume fractions, where fiber spacing is small, deformation of both the matrix and fibers is restricted. Such effects cause the creation of complex triaxial stresses that give rise to ultimate composite strengths in excess of these predicted by the rule of mixtures [37].

Fiber Spacing and Overlap

The preceding discussion has taken into account the normal stress distribution in the fibers to arrive at an expression for the ultimate strength of a whisker-reinforced composite. However, it has not considered the stress concentrating effects that arise in the matrix when discontinuous fibers or whiskers are brought into close proximity of one another. Nonuniform distributions of whisker spacing and overlap, caused by the composite fabrication process, may create serious stress concentrations which can exert a controlling influence on the properties of the whisker composite. These effects have been studied photoelastically in a discontinuous boron fiber-epoxy resin system [32]. Two boron fibers were accurately positioned in each specimen so that the stress concentrations could be determined as a function of fiber spacing and overlap. In order to control the effects of aspect ratio in these experiments, supercritical length to diameter ratio boron fibers ($l/d \simeq 110$) were used. In this system $(l/d)_c$ is 100. Regions of high stress concentration were observed for close spacings and 100% overlap, that is, fibers lying parallel and adjacent to one another (Figs. 11.4 and 11.5). For this case, and the zero percent overlap situation, the maximum

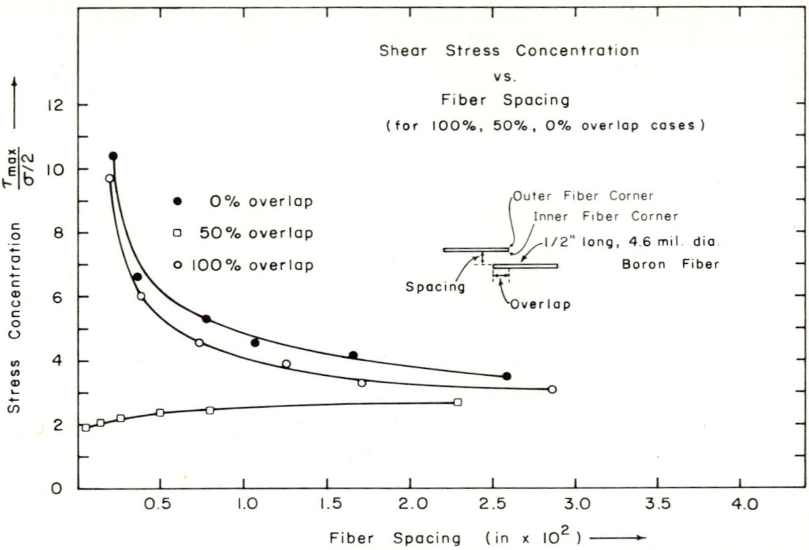

Figure 11.4 Interfacial shear stress concentration versus fiber spacing for the 100%, 50%, and 0% overlap cases; 0.0046 in. diameter, 0.5 in. long boron fibers aligned in the load direction; specimen thicknesses 0.01 in.

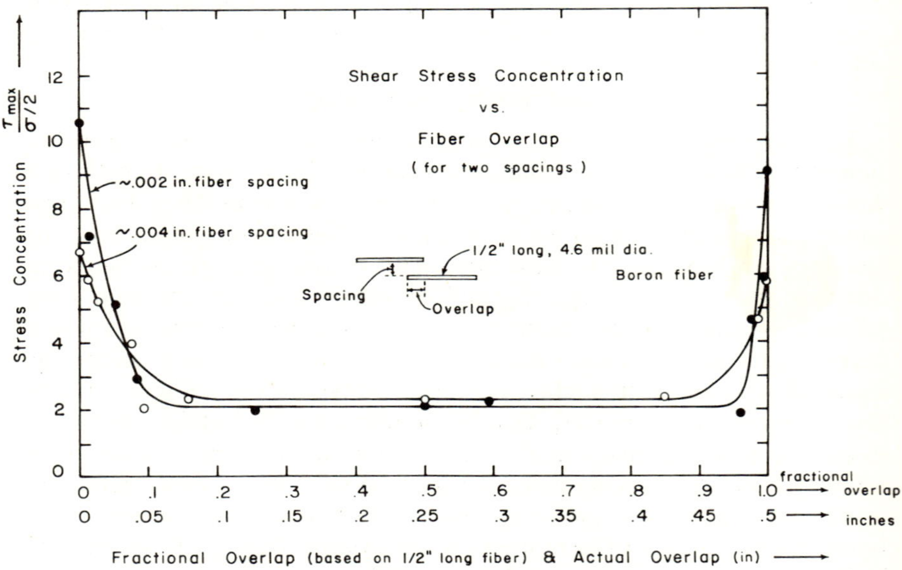

Figure 11.5 Interfacial shear stress concentration versus fiber overlap for spacings of ≈0.002 and ≈0.004 in.; 0.0046 in. diameter, 0.5 in. long boron fibers aligned in the load direction; specimen thicknesses 0.01 in.

shear stress concentrations were substantially greater than those reported for a single fiber [19]. It should be noted that at intermediate overlaps and close spacing, the stress concentrations were reduced.

The variations of maximum shear stress concentration with fiber spacing and overlap are presented in Figures 11.4 and 11.5, respectively. The data were recorded at applied loads of the order of 500 psi. The relatively low applied stresses were necessary so that yielding and fiber debonding did not occur as a result of the stress magnifications. It may be observed from Figures 11.4 and 11.5 that the stress concentrations occurring in the matrix are sensitive functions of fiber spacing and overlap, and stress concentrations in excess of ten were observed. Upon loading to higher stresses (approximately 2500 psi), yielding and fiber debonding occurred in the regions of high stress concentration. Thus, when irregularities in the fabrication process cause fibers to be brought into close proximity to one another, or even touching, serious stress concentrations can result. In a ductile matrix, these will be the regions where yielding and fiber debonding are initiated. In more brittle systems, these will be the areas where cracks are initiated and propagated. Depending on the mechanical properties of the fiber and matrix, the conditions just described may strongly influence the properties of the whisker composite, especially where fatigue strength and fracture toughness are of interest.

TENSILE PROPERTIES

Data on the engineering properties of whisker-reinforced composites are rather limited at the time of this writing. This is due to the fact that whiskers are still relatively expensive (although costs have been decreasing steadily with increased production [38]), and because handling techniques and methods for the incorporation of whiskers into suitable engineering materials are still in the process of development. Furthermore, current and interesting information is sometimes withheld from the engineering community for proprietary reasons.

However, a great deal of research has been performed on model whisker composites and in systems where continuous filaments have been cut into short lengths to approximate the behavior of whiskers. These studies have served to illuminate the mechanisms active in whisker-reinforced composites and to pinpoint potential problem areas. It has also been shown that under certain conditions, notably for high aspect ratio, the behavior of whisker composites may be fairly well represented by continuous filamentary composites, where a great wealth of information exists. An attempt has been made to draw illustrative examples from the research on whisker and other discontinuously reinforced fiber composites. When no such examples are available, and where applicable, the data from continuous composite studies are presented.

Stress-Strain Behavior

It has been observed in the tungsten wire-copper matrix system that the stress-strain curves for both continuous and discontinuous reinforcement are quite similar [4, 8, 39]. That this is so, when the discontinuous fibers are aligned in the load direction and possess a sufficiently high aspect ratio, is shown in Figure 11.6. Here the influence of increasing fiber content on the stress-strain behavior for several continuously reinforced specimens is shown with the results for a 32.4 vol % discontinuously reinforced composite. Deformation, in the case of a highly deformable matrix, usually proceeds in four stages [8]: (1) elastic deformation of both fibers and matrix, (2) fibers continue to deform elastically and matrix begins plastic deformation, (3) both phases deform plastically, and finally (4) the fibers fracture with subsequent failure of the composite. In the fourth stage, fracture of the fibers immediately causes failure of the composite when the fiber volume fraction is in excess of the critical volume fraction as discussed in the section on the rule of mixtures.

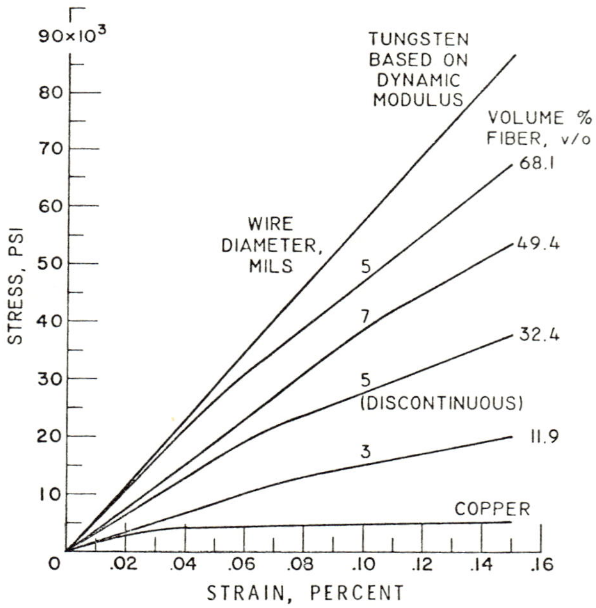

Figure 11.6 Stress-strain curves (low strain region) for tungsten, copper, and tungsten-reinforced copper composites [8].

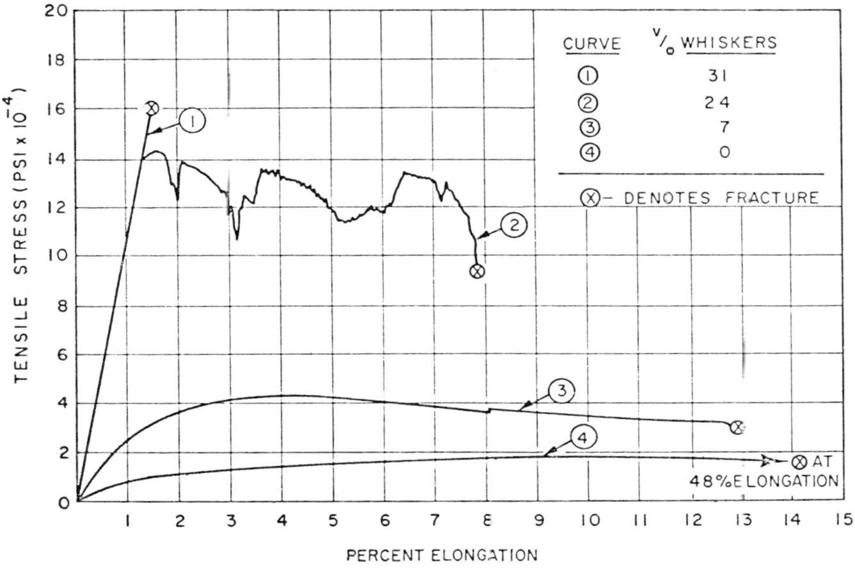

Figure 11.7 Stress versus elongation for silver and composites of silver and sapphire whiskers [40].

The third stage (plastic deformation of both fibers and matrix) is not observed in the case of brittle fibers. Sutton and Chorné [40] have reported the stress-elongation curves for pure silver and for silver reinforced with several different volume fractions of sapphire whiskers (Fig. 11.7). In this case there is no plastic deformation of the fibers. The effect on the stress-strain behavior of progressive failure of the whiskers is most apparent for the 24 vol% curve shown in Figure 11.7. Here the serrated appearance is due to successive fiber failures, the peaks corresponding to fracture of whiskers and then subsequent redistribution of load to other whiskers.

During the first stage of deformation (both phases deforming elastically), McDanels, Jech, and Weeton [8] found that the modulus, E_c, of an aligned tungsten wire-reinforced copper composite could be approximately represented by a relation similar to the rule of mixtures, that is,

$$E_c = E_f V_f + E_m V_m \tag{8}$$

where E_f and E_m are the moduli of the fibers and matrix, respectively. However, since the stress in a discontinuous fiber builds up from either end (see Fig. 11.2), short whiskers contribute something less than $E_f V_f$ to the modulus of the composite. This effect should be negligible if long enough fibers are used, that is, if $l/d \gg (l/d)_c$.

The influence of length to diameter ratio on the stress-strain behavior of a discontinuously reinforced copper-tungsten wire composite has been observed by Kelly and Tyson [41]. Figure 11.8 shows the effect of increasing aspect ratio, at the same fiber volume fraction (0.53), on the stress-strain curve. It may be observed that for increasing length to diameter ratio, the stress-strain curves approach the behavior of a continuously reinforced composite ($l/d = \infty$).

Second stage behavior (fibers elastic, matrix plastic) is an extremely important part of the stress-strain curve. It is in this region that the matrix transmits increasingly larger portions of the applied load to the fibers or whiskers. As the matrix begins to deform plastically, the modulus of the matrix, E_m, is effectively reduced. This tends to increase the fiber-to-matrix modulus ratio, E_f/E_m, thus increasing the fraction of the total load being absorbed by the fibers. McDanels, Jech, and Weeton [8] have presented a typical stress-strain curve emphasizing the effect of elastic fiber, plastic matrix conditions (Fig. 11.9). In this region the modulus of the composite, E_c, may be represented as

$$E_c = E_f V_f + \left(\frac{\sigma'_m}{\epsilon}\right) V_f \tag{9}$$

where (σ'_m/ϵ) is the slope of the matrix stress-strain curve at a given strain.

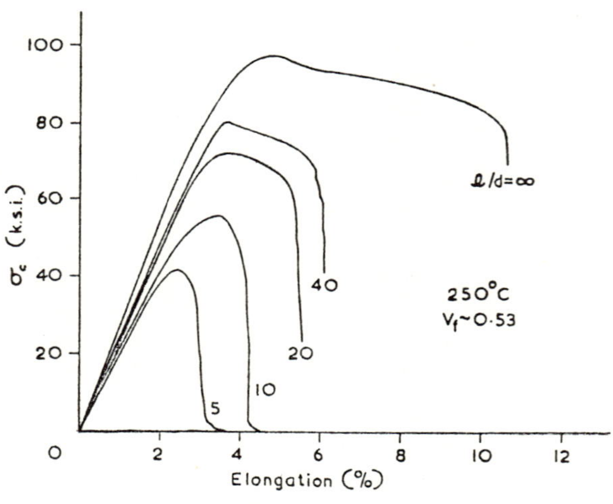

Figure 11.8 The effect of aspect ratio on the stress versus elongation curves for copper reinforced with tungsten wires at 250°C, approximately 53% by volume tungsten [41].

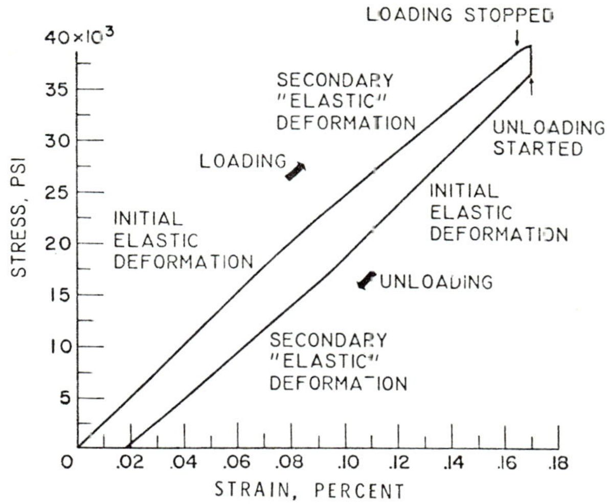

Figure 11.9 Load-unload cycle for a copper-tungsten wire composite; 18.7% by volume of 5 mil diameter tungsten wire [8].

For metal matrices that do not work-harden, or have very low work-hardening coefficients, the composite modulus is fairly well approximated by

$$E_c \simeq E_f V_f \tag{10}$$

Baker and Cratchley [42] have reported similar loading and unloading curves in the aluminum-silica fiber system. They also observed that the first complete load-unload cycle loop remained open, but upon subsequent cycling to the same load, the loop closed, usually within one or two additional cycles. They termed the material "stabilized" once the loop had been closed, but noted that cycling to a higher stress tended to reopen the loop. The situation after the first cycle (open loop, Fig. 11.9) arises since the fibers are left in tension and the matrix in compression when the applied load is removed. The hysteresis loops obtained by Mehan and Jakas [43] (Fig. 11.10) for 20 vol% aluminum oxide whiskers in an aluminum matrix are also due to the same effect. The primary and secondary elastic moduli are clearly shown.

Thus far we have only considered the stress-strain behavior of uniaxially aligned fiber or whisker composites. This may prove to be a rather difficult condition to satisfy with commercial fabrication techniques. Parratt [44, 45] has incorporated 15 vol% silicon nitride whiskers into a silver matrix by hot pressing. The Si_3N_4 whiskers were randomly oriented in a plane causing approximately a one-third dilution of their influence in that plane. Figure 11.11 shows that significant strengthening has occurred even though only a relatively low and random fiber content had been employed. Koppenaal and Parikh [46] investigated the microstrain behavior of a variety of discontinuous metal fibers in silver matrices. They found that the strength of the composite increased with the decreasing fiber diameter since this also served to reduce the interfiber spacing. The stress required to initiate plastic deformation was not dependent on fiber content, but as deformation progressed the fiber-fiber spacing became important. This indicates a mechanism of fiber reinforcement to which we have not alluded until now. In addition to load transfer to the fibers, dispersion strengthening and higher rates of work-hardening of the matrix can result when very fine whiskers are closely spaced in a matrix. This possibility is even more enticing when we realize that the strongest whiskers are those possessing the smallest diameters

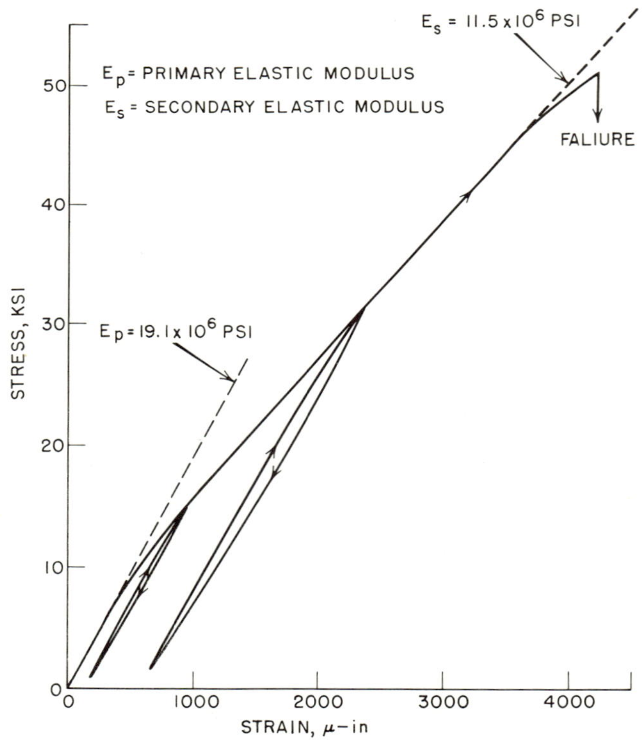

Figure 11.10 Stress-strain curve obtained at room temperature with a strain-gauged 20 vol% Al/Al_2O_3 specimen. The strain rate was 0.002 in./in./min.

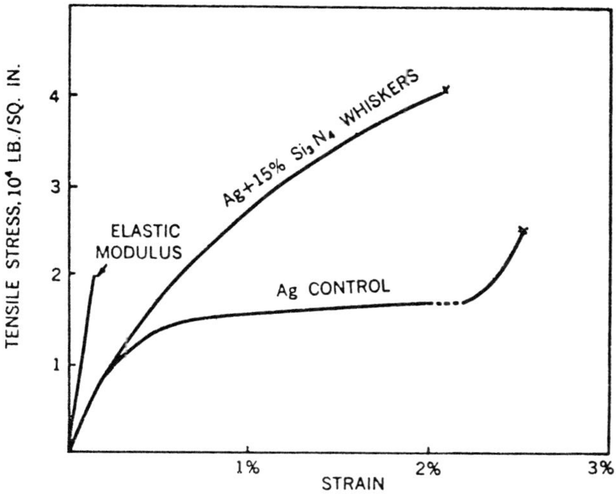

Figure 11.11 Stress-strain curves for silver and silver reinforced with 15% by volume Si_3N_4 whiskers at 20°C [43].

(whisker strength increases with decreasing diameter [1]). In this case, the matrix would be reinforced by dispersion-strengthening effects as well as load transfer to the strongest possible whiskers.

One method of whisker composite fabrication which circumvents many of the inherent fabrication problems is that of unidirectional solidification of eutectic alloys. In this method, the whiskers are produced *in situ*, thus eliminating the problems of whisker alignment, fiber/matrix bonding, and the difficulties of handling and sorting whiskers. Unidirectional solidification is the subject of an earlier chapter [10], but it is worth noting here that significant advances have been made with this approach [47–49]. Of the many alloys investigated [49], the stress-strain behavior for the Ta-Ta_2C system [48] is presented in Figure 11.12. The increased strength, produced by unidirectional solidification yielding "needlelike" whiskers, is apparent from the figure. Here the composite consists of 29 vol% Ta_2C whiskers in a tantalum matrix. A strength of 155,000 psi was achieved in this system. The decreased ability of the "bladelike" whiskers to reinforce the matrix was attributed to their imperfection and short length relative to the "needlelike" whiskers.

Ultimate Strength

The ultimate tensile strength of a uniaxially aligned whisker-reinforced composite is given to a fair degree of accuracy by the rule of mixtures

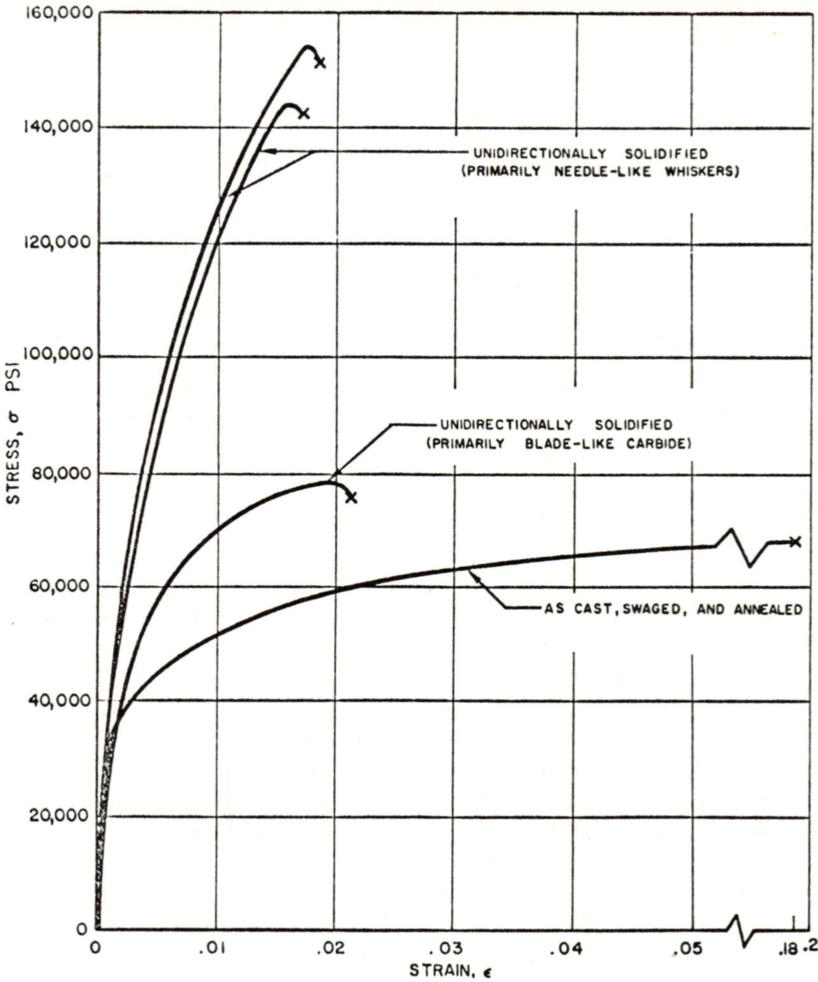

Figure 11.12 Stress-strain curves for unidirectionally solidified Ta-Ta_2C eutectic alloys [47].

relation,

$$\sigma_c = \sigma_f V_f + \sigma'_m(1 - V_f) \tag{5}$$

provided:

1. The average strength (or strength distribution) of the whiskers is known (Chapter 7).
2. The minimum volume fraction, as discussed in the rule of mixtures section is exceeded.

3. The dispersion of whiskers in the matrix is free of nonuniformities resulting from the fabrication process which would give rise to stress-concentrating effects as discussed in the section on fiber spacing and overlap.
4. The whiskers possess sufficient length or aspect ratio [$l/d \gg (l/d)_c$, as discussed earlier] (Chapters 7 and 9).
5. High-strength, continuously bonded fiber-matrix interfaces exist.

The importance of good continuous bonding has been demonstrated in the silver-sapphire whisker system [50] (Fig. 8.10) and later by studies in the aluminum-sapphire whisker system [51]. The composite tensile strength is plotted as a function of volume fraction Al_2O_3 whiskers for the two cases of a poorly bonded and a well-bonded whisker/matrix interface. Fibers of much greater aspect ratio would be required to permit sufficient load transfer to raise the poorly bonded case up to the well-bonded situation. Attempts to achieve favorable bonding conditions have recently been treated [52], and that subject along with other aspects of whisker composite development and fabrication are dealt with in an earlier chapter.

As would be expected from the discussion of stress-strain behavior, the ultimate tensile strength is dependent on fiber aspect ratio. Cratchley [5] has studied this effect in stainless steel fiber-reinforced aluminum prepared by hot pressing. Figure 11.13 shows the composite ultimate strength as a function of volume percent fiber for several different aspect ratios. It is significant that increasing the aspect ratio ultimately produces smaller and smaller increases in the ultimate strength. Kelly and Tyson [4] have predicted that once the aspect ratio is ten times that of the critical aspect ratio, discontinuous reinforcement yields 95% of the strength produced by continuous reinforcement (Fig. 11.14). It is worth noting that the critical aspect ratio for metal [41] matrices is of order 10 and for resin [32] matrices of order 100. Whiskers are available [38, 56] possessing aspect ratios several times greater than these values. The problem is then one of bonding the whiskers into the matrix while preserving their length-to-diameter ratio through the fabrication process. Hahn and co-workers [53] have recently reported progress in the preparation and working of whisker-reinforced metal composites. Results have been achieved by hot pressing aluminum and sapphire whiskers to produce a 60% increase in tensile strength and a 70% increase in modulus. At elevated temperatures they report that a 10%-by-volume addition of sapphire whiskers to aluminum produced a composite which is stronger than the strongest wrought aluminum alloy.

Sutton and Chorné [6] have presented striking results in silver-sapphire whisker composites (Fig. 11.15). The added effect of using finer whiskers, which are stronger, is shown in Figure 11.15. It is worthy of note that a specimen containing 24% whiskers by volume (fine whiskers) produced a

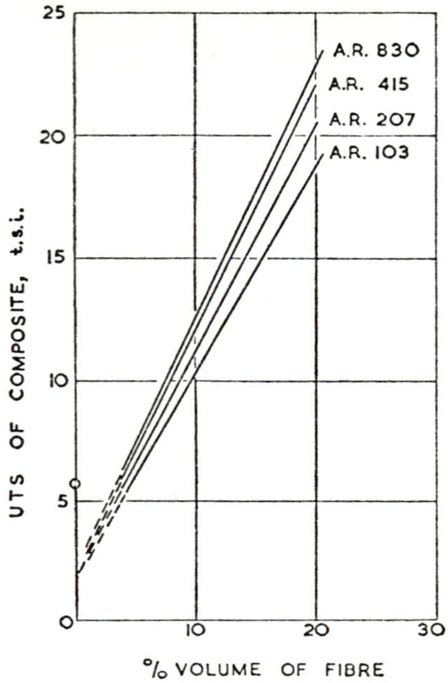

Figure 11.13 The effect of aspect ratio on the UTS of aluminum reinforced with 0.002 in. diameter stainless steel fibers [5].

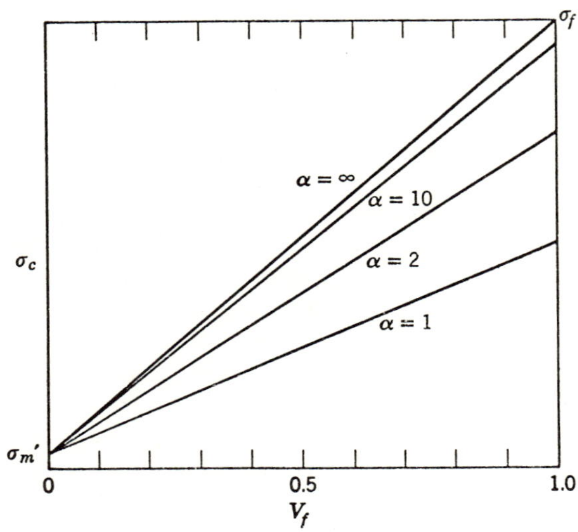

Figure 11.14 Predicted composite strength dependence on volume fraction, discontinuous fibers with $\alpha = l/l_c$ [4].

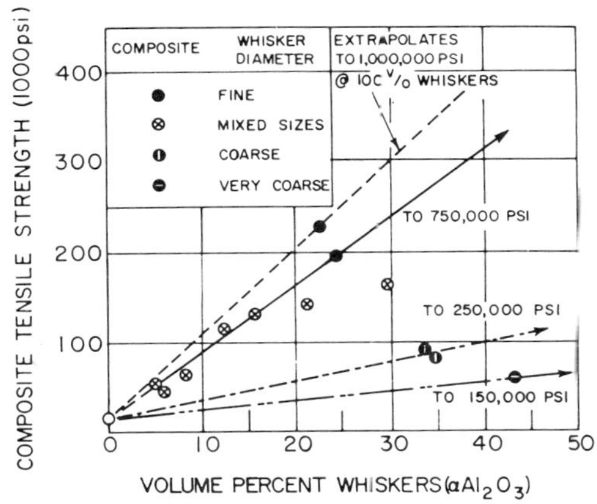

Figure 11.15 Tensile strength versus volume percent for silver-reinforced sapphire whisker composites [6].

tensile strength of 232,000 psi, which is twice that of other methods used to increase the strength of silver.

Results for nichrome reinforced with sapphire whiskers were reported by Raynes and Kelsey [54]. Their results followed a linear rule of mixtures relation (Fig. 11.16), and demonstrated strengths superior to nichrome wire produced under similar conditions. Kelsey [55] has also reported on the strengthening of iron with sapphire whiskers where significant reinforcement was observed.

Most of the preceding discussion has concerned itself with metal matrix composites. Whiskers can also produce pronounced effects on the strength and modulus of resin matrices [56, 57]. Indeed, if it were not for the current high cost and production limitations, whiskers would probably replace fiber glass in many applications. Sutton, Rosen, and Flom [58] have demonstrated the potential improvements of whisker reinforcement in an epoxy (PJ-122) resin matrix. An addition of 14% by volume of aligned sapphire whiskers to the epoxy, yielded a strength of 113,000 psi compared with 73,100 psi for a 49% addition of aligned quartz fibers in the same matrix.

The mechanism by which load is transferred to the reinforcing phase in a discontinuous fiber-reinforced polymer composite has been discussed by Kelly [59]. The essential point here is that load is transferred from matrix to fiber by frictional forces acting at the interface. The relation between the critical aspect ratio, $(l/d)_c$, and the breaking strength of the fiber, σ_{fb}, is similar to Eq. 3,

$$\left(\frac{l}{d}\right)_c = \frac{\sigma_{fb}}{\mu n} \tag{12}$$

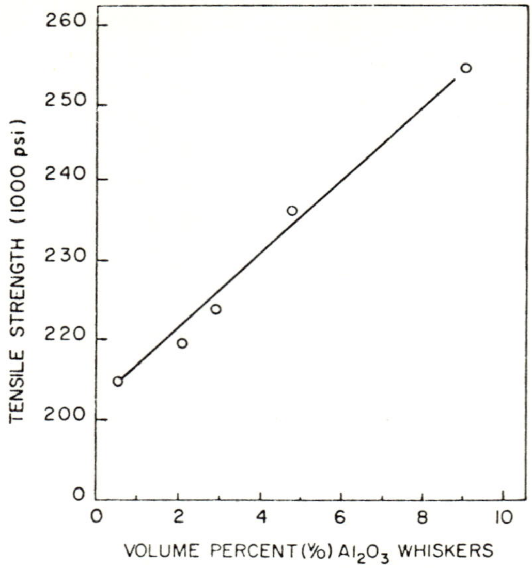

Figure 11.16 Tensile strength versus volume percent fiber for nichrome reinforced with sapphire whiskers [52].

Here, τ_y, the shear strength of the matrix has been replaced by μn, where μ is the coefficient of friction and n is the radial pressure on the fiber due to shrinkage of the resin. Since μn is about an order of magnitude less than the shear strength of metals, we would expect critical aspect ratios in resin systems to be an order of magnitude greater than in metal matrices. Although greater aspect ratios are required in resin matrix composites, such whisker aspect ratios are attainable [38].

ELEVATED TEMPERATURE PROPERTIES

Of the various uses proposed for whisker-reinforced materials, there is greatest hope for improvements in high-temperature applications. As most metals rapidly lose strength with increasing temperature, reinforcement with high-strength refractory whiskers (Al_2O_3, SiC, B_4C, Si_3N_4, C) appears quite promising.

Most investigations reported to date have been concerned with high-temperature tensile strength, with relatively few studies presenting data on the creep or stress rupture of whisker composites. Sutton and Chorné [6] have reported on the short time tensile strength of silver reinforced with 20 to

40 vol% sapphire whiskers as a function of temperature (Fig. 11.17). The advantages of increased length-to-diameter ratio are apparent from Figure 11.17, and are as would be expected from the discussion of room-temperature properties. Figure 11.17 also indicates that significant strengthening was achieved at temperatures as high as 98% of the melting point of the silver matrix. Parratt [44], using powder methods to incorporate a random array of silicon nitride whiskers in a silver matrix, has also reported substantial increases in hot strength of the composites. The tensile strength versus temperature of $\frac{1}{4}$-in. diameter composite rods of 20 vol% sapphire whiskers in aluminum is shown in Figure 11.18 [60]. These data are for specimens prepared with Fe or NiFeCr coatings on the whiskers prior to infiltration with molten aluminum. Again, it may be seen that these composites retain a large percentage of their room-temperature strength.

The elevated-temperature behavior of whisker composites prepared by unidirectional solidification has been studied by workers at United Aircraft Research Laboratories [49, 61]. Results of their work, on aluminum reinforced with Al_3Ni whiskers and $CuAl_2$ blades [49], are shown in Figure 11.19. The results are compared to a precipitation hardened aluminum alloy (7075-T6) and the superior behavior of the whisker composites at elevated temperatures may be observed from the figure. These whisker composites, produced by unidirectional solidification, also display surprising chemical and microstructural stability at elevated temperatures as indicated in Chapter 10. This stability may be attributed to their method of fabrication which provides nearly equilibrium conditions and produces lower energy interfaces.

It is necessary to be wary of chemical or metallurgical interactions between fibers and matrix when interpreting the results of long-time stress rupture or

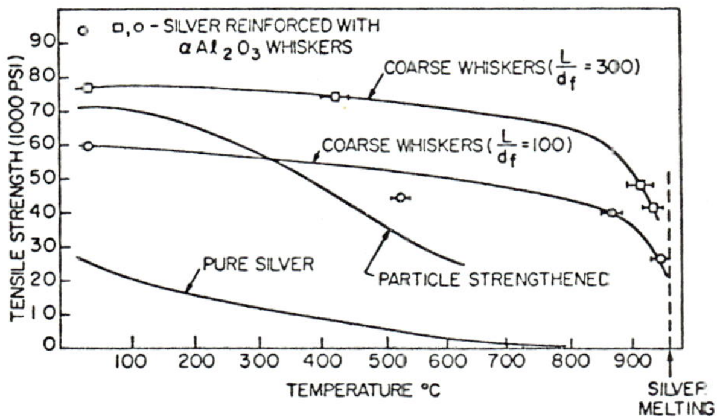

Figure 11.17 Tensile strength versus temperature for silver and silver reinforced with Al_2O_3 particles and with Al_2O_3 whiskers [6].

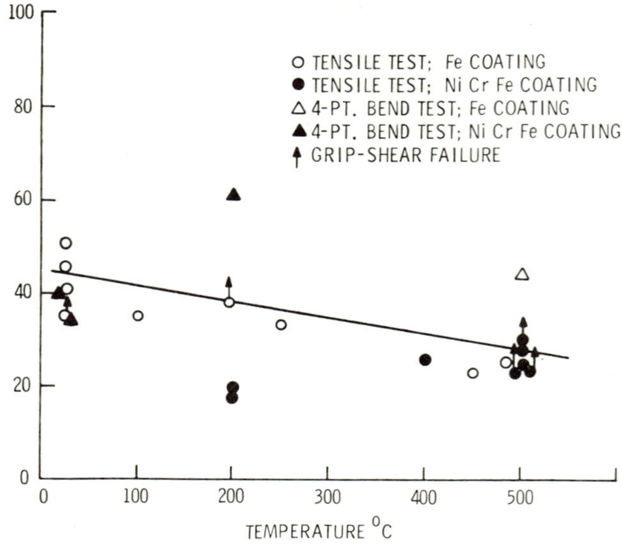

Figure 11.18 Strength of $\frac{1}{4}$-in. diameter 20 vol% Al/Al_2O_3 composite specimen versus temperature.

creep tests at elevated temperatures. Petrasek [24] has investigated the effects of alloying elements in the copper-tungsten fiber system and correlated the elevated-temperature tensile properties of the composite with the depth of penetration of alloying elements into the tungsten fibers. In another NASA study [62] of the copper-tungsten fiber system, discontinuous fiber orientation to the load direction had large effects on tensile strength and mode of failure at elevated temperatures. Namely, for misalignments of as little as 3°, shear failures occurred at 1500°F with much lower tensile strengths than were observed for composites containing well-aligned fibers. In addition, elevated-temperature composite tensile strength was found to increase with increasing aspect ratio, as had been found in the silver-sapphire whisker-reinforced system [6].

The necessity for increased length to diameter ratios at elevated temperatures arises because of the decreased shear strength of the matrix. This may be most easily understood from a consideration of Eq. 3:

$$\left(\frac{l}{d}\right)_c = \frac{\sigma_{fb}}{2\tau_y} \tag{3}$$

Here the critical aspect ratio is shown to have an inverse dependence on the shear strength of the matrix. As τ_y decreases, the critical aspect ratio increases and higher l/d ratio fibers are required to achieve effective reinforcement. Therefore, at any one given value of aspect ratio and volume percent fiber, the high-temperature strength of the composite is governed by the shear strength of the matrix. Petrasek, et al. [62] have shown this to be true by varying the shear strength of a copper matrix reinforced with discontinuous tungsten fibers through chromium alloying additions to the copper. The chromium additions were found to enhance the high-temperature properties of the composite by increasing the shear strength of the matrix even though it reacted with the tungsten fibers.

Kelly and Tyson [41] have studied reinforcement of a copper matrix with discontinuous tungsten fibers and discontinuous molybdenum fibers. Their investigations extended up to 900°C and they observed that the composite strength increased linearly with fiber volume fraction over the temperature range investigated. The ultimate strength of the composite, σ_c, was dependent on the properties of the fiber and matrix at the temperature under consideration and varied with fiber volume fraction, V_f, as

$$\sigma_c = V_f \sigma_{fb}\left(1 - \frac{l_c}{2l}\right) + (1 - V_f)\sigma'_m \tag{13}$$

for $l \geqslant l_c$, and

$$\sigma_c = V_f \tau_y \frac{l}{d} + (1 - V_f)\sigma_{mu} \tag{14}$$

for $l \leqslant l_c$, where σ_{mu} is the ultimate strength of the matrix.

We have discussed the short-time tensile strength of discontinuously reinforced composites and found that their behavior at elevated temperatures is dependent on aspect ratio and volume fraction in much the same way as at

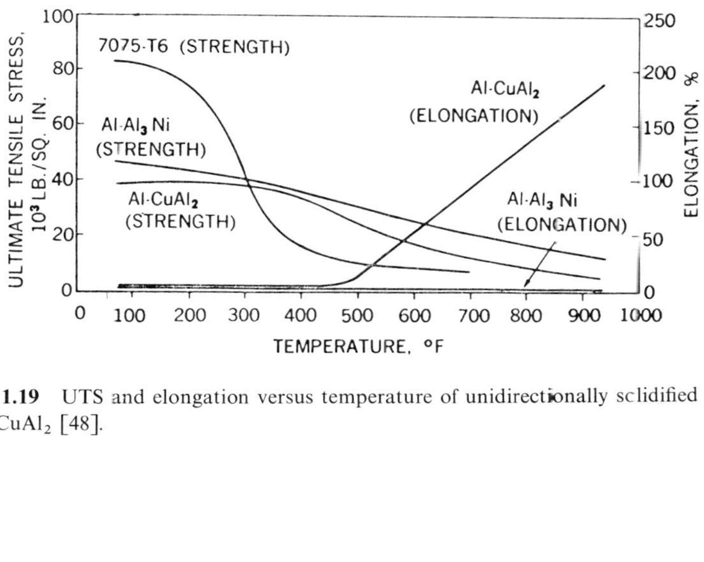

Figure 11.19 UTS and elongation versus temperature of unidirectionally solidified Al-Al$_3$Ni and Al-CuAl$_2$ [48].

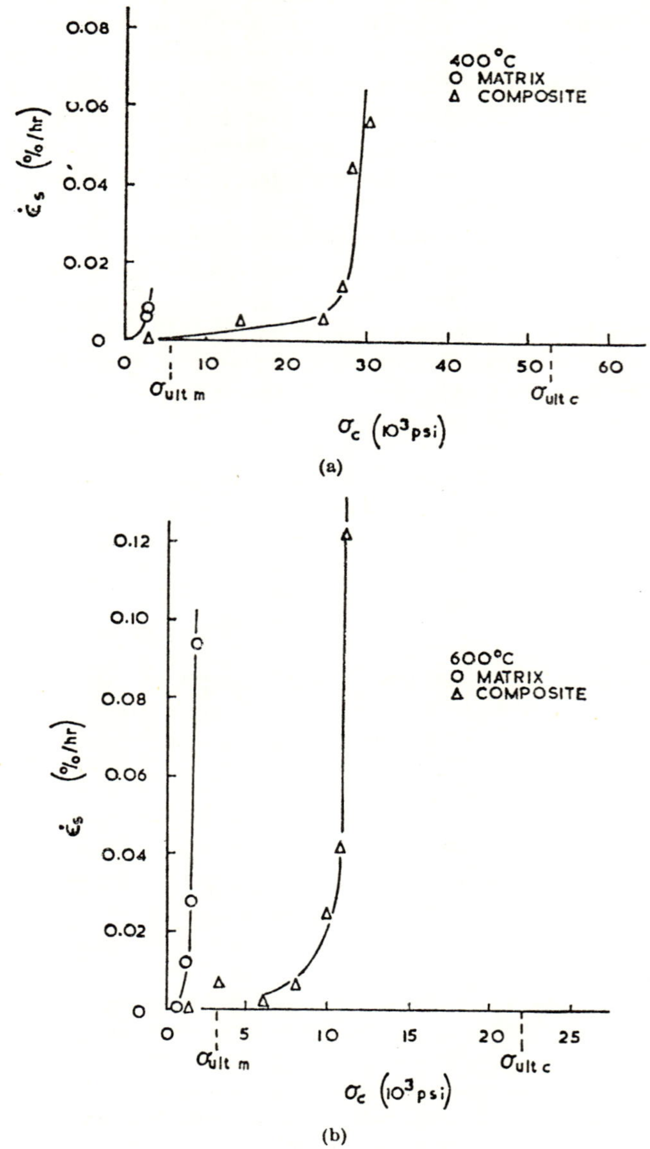

Figure 11.20 Minimum creep rate versus stress for a silver-tungsten composite as compared to the silver matrix alone; (a) 400°C, (b) 600°C [63].

room temperature. Long time properties, such as creep and stress rupture of composites, have also been studied [61, 63–69]. Alloys of titanium reinforced with discontinuous molybdenum fibers have shown increases in stress-rupture life at elevated temperatures [68]. Nickel alloys reinforced with continuous tungsten fibers have also shown longer lives in stress-rupture tests [63, 64]. Investigators at United Aircraft [61, 69] have examined the creep behavior in Al_3Ni whisker and $CuAl_2$ platelet reinforced aluminum prepared by unidirectional solidification (Chapter 10). The whisker composite exhibited a small minimum creep rate with a rapid tertiary creep stage and small amounts of rupture ductility. Creep behavior was further observed to be dependent on the variables of solidification, with creep resistance being greater for alloys solidified at higher rates.

Ellison and Harris [65] reported longer stress rupture life at 1200°F for specimens of Inconel 600 reinforced with small volume percentages (7%) of short 0.003-in. diameter tungsten wires. They also found that discontinuous fibers improved creep properties of the Inconel 600 matrix, but there were chemical interactions between the fibers and the matrix after long times at 1800°F. Kelly and Tyson [66] studied creep in the model silver-tungsten system in an attempt to reduce the effects of chemical interaction. They found the typical creep curve at 500°C for silver reinforced with about 40% by volume discontinuous, aligned tungsten fibers ($l/d = 30$) to possess three regions. There was an initial transient region, followed by a stage of constant creep rate, with a tertiary stage of accelerating creep leading rapidly to failure. The critical aspect ratio in the silver-tungsten system was found to be 22.5 at 400°C and 67 at 600°C. Therefore, measurements of the minimum creep rate for discontinuously reinforced composites of aspect ratio 30, at both of the above temperatures of 400 and 600°C, gave indications of the effect of aspect ratio above and below the critical ratio (Fig. 11.20). From Figure 11.20, it is observed that the composite creep rate is significantly reduced by fibers of both less than and greater than the critical aspect ratio. It was further concluded [66] that the creep of a discontinuously reinforced fiber composite was governed by the rate of creep in shear of the matrix in immediate proximity to the fibers. This is the region of the matrix which is subjected to the highest shear stresses [19, 20, 29], and on this basis, the creep strength of the composite should increase with increasing fiber content, increasing aspect ratio, and increases in the shear strength of the matrix.

Creep rupture data for 20 vol% sapphire whiskers in aluminum at 500°C is shown in Figure 11.21 [60]. This specimen withstood, without failure, 1338 hrs at 500°C at stresses between 12,000 and 14,500 psi. The creep behavior for this specimen (Fig. 11.22) is characterized by very low strain. Metallographic analysis revealed intact whiskers, with no edge dissolution, well distributed in the aluminum matrix [60].

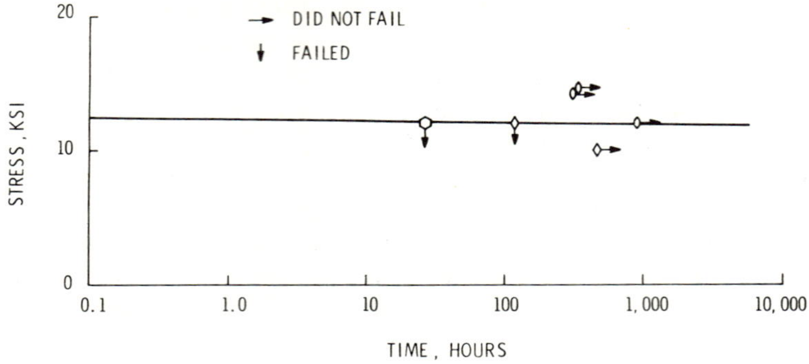

Figure 11.21 Creep-rupture data of 20 vol% Al_2O_3/Al at 500°C [60].

In summary, provided whiskers can be kept from reacting with the matrix at elevated temperatures [70], it appears that substantially improved high-temperature materials can be attained through the use of discontinuous fiber or whisker reinforcement. Both long- and short-time properties will be dependent on fiber volume fraction, with increases in strength occurring at high whisker contents. Short-time tensile strength will depend more strongly on the strength of the fibers than will the long-time properties. On the other hand, creep and stress rupture will be more dependent on fiber aspect ratio and the shear strength of the matrix at the temperature in question.

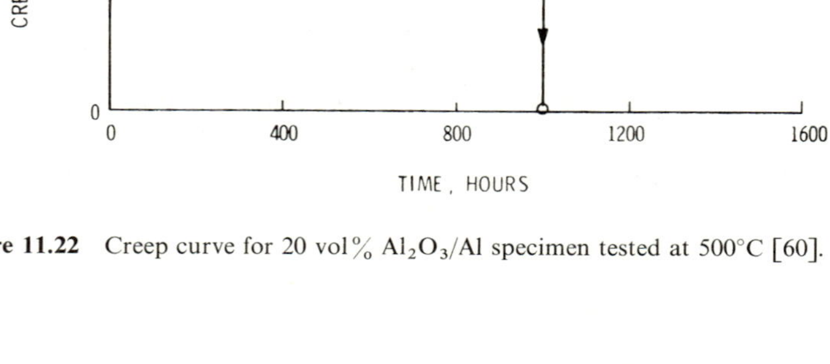

Figure 11.22 Creep curve for 20 vol% Al_2O_3/Al specimen tested at 500°C [60].

FATIGUE

Fatigue remains one of the leading causes of service failure in modern materials applications and so it is natural to inquire into the fatigue properties of whisker-reinforced composites. The fatigue behavior of continuously reinforced materials has only recently come under investigation [71–77], and there have been even fewer studies concerned with fatigue under the conditions of discontinuous or whisker reinforcement [61, 78–79].

Fatigue failure is brought about by the application of repeated loads below the ultimate tensile strength of the material. The oscillating loads cause repeated excursions into the range of plastic behavior with the creation of fatigue cracks which propagate to cause failure. Forsyth, George, and Ryder [71], from their studies on aluminum alloys reinforced with steel and with tungsten wires, indicated that the fibers tended to decrease the rate of fatigue crack growth, rather than hinder the initiation of cracks. Specimens of up to 13% by volume wire were prepared by hot rolling wire mesh with alternate aluminum alloy sheets and possessed a central slot perpendicular to the stress direction. The composites were stressed in tension-tension and fiber volume fractions of as low as 2% significantly reduced the rate of crack propagation. Although only a few tests were carried out, the indications were that there was little advantage of continuous over discontinuous fibers on the basis of reduced rate of crack growth. Fiber strength and volume fraction were more important from this point of view.

Williams and O'Brien [72] reported improved fatigue life for aluminum alloys reinforced by continuous steel fibers where the specimens were prepared by casting aluminum around the high strength wires. Aluminum composites containing silica fiber reinforcement have been fabricated by precoating the fibers and hot pressing [73, 76, 78]. Baker and Cratchley [73] in their study of such a continuous silica-aluminum system, under reversed bending, found that the regions of stress concentration where fibers had failed during composite fabrication generated cracks in the matrix. These cracks were not able to pass into adjacent undamaged fibers, but rather, propagated parallel to the fiber until encountering a crack in adjacent fibers. Baker [78], in a later study on the same silica-aluminum system, but which included discontinuous fibers, observed fatigue damage to occur at fiber ends because of the high shear stress concentrations existing at such locations (see the section on fiber/matrix interaction). Cracks were initiated at such points which then acted as a path for easy fatigue crack propagation. For the fiber volume fractions investigated (≈ 0.30) and for low plastic strains, the cracks could not penetrate undamaged fibers and fatigue life was determined by matrix damage and the properties of the fiber/matrix interface. At higher plastic strains, the fibers were unable to hinder direct propagation of the cracks. Measurements made as a function of length indicated improved fatigue resistance was obtained with longer fibers. However, the analogy to actual whisker reinforcement may be questioned because of the relatively large diameters ($\approx$ 2 mils) employed, and because of relatively weak intercoating boundaries which arose from the fabrication process.

In work on continuous steel wire-reinforced aluminum, prepared by precoating and hot pressing the fibers, Baker [74] was able to vary the fiber/matrix interfacial bonding. This was accomplished by varying the hot-pressing conditions to achieve the desired amount of fiber-matrix reaction. Reversed bending was used, and greater fatigue resistance was observed in specimens where some limited amount of interfacial reaction had occurred. This was due to the increased bond strength which was created in those specimens without a degrading of fiber properties. Fiber volume fraction was also considered and, since fatigue life may usually be increased by decreasing the plastic strain range, increased fiber contents produced enhanced fatigue properties.

Continuous, brittle tungsten fibers (up to 23% by volume) in a copper matrix gave rise to only minor increases in fatigue strength at 10^6 cycles. Ham and Place [75] ascribed these results to stress concentrations at the tips of cracks in the fatigued matrix and indicated that reinforcement against fatigue would be more effective in matrices that did not fatigue harden.

The fatigue behavior in tension-tension of a model silver matrix reinforced with continuous steel and continuous or discontinuous tungsten fibers was rather thoroughly investigated by Morris and Steigerwald [79]. They observed significant increases in fatigue strength for all systems (both continuous and discontinuous) studied. Fatigue properties of the composite were found to depend on fiber volume percent (increasing fiber content producing increased fatigue strength) with the failure mode being determined by the relative strength and fatigue properties of the two components present. The increases in fatigue strength, for silver reinforced with 30% by volume of both continuous and discontinuous ($l/d \approx 100$) tungsten fibers, over that of pure silver are shown in Figure 11.23. The increases in fatigue strength are greater for the continuously reinforced case but are still significant where discontinuous fibers have been employed. There was a sharp decrease in the ratio of endurance limit to ultimate tensile strength of the composite at about 10% fiber by volume. This change was accompanied by a change in the mode of failure at about the same fiber content. Below 10 vol% extensive fiber failure was observed throughout the specimen and at greater fiber contents the fracture of the fibers was found only along the main path of fatigue failure. Morris and Steigerwald [79] attributed this effect to the interaction of stress concentrations at the higher fiber contents, similar to those discussed earlier.

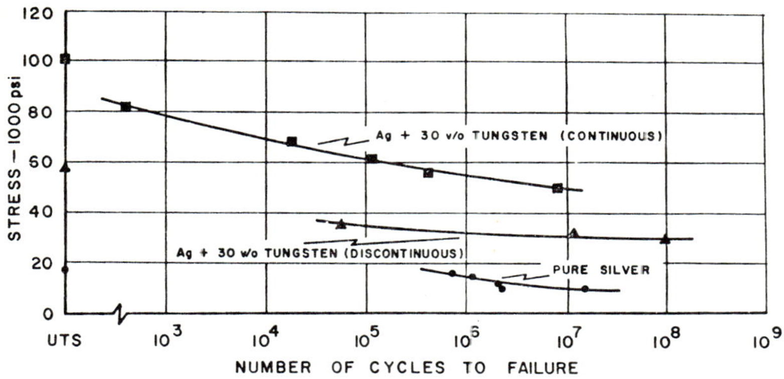

Figure 11.23 Comparison of the fatigue behavior of silver, silver plus 30 vol% continuous aligned tungsten, and silver plus 30 vol% discontinuous aligned tungsten fibers [76].

Differences in the fatigue behavior of the composites containing tungsten versus steel reinforcement are thought to arise from the differences in moduli of these two materials. Since the plastic strain range of the matrix is related to the modulus of the reinforcement, greater fatigue resistance was achieved with the tungsten fibers which possess a higher modulus and strength than the steel reinforcement.

Data by Mehan, Jakas, and Bruch [51] on the low-cycle fatigue behavior of aluminum reinforced with 20 vol% sapphire whiskers are shown in Figure 11.24. The curve is quite flat and in this respect resembles the elevated-temperature rupture behavior of this composite. Although the low fatigue strength (11,000 psi, defined in terms of the alternating stress) is not fully understood, they attribute it in part to unclassified and poorer quality sapphire whiskers. Further studies are under way to explain this effect. However, it may be anticipated that higher fatigue strengths will be attained as whisker selection and quality are improved.

Fiber shape has also been found to be important with regard to the fatigue resistance of a whisker composite. Low-cycle fatigue studies at United Aircraft [61] showed the platelike reinforcement of aluminum with $CuAl_2$, prepared by unidirectional solidification, to be more effective in promoting fatigue strength than the rodlike reinforcement of Al_3Ni.

In summary, we have seen in previous sections that a useful range of composite application occurs when the fibers deform elastically and the matrix is just starting to deform plastically (see section on stress-strain behavior). Such effects, however, give rise to fatigue crack initiation and propagation in the composite. The function of the fibers or whiskers is then

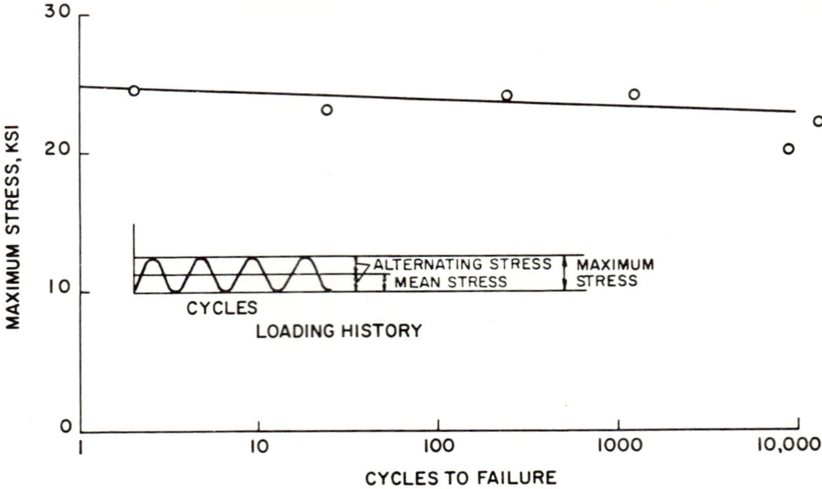

Figure 11.24 Fatigue behavior of 20 vol% Al/Al_2O_3 composites at room temperature [51].

to inhibit the growth or propagation of fatigue cracks. In addition, fibers should assume large fractions of the applied loads, thereby strengthening and stiffening the matrix so that fatigue cracks are not initiated. To achieve this, the plastic strain range in the matrix should be kept as small as possible by using whiskers of high aspect ratio and modulus and the largest whisker contents possible.

FRACTURE TOUGHNESS

We have seen that the incorporation of high-strength and high-modulus whiskers into various matrices can have very beneficial effects on tensile properties and high-temperature strength. These materials are intended for use in engineering applications where toughness is required. It is necessary then to inquire into the effects of brittle fibrous inclusions on fracture toughness and notch sensitivity.

As might be imagined, the addition of larger and larger amounts of the more brittle fibers to a ductile matrix tends to reduce the fracture energy. Parikh [80] has reported such an effect in an aligned steel fiber-silver composite (Fig. 11.25). Similar results have also been observed in copper reinforced with continuous, brittle tungsten fibers [81] under impact conditions.

Cottrell [82] has suggested that the impact resistance of a fiber-reinforced composite may be increased by the use of aspect ratios just slightly less than critical. In addition, the fiber/matrix interface should be weak in shear so that the critical aspect ratio is as great as possible. In this analysis, the fibers pull out of the matrix instead of being broken by the advancing crack. This tends to absorb more energy and slow the growth of the crack. Further, if discontinuous fibers of length approximately equal to the critical length are used, the fibers surrounding a broken whisker will not fracture in the same plane. Load builds up from the ends of short whiskers, and therefore their most highly stressed areas will not necessarily be in the same cross-section. The whiskers then probably break at points close to their own centers rather than in the plane of a broken neighboring fiber. The use of short fibers (aspect ratios less than or approximately equal to the critical aspect ratio) presents a disadvantage in that the strength of the whiskers will not be fully utilized. The average stress over the full length of a critical aspect ratio fiber is approximately one-half that of a much longer fiber. These effects have been discussed in section on tensile properties, and the strength of the composite is reduced by the use of such short whiskers. Depending on the application, some compromise between strength and toughness is required, and research is necessary to determine the parameters for the optimum synthesis.

The fracture mechanics of a copper matrix reinforced with either brittle tungsten or silica fibers have been investigated by Cooper and Kelly [83]. Although their studies were performed with continuous filaments, the results are applicable to whisker composites where whiskers greater than the critical aspect ratios are used. In this work, fibers were observed to fracture at some distance ahead of the advancing crack, that is, three or four fibers beyond the root of the crack. The indication here is that, although the toughness of a composite material is governed by the properties of the fiber,

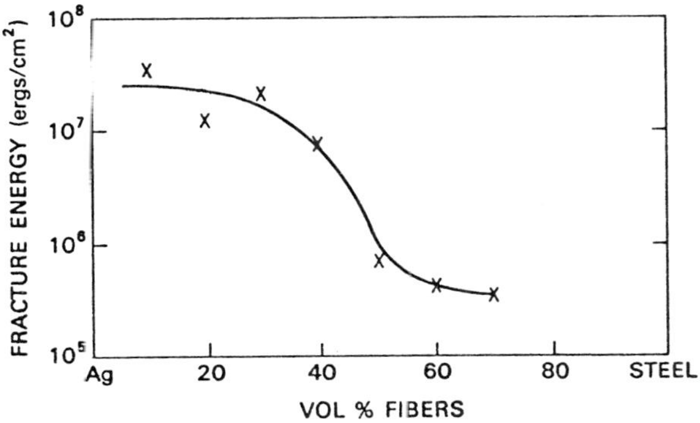

Figure 11.25 Fracture energy versus volume percent fiber for silver matrix composites containing short aligned 0.002 in. diameter steel fibers [80].

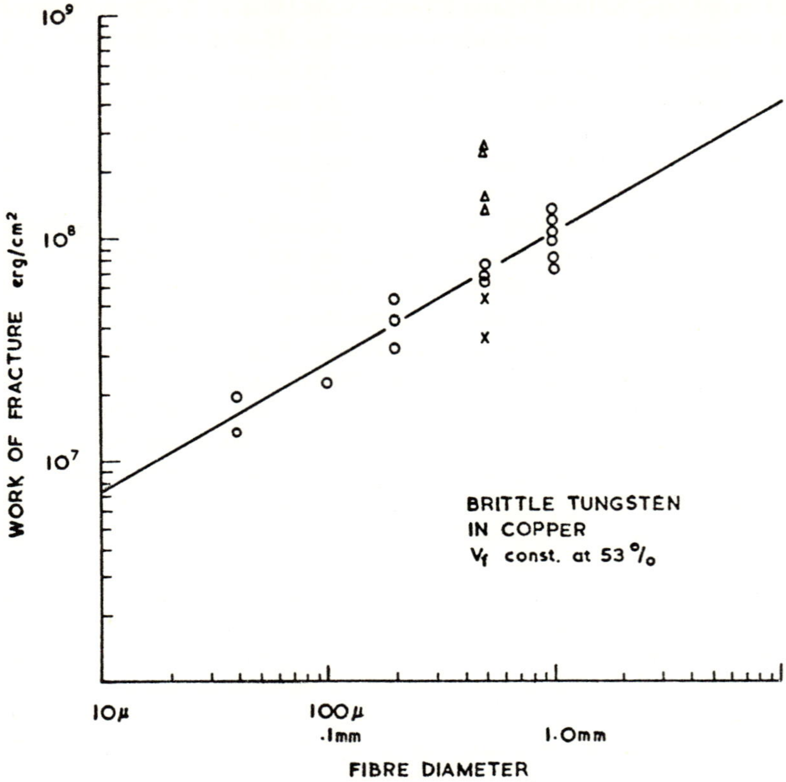

Figure 11.26 Work of fracture versus fiber diameter for vacuum cast copper-tungsten composites containing 53 vol % tungsten [80].

matrix, and the interface, the major contribution to the energy of fracture arises from the matrix. The energy to fracture was determined as a function of fiber diameter for a constant fiber volume fraction of 0.53, and is shown to increase with increasing diameter in Figure 11.26. This may be attributed to the advancing crack having to drive a larger plastic zone for the larger diameter fibers. It is significant to note that the work of fracture is less than 10^7 ergs/cm^2 at 20 μ. The work of fracture of most engineering metals and alloys is usually about an order of magnitude greater ($\approx 10^8$ ergs/cm^2). Since the strongest whiskers possess diameters considerably less than 20 μ, again we see that some compromise is required to achieve reasonable levels of both strength and fracture toughness.

Both Cooper and Kelly [83] and Cratchley and Baker [42] have reported that a fiber composite may be rendered notch insensitive by the presence of weak interfaces perpendicular to the advancing crack. In the silica fiber-aluminum system [42], the weak interface appeared at the intercoating boundaries where the precoated fibers had been hot-pressed together. In the tungsten fiber-copper system [83], the weak planes that served to deflect the cracks parallel to the fibers were at the fiber/matrix interface. The idea that a crack may be deflected by a weak interface, thus rendering the material notch insensitive, is dependent on the fact that a tensile stress exists normal to the applied tensile load. Cook and Gordon [84] calculated the tensile stress parallel to the plane of the crack to be about one-fifth of the maximum stress concentration causing propagation at the crack tip. Figure 11.27 shows the stress parallel to the plane of the crack, σ_x, and the stress

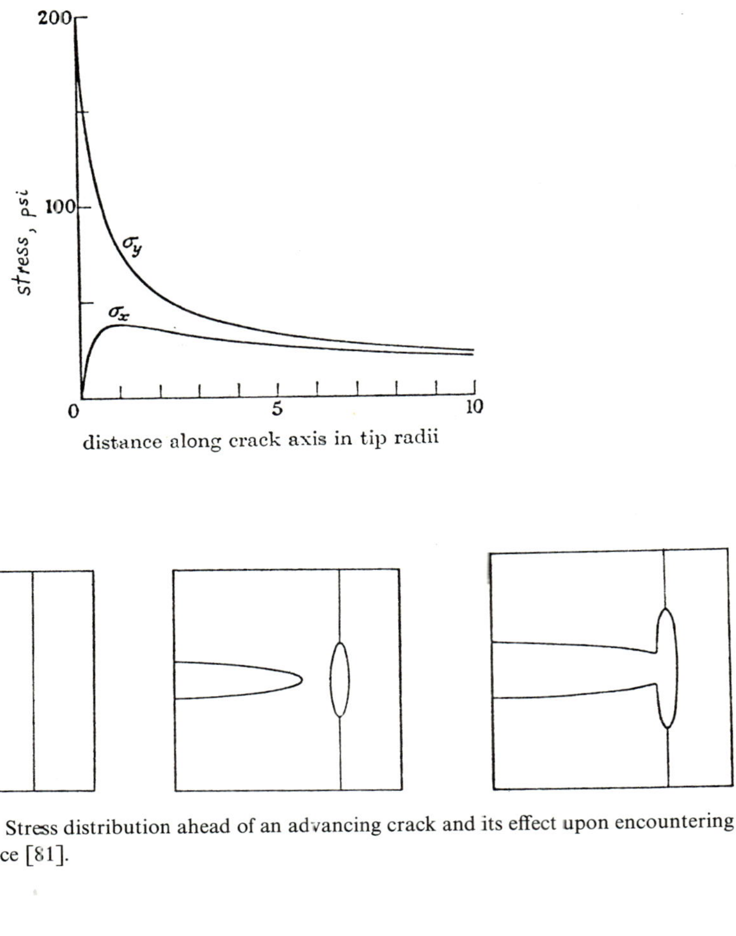

Figure 11.27 Stress distribution ahead of an advancing crack and its effect upon encountering a weak interface [81].

driving the crack, σ_y. σ_x increases from zero, at the crack root, to a maximum at a slight distance (approximately one crack tip radius) ahead of the crack. The schematic situation by which the crack is deflected, because of the above considerations coming to bear at a weak interface, is shown in Figure 11.27. Here we see that the effect of the leading tensile stress is to open up the interface ahead of the crack. The stress concentration is thereby relieved and the crack is divested of its driving force.

The preceding discussion has illustrated that the fracture toughness of a composite is related to the properties of not only the whiskers and matrix, but also the interface between them. In addition, whisker composites will show extremely anisotropic impact behavior [85] which will have to be taken into account when the trade-off between strength and toughness is made.

CONCLUDING REMARKS

The properties of whisker-reinforced composites have been reviewed from the point of view of understanding the principles or mechanisms giving rise to their behavior. The potential of these materials has been amply demonstrated and is described in the literature. However, optimizing the parameters relevant to any particular application and fabrication of whisker

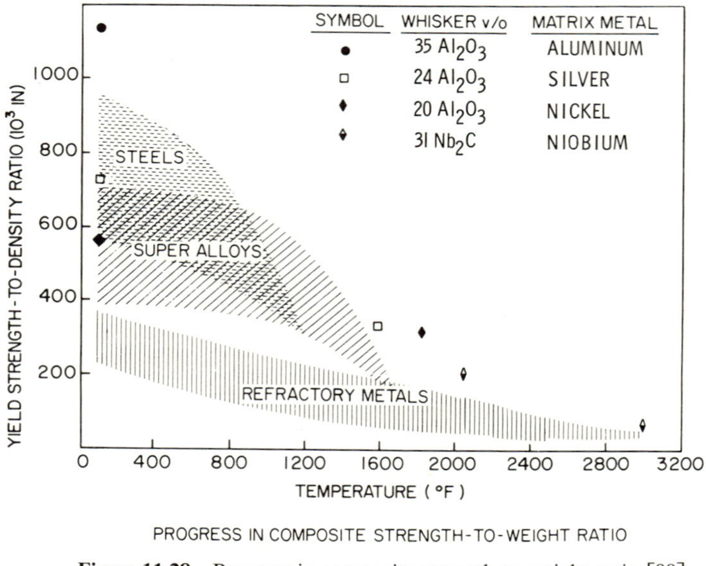

Figure 11.28 Progress in composite strength-to-weight ratio [88].

composites into useful engineering structures, without degrading of whisker properties, will require continued effort. The current research effort is extensive [86, 87] and increases in strength have been achieved and further increases are expected. Figure 11.28 summarizes recent progress in strength-to-density ratio for whisker-reinforced metals and illustrates the high-temperature superiority of these composites over currently available superalloys and refractory metals [88]. The improvement in such parameters as strength-to-weight and stiffness-to-weight ratios should be even more striking. Many unusual systems and novel fabrication techniques [89] are currently under investigation and new developments are constantly emerging.

ACKNOWLEDGMENTS

The authors gratefully acknowledge M. Moss and D. R. Adolphson, both of Sandia Corporation, for their helpful criticisms of the original manuscript.

REFERENCES

[1] R. V. Coleman, "The Growth and Properties of Whiskers," *Met. Rev.*, **9,** 261 (1964).
[2] S. S. Brenner, *Fiber Composite Materials*, American Society for Metals, Metals Park, Ohio, 1965, p. 11.
[3] L. R. McCreight, H. W. Rauch, Sr., and W. H. Sutton, *Ceramic and Graphite Fibers and Whiskers*, Academic Press, New York, 1965.
[4] A. Kelly and W. R. Tyson, "Fiber Strengthened Materials," *High Strength Materials*, V. F. Zackay, Ed., John Wiley and Sons, New York, 1965, p. 578.
[5] D. Cratchley, "Factors Affecting the UTS of a Metal/Metal Fiber Reinforced System," *Powder Met.*, **11,** 59 (1963).
[6] W. H. Sutton and J. Chorné, "Potential of Oxide-Fiber Reinforced Metals," *Fiber Composite Materials*, American Society for Metals, Metals Park, Ohio, 1965, p. 173.
[7] R. W. Jech, D. L. McDanels, and J. W. Weeton, "Fiber Reinforced Metallic Composites," paper presented at Sixth Sagamore Conference, Racquette Lake, New York, August 19, 1959.
[8] D. L. McDanels, R. W. Jech, and J. W. Weeton, "Analysis of Stress-Strain Behavior of Tungsten-Fiber-Reinforced Copper Composites," *Trans. AIME*, **233,** 636 (1965).
[9] R. H. Krock, "Some Comparisons Between Fiber-Reinforced and Continuous Skeleton Tungsten-Copper Composite Materials," *J. Materials*, **1,** 278 (1966).
[10] N. F. Dow, "Study of Stresses Near a Discontinuity in a Filament Reinforced Composite Metal," General Electric Co., Report R63 SD61 (1963).
[11] S. Umekawa, R. Kotfila, and O. D. Sherby, "Elastic Properties of a Tungsten-Silver Composite Above and Below the Melting Point of Silver," *J. Mech. Phys. Solids*, **13,** 229 (1965).
[12] B. Paul, "Prediction of Elastic Constants of Multiphase Materials," *Trans. Met. Soc. AIME*, **218,** 36 (1960).
[13] Z. Hashin and B. W. Rosen, "The Elastic Moduli of Fiber-Reinforced Materials," *J. Appl. Mech.*, **31,** 223 (1964).

[14] B. D. Coleman, "On the Strength of Classical Fibers and Fiber Bundles," *J. Mech. Phys. Solids*, **7**, 60 (1958).

[15] B. W. Rosen, "Tensile Failure of Fibrous Composites," *AIAA Journal*, **2**, 1985 (1964).

[16] J. Gurland, "Some Aspects of the Fracture of Metallic Composites," U.S. AEC Contract No. A7(30-1)12394, Technical Report No. 18, February, 1966.

[17] P. W. Jackson and D. Cratchley, "The Effect of Fibre Orientation on the Tensile Strength of Fibre-Reinforced Metals," *J. Mech. Phys. Solids*, **14**, 49 (1966).

[18] G. A. Cooper, "Orientation Effects in Fibre-Reinforced Metals," *J. Mech. Phys. Solids*, **14**, 103 (1966).

[19] D. M. Schuster and E. Scala, "The Mechanical Interaction of Sapphire Whiskers with a Birefringent Matrix," *Trans. Met. Soc. AIME*, **230**, 1635 (1964).

[20] T. F. MacLaughlin and M. J. Salkind, "Effect of Fiber Geometry on Stress in Fiber-Reinforced Composite Materials," Watervliet Arsenal Report No. WVT-6521, May, 1965.

[21] J. Milewski, "Methods for Incorporating Whiskers in Various Matrices for Reinforcement," presented at Am. Cer. Soc. Meeting, May, 1965.

[22] W. H. Sutton, "Investigation of Bonding in Oxide-Fiber (Whisker) Reinforced Metals," General Electric Co. Contract DA 36-034-PRD-3768RD, Final Report, June, 1964.

[23] D. W. Petrasek and J. W. Weeton, "Effects of Alloying on Room Temperature Tensile Properties of Tungsten-Fiber-Reinforced-Copper-Alloy Composites," *Trans. Met. Soc. AIME*, **230**, 977 (1964).

[24] D. W. Petrasek, "Elevated-Temperature Tensile Properties of Tungsten Fiber Composites," *Trans. Met. Soc. AIME*, **236**, 887 (1966).

[25] A. P. Levitt, "Recent Advances in Alumina Whisker Technology," *Mat. Res. Stds.*, **6**, 64 (1966).

[26] S. M. Wolf, A. P. Levitt, and J. Brown, "Whisker-Metal Matrix Bonding," *Chem. Eng. Prog.*, **62**, 74 (1966).

[27] W. R. Micks, "Composite Materials—Considerations for Future Research," Presented at the IAS National Summer Meeting, Paper No. 62–167, June 1962.

[28] R. H. Krock, "Whisker-Strengthened Materials," *Int. Sci. Tech.*, **59**, 38 (1966).

[29] W. R. Tyson and G. J. Davies, "A Photoelastic Study of the Shear Stresses Associated with the Transfer of Stress During Fiber Reinforcement," *Brit. J. Appl. Phys.*, **16**, 199 (1965).

[30] L. J. Cohen and J. P. Romualdi, "The Stress, Strain and Displacement Fields in a Composite Reinforced with Discontinuous Fibers," Douglas Missile and Space Systems Division, Douglas Paper No. 4128, 1966.

[31] H. L. Cox, "The Elasticity and Strength of Paper and Other Fibrous Materials," *Brit. J. Appl. Phys.*, **3**, 72 (1952).

[32] D. M. Schuster and E. Scala, "Mechanical Interactions in Fiber Reinforced Photoelastic Composites," Proceedings of an Air Force Conference, *Fundamental Aspects of Fiber Reinforced Plastic Composites*, Dayton, Ohio, May 1966.

[33] B. W. Rosen, "Mechanics of Composite Strengthening," *Fiber Composite Materials*, American Society for Metals, Metals Park, Ohio, 1965, p. 37.

[34] L. J. Broutman, "Glass-Resin Joint Strengths and Their Effect on Failure Mechanisms in Reinforced Plastics," *Polymer Eng. Sci.*, (July 1966), p. 263.

[35] F. J. McGarry, "Resin-glass Bond Characteristics," *ASTM Bulletin* (January 1959), 63.

[36] A. Kelly and G. J. Davies, "The Principles of Fiber Reinforcement of Metals," *Met. Rev.*, **10**, 1 (1965).

[37] H. R. Piehler, "Plastic Deformation and Failure of Silver-Steel Filamentary Composites," *Trans. Met. Soc. AIME*, **233**, 12 (1965).

[38] J. V. Milewski, R. G. Shaver and S. H. Morgan, "How Whiskers Differ from Fiberglass and Boron in Characteristics and Processing Techniques, "Society of Plastics Engineers 22nd Annual Technical Conference, Montreal, Canada, March 10, 1966.

[39] J. W. Weeton and R. A. Signorelli, "Fiber-Metal Composite Materials," NASA Technical Note (D-3530), 1966.

[40] W. H. Sutton and J. Chorne, "Development of High-Strength, Heat-Resistant Alloys by Whisker Reinforcement," *Met. Eng. Quart.*, **3,** 1 (1963).

[41] A. Kelly and W. R. Tyson, "Tensile Properties of Fibre-Reinforced Metals: Copper/ Tungsten and Copper/Molybdenum," *J. Mech. Phys. Solids*, **13,** 329 (1965).

[42] A. A. Baker and D. Cratchley, "Stress-Strain Behavior and Toughness of a Fibre-Reinforced Metal," *Appl. Mat. Res*, **5,** 92 (1966).

[43] R. L. Mehan and R. P. Jakas, "Behavior Study of Sapphire Wool Aluminum Composites," presented in "Review of Research," General Electric Space Sciences Laboratory under Air Force Materials Laboratory Contract No. F 33615-67-C-1308.

[44] N. J. Parratt, "Reinforcing Effects of Silicon-Nitride Whiskers in Silver and Resin Matrices," *Powder Met.*, **7,** 152 (1964).

[45] N. J. Parratt, "Whisker Reinforced Plastics and Metals," *Chem. Eng. Prog.*, **62** 61 (1966).

[46] T. J. Kopperaal and N. M. Parikh, "Microstraining in Fiber-Reinforced Silver," *Trans. Met. Soc. AIME*, **224,** 1173 (1962).

[47] R. W. Hertzberg, "Potential of Unidirectionally Solidified Eutectic Alloys as Reinforced Composites," *Fiber Composite Materials*, American Society for Metals, Metals Park, Ohio, 1965, p. 77.

[48] F. D. Lemkey and M. J. Salkind, "The Growth and Properties of Carbide Whisker-Reinforced Refractory Metals from the Melt," *Crystal Growth*, H. S. Peiser, Ed., Pergamon Press, New York, 1967, p. 171.

[49] M. J. Salkind, F. D. George, F. D. Lemkey, B. J. Bayles, and J. A. Ford, "Whisker Composites by Unidirectional Solidification," *Chem. Eng. Prog.*, **62,** 52 (1966).

[50] W. H. Sutton and J. Chorné, "Factors Affecting the Tensile Strength of Metals Reinforced with Strong Fibers," *Strengthening Mechanisms*, Syracuse University Press, 1966, p. 549.

[51] R. L. Mehan, R. P. Jakas, and C. A. Bruch, "Behavior Study of Sapphire Wool Aluminum Composites," -AFML-TR-68-100, May 1968, Space Sciences Laboratory, General Electric Co., under Air Force Contract No. F 33615-67-C-1308.

[52] D. Cratchley, "Experimental Aspects of Fibre-Reinforced Metals." *Met Rev.*, **10,** 79 (1965).

[53] H. Hahn, P. Lare, and A. Divecha, "Progress in Preparation and Working of Whisker-Reinforced Metal Composites," ASME Design Engineering Conference, New York, N.Y., Preprint No. 67-DE-35, May, 1967.

[54] B. C. Raynes and R. H. Kelsey, "Studies of the Reinforcement of Metals with Ultra High Strength Fibers (Whiskers)," Horizons, Inc., Final Report, Contract NOw 62-0235-C, Oct. 1962.

[55] R. H. Kelsey, "Reinforcement of Nickel Chromium Alloys with Sapphire Whiskers," Horizons, Inc., Final Report, Contract NOw 63-0138-C, Oct. 1963.

[56] R. G. Shaver, "Whisker-Strengthened Plastics," ASME Design Engineering Conference, New York, N.Y., Preprint No. 67-DE-9, May, 1967.

[57] J. J. Shyne and R. G. Shaver, "Whisker Composite Technology," presented at the 10th National SAMPE Symposium, San Diego, Calif., Nov. 1966.

[58] W. H. Sutton, B. W. Rosen, and D. G. Flom, "Whisker-Reinforced Plastics for Space Applications," *SPE Journal*, **20,** 567 (1964).

[59] A. Kelly, *Strong Solids*, Clarendon Press, Oxford, 1966, p. 134.

[60] R. L. Mehan, "Behavior Study of Sapphire Wool-Aluminum and Aluminum Alloy Composites," Quarterly Prog. Report No. 3 Oct. 1968 General Electric Space Sciences Laboratory, Air Force Materials Laboratory Contract No. 33615-67-C-1308.

[61] M. J. Salkind, F. D. George, F. D. Lemkey, and B. J. Bayles, "Investigation of the Creep, Fatigue, and Transverse Properties of Al_3Ni Whisker and Cu αl_2 Platelet Reinforced Aluminum," United Aircraft Research Laboratories, Final Report Contract NOw 65-0384-d, May 11, 1966.

[62] D. W. Petrasek, R. A. Signorelli, and J. W. Weeton, "Metallurgical and Geometrical Factors Affecting Elevated Temperature Tensile Properties of Discontinuous Tungsten Fiber Reinforced Composites," NASA TM X-52232, presented at Metals Congress, Chicago, Ill., November 1966.

[63] A. V. Dean, "The Development of Composite Materials for Service at Medium and Elevated Temperatures," *Appl. Mat. Res.*, **3,** 195 (1964).

[64] A. V. Dean, "Reinforcement of Nickel Base Alloys with High Strength Tungsten Wires," *J. Inst. Metals*, **95,** 79 (1967).

[65] E. G. Ellison, and B. Harris, "The Elevated Temperature Properties of a Nickel Alloy Reinforced with Tungsten Wires," *Appl. Mat. Res.*, **5,** 33 (1966).

[66] A. Kelly and W. R. Tyson, "Tensile Properties of Fibre-Reinforced Metals—II: Creep of Silver-Tungsten," *J. Mech. Phys. Solids*, **14,** 177 (1966).

[67] A. B. Lomax, "A Model System to Simulate Creep of Composites," *Metals and Materials*, **1,** 117 (1967).

[68] R. W. Jech, E. P. Weber, and A. D. Schwope, "Development of Titanium Alloys for Elevated Temperature Service by Powder Metallurgical Techniques," Final Report, Contract No. NOas 55-953-C, July 1957.

[69] M. J. Salkind, F. D. Lemkey, F. D. George, and B. J. Bayles, "Eutectic Composites by Unidirectional Solidification," Presented at the 10th National SAMPE Symposium, San Diego, Calif, Nov. 1966.

[70] E. H. Andrews, "Destruction of Silicon Nitride Whiskers by Reaction with Metals at High Temperatures," *J. Mat. Sci.*, **1,** 377 (1966).

[71] P. J. E. Forsyth, R. W. George, and D. A. Ryder, "Some Preliminary Tests on Aluminum Alloy Sheets Reinforced with Strong Wires," *Appl. Mat. Res.*, **2,** 223 (1963).

[72] R. V. Williams and D. J. O'Brien, "The Reinforcement of Metals with Metal Fibers," *Appl. Mat. Res.*, **3,** 148 (1964).

[73] A. A. Baker and D. Cratchley, "Metallographic Observations on the Behavior of Silica Reinforced Aluminum Under Fatigue Loading," *Appl. Mat. Res.*, **3,** 215 (1964).

[74] A. A. Baker, "The Effect of Fibre Volume Fraction and Interfacila Bond on the Fatigue of Aluminum Reinforced with Stainless Steel Wires," *Appl. Mat. Res.*, **5,** 143 (1966).

[75] R. K. Ham and T. A. Place, "The Failure of Copper-Tungsten Fibre Composites in Repeated Tension," *J. Mech. Phys. Solids*, **14,** 271 (1966).

[76] A. A. Baker, J. E. Mason, and D. Cratchley, "High-Strain Fatigue Studies of a Composite Material," *J. Mat. Sci.*, **1,** 229 (1966).

[77] T. H. Courtney and J. Wulff, "Matrix Limited Fatigue Properties in Fibre Composite Materials," *J. Mat. Sci.*, **1,** 383 (1966).

[78] A. A. Baker, "The Effect of Fibre Diameter and Discontinuous Fibres on the Fatigue of a Fibre-Reinforced Metal," *Appl. Mat. Sci.*, **5,** 210 (1966).

[79] A. W. H. Morris and A. E. Steigerwald, "An Investigation of the Fatigue Behavior of Tungsten-Reinforced and Steel-Reinforced Silver Composites," *Trans. Met. Soc.*, **239,** 730 (1967).

[80] N. M. Parikh, "Deformation and Fracture Modes," *Fiber Composite Materials*, American Society for Metals, Metals Park, Ohio, 1965, p. 115.

[81] A. Kelly, "The Strengthening of Metals by Dispersed Particles," *Proc. Roy. Soc. A.*, **282,** 63 (1964).

[82] A. H. Cottrell, "Strong Solids," *Proc. Roy. Soc. A*, **282,** 2 (1964).

[83] G. A. Cooper and A. Kelly, "Tensile Properties of Fibre-Reinforced Metals: Fracture Mechanics," *J. Mech. Phys. Solids*, to be published.

[84] J. Cook and J. E. Gordon, "A Mechanism for the Control of Crack Propagation in All-Brittle Solids," *Proc. Roy. Soc. A.*, **282** 508 (1964).

[85] M. J. Salkind and F. D. George, "Investigation of the Impact Resistance of Al_3Ni Whisker Reinforced Aluminum Prepared by Unidirectional Solidification," United Aircraft Research Laboratories, Technical Report AFML-TR-66-234, August, 1966.
[86] G. A. Hoffman, "Survey of Developments in Whisker Composites in the United States," Rand Corporation Report P-3129, April, 1965, AD 614 991.
[87] M. J. Salkind, "Composites, Where Do They Stand?" *Metals/Materials Today*, **40,** 29 (July 1967).
[88] W. H. Sutton, private communication.
[89] J. E. Gordon, "Some Considerations in the Design of Engineering Materials Based on Brittle Solids," *Proc. Roy. Soc. A.*, **282,** 16 (1964).

Ernest G. Wolff and Thomas Vasilos

12 Applications of Whisker Composites

ABSTRACT

Potential applications of whisker composites include turbine and other aerospace components, deep submergence structures, armor plate, dental ceramics, and electronic and wear-resistant materials. Current development efforts in the use of whisker additions to metals, plastics, ceramics, and other composites are also reviewed in order to assess the competitive potential and future impact of whisker composites on materials technology.

INTRODUCTION

The commercial use of whisker composites received its initial impetus from the discovery at the Bell Laboratories in the 1950s of the exceptionally high strength of single-crystal fibers [1]. Since that time to about 1958 the principal activity in whisker technology was the study of their growth mechanisms and properties. This was followed by interest in whiskers as reinforcing agents for a variety of matrix materials. Theoretical studies were initiated for the prediction of the mechanical properties of fiber-reinforced composites [2]. Model systems were investigated and problem areas were identified. For example, the critical dependence of the whisker-matrix interface, the effective whisker strength, the degree of alignment, and the residual stresses due to a particular fabrication technique make it difficult to predict, apart from a general law-of-mixtures approximation, the mechanical properties of a given composite system. The properties of whiskers themselves after handling, coating, grading, aligning, and subsequent incorporation into a matrix are unknown in most cases with the possible exception of unidirectional solidification. A composite may contain millions of whiskers, but only a few hundred can be adequately characterized, in terms of tensile properties alone, in a year.

It is the object of this chapter to review both specific and general areas where whisker-reinforced composites are being or can be seriously considered for application and where research and development efforts are going on. The status of whisker technology is assessed, and an attempt is made to indicate why whisker composites are likely to demonstrate a competitive potential in the areas cited. For example, the proposed applications are motivated not only by unidirectional strength of the whisker reinforcements, but by a combination of properties, such as composite stiffness-to-weight ratio. The uses of whiskers by themselves (e.g., insulation, paper, felts, filters, dipoles for light modulation) are beyond the scope of this chapter. In the concluding remarks, an attempt is made to assess the future impact of whisker composites on materials technology.

SPECIFIC APPLICATIONS

Turbine Engine Components

The successful application of "Hyfil"—a carbon fiber composite—as a compressor blade material in the Rolls Royce RB211 engine for the Lockheed L-1011 airbus demonstrates a substantial back-up technology for wider application of composite materials [3]. Whiskers are likely to be considered chiefly for application in the high-temperature engine components, such as blades or vanes, for several reasons: (1) the obvious competition from Hyfil and other fiber composites; (2) the excellent retention of whisker strength to high temperatures, where many continuous fibers recrystallize; (3) the operating efficiency improvements obtainable in extending the capabilities of turbine materials (above about 1800°F); and (4) the availability of alternative, oxidation-resistant alloys for use below 1800°F.

A turbine blade is the most critical component in the gas turbine engine. It is subject to high temperatures (1700–2000°F) and temperature gradients from the root to the airfoil tip (as much as 1000°F). It must withstand static and dynamic stress systems, thermal fatigue, impact/abrasion, and severely oxidizing environments, and must exhibit long time (1000 hrs and more) resistance to these conditions. To date the limiting factor in the adoption of higher turbine inlet temperatures (temperatures of combustion) has been the high-temperature materials capability of the hot end components. It is probable that cobalt- and nickel-base superalloys, now extensively used, have reached their limit of application. Protective coatings, new cooling methods, and dispersion-strengthened materials represent possible routes to

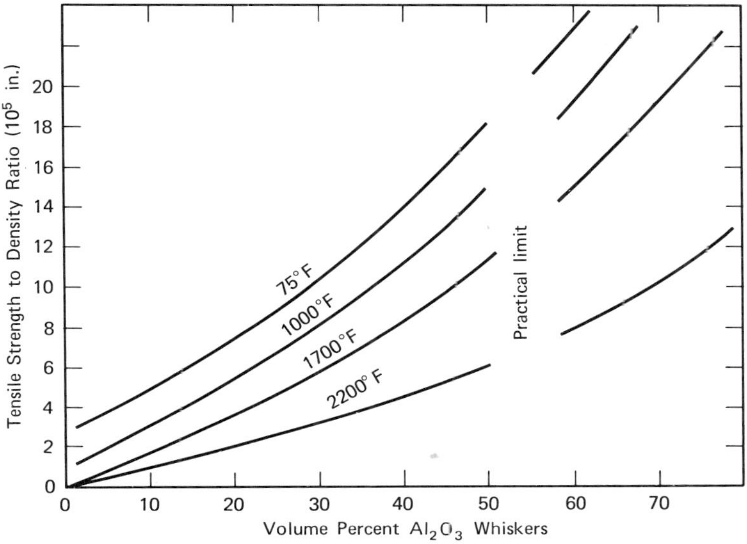

Figure 12.1 Theoretical strength to density ratio of nickel-alumina whisker composites as a function of volumetric loading and temperature.

marginal performance improvements. The combination of high-temperature strength (over 10^5 psi at 3000°F for Al_2O_3 whiskers [4]) and expected long-term stability of these whiskers with nickel [5] suggests that Al_2O_3/nickel alloy whisker composites have considerable practical potential as turbine components. Figure 12.1 illustrates the theoretical tensile strength-to-density ratio of nickel/alumina whisker composites as a function of volumetric whisker loading and temperature. This composite system has been studied at a number of laboratories including Avco [6, 7], General Electric [8], Horizon's [9], Melpar [10, 11], Tyco [12, 13], United Aircraft [14–16], Watervliet Arsenal [17] in the United States, the Atomic Weapons (AWRE) and Explosives (ERDE) Research Establishments in England [18, 19], and the Compagnie Francaise Thomson-Houston in France [20].

Workers at Horizon's Inc. [9] carried out most of the early research on the reinforcement of nichrome (and iron) with Al_2O_3 whiskers. Much of this effort was directed toward improving the bond at the metal-whisker interface, which was found to be strongly influenced by the oxygen present during hot pressing and melting.

The most extensive effort in this area has been carried out by W. H. Sutton and co-workers at General Electric [8]. Fabrication techniques such as hot pressing, electroforming, and liquid infiltration were explored and numerous

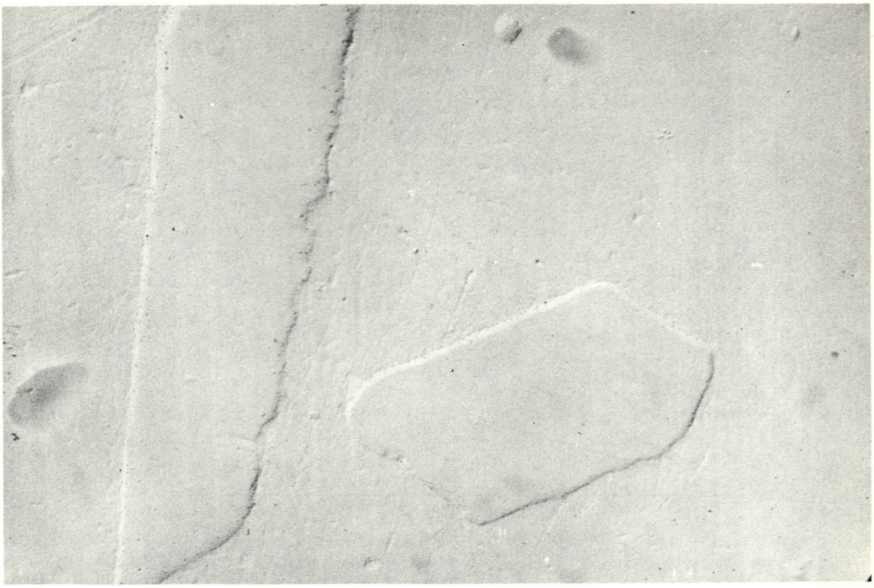

Figure 12.2 Ni/Al_2O_3 whisker composite (mag. 7500X).

problem areas defined, for example, wetting, bonding, whisker degradation, and alignment. In general hand lay-up techniques were used to achieve alignment and to demonstrate the potential of whisker composites, for example, 100,000 psi tensile strength and strength-to-density ratios of over 3×10^5 in. at temperatures to at least 1850°F. Problems still exist in control of interface conditions during processing, and in long-term, high-temperature exposure, and in developing a reproducible and economic fabrication technique yielding high-tensile strengths.

The effort at United Aircraft [14] toward composite turbine blade materials stems from considerable experience with unidirectional solidification of binary eutectic alloys, particularly the Al-Al_3Ni, Cu-Cr, and Al-$CuAl_2$ systems. Ingots of 3 in. in diameter and over 1 ft long have been made and experiments are under way to determine the potential of the method for production of complex shapes, such as a turbine blade. A drawback of this process is the relatively slow solidification rate and the relatively high purity of starting materials required. Continuous casting was suggested as a future modification. More recently, systems such as Nb-Nb_2C, Ta-Ta_2C, and Ni-NiMo have been studied [15, 16].

Liquid-phase hot pressing [11] and liquid infiltration [8] have been studied but present considerable difficulties, especially with deleterious chemical reactions. Electrodeposition of the nickel matrix with codeposition of the whiskers avoids many of these problems and has been studied at several laboratories [7, 8, 11, 17, 20]. Problems of low volumetric loading due to settling limitations [17] can be overcome by plating fluid flow control [7, 20]. Heat treatments mitigate effects of residual stresses and grain boundary precipitates [8]. Although further clarifications of alloying possibilities and interfacial bond requirements are needed, the process does yield intimate whisker-matrix contact with no degradation (Fig. 12.2). It can be scaled up to provide large sheets (possibly curved) containing aligned discontinuous fibers of at least 27 vol% loading (Fig. 12.3) or whiskers. Figure 12.4 shows that foils containing whiskers can be hot forged into a thicker plate.

Apart from Al_2O_3 (and very likely ThO_2), few potential nonmetal fiber reinforcements (e.g., B, SiC, BN, B_4C) are in stable equilibrium with nickel-base alloys at high temperatures [19, 21]. (Silicon nitride whiskers, if oxidized, were found to dissolve very slowly in nickel at 1100°C [21].) The use of metal carbide, boride, silicide, or oxide whiskers was recommended

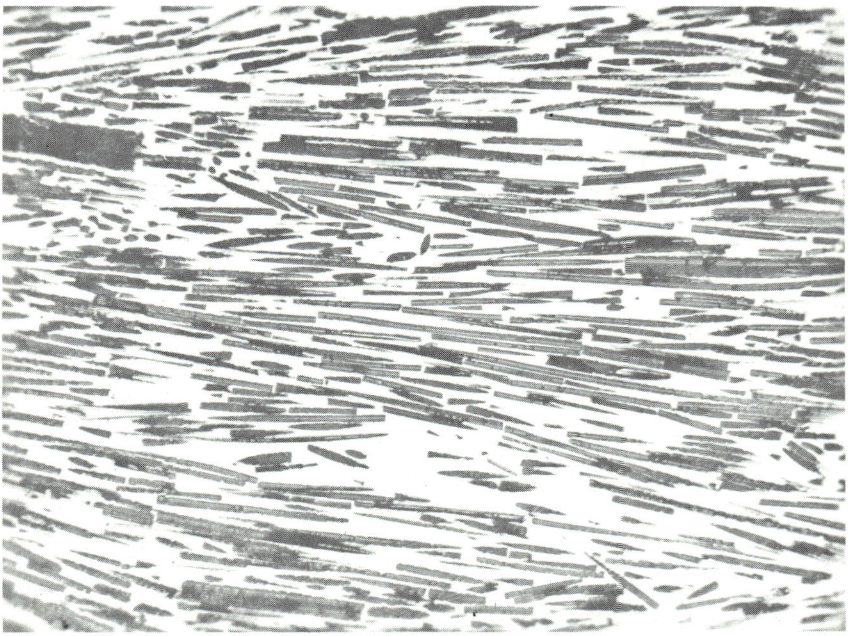

Figure 12.3 Discontinuous alumina fiber-nickel matrix composite made by electroforming and hydrodynamic alignment (mag. 50X).

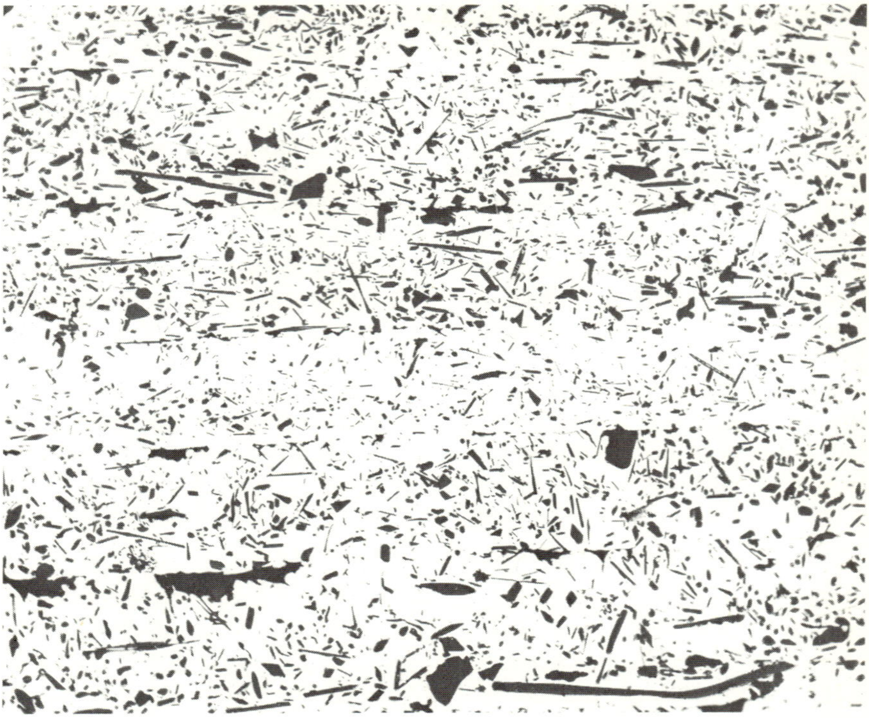

Figure 12.4 Cross-section of hot-forged stack of Ni/Al_2O_3 whisker foils, near edge of stack (mag. 100X).

by Tyco [12]. The alternative is to precoat the whiskers and choose a kinetically favorable, though basically unstable, system (e.g., Ti-SiC or possibly certain Ni-Cr-SiC mixtures). Single-crystal fibers of alpha-alumina, about 4 mils diameter by 14–18 in. long have been grown by a modified Czochralski melt pulling technique at Tyco [12]. Tensile strengths up to 500,000 psi have been measured and studies are under way to incorporate these filaments into Ni-Cr alloys for potential use in an SST high-temperature turbine blade, where the design goals are 3000 hrs stress rupture life at 2200°F with 20,000 psi load.

It is apparent that much effort has been expended on the application of whisker composites to turbine components and that, while promising approaches have been demonstrated, considerable work still needs to be done.

Rocket Nozzles

The use of Al_2O_3 whisker wool and needles (3–10% loading) to reinforce W or Mo rocket nozzles has been reported by Thermokinetic Fibers, Inc. [22]. These whiskers were introduced by a powder metallurgy process and a doubling of the 3000°F strength was reported. Phenolic resins containing refractory materials, including alumina fibers, have been successfully tested as rocket-thrust chamber materials (wall and nozzle) in low-pressure (<200 psia) N_2O_4-hydrazine propellant systems at the Aeroneutronic Division of Philco [23]. Thermodynamic considerations were presented for prediction of suitable combinations of resin and reinforcement for a given propellant chemistry.

The use of cluster mixes of 30–60% MgO or Al_2O_3 whiskers has been considered for potential use as a filler in epoxy and novalac resin ablative throat inserts (and leading edges). The whiskers would ablate appreciably at several hundred degrees above the use temperatures of silica fibers, which are currently employed for such purposes.

Aircraft and Helicopter Structures

Composite aerospace structure prototypes that have been studied to data (e.g., the F-111 horizontal stabilizer midsection, the T-30 wing box center section, or helicopter rotor blades) have utilized continuous (boron) fibers [24]. Although these components do not presently envisage the use of whiskers, one can foresee a definite future potential for whiskers in these areas. The experience gained in these programs will be of significant help in the evaluation of the potential of whiskers.

Diffusion bonding has been found to be an attractive fabrication technique for fiber-reinforced composites because it can be used to fabricate large sheet structures. Figure 12.5 illustrates a plate that was diffusion bonded in a graphite mold from a special mixture of aluminum foil, high-purity Al powder, and SiC whiskers. The whiskers were aligned unidirectionally in the plane of the plate. This alignment was further enhanced by subsequent cold rolling in that direction. There is a significant improvement in yield and tensile properties over the control sample (where additional powder was used in place of the SiC whiskers), in both the diffusion-bonded and the cold-rolled state.

Work at Melpar is directed to reinforcing aluminum alloys with SiC or Al_2O_3 whiskers to produce sheet structures by hot rolling. The initial fabrication technique was that of liquid-phase hot pressing [11]. Aluminum and potentially magnesium matrix composites called "Whiskerloy" are extruded composites containing up to 40 vol% aligned whiskers in rod, ribbon, and right-angle shapes. Ribbons and thin sections may find use in

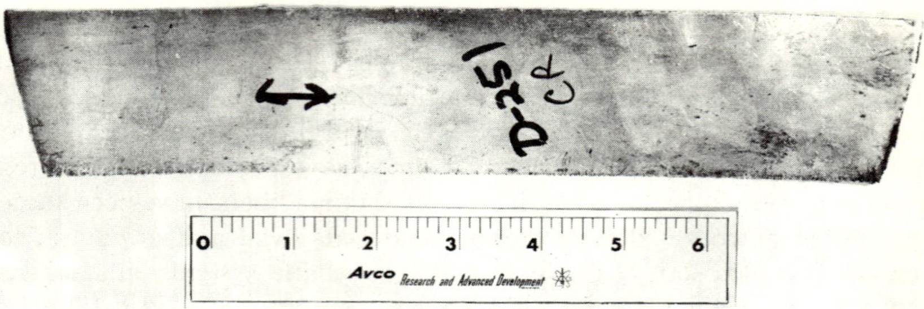

Figure 12.5 Hot-pressed and cold-rolled plate of aluminum plus 6.3 vol% aligned SiC whiskers (mag. 0.4 X).

laminates and right-angle structures as stiffeners for weight-sensitive aerospace structures [25].

Reentry Vehicles and Leading Edges

A current effort in the application of composite materials to reentry vehicles is an Air Force sponsored program to fabricate a cylindrical midsection from boron and glass fibers in a plastic matrix [24]. As in aircraft structure developments, the experience gained in these initial stages should be of ultimate benefit to a consideration of whisker reinforcement. The potential advantages of integral fabrication of both heat shield and substructure suggest the further use of a whisker-reinforced structure for the entire reentry vehicle. A limited effort has been carried out at Avco to improve the radial strength of boron filament integrated heat shields by whisker additions [26]. In this work 5 wt% SiC and also Al_2O_3 whiskers were added to tensile specimens of epoxy resin. There was not tensile strength improvement at this loading, but a slight modulus improvement was obtained, suggesting a need for whisker alignment.

Biological Materials

There are several dental schools working in whisker- or fiber-reinforced dental porcelain, resins, cements, amalgams, and dental alloys. The important material requirements include favorable tensile, impact abrasion, water absorption, and thermal expansion characteristics.

Preliminary work was carried out by R. N. Boyd of New York University on the problems of the weakening of porcelains after pigment additions [27]. Small additions of whiskers added to this pigmented ceramic porcelain restored it to its original (unpigmented) strength. Similar results with short whiskers in reinforcing castable plastics were also obtained.

Grant and Greener of Northwestern University dispersed whiskers (Al_2O_3, SiC) and fibers (BN, steel) in a polymer prior to adding a monomer and sometimes also a silane coupling agent [28]. The autopolymerizing material was then packed into a tin-lined gypsum investment mold and held under pressure during polymerization. Samples were tested in simple three-point loading to determine stress-strain behavior. Whiskers (e g., 8–13% sapphire) tended to markedly increase both the flexure modulus and ultimate bending strength. The conventional plastic behavior of the polymethyl methacrylate changed to a behavior somewhat analogous to that of a ductile metal. The use of silane appeared to improve the whisker-matrix bonding. Impact properties were not necessarily improved, however, and a laminated structure to prevent whisker protrusion was recommended. In subsequent work, both in vitro and in vivo testing of whisker- and fiber-reinforced biological materials is planned, including electrochemical potential studies and rabbit and monkey implantation [29]. Surgical and histopathological techniques have been developed to permit studies of the interaction of implanted composites and biological hosts.

There are as yet no whisker-reinforced composites presently on the biological market although glass fibers in epoxy-methyl methacrylate copolymer bases may be forerunners [30].

Armor Plate

Kliman [31] has reported on the impact strength of alumina whisker-ceramic matrix composites. Work has been carried out at Melpar [32] on the use of sapphire additions to ceramics and glasses for opaque and transparent armor plate. The matrix materials include originally opaque materials such as alumina or mullite, as well as special glasses. The composite optical transparency, with the latter, is hopefully achieved by matching the index of refraction of the glass to that of the sapphire whiskers. Preliminary results indicated that a 10 wt% randomly oriented whisker addition increased the as-hot pressed flexural strength for opaque specimens of equivalent density.

Electronic Applications

Whiskers, added to slip castings, could result in composite thin-walled, high dielectric insulators. Whisker materials have been considered as substrates for the deposition of electronic circuits [33]. The exploitation of SiC whiskers in miniaturized electronic components (e.g., single-crystal thermistors) and micro-circuits has also been suggested [34].

The unique properties of unidirectional solidified alloys may lead to some interesting nonstructural applications [35–37]. These are based on unique electronic, magnetic, thermoelectric, thermal conductive, and piezoelectric properties of these anisotropic structures. Devices include contactless switches or variable resistors, brushless DC motors, probes for mapping magnetic fields (e.g., in cyrogenic magnets), infrared polarizing devices, better ferromagnetic materials, electronic image dissectors, target emitters, contact modulators, optical devices (e.g., for imaging), electron beam to optical converters, thermal conductors, improved thermoelectric materials, and possibly hard superconductors. Finally, unique filters could be made by blowing out liquid "fibers" to leave a structure with a controlled porous structure.

High-strength whiskers, aligned in a copper or other highly conductive matrix, could be used to strengthen electric cables, especially at higher temperatures. By being aligned in the direction of current flow, the loss of conductivity would be minimized. Other electrical applications sought include reinforced electrical contacts [38], better magnets, and improvements in the general field of fiber optics. Unidirectionally solidified Al_2O_3-TiO_2 structures have also been considered at Electricité de France for use in magneto-hydrodynamic generator channels [20].

Wear Resistant Applications

Hahn [39] has suggested that whisker composites may be ideal materials for bearing and moving parts in space applications where normal lubrication against wear is impossible. Sapphire whiskers could conceivably be aligned to stand up along the top of the bearing material to give greater wear.

Composite electrochemical materials (CEM) consisting of refractory powders dispersed in electroformed metals such as nickel have a wide range of potential applications (Table 12.1). Further improvements in the performance of these materials could be achieved by the replacement, or addition, of whiskers to the powder particles. The high hardness of SiC whiskers, for example, suggests their use in gear boxes, clutch plates, heavy-duty machine parts, and catapult guides on aircraft carriers.

Deep Submergence Structures

The increasing use of deep submergence structures for undersea oil or food exploration, or rescue work, represents a potential application area for whisker composites. Low-weight, high strength, and long fatigue life in the corrosive ocean environment are required material properties. Whisker materials, such as alumina, have many of these properties, such as high compressive strength, high rigidity, and resistance to seawater corrosion. Unlike aerospace materials, undersea structures can be designed for simpler loading conditions (e.g., simple hydrostatic pressure), or principally compressive or bending stresses. The present high cost of whiskers may preclude their immediate use in direct hull structures, but there are numerous associated components that could benefit from the high whisker strength

TABLE 12.1 APPLICATIONS ON WHICH THE COMPOSITE ELECTROCHEMICAL MATERIALS (CEM) PROCESS HAS BEEN USED TO REDUCE WEAR OR IMPART INDICATED PROPERTY[a]

Spinning reel parts
Thread guides—textile industry
Diamond dressing wheels (abrasion)
Film guides for movie projectors
Mold cavities (anti-stick and wear agent for molten glass)
Grass shears (self-sharpening)
Cloth-cutting circular knives (self-(sharpening)
Domestic knives (self-sharpening)
Scissors and tin snips (self-sharpening)
Air gauges—machine operated
Nuclear control rods (boron carbide particles absorb neutrons)
Chain sprockets
Hammer mill hammers and separating screens
Fan blades
Acoustic tile drills (self-sharpening)
Screw driver bits
Diamond hones and nail files (abrasion)
Side coat on cement cutting saws
Plane irons (self-sharpening)
Screens for pulverized paint pigment
Twist drills—low speed
Oil well drill bit carriers
Plug gauges
Lawn mower reels and cutter bars (slight abrasiveness)
Metal saw blades for specific applications
Mixer paddles
Pump shafts—oscillatory
Boring bar guide surface
Piston ring depressors (mass production applications)
Various electroforms (mold cavities, ring gauges, nozzles)
Gate valves for petroleum industry
Cams
Conveyor parts
Compression rings
Expanding mandrels for handling c.i. cylinder sleeves
Alignment pins for production machines
Embossing
Textile machinery parts

[a] Courtesy Platecraft of America.

to weight ratio and also the resulting inhibition of crack propagation. Potential applications may include ring stiffeners, hatches, cabling, fittings, piping, window components, billets, and undersea instrument housings. Hemispherical end sections for tubes or small hull ends could be made from hot pressed powder/whisker mixture in special dies. Since compressive strength depends on high interlaminar shear strength [40], the addition of whiskers to glass filament wound plastics may be helpful. These composites are being considered for improved buoyancy ratio hulls, due to their high strength-to-weight ratios. Potential, and largely unexamined, problems include ports and openings. Stress corrosion and low-cycle fatigue properties of candidate materials need extensive investigation [41, 42].

Miscellaneous Applications

Miniature rockets (e.g., of the 1-lb pound thrust level) could be usefully reinforced with whiskers. Both the general housing and the combustion chamber strength-to-weight ratio, especially at higher temperatures, could be improved over presently used refractory metals. Temperatures in the combustion chamber range as high as several thousand degrees and a common failure mode is bursting due to excessive pressure build-up. Reinforcement with refractory whisker materials would very likely increase rocket reliability.

Fuse parts and high strength molded items based on phenolic resins have 20 to 50% greater strength with small additions of whisker wool. The addition of whiskers to metal castings should reduce brittleness and increase the toughness of items such as valves, pumps, and nozzle components. Welding rods and cutting tools incorporating whiskers are now being evaluated [22]. The use of whisker-reinforced composites is also suggested for cryogenic pressure vessels, where the toughness of aluminum at low temperatures and the strength of materials such as titanium or steel could be usefully combined. Al_3Ni whisker-reinforced aluminum made by eutectic solidification has exhibited good toughness even at liquid nitrogen temperatures.

Work at Horizon, Inc., has indicated that 8 wt% whisker additions can increase the tensile strength of iron wire by 34%, and 20 wt% additions gave 280% greater tensile strength.

The addition of whisker mats to phenolic and epoxy resins would permit construction of sheet laminates such as high-strength breadboards for computer systems. The reinforcement of weld joints and glass to metal seals is also promising. Shyne and Shaver [43] report that whiskers have been successfully passed through an electric arc in welding studies without damage to their properties. Similarly, whisker-reinforced soldering tapes should simplify joining and repairing of metal components. The whiskers would hold the solder in the joint area by capillary action, as well as strengthen the final joint.

Research at National Beryllia [22] showed that whisker mats can be plated to produce a strong whisker-reinforced porous material suitable for "breathing" and/or gas-cooled structures.

Silicon carbide whiskers, because of their excellent resistance to hydrofluoric acid, appear to be likely reinforcements for polyethylene pipe and propulsion devices.

Whiskers have been suggested for use as catalyst supports in chemical reactors at high temperatures. By replacing powder substrates, the fluid properties in the reactor bed could be suitably modified [20].

The crystal fiber process developed at Tyco [12], (chapter 4) suggests potential applications for special single-crystal sapphire shapes. Available shapes include ribbons, rods, tubes, and arbitrary cross-sectional shapes with the long axis in the *c*-direction. One or more arbitrarily shaped holes can also be furnished. Possible applications include integrated circuits, where the sapphire is expensively machined to a thin layer. More efficient ruby lasers could be made by this technique. A thin strip of sapphire, which is very hard, might be used as a razor blade. High temperature windows, heating filament holders, and other optical devices could now be made from small single-crystal alumina with controlled cross sections.

A more exotic use in the future might be in ferrofluids, where extremely fine magnetic whiskers, $\leqslant$ 200 Å long could be used to impart anisotropic and other unusual properties to magnetic liquids. Adequate dispersion of whiskers is a major problem area for this application.

GENERAL APPLICATIONS

Whiskers are being added to metals, plastics, ceramics, and other composites for a variety of reasons. Initially, the common purpose is to improve the properties of the matrix material and to increase the performance in those applications where previously only the whisker-free materials were used. The results of the studies described here are, therefore, of general validity and will contribute to the choice and design of materials for many presently unspecified applications. With time it is hoped that certain whisker composites may be considered for use on the basis of their overall, and unique, properties alone.

Metal Matrix Composites

Ionic compound whiskers generally possess lower surface energies than do common metals. There is consequently a high equilibrium contact angle at the interface and poor wetting of the whiskers by the metal occurs. A contribution to overcoming this problem was made by Rosenberg of MITRON who evaluated whisker additions to casting alloys, such as an Al-4.5% Cu alloy [44]. The technique is basically as follows.

1. Introducing ceramic whiskers (e.g., Al_2O_3 or SiC) into the molten alloy by using a "master slug."
2. Distributing them in the melt to a 0.2 wt% loading.
3. Pouring the whisker-containing alloy into a mold cavity to form a cast shape.

The "master slug" consists of a proprietary metal matrix composite containing about 2 vol% alumina whiskers. This dissolves readily in the molten alloy (e.g., alloy 195), at temperatures in the range 1300 to 1400°F, thereby overcoming inherent problems in matrix-whisker wettability. The whiskers produce "stiching" patterns across the primary grain boundaries as a result of greater wetting of the whisker with the copper rich eutectic phase at the grain boundaries than with the aluminum phase (see Fig. 12.6).

Figure 12.6 View of 10–30 μ sapphire ribbon fibers at grain boundaries of cast aluminum-copper alloy. Whiskers were introduced into melt prior to solidification (mag. 800X). (Work done at Mitron RD, Waltham, Massachusetts. Fiber type 3B from Thermokinetic Fibers, Inc., Nutley, New Jersey.)

Hot tearing during the final stages of solidification is minimized through a transferral of shrinkage stresses to the whiskers. The whiskers cause a 10% increase in the yield and ultimate tensile strength, as well as increased ductility. It is questionable whether or not the mechanism is one of true fiber reinforcement, but it is likely that it is not dispersion hardening. There is, doubtless, some interference with grain boundary sliding and this would help to explain the improved tensile properties. The mechanism of reinforcement as such appears to be an interesting new development in composite materials.

The method helps to avoid cracking at sharp corners in a casting. Billets and blooms can be cast using this technique, followed by working or forging to wrought products. Both the cast and wrought products can be welded with a whisker-reinforced filler rod without destroying the integrity and distribution of the fibers. A logical extension of this technique could be made to improve the properties of nickel- and cobalt-base superalloys, which are commonly formed by a vacuum casting technique.

An approach to overcoming wetting problems is that of giving the whiskers a metallic coating. The advantages include protection during handling, elimination of whisker-whisker contacts, possible high-temperature diffusion barriers and, if the coating is magnetic, a means of alignment. Electroless, electrolytic, and evaporated coatings have been studied at Melpar [11] and at Avco [45], and sputtered coatings at General Electric [8].

The effectiveness of metal whiskers grown in a metallic matrix has been explored by Herzberg [46]. Provided dislocations do not pass easily into them, these whiskers can provide substantial reinforcement. Mollard and Fleming have shown that any alloy of a eutectic system (which solidifies into the two-phase field at equilibrium) can be solidified with a eutectic-like structure [47]. Growth of fibrous structures from a melt is also possible by dendritic growth, and by precipitation, although the latter method is likely to result in a structure which is unstable at high temperatures. Growth of sapphire whiskers in molten aluminum is an interesting development in high pressure technology [48].

Liquid-phase hot pressing (e.g., of Al_2O_3 whiskers in an Al-Si alloy matrix [11]) leads to possible nonuniform fiber distribution, porosity, and whisker degradation. However, it does present a potential advantage in being able to achieve high whisker loadings (e.g., over 20 vol%). Melpar also found that whiskers aligned perpendicularly to the line of rolling direction do not break significantly. Extrusion is another promising whisker alignment technique [25, 49]. A slurry of metal (e.g., Al), matrix powder, (SiC) whiskers, and an organic plasticizer solution are coextruded for alignment, and the mixture can then be densified by hot working.

Since most whiskers are highly refractory, a major potential benefit is that of significant high-temperature strength improvement. It is important, however, to maintain favorable interface conditions between the metal and whisker during fabrication, and to be aware of possible effects of environment during use—for example, oxygen diffusion. Calow and Barclay [21] have shown that whiskers can be unstable in a metal for a variety of reasons, including deformation due to fabrication, whisker phase changes, surface undulations, and effects of foreign (or adsorbed) elements at the interface. Due to their small size, whiskers have not only a higher surface-to-volume ratio than continuous fibers (and therefore will have more adsorbed atoms per given volume) but the curvature of the surface is higher and this increases the surface energy, and hence whisker reactivity and instability. The use of a whisker/metal composite at high temperatures, therefore, requires more careful attention to interface relationships than a similar fiber/metal composite.

It may be concluded that in metals, whiskers can serve a variety of functions. These include modification of the microstructure, control of nucleation of alloys, conventional fiber reinforcement, and possibly dispersion strengthening (if the whiskers are very close together). An additional possibility is that of providing improved oxidation resistance.

The use of oxide whiskers to improve the oxidation resistance of refractory metals has been investigated at Avco [50]. This work was based on a recognition of the considerable whisker strength retention at high temperatures. Brenner [51] has shown that at 20°C below the melting point, the average strength of sapphire whiskers is greater than the fracture strength of bulk crystals at room temperature. This suggested the use of these whiskers for incorporation into refractory metals (e.g., molybdenum) where strength drops off sharply above about 1100°C [52].

Composites were fabricated by cold pressing and sintering molybdenum powder-Al_2O_3 whisker mixtures, both precoated with nickel (0.75% addition) so that sintering could be completed at relatively low temperatures by the activated sintering technique [53], and to promote a sound bond between the alumina and the metal matrix. Sintering was conducted at 2550°F in H_2 for 4 hours after a 30 minute pre-sintering treatment at 1830°F in H_2. This resulted in 90–92% theoretically dense pure Mo samples. Composites containing 10–30 vol% Al_2O_3 whiskers were similarly prepared.

Oxidation studies were conducted by heating samples at 1650°F in still air for periods up to 3 hrs. Measurements of the surface recession are given as follows.

Material	Recession at 1650°F in Air	
	After 30 Minutes	After 90 Minutes
Mo	0.030 cm	0.084
Mo + 10% Al_2O_3	0.034	0.073
Mo + 20% Al_2O_3 whiskers	0.015	0.054
Mo + 30% Al_2O_3 whiskers	(0.008)*	0.004

* This sample gained weight initially, therefore, the dimensions increased at 30 minutes.

The results indicated that there is no practical difference in the oxidation behavior of Mo and Mo-10% Al_2O_3 whiskers. However, the addition of 20 and 30 vol% Al_2O_3 whiskers did decrease the surface recession rate relative to that of pure Mo.

Reinforced Plastics

The potential improvements in mechanical properties obtainable by whisker additions to plastic matrices have been demonstrated. For example, the fracture strength of an epoxy (PJ122) was increased tenfold (from 11,200 psi to 113,000 psi) when 14.2 vol% discontinuous alumina whiskers were added to it [54]. This strengthening is over 50% higher than the same epoxy reinforced with 49.1 vol% continuous quartz fibers (Fig. 12.7). A principal requirement is good alignment of the whiskers. Work in France

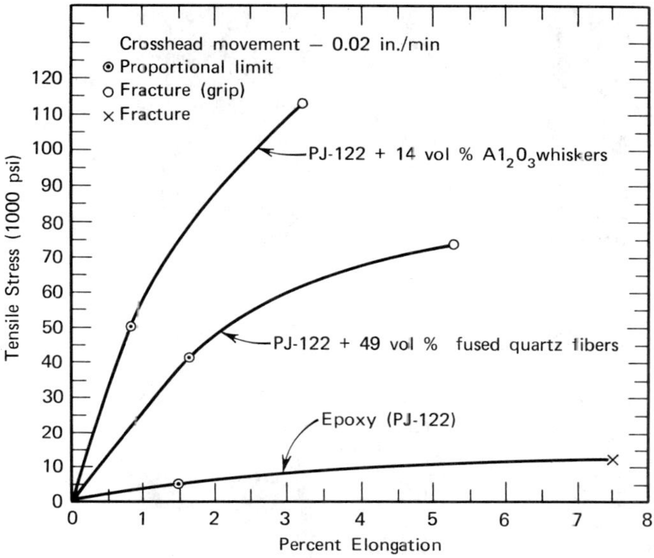

Figure 12.7 Stress-elongation curves for epoxy (PJ-122) and for epoxy reinforced with continuous fused quartz fibers and with discontinuous Al_2O_3 whiskers (Sutton et al. [54]). Permission Society of Plastic Engineers, Inc.

has shown that $\geqslant 3\%$ Al_2O_3 whiskers stop crack propagation in resins, and $\geqslant$ 10 vol% whiskers causes significant structural reinforcement [20]. At this loading, it is possible to obtain a threefold improvement in UTS and an eightfold improvement in elastic modulus.

Research has been carried out at Avco with oriented SiC whiskers in organic matrices (Fig. 12.8). In this case whiskers are used to form filaments which in turn are converted to whisker composites. Similar work has been carried out by Parratt [55] who converted silicon nitride whiskers into aligned yarns by a fluid spinning process. Carborundum Company has developed continuous, composite filaments consisting of 9–11 wt% of oriented whiskers in a resin bonded matrix [34]. These are marketed commercially to give, after removal of the polyacrylonitride copolymer binder by heating in air, an oriented whisker yarn that can be further infiltrated with the desired matrix.

General Technologies Corporation has also drawn silicon carbide wool (e.g., TKF Grade 5A SiC) into yarns and infiltrated the structure with epoxy resin. The whisker-spinning operation is difficult if the individual whiskers are less than $\frac{3}{8}$ in. long [43]. A 25 vol% whisker yarn increased the strength of the resin from 8000 to 125,000 psi. A modulus value of 30 million psi has

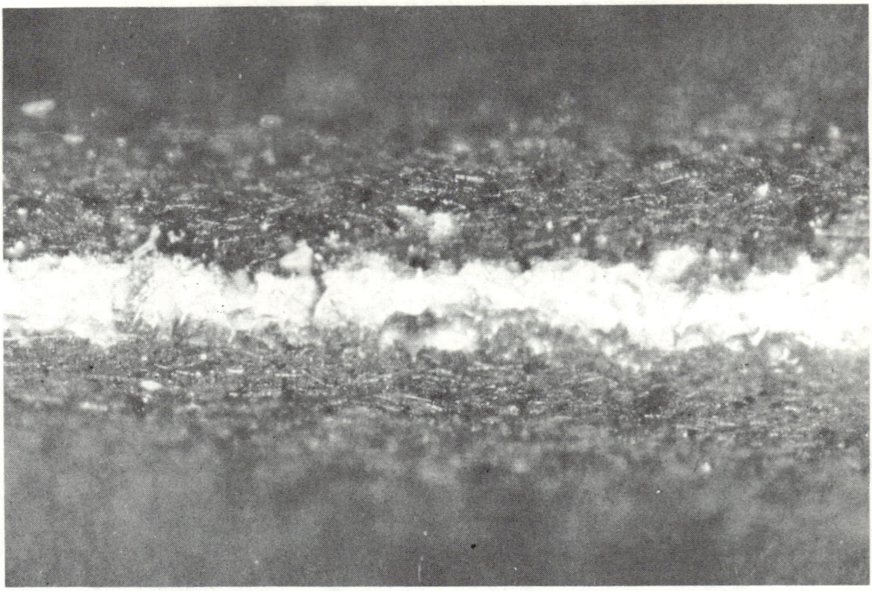

Figure 12.8 Longitudinal section of SiC whisker containing extruded fiber (resin binder) (mag. 150X).

been measured on one of these aligned composites. It is considered important to minimize the twist relative to yarn diameter in order to achieve alignment and densification. This would require considerable modification of conventional textile techniques before these can be used extensively for advanced whisker applications.

There is also considerable potential for a planar orientation of whiskers. Whisker mats, impregnated with resins, lead to bidirectionally reinforced composite structures. For example, asbestos laminates are now being made that are 45% filled and that have tensile strengths in the 50,000 psi range. These are made from fibers having strengths of 200,000 to 300,000 psi [56]. One can envisage that the use of 2–3 $\times$ 10^6 psi whiskers would also result in a tenfold improvement in laminate tensile strength—to about 500,000 psi. Milewski and Shyne [57] suggested uses for different types of whiskers in plastic matrices. Unidirectional, high l/d type whiskers would be applicable to beams and plastic wire type applications. Bead or ribbon whiskers, in a planar mat configuration, would be applicable to bidirectional laminates such as breadboards or component boxes, while whisker clusters would be ideally suited for precision molded parts such as small motors, gyros, and computer parts.

Epoxies such as the diglycidyl ethers of bisphenol, epoxy novalacs, and cycloaliphatic types are excellent choices of matrix materials because they are intrinsically strong, have good adhesive properties, and shrink very little when cured. Phenolic resins are limited by the high pressures required to mold them. A great deal of experience and data have been gathered on glass fiber-reinforced polyesters, and this information should be of help in working with whiskers. Other resins that might be used as matrices for whisker reinforcements are polyurethanes, alkyds, polybenzimidazole, and polyimides.

The combination of whiskers with new high-strength, high-temperature polymers should result in excellent high-temperature molding plastics. Potential properties appear to be 100,000 psi at 100–1500°F [22]. Particularly promising reinforcements for plastics are boron carbide whiskers, because of their especially high strength and modulus to density ratios.

Glass and Ceramics Composites

Whiskers have been found to improve the impact strength, elastic modulus, thermal shock resistance, hardness, and compressive strength of glasses and ceramics [22, 58]. They have also been used to yield finer grain structures of hot pressed ceramics and improve the strength of slip castings [58].

Fusion processes involving ceramic matrices are subject to, in many instances, whisker dissolution, or at least severe degradation. (This is also true of many whisker systems.) Single crystals of Al_2O_3, for example, have been shown to be severely degraded in a wide range of molten glasses [59]. However, much depends on the glass system utilized.

Research at R.P.I. has involved small whisker additions in calcium aluminate glass, which produced a significant increase in the glass hardness. Excellent wetting and bonding was obtained at high fusing temperatures with no whisker degradation [22]. Silicon carbide whiskers have been added to molten silica boules which have been drawn out into filaments to form a fiber core structure. An alignment of the whiskers has been obtained as well as a small increase in the modulus and tensile strength of the filament [43]. In other work [28], a 10% whisker addition to a zinc oxide ceramic was found to increase the compressive strength from 2500 to 7500 psi. Sapphire whiskers also increased the compressive strength of hydraulic cements by a factor of four.

Silicon carbide whisker additions are also being studied as an additive to a low-density porous silica structure; in one case a 5% whisker addition raised the modulus of the fused silica structure from 4.5 to 6 million psi [22]. Work at the Naval Ordnance Laboratory [22] showed that the compressive strength of glass could be increased by 59% with the addition of about 1 wt %

randomly oriented whisker additions. At Alfred University whiskers were included in alumina and mullite, and hot pressed and fired to high temperatures with no whisker recrystallization and very little whisker damage from pressure. Matrix grain growth was arrested at the interface, yielding a finer-grained structure.

Thermokinetic Fiber's sapphire paper now presents the possibility of high-strength ceramic laminates which should have increased toughness through the use of ceramic impregnation. When coated and bonded with a phosphate, silica, or other compatible solution, this paper should make a good reinforcement for ceramic slip-casting applications.

Multiple Reinforcements

Whiskers have been used to increase the shear strength of the matrix of continuous filament composites [22, 26, 34, 60, 61]. Typical systems include SiC or Al_2O_3 whisker-reinforced glass or boron fiber-resin matrix composites. For example, whiskers have been added to glass filament wound rings at the Naval Ordnance Laboratory [22]. A 4% addition of loose whiskers of various sizes and types added to the resin prior to wet winding resulted in a 38% improvement in interlaminar shear strength. Silicon carbide whiskers were found to increase the interlaminar shear strength of unidirectional boron fiber composites by increasing the matrix modulus (hence matrix shear strength) and by reducing crack propagation through crack dissipation [34]. Two types of whisker mixtures were used to reinforce pin-bearing boron-resin tape composites at Douglas [60]. These were 1 or 2% of mixed SiC-Al_2O_3 whiskers or mixed AlN and gave 5 to 24% improvements. Although the whiskers did not appear to improve the pin-bearing properties, they did preclude a shearout mode of failure in the composite specimens. Additions of 5 to 20% whiskers to tape windings generally afford local reinforcement of highly stressed areas of flanges and bosses of filament-wound structures and a chief benefit is the prevention of crazing [20].

Studies at Avco [26] indicated that care is required to eliminate all porosity during incorporation of small, high surface area SiC whiskers into a boron filament, epoxy resin composite. The effectiveness of whiskers in this class of applications is also limited by the tendency for whiskers to agglomerate. (The reader is also referred to related remarks by Wolf and Levitt in Chapter 8.) A development from Thermokinetic Fibers, Inc. [61], is a process called "whiskerizing," which allows whiskers to be grown on virtually any surface able to withstand elevated temperatures for short times. The first products from this process involve β-silicon carbide whiskers on graphite filaments, yarns, and cloth. Since the whiskers are oriented mainly at right angles to the graphite filament, they help to improve the interlaminar shear strength in a resin matrix (see Fig. 12.9). This development may also help to solve a difficulty in the fabrication of this type of composite, namely uniform dispersion of the whiskers in the resin.

Unpublished work at Avco has shown that a thin coating of a whisker-organic slurry on aluminum foil layers helps to promote bonding in diffusion-bonded boron filament-aluminum matrix composites.

At the present time, there is no reported research on the combination of whisker reinforcement with composites such as dispersion-hardened metals or honeycomb or other macrolaminate structures. The use of whisker reinforcement in age-hardening alloys represents a multiple use of metal-strengthening mechanisms. The authors are not aware of studies to investigate any mutual interactions of fibers with other simultaneous strengthening mechanisms. This is clearly an area for future basic research of direct applicability to advanced whisker composite materials.

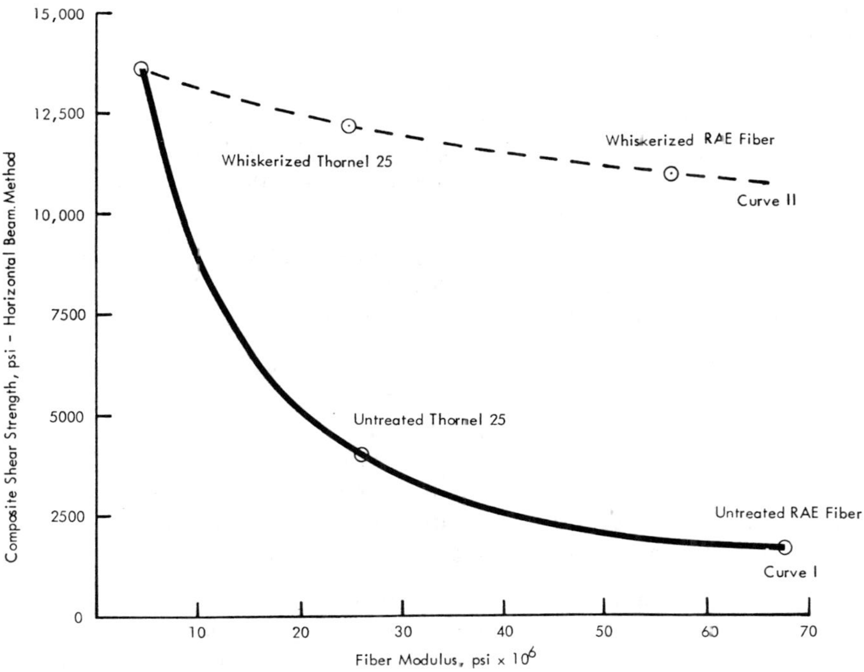

Figure 12.9 Carbon fibers composite shear strength versus fiber modulus (epoxy resin, amine cure). Data of Naval Ordnance Laboratory.

DISCUSSION

Whiskers can be used in a variety of methods and concentrations to produce different effects. In concentrations of 1–5%, the prime purpose of whisker additions will be to produce major changes in the grain structure of the matrix; for example, in dispersion hardening, grain refinement, or epitaxial crystallization. Several laboratories have initiated investigation of whisker effects in hot-pressed aluminum oxide and beryllium oxide bodies, glasses, and ceramics. The major findings reported so far are the following.

1. High-temperature grain refinement of matrix ceramics.
2. No recrystallization of whiskers in matrix at elevated temperatures for long periods of time.
3. Good wetting and bonding to glass without dissolving.
4. Increased hardness in glass ceramics.

Most investigators are working with 5 to 25% additions depending upon the type of effect they wish to produce. The following percentages of whiskers are suggested for different effects [22]:

1. 0 to 5% for grain refinement and crystallization control.
2. 5 to 25% for major improvement of mechanical properties.
3. 25 to 90% for development of super composites with multiple improvements in mechanical, electrical, and thermal properties. (High percentages represent an advanced state of the art of whisker technology.)

One or more of the following fabrication process problems are common to all potential applications. These are listed in terms of goals to be achieved before the full potential of a whisker-reinforcement approach is realized:

1. An adequate volume fraction of high-strength whiskers.
2. A uniform or controlled whisker dispersion.
3. Orientation for maximum effectiveness.
4. Maximum whisker-matrix bond strength.
5. Minimum whisker degradation due to handling.
6. Minimum whisker degradation due to reaction with matrix.
7. Competitive materials and manufacturing costs.

The stringent requirements of a successful whisker composite may be expected to require a combination of two or more fabrication techniques from a formidable list of possibilities. These may include cold press and sintering, hot pressing, casting, liquid impregnation, diffusion bonding, plasma spraying, spinning, slip casting, compression molding, electroforming, and extrusion or forging. Additional refinements to each of these processes are required to achieve the desired alignment of whiskers.

In addition to manual alignment, semi-automatic methods investigated to date include the use of electrostatic and magnetic fields, extrusion, flocking, controlled solidification, vibratory methods, fluid spinning, combing, hydrodynamic alignment, centrifugal casting, and combinations of these. Further refinements are needed to adopt these to an automatic, commercially feasible process.

The limitations of whisker composites lie in the areas of geometries available, volume use, costs and, to some extent, whisker materials available. New developments, such as preformed whisker tapes can be expected to extend the range of application of whiskers and lower fabrication costs. There will always be competition from fiber-reinforced materials, especially for structural applications where there are few and well-defined stresses to which continuous fibers can be easily oriented. It has been noted, however, that unidirectional alignment is often undesirable. Sheet or cylindrical structures are usually bidirectionally stressed and moldings or castings are subject to a large variety of mechanical and/or thermal stresses, including compressive stresses.

A matter of continuing concern for the maximum effectiveness of whisker composites is that of separating the really high strength whiskers from the total number harvested in the whisker growth process. For example, Al_2O_3 whiskers having cross-sectional areas of greater than about 20–50 μ^2 tend to have strengths of less than 1 million psi [8]. Also, unless the weaker ones are removed, the average bundle strength of the whiskers decreases since the scatter in the individual tensile strength increases. This is reflected in a decrease in composite strength.

The discussion has been concerned mainly with refractory compounds or nonmetal whiskers such as the more common Al_2O_3 or SiC whiskers, and also C, BeO, B_4C, MgO and Si_3N_4 whiskers. In addition, metallic whiskers (e.g., Fe, Cu, Cr, and Au) are presently available in small quantities, but little composite research has been done with these. It is probable that metallic whiskers will find structural use mainly at low temperatures, for example, in resin matrices. Kelly [62] shows that there is no advantage in considering whiskers of the face-centered cubic transition metals because their theoretical shear strengths are low and the presence of dislocations would lead to rapid loss of strength. Chemical and physical properties may, however, be of interest (e.g., magnetic properties).

A major future problem area in the utilization of whisker composites will be that of joining. Grimes [63] has pointed out that, particularly for components, joining will almost always be required. Basic parameters that need to be considered for composite joining design include joint function and environment, the type and magnitude of the loads to be transferred, the strain compatibility of discrete elements of the joint, and space and cost requirements. Considerably more research needs to be done in this area.

It may also be expected that the application of whiskers to high-temperature systems will be limited because of the chemical compatibility problems. The utilization of SiC and C whiskers, where the strength may actually be higher at the desired use temperature than at room temperature, in metallic matrices, presents a real challenge to those engaged in expanding the field of whisker applications.

CONCLUSIONS

A number of prophecies have been attempted regarding the future exploitation of whisker-reinforced composites [54, 63–66]. Most of the basic problems associated with the attainment of high-strength composites were identified by the Engineering Foundation Research Conference on Composite Materials in 1962 [63]. This included areas for investigation that are still valid today, for example, bundle strengths, interface reactions, load transfer, failure mechanisms, fatigue, and micromechanics. Hibbard [64] in 1963 predicted that if these areas are more widely understood, "We may see whiskers on the man in the moon in 1975." Proper orientation was stressed by West [65] in 1963 as a requisite for structural whisker composites. At that time it appeared doubtful if whiskers could actually be used to improve interlaminar shear strength of resin-based composites, but there are now indications that, with favorable distribution and alignment, shear strength can be improved. Principal areas for applications of whisker composites were likely to include "gas turbine blades, SST structural parts, and hulls for deep submergence vessels" [65].

A detailed time scale for the engineering application of whisker composite materials was prepared by W. H. Sutton [54] and, as this is based on a retroactive review as well as a hopeful anticipation, the "prospectus" is reproduced as follows:

1952	Discovery of inherently high strength of whiskers reported.
1951–1959	Scientific investigations conducted on growth and properties of whiskers. Several patents on the growth of whiskers issued.
1958	*In situ* growth of whiskerlike fibers by unidirectional solidification from alloy melts.
1960	Whiskers available in tenths of grams for exploration of composite materials.
1961	First whisker composites to show marked improvement in strength.
1960–1965	Research on the growth of whiskers, development of whisker handling and processing methods, development of composite fabrication techniques, and testing and

	evaluation of composite performance. Silicon carbide whiskers available in tonnage quantities.
1965–1970	Scale-up of whisker growth and whisker types; scale-up of exploratory composite materials for testing and performance evaluation.
1970–1975	Early applications of whisker-reinforced composites in special aerospace components.
1975	Applications in jet engines, SST structures, deep-submergence structures, as special impregnated fabrics, wires, cables, sheets, etc., and eventually in other commercial areas, such as gas/steam turbines and nuclear structural components.

Similarly, the ASM Advisory Technical Awareness Council [66] predicts that in the 1970–1980 period filaments with ultimate tensile strengths of 1 to 2 million psi will be utilized in composites. Composites in turn will supplement many metals in traditional forms. For the 1980–1990 period it is predicted that composites will efficiently combine strength and low cost of steels with desirable properties of plastics, ceramics, and nonferrous materials. In addition, computers will be used to formulate composites, design parts, and direct production with tapes. Standard computer models of composition, microstructure, and design will permit tailoring of microstructure for the function. Work will continue on forming, machining, and joining of novel composites. Applications exploiting the controlled directionality of properties will accelerate rapidly and composites and filamentary potentials will be introduced in consumer goods. By the years 2000–2010 it is predicted that designs based on single monolithic materials will not be able to compete with composites in material utilization and structural efficiency.

The foregoing predictions are certainly optimistic, for few practical applications of whisker composites exist at the present time. However, although whisker composites are currently in the development stage, they do represent a follow-up to the already successfully employed fiber composites such as fiber glass-plastics and graphite or carbon filament-plastic materials. Whiskers do exist with several million psi tensile strength and much has been learned regarding their effectiveness in composites. It is concluded therefore, that the next several years will be directed to extensive testing and evaluation of somewhat standardized ranges of whisker composites with a view to generating extensive structural design data. On the speculative side, one can predict that in the more distant future, composites with 1 million psi tensile strength will be developed and the applications for these should be very extensive, especially after the cost has been brought down to a competitive basis.

REFERENCES

[1] C. Herring and J. K. Galt, *Phys. Rev.*, **85** 1060 (1952).

[2] A. Kelly and G. J. Davies, *Metals Rev.*, **10,** 37 (1965), 1–77.

[3] M. Gunton, article in *Metalworking News*, **9,** 402 (April 8, 1968).

[4] S. S. Brenner, *J. Appl. Phys.*, **33,** 33 (1962).

[5] H. M. Burte and C. T. Lynch, Paper No. 1, 70th Annual Meeting, ASTM, June 25–30, 1967.

[6] E. G. Wolff, Paper EM67-687, Engineering Materials Seminar, ASTME, San Francisco, Calif., Nov. 28–29, 1967.

[7] E. G. Wolff, "Hydrodynamic Alignment of Discontinuous Fibers in a Metal Matrix," *Fiber Science and Technology*, Elsevier Publishing Co., Limited, Essex, England, March 1969.

[8] J. Chorne et al., Final Reports, Contract NOw-65-0176-c, March 1966, Contract NOw-66-0443-d, January 1967, and NOOO19-67-C-0243, January 1968.

[9] B. C. Raynes et al., Reports under Contracts NOw-62-0235-c, NOw-63-0138-c and 64-0125-c, Final Report October 27, 1964.

[10] B. H. Dennison et al., presented at 11th Meeting, Refractory Composites Working Group, Los Angeles, Calif., January 24–28, 1966.

[11] H. Hahn et al., Final Report, Contract NOw-65-0387-8, April 30, 1966. Also preliminary work under Contracts NOw-66-0616-c, and NOOO19-67-C-0404.

[12] High Temperature Sapphire Filaments (Tyco), see for example, *Metalworking News* (Nov. 6, 1967), p. 14 or the *N.Y. Times* (October 22, 1967).

[13] H. E. Bates, F. Wald, and M. Weinstein, presented at 10th National Symposium, SAMPE, San Diego, Calif., November 9–11, 1966.

[14] R. W. Hertzberg, "Fiber Composite Materials," *Am. Soc. Metals*, 77–87 (1965).

[15] M. J. Salkind et al., *Chem. Eng. Prog.*, **62,** 52 (1966).

[16] E. R. Thompson, 12th Meeting, Refractory Composites Working Group, Denver, Colo., October 17–19, 1966.

[17] I. Ahmad, V. P. Greco, and J. M. Barranco, *J. Composite Materials*, **1,** 1 (1967), 19.

[18] See L. R. McCreight, H. W. Rauch, and W. H. Sutton, AFML-TR-65-105, May 1965.

[19] L. A. Warnes, ERDE, Res. Report 17/R/65 (AD477233), July 16, 1965.

[20] P. Grenier and A. Kelly, "Plastic Resin Reinforcement with Alumina Trichites," *Comptes Rendus*, Academy of Science, Paris, France V 266 B, No. 14 (1968), 944–947.

[21] C. A. Calow and R. B. Barclay, *J. Mat. Sci.*, **2,** 404 (1967).

[22] J. V. Milewski and J. J. Shyne, paper presented at ACS Meeting, May 3, 1965. See also Sheets and Bulletins, Thermokinetic Fibers, Inc.

[23] A. M. Saul, W. L. Smallwood, and H. M. Blaes, 10th Meeting Refractory Composites Working Group, Atlanta, Ga., April 1965.

[24] L. D. Whipple, 10th National SAMPE Symposium, San Diego, Calif., November 9–11, 1966.

[25] "Whisker Reinforced Shapes Made to Order," *Aerospace Technology*, February 12, 1968, 48.

[26] B. Coplan et al., Contract AF 33(615)-2120, AFML-TR-66-295 (1966).

[27] R. N. Boyd, Contract Report to Thermokinetic Fibers, Inc., April 1965.

[28] A. A. Grant and E. H. Greener, *Australian Dental J.*, **12,** 1 (1967), 29–33. Also internal reports from Northwestern University, Chicago, Ill., 1966, 1967.

[29] E. P. Lautenschlager, Northwestern University, Dept. of Biological Materials. Private communication.

[30] "Addent" 3M Company and Dakor, Caulk Division of Dentist's Supply Co. of New York.

[31] M. I. Kliman, Watertown Arsenal Report WAL-TR-371/51, D/A Proj. 59332007, November 1962.

[32] A. P. Divecha, P. J. Lare, and H. Hahn, 12th Meeting Refractory Composites Working Group, Denver, Colo., October 17–19, 1966.

[33] O. Y. Ang, and H. M. Manasevit, *Electrochem. Soc.*, 9–13 May, 1965.

[34] L. C. Wohrer and J. Economy, 10th National SAMPE Symposium, San Diego, Calif., November 9–11, 1966. See also Carborundum Brochures on SiC Whiskers (Bull. #202, 1966 and No. A-1907).

[35] H. Weiss, *Electron. News* (October 12, 1964).

[36] F. S. Galasso, *J. Metals* (June 1967), 17.

[37] M. J. Salkind, F. D. Lemkey, and F. D. George, "Whisker Composites by Eutectic Solidification," Chapter 10 in this book.

[38] H. Schreiner, *Z. Metallkunde*, **58,** 600 (1967).

[39] H. Hahn, Report in *Metalworking News*, February 27, 1967, 15.

[40] L. J. Broutmann and R. H. Krock, *Modern Composite Materials*, Addison-Wesley, Reading, Mass., 1967.

[41] "Materials Survey for the Rescue and Search Vehicles of the Deep Submergence Systems Project," David Taylor Model Basin, Navy, March 1965, AD614421D.

[42] B. T. Woodland, *Space/Aeronautics*, **47,** 3 (March 1967), 100.

[43] J. J. Shyne and R. G. Shaver, 10th National Symposium, SAMPE, San Diego, Calif., November 9–11, 1966.

[44] H. Rosenberg, MITRON R & D Corporation, private communication. MITRON Data Sheets (1966).

[45] R. H. Duff, F. T. Mansur, and E. G. Wolff, paper presented at 10th National SAMPE Sympsoium, San Diego, Calif., November 1966.

[46] R. W. Herzberg and R. W. Kraft, *Trans. Am. Inst. Min. Metall. Petrol. Engrs.*, **227,** 580 (1963).

[47] F. R. Mollard and M. C. Flemings, *Trans. AIME*, **239,** 1562 (October 1967).

[48] "Crystallization Innovations," e.g., Bendix Corp., see *Metalworking News*, March 20, 1967, p. 10.

[49] H. N. Barr, 3rd Quarterly Progress Report, Contract AF 33(615)-5375, January 1, 1967—March 31, 1967, Hittman Associates, Baltimore, Md.

[50] E. G. Wolff, T. Sturiale, and T. Vasilos, 11th Meeting Refractory Composites Working Group, Los Angles, Calif., January 1966.

[51] S. S. Brenner, chapter in *Fiber Composite Materials*, ASM, Ohio, 1965.

[52] F. F. Schmidt and H. R. Ogden, DMIC Report 190, September 20, 1963.

[53] J. T. Smith and D. P. Wilde, *J. Appl. Phys.*, **36,** 2 (1965), 595.

[54] W. H. Sutton, *Astronautics and Aeronautics* (August 1966), 46. See also *SPE Journal*, **20,** 1203 (1964).

[55] N. J. Parratt, *Chem. Eng. Prog.*, **62,** 3 (1966), 61. Also Preprint 18D, Whisker Technology Symposium AICE, 58th Annual Meeting, Philadelphia, Pa. December (1965).

[56] D. V. Rosato, *Asbestos, Its Industrial Applications*, Reinhold Publishing Corp., New York, 1959.

[57] J. V. Milewski and J. J. Shyne, presented at SPE Meeting, Atlantic City, New Jersey, January 1964.

[58] W. J. Corbett, A. T. Sales, and J. D. Walton, Sandia Corp. Contract 16-2092 to Georgia Inst. Tech., Report, August 6, 1965.

[59] T. S. Busby and J. Eacles, *Glass Technology*, **5,** 3 (1964), 115.

[60] C. E. Anderson, D. G. Smillie, G. M. Lehman, Progress Reports under Contract No. F 33615-67-C-1582, Douglas Aircraft, 1967–68.

[61] GTC Bulletin 105, "Whiskerizing," 1967.
[62] A. Kelly, *Strong Solids*, Clarendon Press, Oxford, 1966, p. 179.
[63] Engineering Foundation Research Conference on Composite Materials, August 1–3, 1962, Tilton, N. H.
[64] W. R. Hibbard, *Chem. Eng.* (November 11, 1963), 203.
[65] P. West, Mat. in *Design Eng.*, **61,** 6 (1965), 112.
[66] American Society for Metals, Advisory Technical Awareness Council, p. 60 in *Met. Prog.*, **92,** 6 (December 1967).

Index

In this text the terms Al_2O_3, α-Al_2O_3, Alumina, Aluminum oxide, and Sapphire are equivalent when used to describe whiskers. Each of these terms has been used or cross-referenced in this index where deemed appropriate.